NATURAL HAZARDS

NATURAL HAZARDS

Edward Bryant

The right of the University of Cambridge to print and sell all manner of books was granted by Henry VIII in 1534. The University has printed and published continuously since 1584.

CAMBRIDGE UNIVERSITY PRESS
CAMBRIDGE NEW YORK PORT CHESTER MELBOURNE SYDNEY

Published by the Press Syndicate of the University of Cambridge
The Pitt Building, Trumpington Street, Cambridge CB2 1RP, UK
40 West 20th Street, New York, NY 10011, USA
10 Stamford Road, Oakleigh, Melbourne 3166, Australia

First published 1991

Printed in Hong Kong by Colorcraft

National Library of Australia cataloguing-in-publication data:

Bryant, Edward, 1948–
Natural Hazards.

Bibliography.
Includes index.
ISBN 0 521 37295 X.
ISBN 0 521 37889 3 (pbk.).

1. Hazardous geographic environments. 2. Natural disasters. I. Title.

551

British Library cataloguing-in-publication data:

Bryant, Edward, 1948–
Natural disasters.
1. Natural disasters.
I. Title

904.5

ISBN 0–521–37295–X
ISBN 0–521–37889–3 pbk

Library of Congress cataloguing-in-publication data:

Bryant, Edward, 1948–
Natural hazards/Edward Bryant.
p. cm.
Includes bibliographical references and index.
ISBN 0–521–37295–X: $60.00 (est.). — ISBN 0–521–37889–3 (pbk.): $25.00 (est.)
1. Natural disasters. I. Title.
GB5014.B79 1991
363.3'4—dc20

To Dianne, Kate and Mark

Contents

Illustrations

Note: World maps used in this book are Hoelzel's planispheres. Other maps are modified conical projections.

Tables

Preface

I am sitting in my office perplexed but stimulated by the climatic events of the last three years, stimulated because they have occurred before, and perplexed because some scientists have attributed the extreme events of recent years to evidence of Greenhouse warming. The 1988–89 wet summer in Australia was predictable as far back as May 1986. That was the beginning of the 1986–87 El Niño-Southern Oscillation event. My reputation for climatic predictions over the last three years was enhanced by that event. I had the good fortune of being in Canada during the winter of 1986–87, which was one of the warmest on record. Canadian friends were puzzled when I mentioned this strange phenomenon called 'El Niño' in the South Pacific. They hastily distanced themselves from me when I explained how climatic behaviour, triggered by warm water movement from northern Australia, was directing their winter weather. As El Niño ran its course and the strongest La Niña event in 40 years developed, I began warning people that eastern Australia would have record floods in the summer of 1988–89. Acquaintances scoffed at my foolery during a seven-week drought in September and October, but I confidently forecast that the deluge would begin on 15 November. It did: and rain fell for the next eight months over most of eastern Australia, breaking records for consistency and amount. Some desert lakes filled for the first time in recorded history, and most rivers entered an extreme flood stage. When friends queried me as to when the 'big wet' would stop, I confidently stated months in advance that it would finish on 30 June. Since then, the rain has ceased and a winter's sun now warms me as I type this preface.

The rainfall did have its benefits for me. I had three months' recreational leave from work during which I intended drawing diagrams for this book, as well as relaxing by bushwalking and visiting cultural centers in Sydney. The diagrams became more onerous than foreseen and took most of my free time; but then the deluge prevented any diversionary recreational activities.

My perplexity still remains. This book is not about Greenhouse. It is about everyday climatic and geological hazards that can, and will be predicted given substantial research effort. Some of the optimism portrayed in the book may be viewed as incorrect, but hopefully the reader will be imbued with a sense of excitement that we can do something now about many of the hazards that afflict us. Greenhouse tends not to do that. The predictions are 40–50 years in front of us. Even my young children, who cannot yet understand what is meant by the Greenhouse effect, will be old by then. The way that Greenhouse is being portrayed in the media is depressing, and is evoking in young people a nihilistic sense of despair and futility. Our existing environment has not so far succumbed to Greenhouse, and may not—at least for a few years yet. With effort, we can understand, predict and mitigate the hazards around us. By doing this, we can hopefully cope with Greenhouse when and if it arrives. That is what this book is about—natural hazards that are within the present realm of possibility, that we have lived with, will continue to experience and, hopefully, can survive.

In order to convey this point of view clearly in the book, adherence to academic referencing has been kept to a minimum. Usually each section begins by listing the major papers or books on a topic that have influenced my thinking and writing. Full reference to these publications can be found at the end of each chapter. I apologize to anyone who feels that their crucial work has been ignored; but the breadth of coverage in this textbook precluded a complete review of the literature on many topics.

I should also add that, in discussing the impacts of particular disasters, I have frequently given monetary assessments of the damage caused or the cost to the community. Unless specifically stated, monetary values given refer to values at the time the disaster occurred and have not been adjusted.

Ted Bryant
10 July 1989

Acknowledgments

The sections on beach erosion and astronomical influences on the Southern Oscillation were based upon research funded by the University of Wollongong and its Geography Department. The following individuals, groups or organizations provided data (much of it free of charge!) for this research, for which I am extremely grateful: NOAA, Boulder, Colorado, for climatic data for the world; the Permanent Service for Mean Sea Level, Bidston Observatory, United Kingdom, for world monthly sea-level data; the Maritime Services Board of New South Wales for Sydney sea-level data; the CSIRO Division of Oceanography, Cronulla, New South Wales, for sea surface temperature data offshore from Sydney; Dr A. B. Pittock, CSIRO Atmospheric Research, Melbourne, for Southern Oscillation and Hadley cell data; more than 60 private citizens, organizations, semi-autonomous government bodies, and state departments who supplied me with 154 old photographs of Stanwell Park beach. Several people generously provided data from their own research as follows: Dr Bob Wasson and Ms Wendy Gallagher, CSIRO, Canberra, for figure 5.5; Jane Cook, Faculty of Education, University of Wollongong, for her Kiama landslip data in figure 5.9; and Professor Eric Bird, Department of Geography, University of Melbourne, for updating table 4.2. In addition, the following people gave me leads to materials that were incorporated into the book: Professor Bruce Thom, Department of Geography, University of Sydney, for information on east-coast lows; Dr Geoff Boughton, Department of Structural Engineering, Curtin University of Technology, Perth, for information on the lower limits of seismically induced liquefaction; Col Johnson, New South Wales South Coast State Emergency Services, for pointing out the importance of Critical Incidence Stress Syndrome amongst rescue workers; Rob Fleming, Australian Counter Disaster College, Mount Macedon, Victoria, for access to photographic files on hazards in Australia; and Bert Roberts, Department of Geography, University of Wollongong, for pointing out the relative literature on uniformitarianism.

The following people or publishers permitted diagrams or tables to be reproduced: Methuen, London, for figures 2.18 and 2.20; Australia and New Zealand Association for the Advancement of Science, *Search*, for figure 2.22; Geographical Society of New South Wales, *Australian Geographer*, for figure 4.11; CSIRO Editorial and Publishing Unit, Melbourne, for figures 4.12 and 4.13; the Chemical Rubber Company, Boca Raton, Florida, for figure 4.17; Elsevier Science Publishers, Physical Sciences and Engineering Division, *Marine Geology*, for figure 4.19; Gebrüder Borntraeger, *Zeitschrift für Geomorphologie Supplement*, for figure 5.2; University of Wisconsin Press, for figure 5.3; Harcourt Brace Jovanovich Group, Academic Press, Sydney, for figures 10.15, 12.4 and 12.9; Prentice Hall, New Jersey, for figure 11.17 and permission to reproduce figures 2.24, 11.12 and 11.13; Butterworths, Sydney, for figures 13.2 and 13.12; and Dr John Whittow, Department of Geography, University of Reading, for figures 13.11 and 13.16. Acknowledgment is made to the following organizations who did not require permission on copyrighted material: the United States Geological Survey, Reston, for figures 13.7 and 13.14; the United States Marine Fisheries Service, Seattle, for table 5.1; and the American Geophysical Union for table 5.2. All of the above diagrams and tables are acknowledged fully in the text.

The following people and organisations provided photographic material or assisted in searching for it: John Fairfax and Sons Limited, *Sydney Morning Herald*, for figure 2.12; Dr Geoff Boughton, Department of Structural Engineering, Curtin University of Technology, Perth, for figure 2.15; Esther Fell, Bureau of Meteorology, Melbourne, for figure 3.7; John Telford, Fairy Meadow,

New South Wales, for figure 4.4; Dr Ann Young, Department of Geography, University of Wollongong for figure 4.16 (second photograph); Bob Webb, Mal Bilaniwskyj and Charles Owen, New South Wales Department of Main Roads, Wollongong and Bellambi offices, for figure 5.10; Dr Gerald Nanson, Department of Geography, Wollongong University, for figure 8.7; the State Library of Victoria, who kindly permitted the reproduction of William Strutt's painting, *Black Thursday*, appearing as figure 9.1; N. P. Cheney, Director, and A. G. Edward, National Bushfire Research Unit, CSIRO Canberra, for figures 9.5 (left); and 9.6 respectively; Jim Bryant, Ontario Department of the Environment, for figure 9.5 (right); Pictorial Department of the *Age* newspaper, Melbourne, for figures 9.10 and 14.1; the United States Geological Survey, for figures 11.7, 11.10, 11.14, 11.15, 12.2, and 12.3; the Institute of Geological Science, London, United Kingdom, for figure 12.7; J. L. Ruhle and Associates, Fullerton, California, for figure 12.10; A. G. Black, Senior Soil Conservator, Hawke's Bay Catchment Board, Napier, New Zealand, for figure 13.10; Dr Peter Lowenstein, for figure 13.13, courtesy C. O. McKee, Principal Government Volcanologist, Rabaul Volcanological Observatory, Department of Minerals and Energy, Papua New Guinea Geological Survey; and Dave Cunningham, New South Wales National Parks, for figure 13.15.

The following people reviewed parts of the manuscript or contributed valuable editorial comments: from the Department of Geography, University of Wollongong, Dr Gerald Nanson reviewed the introduction and the section on floods and historical changes to rivers, Dr Robert Young critically read the introduction and commented on the organization, and Dr Ann Young substantially reviewed the chapter on land instability; Colin Johnson from the New South Wales State Emergency Services, South Coast branch, commented upon and contributed to the content dealing with response. Finally, I am indebted to Professor Russell Blong, School of Earth Sciences, Macquarie University, who methodically read the complete manuscript and made many constructive comments that greatly enhanced its final version. Robin Derricourt, of Cambridge University Press, Sydney, guided the organization of the textbook and its final production. Roderic Campbell, Sydney subeditor, greatly improved the continuity of the completed manuscript. Robyn Johnson, now of Robyn's Photographics, Wollongong, and Simone Rose, Centre for Teaching Development, University of Wollongong, processed some of the photographic material. The majority of this text has been presented as lecture material. I would like to thank many students in my 2nd year Environmental Hazards classes who challenged my ideas or rose enthusiastically to support them. Finally, I would like to thank my wife, Dianne, who took on the onerous task of proofreading the text, and who questioned its content from the point of view of an undergraduate student.

1

Introduction to Natural Hazards

Rationale

The field of environmental studies is usually introduced to students as one of two themes. The first theme examines human effects upon the Earth's environment, and is concerned ultimately with the question of whether or not people can irreversibly alter that environment. Such studies include the effect of human impact on climate, of land-use practices on the landscape in prehistoric and recent times, and of nuclear war upon the Earth's environment. The second theme totally disregards this question of human impact on the environment. It assumes that people are specks of dust moving through time subject to the whims of nature. In this sense, calamities are 'acts of God'—events that make the headlines on the evening news; events you might wish on your worst enemy but would never want to witness yourself. University and college courses dealing with this theme usually treat people as living within a hostile environment, over which they have little control. Such courses go by the name of 'Natural' or 'Environmental' Hazards. While this book is concerned foremost with the various ways that nature can wreak havoc upon humans, the message is not one of total doom and gloom. Also discussed are the various mechanisms by which we can predict and avoid natural calamities such as landslide, cyclones, earthquakes and drought. For those hazards that cannot be avoided by humans, the book examines how people can minimize their effects and rectify their negative consequences.

The book is organized around climatic and geological hazards. Intense storms and winds, oceanographic factors such as waves, ice and sea-level changes, together with extreme precipitation phenomena, are all investigated under the heading of climatic hazards. Under the topic of geological hazards, earthquakes, volcanoes and land instability are examined. This book is not concerned with biological hazards such as plague, disease and insect infestations; but one should realize that death and property loss from these latter perils have been just as severe as, if not worse than, those generated by geological and climatic hazards. A final section looks at the social impact such hazards are likely to have.

Historical background

The world of myths and legends (Holmes, 1965; Day, 1984; Myles, 1985; Milne, 1986)

Myths are traditional stories focusing on the deeds of gods or heroes often to explain some natural phenomenon, while legends refer to some historical event handed down usually by word of mouth in traditional societies. Both incorporate natural hazard events. Prehistoric peoples viewed many natural disasters with shock and as a threat to existence. Hazards had to be explained and were often set within a world of animistic gods. Such myths remain a feature of belief systems for many peoples across the world—from Aborigines in Australia and Melanesians in the South Pacific, to the indigenous inhabitants of the Americas and the Arctic Circle (see figure 1.1 for the location of major place names mentioned in this chapter). With the development of written language, many of these oral myths and legends were incorporated into the writings of the great religions of the world. Nowhere is this more evident than with stories about the great flood. Along with stories of Creation, the Deluge is an almost ubiquitous theme. The biblical account has remarkable similarities to the

Babylonian Epic of Gilgamesh. In the Babylonian epic, Noah is replaced by a man called Utnapishtim, who actually existed as the tenth king of Babylon (Noah was the tenth descendant of Adam). The biblical account parallels the epic, which tells of the building of an ark, the loading of it with different animal species, and the landing of the ark upon a remote mountaintop. Babylonian civilization was founded in what was called Mesopotamia, situated in the basin of the Euphrates and Tigris rivers, which historically have been subject to cataclysmic flooding. Clay deposits in excess of 2 m have been found at Ur, suggesting a torrential rain event in the headwaters of the valley that caused rivers to carry enormous *suspension loads* (sediments carried downstream) at high concentrations. Such an event was not unique. The epic and biblical accounts are undoubtedly contemporaneous and, in fact, a similar major flood appears in the written accounts of all civilized societies throughout Middle Eastern countries. For example, broken tablets found in the library of Ashurbanipal I (668–626 BC) of Assyria, at Nippur in the lower Euphrates valley, also recount the Babylonian legend.

The Deluge also appears worldwide in verbal and written myths. The Spanish Conquistadores were startled by flood legends in Mexico that bore a remarkable similarity to the biblical account. Because these staunchly Catholic invaders saw the legends as blasphemous, they ordered the destruction of all written native references to the Deluge. The similarities were most remarkable with the Zapotecs of Mexico. The hero of their legend was Tezpi, who built a raft and loaded it with his family and animals. When the god Tezxatlipoca ordered the flood waters to subside, Tezpi released firstly a vulture and then other fowl, which failed to return. When a humming-bird brought back a leafy twig, Tezpi abandoned his raft on Mount Colhuacan. A Hawaiian legend even has a rainbow appearing, which signals a request for forgiveness by their god Kane for his destructiveness. Almost all legends ascribe the flood to the wrath of a god, but others state that it just happened. The Aborigines have a story about a frog named Tiddalik drinking all the water in the world. To get the water back, all the animals try unsuccessfully to get the frog to laugh. Finally, only a cavorting eel manages the task. However, the frog disgorges more water than expected and the whole Earth is flooded. In some myths floods are attributed to animals urinating after being picked up. Many myths are concerned with human taboo violations, women's menstruation, or human

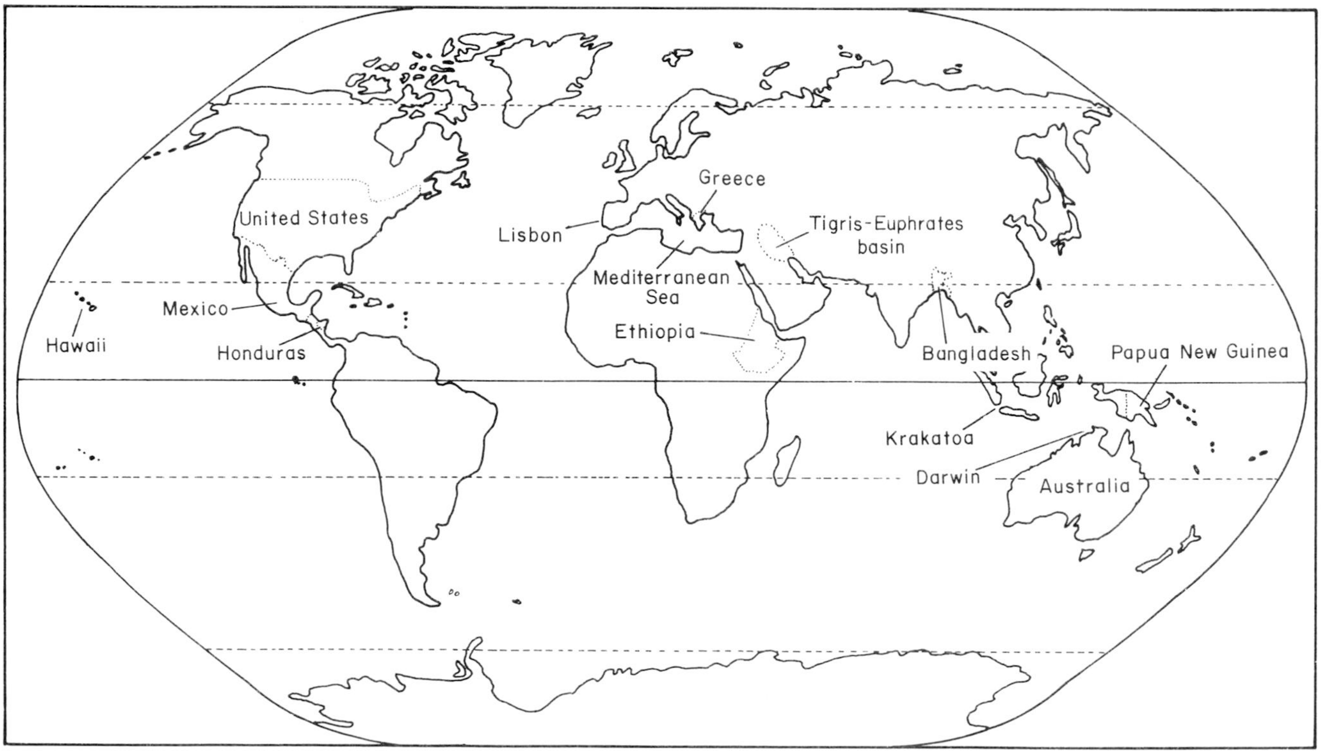

Fig. 1.1 Location map

carelessness. Almost all stories have a forewarning of the disaster, the survival of a chosen few and a sudden onset of rain from above. However, the Bible has water issuing from the ground, as do Chinese, Egyptian and Malaysian accounts. The latter aspect may refer to earthquakes breaking reservoirs or impounded lakes. Many Pacific island and Chilean legends recount a swelling of the ocean, in obvious reference to *tsunami*. Despite the similarities of legends globally amongst unconnected cultures, there is no geological evidence for contemporaneous flooding around the world, or over continents. It would appear that most of the flood myths recount localized deluges, and are an attempt by human beings to rationalize the flood hazard.

Almost as ubiquitous as the legends about flood are ones describing volcanism or the disappearance of lost continents. North American natives refer to sunken land to the east; the Aztecs of Mexico believed they were descended from a people called Az from the lost land of Aztlán. The Mayas of Central America referred to a lost island of Mu in the Pacific, which broke apart and sank with the loss of millions of inhabitants. Chinese and Hindu legends also recount lost continents. Perhaps the most famous lost continent legend is the story of Atlantis written down by Plato in his *Critias*. He recounts an Egyptian story, which has similarities with Carthaginian and Phoenician legends. Indeed, if the Aztecs did migrate from North Africa, the Aztec legend of a lost island to the east may simply be the Mediterranean legends transposed and reorientated to Central America. The disappearance of Atlantis probably had its origins in the catastrophic eruption of Santorini around 1628 BC (dated on attenuated tree-ring widths from the western United States and Ireland). Present-day Santorini is the largest of a small group of islands, situated in the southern Aegean Sea north of Crete (figure 1.2), which remain at the location of the volcanic eruption in 1628 BC. This eruption terminated Minoan civilization in the eastern Mediterranean Sea. Certainly, Greek flood myths refer to this or similar events that generated tsunami in the Aegean. It is quite possible that this event entered other tales in Greek mythology involving fire from the sky, floating islands and darkening of the sky by Zeus.

The timing of the Santorini eruption also fits into the description of the plagues of Egypt (Exodus 6:28–14:31). Accounts of the three days of darkness possibly refer to *tephra* clouds blowing south across Egypt at the beginning of the eruption. The darkness was described as a 'darkness which could be felt'. Volcanic shards have been found in the soils of the Nile delta with the exact same chemical composition as tephra on the Santorini islands. The 'river of blood' may refer to pink pumice from the eruption preceding the explosion. This pumice, after it was deposited, would easily mix with rainwater and flow into any stream or river, colouring it red. The parting of the Red Sea probably occurred in the marshes at its northern end. The Bible attributes the parting to wind (Exodus 14:21); however, the event is similar to the theorized behaviour of a tsunami in the Red Sea generated by the Santorini explosion. The eruption of Krakatoa in 1883 generated tsunami waves in distant oceans and inland lakes by the transfer of energy from the passage of an atmospheric pressure-wave. The Santorini eruption was larger, and its atmospheric-induced pressure wave could easily have generated wind, and set up *seiching* (rhythmic oscillations in enclosed basins of water) in and around the Red Sea. The parting of the waters is characteristic of the initial drawdown of sea-level by some tsunami, while the subsequent swamping of the Egyptian army and cavalry represents the following inundation of the shoreline by one or more large tsunami

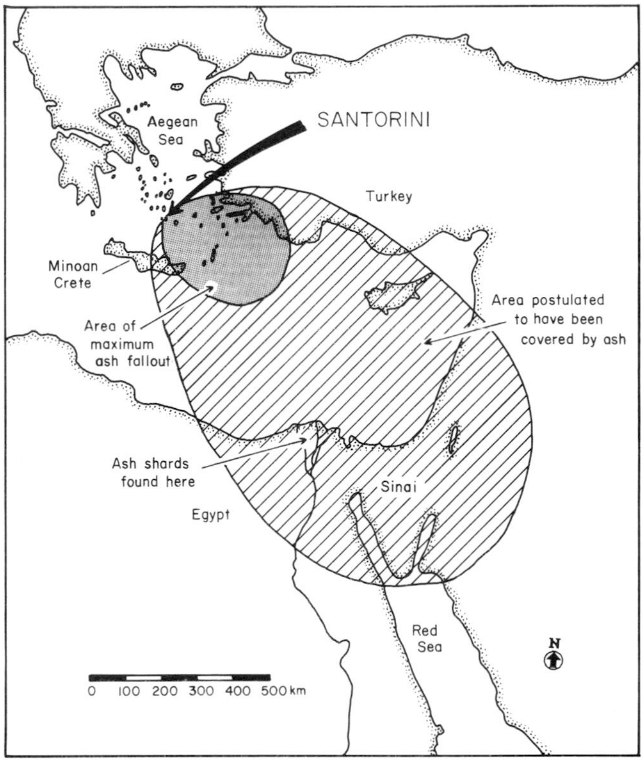

FIG. 1.2 Location of Santorini and extent of ash fallout following the eruption of 1628 BC

waves. In shallow embayments, this inundation can occur inland over long distances because of the long wave-length of the tsunami wave.

The Krakatoan eruption of 1883 also spawned many legends in surrounding islands. Blong (1982) describes modern legends in Papua New Guinea that can be related to actual volcanic eruptions in the seventeenth century. The Pacific region, because of its volcanic activity, abounds in eruption mythology. For example, Pele, the fire goddess of Hawaii, is chased by her sister from island to island eastwards across the Pacific. Each time she takes refuge in a volcano, she is found by her sister, killed, and then resurrected. Finally, Pele implants herself triumphantly in the easternmost island of Hawaii. Remarkably, this legend parallels the temporal sequence of volcanism in the Hawaiian islands.

Earthquakes also get recounted in many myths. They are attributed by animistic societies in India to animals trying to get out of the Earth. Sometimes the whole Earth is viewed as an animal. Ancient Mexicans viewed the Earth as a gigantic frog that twitched its skin once in a while. Timorese thought that the Earth was supported by a huge giant, who shifted it from one shoulder to the other whenever it got too heavy. Greeks did not ascribe earthquakes to Atlas, but to Poseidon disturbing the waters of the sea, which set off earth tremors. The oscillating water levels around the Greek islands during earthquakes were evidence of this god's actions. Earthquakes in the Bible were used by God to punish humans for their sins. Sodom and Gomorrah were destroyed because they refused to give up sins of the flesh (Genesis 19:23–25). The fire and brimstone from heaven, mentioned with these cities' destruction, were most likely associated with the lightning that is often generated above earthquakes by the upward movement of dust. The Bible also viewed earthquakes as divine visitation. The arrival of the angel sent to roll back Christ's tombstone was preceded by an earthquake (Matthew 28:2).

Until the middle of the nineteenth century, western civilization regarded hazards as 'acts of God' in the strict biblical sense, as punishment for people's sins. The Black Death, which swept through Europe in the fourteenth century, was viewed in these terms, and gave rise to fanatical religious fervour wherever it struck. The Lisbon earthquake of 1 November 1755, which completely destroyed that city, a major center of European civilization at the time, had a major impact on the then developing European Enlightenment. The earthquake struck on All Saint's Day, when many Christian believers were praying in church. Wesley viewed the Lisbon earthquake as God's punishment for the licentious behaviour of believers in Lisbon, and retribution for the severity of the Portuguese Inquisition. Kant and Rousseau viewed the disaster as a natural event, and emphasized the need to avoid building in hazardous places. The Lisbon earthquake also awakened scientific study of geological events. Mitchell in 1760 documented the spatial effects of the earthquake on lake levels throughout Europe. He found that no seiching was reported closer to the city than 700 km, nor farther away than 2500 km. The seiching affected the open coastline of the North Sea and also occurred in Norwegian fjords, Scottish lochs, Swiss Alpine lakes and rivers and canals in western Germany and the Netherlands. He deduced that there must have been a progressive, wave-like tilting movement of the Earth outwards from the center of the earthquake, which was different from the type of wave produced by a volcanic explosion.

Catastrophism vs uniformitarianism (Holmes, 1965; Gould, 1965; Woodcock and Davis, 1978)

Mitchell's work in 1760 on the Lisbon earthquake effectively represents the separation of two completely different philosophies for viewing the physical behaviour of the natural world. Before the 1800s, geological, but not scientific methodology was dominated by the *Catastrophists*. This word should be treated with caution because it does not now mean what it did before 1800. A Catastrophist was a person who believed that the shape of the Earth's surface—the stratigraphic breaks evidenced in rock columns—and the large events which were associated with observable processes, were cataclysmic: more importantly they were acts of God. Lyell, one of the fathers of Geology, sought to replace this 'catastrophe theory' with the idea that geological and geomorphic features were the result of cumulative, slow change, by natural processes operating at relatively constant rates. The term 'natural' implies that the processes would follow laws of nature, as defined by physicists and mathematicians going back to Bacon.

Whewell in a review of Lyell's work coined the term *uniformitarianism*, and subsequently a long-term debate broke out, on whether or not the slow processes we observe at present apply to past unobservable events. Some thought to confuse the issue more by stating that 'the present is the key to the past'. In fact, the idea of uniformitarianism involves two concepts. The first implies that geological processes follow natural laws applicable to science (often termed 'methodological uniformitarianism'): there are no acts of God. This type of uniformitarianism was established to counter the arguments raised by the Catastrophists of the time. The second concept implies constancy in the rates of change and in physical form through time ('substantive uniformitarianism'). The type and rate of processes operative today have remained constant throughout geological time. For example, waves break upon a beach today in the same manner that they would have 100 million years ago (assuming that the Earth has not expanded and, hence, that gravitational constant has remained the same). Few people would object to this concept; however, it is not strictly true. There is no analogy at present to the type of mountain-building processes that formed the Alps, nor to the *mass extinctions* and sudden discontinuities that have dominated the geological record. Also included in this second concept is the belief that the rate of change of physical form (namely, landscape development) is slow. Anyone who has witnessed a faultline being upthrusted during an earthquake would find this view limiting. In fact, the second concept of uniformitarianism is irrelevant. From a philosophical point of view it is simply inductive reasoning, which states that past processes and landforms can be implied from present, rigidly gathered, observations (empirical scientific methodology). The statement, 'the present is the key to the past', does not substitute for inductive reasoning any better than the term 'uniformitarianism' does. There is, thus, no need for the term or debate surrounding uniformitarianism. In one sense, it was thought up to counter the belief that acts of God dictate nature; in another sense, it implies nothing more than inductive reasoning, which any scientist accepts outright.

Catastrophe theory (Zeeman, 1977; Woodcock and Davis, 1978)

While the term 'uniformitarianism' can be dismissed, the term 'catastrophism' has come full circle. *Catastrophe* theory today no longer implies acts of God. The term implies large magnitude, or sudden events, that appear to alter completely processes or landscapes that existed before the event. In this sense, catastrophism can be divided into two concepts. The first implies a large magnitude event which may or may not be rare. A large coastal storm may be of this type, as long as one can still recognize that the beach exists in the same form, and with the same wave and current processes as before the storm. The second concept implies some sudden irrevocable change, or at least an altered state that will not easily revert back to the pre-catastrophe state. An earthquake lifting part of the sea bed above sea-level would be of this type. The landforms, and the processes operating after the event, are not the same as those operating beforehand.

Irrevocable catastrophism was explained succinctly in mathematical models by the French mathematician René Thom in the 1960s. Consider the diagram in figure 1.3, which plots the deformation of an object as stress is applied to it. This graph might represent the movement of a piece of chalk skipping across a blackboard surface, or it could describe the displacement of the Earth's crust along a plate boundary. Segments *a* and *c* can be described mathematically by simple linear equations that are similar in formulation to each other; however, these segments cannot be described together because of the rapid change that is occurring at point *b*. In reality, the earthquake process is not well characterized in figure 1.3. The figure is one-dimensional, in that there is only one controlling or independent factor—stress. The process of deformation also depends upon a second controlling factor—the rate at which stress is applied over time. The deformation process can now be modelled as a surface similar to that shown in figure 1.4. The surface in this diagram is a stability surface, in that the magnitude of deformation can only be plotted on this surface, and not under or above it. The surface, however, contains a fold or area of discontinuity that is two-dimensional. Low stress over a short period of time produces minimal

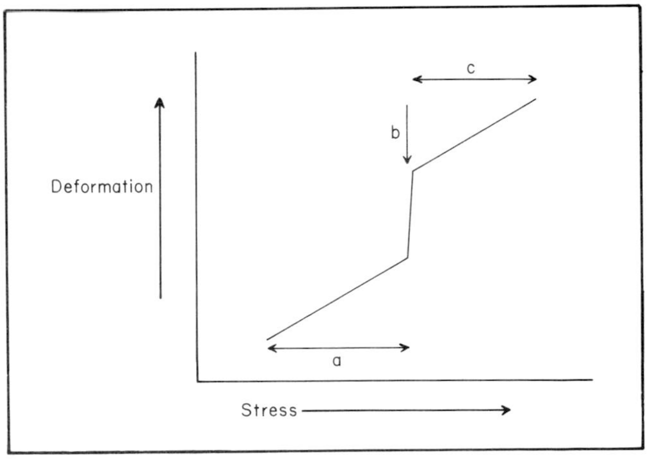

FIG. 1.3 Schematic representation of the relationship between the stress applied to an object and its deformation. Point *b* marks a discontinuity or sudden change.

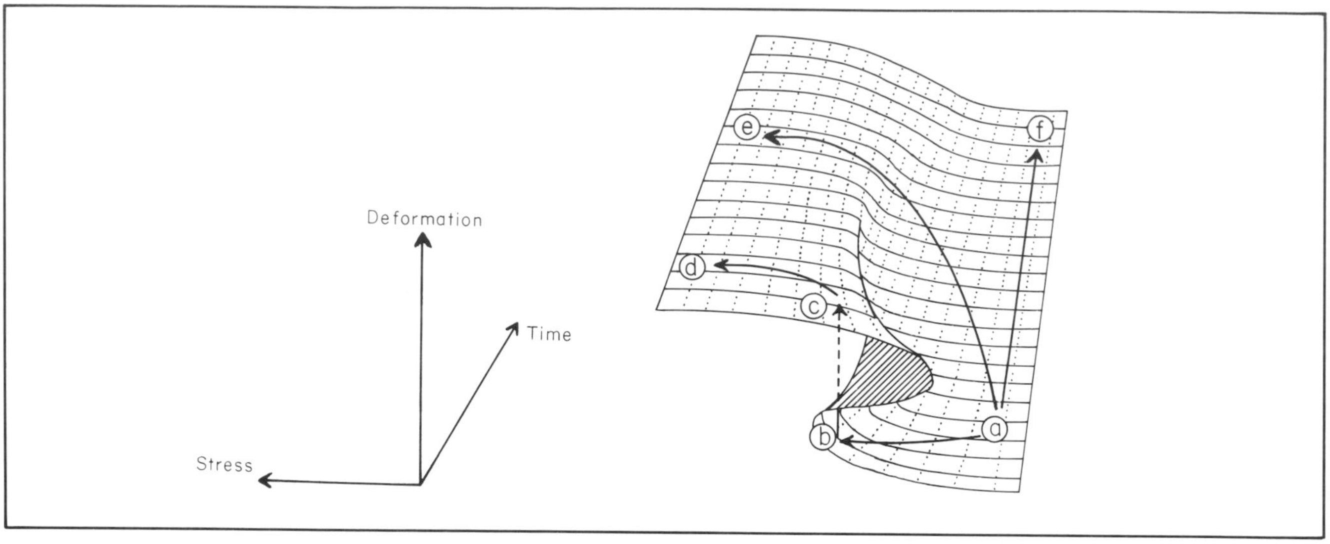

FIG. 1.4 The same relationship as plotted in figure 1.3, with the addition of a second controlling factor, time. The surface of this map plots deformation and represents the elementary cusp catastrophe formulated mathematically by René Thom.

deformation. If this low stress is applied over a long period of time (path *a* → *f*), then some deformation may occur. However, if high stress is applied over a short timespan, deformation may initially occur at a slow rate (*a* → *b*), but then a point is reached whereupon sudden deformation occurs (*b* → *c*). The result is a sudden change in the degree of deformation along the surface. If stress is increased more, greater deformation will result (*c* → *d*) without any further sudden change. If the same degree of stress is applied at a slower rate (path *a* → *e*), no sudden deformation need occur. As long as stress is applied within a certain time period, within the region of the fold (technically labelled a 'cusp' for a surface), then there will be a discontinuous shift in the degree of deformation. Such a process is what happens to toffee candy. If it is twisted slowly, it deforms gradually and behaves as a viscous fluid. If it is twisted sharply, it will snap behaving as a rigid solid.

Figure 1.4 is a geometric representation of a system of behaviour having one behavioural axis and two controlling functions. It is termed a 'cusp catastrophe'. Thom's definition of a catastrophe, in the broadest sense, describes any discontinuous transition that may occur in a system that can have more than one stable state (locations *d* and *e* in figure 1.4), or can follow more than one stable pathway of change (*a* → *b* → *c* → *d*, or *a* → *e* in figure 1.4). At first, catastrophe theory appears elegant. Mathematically there are only seven fundamental or elementary catastrophes, of which only the simplest—the fold (figure 1.3) and the cusp (figure 1.4)—can be visualized completely in two dimensions. These catastrophes seem to represent, ideally, behaviour in the real world, and have been applied successfully in Biological and Social Sciences. However, catastrophe theory has not been easily substantiated in the Physical Sciences. Its application to geological phenomena has been hotly debated. In fact, it is doubtful if figure 1.4 describes earthquake behaviour. This is not to say that catastrophe theory is irrelevant in characterizing natural hazards. It represents a recent development in mathematics, which attempts to simplify, unify and describe mathematically the underlying structure of physical processes in nature. Other fields of mathematics such as *bifurcation theory* and *chaos theory* are presently being used to resolve structure in seemingly random processes, and may be more appropriate than catastrophe theory in characterizing the behaviour and structure of the natural world. While this textbook minimizes the use of mathematics in characterizing natural hazards, the reader should be aware that the above mathematical concepts may some day supplant our present-day emphasis on process studies in natural hazard research.

The relationship between humans and natural hazards (Susman, et al., 1983; Watts, 1983)

The above concepts emphasize the *natural* aspect of natural hazards. Disasters can also be viewed from a sociological or humanistic viewpoint. An extensive chapter at the end of this book deals with this aspect by examining the human response to natural hazards at the personal or group level (part III—Social Impact). While there are great similarities in natural hazard response amongst various cultures, societies and political systems, it will be shown that fundamental differences also occur. This is evident especially in the different ways countries of contrasting political ideology, or of differing levels of economic development cope with drought. Both Australia and Ethiopia were afflicted by droughts of similar severity at the beginning of the 1980s. In Ethiopia drought resulted in starvation followed by massive international appeals for relief. However, in Australia no one starved and drought relief was managed internally.

In dealing historically with the scientific and mathematical description of natural hazards, it is apt to discuss a sociological viewpoint on the subject that has been formulated in the twentieth century. It should be emphasized that, in this book, I am not advocating the following sociological views, nor do I adhere to them in any way; but there are enough people in the world who do, so that the subject cannot be ignored. This sociological viewpoint states that the severity of a natural hazard depends upon who you are, and to what society you belong to at the time of the disaster. It is exemplified and expressed most forcibly by Marxist theory. Droughts, earthquakes and other disasters do not kill or strike people in the same way. The poor and oppressed suffer the most, experience worst the long-term effects, and are more likely to have the higher casualties as a result. In Marxist philosophy, it is meaningless to separate nature from society. People rely upon nature for fulfilment of their basic needs. Throughout history humans have met their needs by utilizing their natural environment through labour input. Humans do not enter into a set contract with nature. In order to produce, they interact with each other, and the intrinsic qualities of these interactions determine how individuals or groups relate to nature. Labour, thus, becomes the active and effective link between society and nature. If workers or peasants are able to control their own labour, they are better adapted to contend with the vagaries of the natural environment. The separation of workers from the means of production implies that their relationship with nature is governed by others, namely those people (capitalists or bosses) who control the means of production. This state of affairs is self-perpetuating. Because capitalists control labour and production, they are better equipped to survive and recover from a natural disaster than are the workers.

The main point about the Marxist view of natural hazards is that hazard response is contingent upon the position people occupy in society in the production process. This social differentiation of reactions to natural hazards is a widespread phenomenon in the Third World, where it is typical for people at the margins of society to live in the most dangerous and unhealthy locations. A major slum in San Juan, Puerto Rico, is frequently inundated at high tide; the poor of Rio de Janeiro live on the precipitous slopes of Sugarloaf Mountain, subject to landslide; the slum dwellers of Guatemala live on the steeper slopes affected by earthquakes; the Bangladeshi farmers of the Chars (low-lying islands at the mouth of the Ganges-Brahmaputra delta in the Bay of Bengal) live on coastal land subject to storm surge flooding. The contrast in hazard response between the Third World and Westernized society can be illustrated by comparing the effects of Cyclone Tracy—which struck Darwin, Australia, in 1974—to the effects of hurricane Fifi—which struck Honduras in September of the same year. Both areas lie in major tropical storm zones, and both possess adequate warning facilities. Both tropical storms were of the same intensity, and destroyed totally about 80 per cent of the buildings in the main impact zones. Over 8000 people died in Honduras, while only 64 died in Darwin. A Marxist would reason that the destitute conditions of underdevelopment in the Honduras accounted for the difference in the loss of life between these two events. Society is differentiated into groups with different levels of vulnerability to the hazard.

One might argue, on the other hand, that many people in Third World countries, presently affected by major disasters, simply are unfortunate to be living in areas experiencing large-scale climatic change or *tectonic* activity. After all, European populations were badly decimated by the climatic cooling that occurred after the middle of the fourteenth century, and everyone knows that the Los

Angeles area is well overdue for a major earthquake. One might also argue that Third World countries bring much of the disaster upon themselves. In many countries afflicted by large deathtolls due to drought or storms, the political systems are inefficient, corrupt or in a state of civil war. If governments in Third World countries took national measures to change their social structure, they would not be as vulnerable to natural disasters. For example, Japan, which occupies a very hazardous earthquake zone, has performed this feat through industrialization this century. One might also argue that Australians or Americans cannot be held responsible for the misery of the Third World. Apart from foreign aid, there is little we can do to alleviate their plight. The Marxist view disagrees with all these views. Marxism stipulates that Western, developed nations are partially responsible for, and perpetuate the effect of disasters upon people in the Third World, because we exploit their resources and cash-cropping modes of production. The fact that we give foreign aid is irrelevant. If our foreign aid is being siphoned off by corrupt officials, or used to build highly technical projects that do not directly alleviate the impoverished, then it may only be keeping underprivileged people in a state of poverty or, it is suggested, further marginalizing the already oppressed, and thus exacerbating their risk to natural hazards. Even disaster relief maintains the status quo, and generally prohibits a change in the status of peasants regarding their mode of production or impoverishment.

The basis of Marxist theory on hazards can be summarized as follows:

- the forms of exploitation in Third World countries increase the frequency of natural disasters as socio-economic conditions and the physical environment deteriorate;
- the poorest classes suffer the most;
- disaster relief maintains the status quo, and works against the poor even if it is intentionally directed to them; and
- measures to prevent or minimize the effects of disasters, which rely upon high technology, reinforce the conditions of underdevelopment, exploitation and poverty.

As stated earlier, Marxist philosophy does not form the underlying basis of this book. It has been presented simply to make the reader aware that there are alternate viewpoints about the effects of natural hazards. The Marxist view is based upon the structure of societies or cultures, and how those societies or cultures are able to respond to changes in the natural environment. The framework of this book, on the other hand, is centered around the description and explanation of natural hazards that can be viewed mainly as uncontrollable events happening continually over time. The frequency and magnitude of these events may vary with time, and particular types of events may be restricted in their worldwide occurrence. However, natural hazards cannot be singled out in time and space only when they affect a vulnerable society. An understanding of how, where and when hazards take place can only be achieved by studying all occurrences of that disaster. The arrival of a tropical cyclone at Darwin, with the loss of 64 lives, is just as important as the arrival of a similar-sized cyclone in the Honduras, with the loss of 8000 lives. Studies on both events contribute to our knowledge of the effects of such disasters upon society and permit us to evaluate how societies respond subsequently.

Chapter outlines

While the book has been written assuming little prior knowledge in Earth Science at the university level, there are occasions where basic terminology (jargon) has had to be used in explanations. More explicit definitions of some of these unfamiliar terms can be found in the glossary at the end of the book (terms appearing in the glossary are *italicised* where they are first discussed in the text). Individual chapters basically describe a specific type of hazard and its associated phenomena. The mechanisms controlling and predicting this particular hazard's occurrence are outlined and some of the more disastrous occurrences worldwide are summarized. Where appropriate, some of mankind's responses to, or attempts at mitigating the hazard are specified.

The hazards covered in this book are summarized in table 1.1 and a chapter reference for each hazard is given in the last column. These hazards are listed in relative order of importance based upon a subjective assessment that reflects the emphasis given to each in the text. Table 1.1 also grades the hazards, on a scale of 1 to 5, as to degree of severity, timespan of the event, total areal extent, total deathtoll, economic consequences, social disruption, long-term impact, lack of prior warning or suddenness of onset, and number of associated hazards.

The hazards are organized in the book under two main parts: climatic hazards and geological

Table 1.1 Ranking of Hazard Events by Characteristics and Impacts

Overall Rank[b]	Event	*Grading of Characteristics and Impacts*[a]									
		Degree of Severity	*Length of Event*	*Total Areal Extent*	*Total Loss of Life*	*Total Economic Loss*	*Social Effect*	*Long-term Impact*	*Suddenness*	*Occurrence of Associated Hazards*	*Chapter Location*[c]
1	Drought	1	1	1	1	1	1	1	4	3	5, 6
2	Tropical Cyclone	1	2	2	2	2	2	1	5	1	2
3	Regional Flood	2	2	2	1	1	1	2	4	3	5, 8
4	Earthquake	1	5	1	2	1	1	2	3	3	10, 11
5	Volcano	1	4	4	2	2	2	1	3	1	10, 12
6	Extra-tropical Storm	1	3	2	2	2	2	2	5	3	2
7	Tsunami	2	4	1	2	2	2	3	4	5	11
8	Bushfire	3	3	3	3	3	3	3	2	5	9
9	Expansive Soils	5	1	1	5	4	5	3	1	5	13
10	Sea-level Rise	5	1	1	5	3	5	1	5	4	4
11	Icebergs	4	1	1	4	4	5	5	2	5	4
12	Dust Storm	3	3	2	5	4	5	4	1	5	4
13	Landslides	4	2	2	4	4	4	5	2	5	13
14	Beach Erosion	5	2	2	5	4	4	4	2	5	4
15	Debris Avalanches	2	5	5	3	4	3	5	1	5	13
16	Creep & Solifluction	5	1	2	5	4	5	4	2	5	13
17	Tornado	2	5	3	4	4	4	5	2	5	3
18	Snowstorm	4	3	3	5	4	4	5	2	4	7
19	Ice at Shore	5	4	1	5	4	5	4	1	5	4
20	Flash Flood	3	5	4	4	4	4	5	1	5	8
21	Thunderstorm	4	5	2	4	4	5	5	2	4	7
22	Lightning Strike	4	5	2	4	4	5	5	1	5	7
23	Blizzard	4	3	4	4	4	5	5	1	5	7
24	Ocean Waves	4	4	2	4	4	5	5	3	5	4
25	Hail Storm	4	5	4	5	3	5	5	1	5	7
26	Freezing Rain	4	4	5	5	4	4	5	1	5	7
27	Localized Strong Wind	5	4	3	5	5	5	5	1	5	3
28	Subsidence	4	3	5	5	4	4	5	3	5	13
29	Mud & Debris Flows	4	4	5	4	4	5	5	4	5	13
30	Air-supported Flows	4	5	5	4	5	5	5	2	5	13
31	Rockfalls	5	5	5	5	5	5	5	1	5	13

[a] Hazard characteristics and impacts are graded on a scale of 1 (largest or greatest) to 5 (smallest or least significant).
[b] Overall rank is based on average grading.
[c] The relevant chapter number in this book for discussion of the specific hazard event.

hazards. Climatic hazards are discussed under the following chapter headings: large-scale storms; wind; oceanographic phenomenon such as waves, ice, sea-level rise and beach erosion; causes and prediction of drought and rainfall events; response to droughts; associated precipitation events; flooding hazards; natural fires. Chapter 2 on large-scale storms examines the atmospheric physics controlling wind speed, and discusses the formation of large-scale tropical vortices known as tropical cyclones. Tropical cyclones are the second most important hazard, generating the greatest range of associated hazard phenomena. Familiar cyclone disaster events are described together with their development, magnitude and frequency, geological significance, and impact. The response to cyclone warnings in Australia is then compared to that in the United States and Bangladesh. Extra-tropical cyclones are subsequently described with reference to major storms in the northern hemisphere, and the May–June storms of 1974 in New South Wales, Australia. The chapter concludes with a discussion on the mechanisms responsible for storm surges produced by the above events. Here, the concepts of probability of occurrence and exceedence are introduced.

Chapter 3 summarizes more minor hazards generated by wind. Attention is given to the description of tornadoes, and to the various theories put forward for their generation. Tornado disasters are summarized, and the geological and ecological significance of such events is presented with some description of the measures used to avert such disasters in the United States. Following this, the circumstances favouring the formation of dust storms is discussed. Here, astronomical cycles, such as sunspots or the 18.6-year lunar tide, are linked for the first time in the book to the occurrence of a natural hazard—namely the dust-storm frequency in the United States. Major dust-storm events worldwide are summarized, and the geological significance of such storms emphasized. The chapter concludes with a description of the mechanics of more minor, localized strong wind events.

Chapter 4 deals with oceanographic hazards encompassing a variety of phenomena. Most of these hazards affect few people but they have long-term consequences, especially if global warming becomes significant over the next 50 years. Because of their areal extent and long period of operation, most of the phenomena in this chapter can be ranked as middle-order hazards. Waves are described first together with the theory for their generation. The worldwide occurrence of large waves is illustrated using ship observation and *SEASAT altimeter* data. This is followed by a commentary on the additive effect of individual waves incorporated within wave spectra. Secondly, a brief description is given on hazardous sea-ice phenomena in the ocean and at shore. Again, the effect of sunspot cycles is noted, this time upon ice occurrence. Thirdly, one of the greatest concerns today is the impending rise in sea-level that is supposedly happening because of melting of icecaps or thermal expansion of oceans within a Greenhouse warming environment. The Greenhouse effect is briefly discussed, with a presentation of some of the restraints upon such a scenario. Worldwide sea-level data are then examined to show that sea-levels may not be rising worldwide as generally believed, but may be influenced in behaviour by regional climatic change. Finally, the chapter discusses the various environmental mechanisms of sandy beach erosion (or accretion) caused mainly by changes in sea-level, rainfall or storminess. This evaluation utilizes an extensive data set of shoreline position and environmental variables collected for Stanwell Park, Australia, between 1943 and 1978. Much of the discussion in this chapter draws heavily upon the author's own research experience.

The above chapters have all alluded to the coincidence of high-magnitude, climatic hazards and peaks in the sunspot cycle. This point is developed further in chapter 5, which discusses the causes and prediction of worldwide drought and flood. Because of its severity and global occurrence, drought is probably the worst hazard to afflict the human race. To begin with, the more obvious causes of drought, such as changes in global circulation patterns, are described. This is followed by an evaluation of people as a major factor instigating, or prolonging droughts in India, the Sahel region of Africa, and Australia. In recent years, a major link has been found between air circulation in the South Pacific region, and the occurrence of drought and heavy rainfall in the southern hemisphere. Intense, tropical, east-west air circulation exists between South America and Australia. This circulation reverses aperiodically producing drastic climatic changes over a 2–7-year timespan. This phenomenon, termed the *Southern Oscillation*, is explained in detail, together with a description of its worldwide *teleconnections*, especially around the Indian Ocean and within Australia. It is proposed that the 11- and 22-year sunspot, and 18.6-year lunar cycles, may affect the behaviour of the Southern Oscillation. These astronomical cycles are defined, and evidence presented to show a correspondence between droughts or periods of heavy rainfall and the 18.6-year lunar cycle worldwide.

Because chapter 5 emphasizes drought events worldwide, it is only logical that it be followed in chapter 6 by a discussion of the world's response to drought conditions. Different responses for Third World African countries are discussed in relationship to responses in developed Westernized countries such as the United States, England and Australia. The chapter continues by describing the way the international community alleviates drought. It is only apt that this section concludes with a description of one of the most successful attempts at disaster relief on a world scale. The inspiration and enthusiasm shown by Bob Geldof and the reactions of individuals to his initiatives, regardless of their economic or political allegiance, gives hope that the effects of even the worst natural calamities affecting humanity can be minimized. The chapter concludes by discussing the long-term effects of drought, which add to the destructiveness of this hazard.

The book then progresses to a discussion of natural hazards associated with extreme precipitation and drought. The minor hazards associated with precipitation are discussed in chapter 7. While many of these hazards are very localized, some—like blizzards and freezing rain—occur frequently enough and are such a major disruption to community activities that they warrant discussion. The chapter systematically examines thunderstorms, hail and lightning, followed by an evaluation of snowstorms, blizzards and freezing rain as hazards in temperate regions. Chapter 8 covers flooding hazards separately. Flash floods and regional flooding are usually severe enough to invoke states of emergency. In addition, regional flooding accounts for some of the largest deathtolls registered by any hazard event, ranking third in severity and impact after drought and tropical cyclones. This chapter begins by introducing the climatic processes responsible for flash flooding. This is followed by presentations of examples of flash flooding events in the United States and Australia. Regional flooding is discussed in the context of the Mississippi River floods of 1973 and the Great Australian Floods of 1974. Finally, the chapter concludes with an evaluation of the geological consequences of flooding and its role in shaping the course of the Mississippi and the Hwang Ho (Yellow) rivers.

Drought-induced bushfires or forest fires are then treated as a separate entity in chapter 9. It is shown that different types of fires are restricted to certain vegetation zones worldwide. The conditions favouring intense bushfires and the causes of such disasters are described. Major natural fire disasters in the world, and in the United States and Australia, are then presented with an emphasis on events in Australia. This chapter concludes with a discussion of the quandary whether or not to use *prescribed burning* as a mitigation method to deal with the encroachment of urban development into forests.

Geological hazards, the second part of the book, are covered under the following chapter headings: causes and prediction of earthquakes and volcanoes, earthquakes, volcanoes and land instability. Chapter 10 presents the worldwide distribution of earthquakes and volcanoes, together with a presentation of the scales for measuring earthquake magnitude. This is followed by an examination of causes of earthquakes and volcanoes under the headings of plate boundaries, hot spots, regional faulting and the presence of reservoirs or dams. Next, the presence of cyclicity in the occurrence of earthquakes and volcanoes is examined. The long-term prediction of volcanic and seismic activity is evaluated, by correlating the occurrence of these geological hazards with variations in daily length of the Earth's rotation, the Southern Oscillation and geomagnetic activity. The chapter concludes with a presentation on the geophysical, geochemical and geomagnetic techniques for forecasting volcanic and seismic activity over the short term.

This introductory chapter is then followed by two chapters that are concerned separately with earthquakes and volcanoes. Both hazards, because of their suddenness and high energy release, have the potential to afflict human beings physically, economically and socially. Both rank as the most severe geological hazards. Chapter 11 first describes types of seismic waves and seismic risk maps. Earthquake disasters for Tokyo and Alaska are then recounted, followed by an assessment of the Californian earthquake hazard. The associated phenomenon of liquefaction or thixotrophy, and its importance in earthquake damage, is subsequently presented. One of the major phenomena generated by earthquakes is a tsunami. This phenomenon is described in detail, together with a presentation of major, worldwide tsunami disasters. The prediction of tsunami in the Pacific region is then discussed, and the tsunami threat for Australia evaluated. Chapter 12, on volcanoes, emphasizes types of volcanoes and associated hazardous phenomena such as lava flows, tephra clouds, pyroclastic flows and base surges, gases and acid rains, lahars (mud flows), and glacier bursts. The major disasters of Vesuvius (AD 79), Mount Pelée (1902), Krakatoa (1883) and Mount St Helens (1980) are then discussed in detail.

The section on geological hazards concludes with a comprehensive treatment of land instability in chapter 13. This chapter opens with a description of basic soil mechanics including the concepts of stress and strain; friction and cohesion; shear strength of soils; pore-water pressure; and rigid, elastic and plastic solids. Land instability is then classified and described under the headings of expansive soils, creep and solifluction, mud and debris flows, landslides and slumps, rockfalls, debris avalanches, and air-supported flows. Many of these hazards rank in severity and impact as middle-order events. For example, expansive soils, while not causing loss of life, rank as one of the costliest hazards found over the long term because of their ubiquitousness. For each of these categories, the type of land instability is presented in detail, together with a description of major disasters. The chapter concludes with a cursory presentation on the natural causes of land subsidence.

Because the emphasis of the previous chapters has been on the physical mechanisms of hazards, the book closes in chapter 14 with a discussion of the social impact of many of these phenomena. Here, the way hazards are viewed by different societies and cultures is emphasized, which is then followed with detailed descriptions of responses before, during and after the event. As much as possible, the characterization of responses under each of these headings is elucidated using anecdotes. It is shown that individuals or families have different ways of coping with, or reacting to disaster. The chapter concludes by emphasizing the psychological ramifications of a disaster on the victims, the rescuers and society.

References

Blong, R. J. 1982. *The time of darkness, local legends and volcanic reality in Papua New Guinea.* Australian National University Press, Canberra.

Further reading

Day, M. S. 1984. *The many meanings of myth.* Lanham, NY.

Gould, S. J. 1965. 'Is Uniformitarianism necessary?'. *American Journal of Science* v. 263 pp. 223–28.

Holmes, A. 1965. *Principles of Physical Geology.* Nelson, London.

Milne, A. 1986. *Floodshock: the drowning of planet Earth.* Sutton, Gloucester.

Myles, D. 1985. *The Great Waves.* McGraw-Hill, NY.

Susman, P., O'Keefe, P. and Wisner, B. 1983. 'Global disasters, a radical interpretation'. In Hewitt, K. (ed.) *Interpretations of calamity.* Allen and Unwin, Sydney, pp. 263–83.

Watts, M. 1983. 'On the poverty of theory: natural hazards research in context'. In Hewitt, K. (ed.) *Interpretations of calamity.* pp. 231–61.

Woodcock, A. and Davis, M. 1978. *Catastrophe theory.* Penguin, Ringwood.

Zeeman, E. C. 1977. *Catastrophe theory: selected papers 1972–1977.* Addison-Wesley, Reading.

I

Climatic Hazards

2

Large-scale Storms as a Hazard

Introduction (Linacre and Hobbs, 1977; Davies, 1980)

Barometric pressure represents the weight of air above a location on the surface of the Earth. When the weight of air over an area is greater than over adjacent areas, it is termed 'high pressure'. When the weight of air is lower, it is termed 'low pressure'. Points of equal pressure across the Earth's surface can be contoured using *isobars*, and on weather maps contouring is generally performed at intervals of 4 *hectopascals* (hPa) or millibars (mb). Mean pressure for the Earth is 1013.6 hPa. Wind is generated by air moving from high to low pressure simply because a pressure gradient exists owing to the difference in air density (figure 2.1). The stronger the pressure gradient, the stronger the wind. This is represented graphically on weather maps by closer spaced isobars.

In reality, wind does not flow down pressure gradients but blows almost parallel to isobars at the surface of the Earth because of *Coriolis force* (figure 2.2). This force exists because of the rotation of the Earth and is, thus, illusionary. For example, a person standing perfectly still at the pole would appear to an observer viewing the Earth from the Moon to turn around in a complete circle every 24 hours as the Earth rotates. However, the same person standing perfectly still at the equator would always appear to be facing the same direction and not circumscribing a circle. Because of Coriolis force, wind tends to be deflected to the left in the southern hemisphere and to the right in the northern hemisphere. Coriolis force can be expressed mathematically by the following equation:

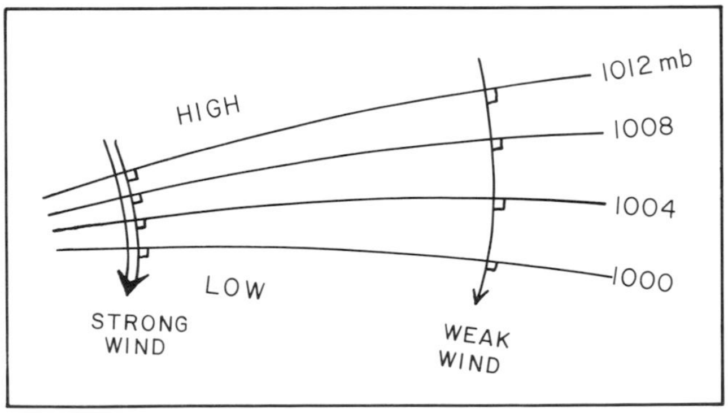

Fig. 2.1 Pressure gradient generation of winds between high and low pressure

$$\text{Force} = 2\omega \sin\phi \, v \qquad (2.1)$$

where ω = rate of spin of the earth
$\sin \phi$ = latitude
v = wind speed

Clearly, the stronger the pressure gradient, the stronger the wind and the more wind will tend to be deflected. This deflection forms a *vortex*. The stronger the wind, the smaller and more intense the resulting vortex. Very strong vortices are known as hurricanes, typhoons, cyclones and tornadoes. The above formula also indicates that Coriolis force is zero at the equator and increases to a maximum value at the pole. As a result, vortices do not develop at the equator and become stronger towards the pole.

This is not a complete picture of vortex formation. Air, besides moving across the surface of the earth, can also rise. This rising motion is a prerequisite for vortex development. As air rises, adjacent air must move in laterally to replace it. The faster air rises, the faster surface winds rush in. There are two significant areas of the world

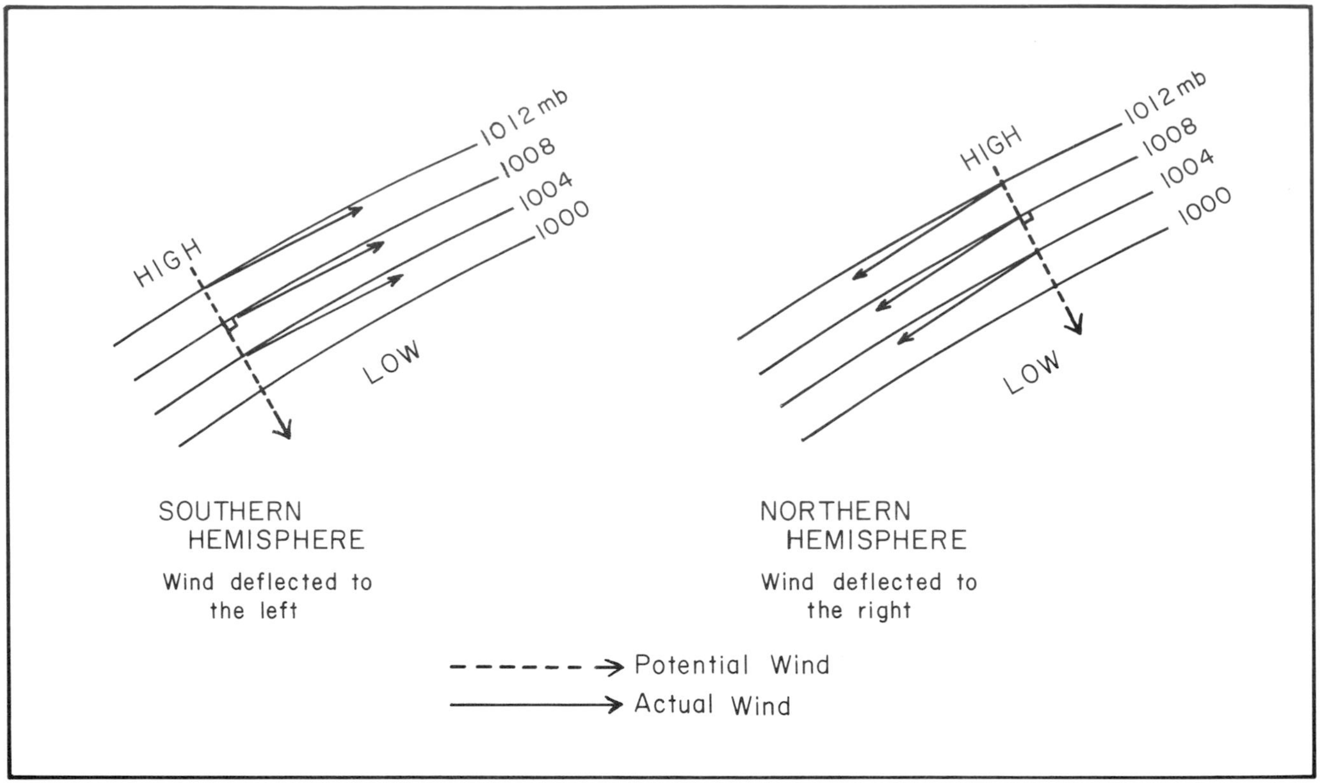

FIG. 2.2 Actual wind movement relative to isobars because of Coriolis force

where rising air motion predominates, leading to the formation of vortices. These areas can be located in relationship to the *general air circulation* pattern that controls air movement in the atmosphere between the equator and poles (figure 2.3). Intense heating by the sun at the equator causes air to rise and spread out polewards in the upper *troposphere*. As this air moves toward the poles, it cools and begins to descend back to the Earth's surface at 20°–30° North and South of the equator, resulting in high pressure known as *Hadley cells*. Upon reaching the Earth's surface, this air either returns to the equator or moves polewards. Because of Coriolis force, equatorial moving air forms two belts of easterly *trade winds* astride the equator (figure 2.4). Over the western sides of oceans these tropical easterlies pile up warm water, resulting in *convection* of air that forms low pressure and intense instability. In these regions, intense vortices known as *tropical cyclones* preferentially develop. At the pole, air cools and spreads towards the equator along the Earth's surface. Where cold polar air displaces upwards warmer subtropical air, a cold *polar front* develops with strong uplift and instability. Tornadoes and strong westerlies can be generated near the polar front over land, while intense extra-tropical storms develop near the polar front especially over water bodies. Two belts of strong wind and storms are thus formed in the zone of the roaring forties, dominated by westerlies on the equator side of the polar front in each hemisphere.

If one lived in Greenland, the Antarctic or downslope from mountain valleys (see figure 2.5 for the location of all major place names mentioned in this chapter), one would realize that other types of strong winds can also develop. These winds exist because of *topographic enhancement* or the flow downslope of colder, denser air under the effect of gravity. In this chapter, only strong regional wind vortices or storms will be examined. More localized wind hazards such as tornadoes, dust storms and topographically enhanced winds will be dealt with in chapter 3. In the present chapter, the process of tropical cyclone development and the magnitude and frequency of such storms will be described first, followed by a description of some of the more familiar cyclone disasters. The geological significance and impact of tropical cyclones will then be discussed. This section will conclude with a comparison of the human response to cyclones in Australia, United States and Eastern Pakistan

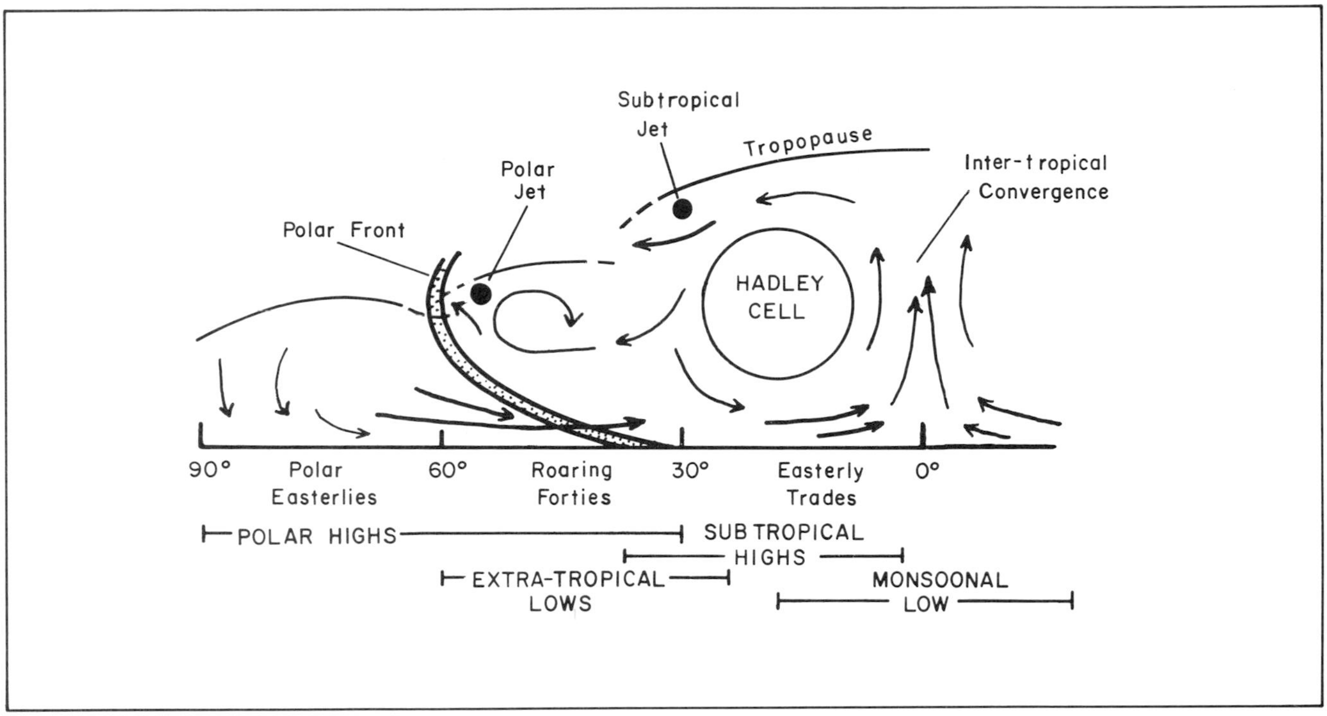

FIG. 2.3 Palmén schematic model of general air circulation between the pole and the equator

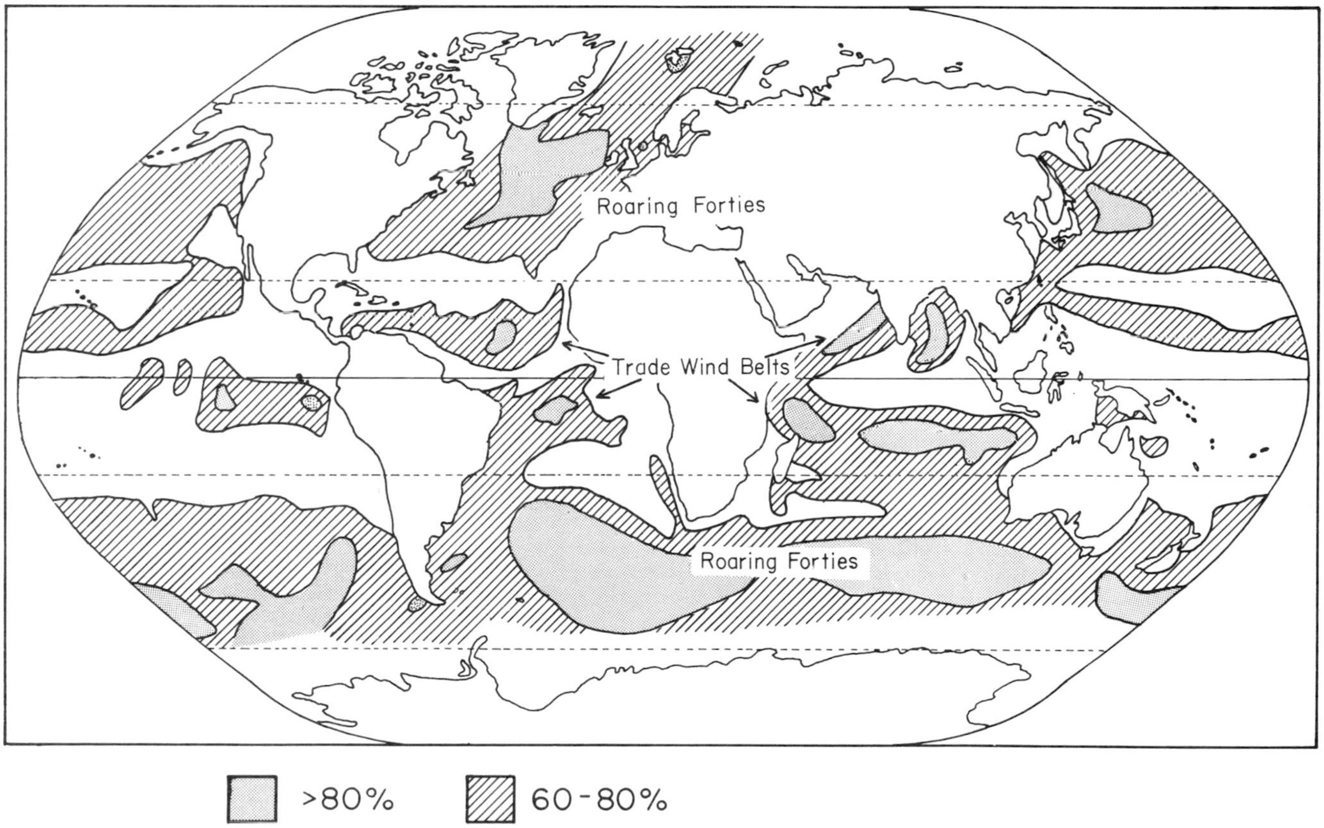

FIG. 2.4 Location of mid-latitude westerly and easterly trade wind belts with winds over 20 km hr^{-1} 60 per cent or 80 per cent of the time in either July or January (based upon Davies, 1980)

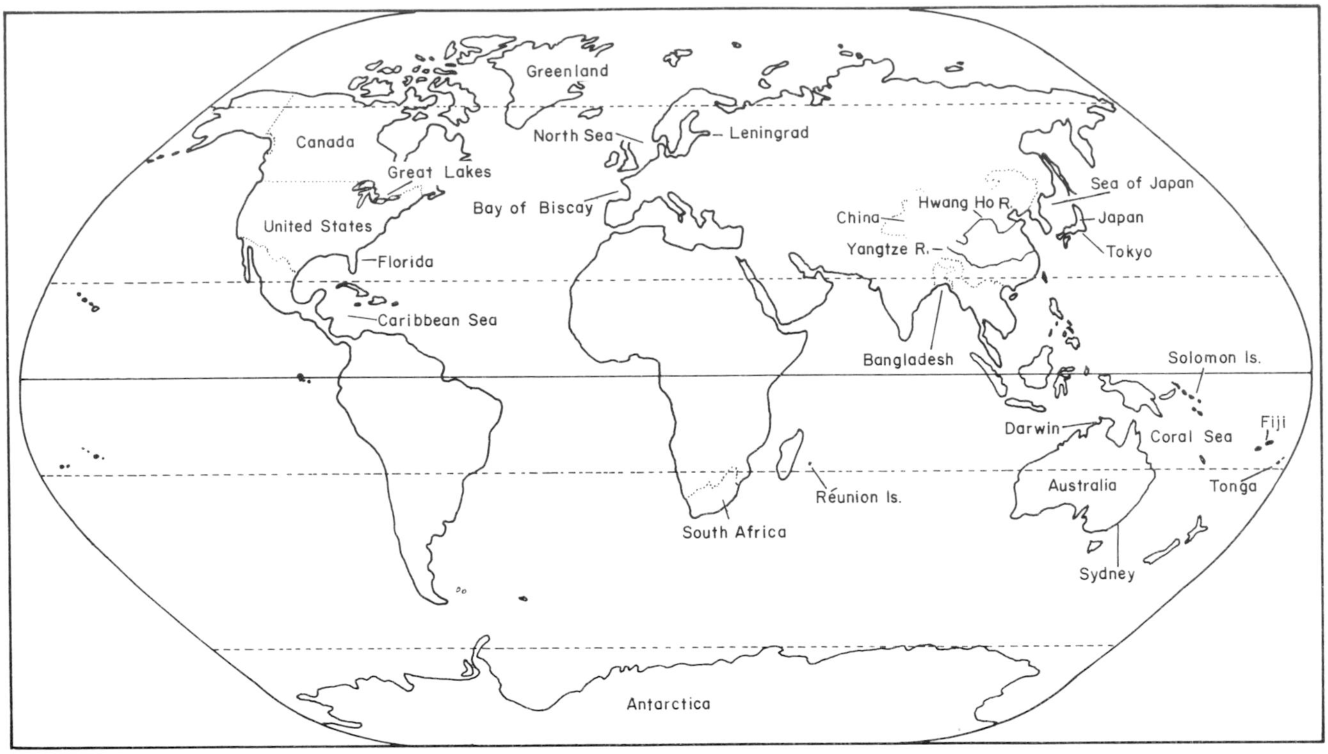

FIG. 2.5 Location map

(now Bangladesh). Extra-tropical cyclones will then be described with particular reference to major storms in the northern hemisphere and exceptional development of lows or *bombs* off the east coast of Australia and, possibly, Japan and the United States. The chapter will conclude with a discussion of the factors controlling the development of *storm surges* generated by the above events.

Tropical cyclones

Introduction (Nalivkin 1983)

Tropical cyclones are defined as intense cyclonic storms that originate over warm tropical seas. In North America, the term 'hurricane' is used because cyclone refers to an intense, counterclockwise rotating, extra-tropical storm. In Japan and south-east Asia tropical cyclones are called 'typhoons'. Figure 2.6 summarizes the hazards relating to tropical cyclones. These can be grouped under three headings: storm surge, wind and rain effects. Storm surge is a phenomenon whereby water is physically piled up along a coastline by low pressure and strong winds. This leads to loss of life through drowning, inundation of low-lying coastal areas, erosion of coastline, loss of soil fertility due to intrusion by ocean salt-water and damage to buildings and transport networks. High-wind velocities directly can cause substantial property damage and loss of life, and constitute the main agent for crop destruction. Surprisingly, strong winds—simply because they are so strong—can also exacerbate the spread of fires in urban and forested areas, even under heavy rainfall. Rainfall is responsible for loss of life, property damage and crop destruction from flooding, especially on densely populated floodplains. Contamination of water supplies can lead to serious disease outbreaks weeks after the cyclone. Heavy rain in hilly or mountainous areas is also responsible for *landslides* or mud flows as floodwaters in stream and river channels mix with excess sediment brought down slopes. The destruction of crops and saline intrusion can also result in famine that can kill more people than the actual cyclone event. This was especially true on the Indian subcontinent during the latter part of the nineteenth century.

Earthquakes are not an obvious consequence of cyclones; however, there is substantial evidence for their occurrence during cyclones. Pressure can vary dramatically in a matter of hours with the passage of a cyclone, bringing about a consequentially large decrease in the weight of air above the Earth's surface. The deloading can be as much as 2–3 million tonnes km^{-2} over a matter of hours. In

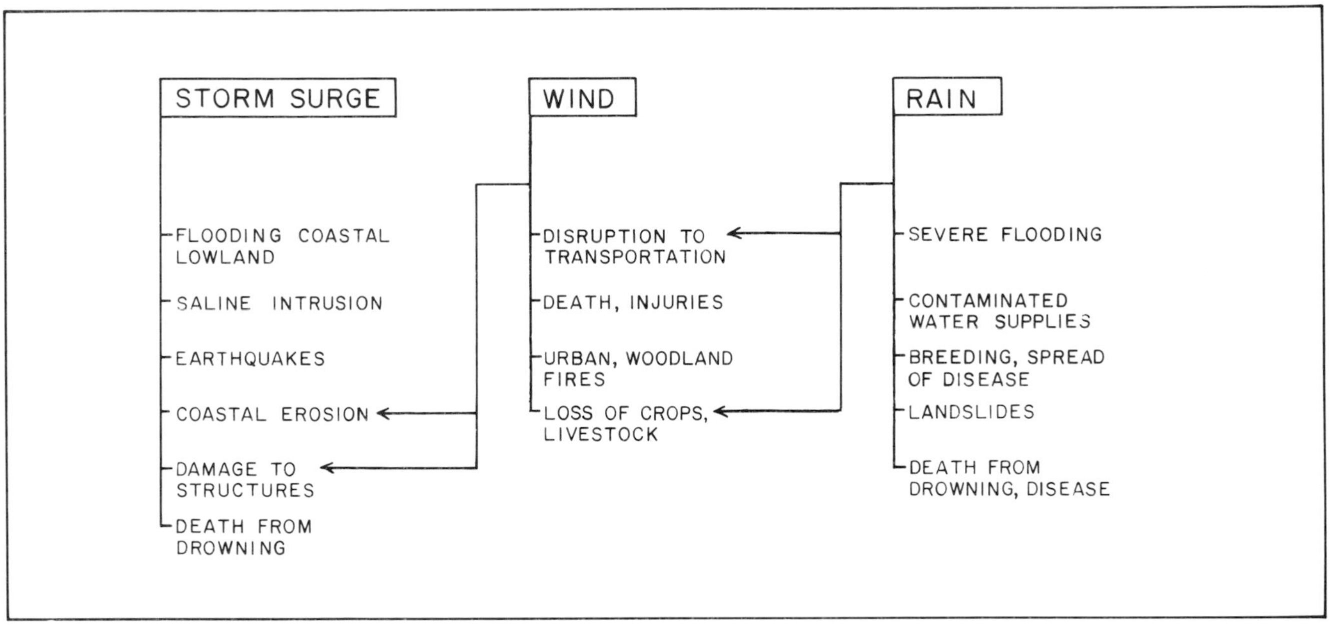

Fig. 2.6 Hazards related to occurrence of tropical cyclones

addition, tidal waves or surges on the order of 10–12 m in height can occur in shallow seas with a resulting increase in pressure on the Earth's surface of 7 million tonnes km^{-2}. In total the passage of a cyclone along a coast can induce a change in load on the Earth's *crust of* 10 million tonnes km^{-2}. In areas where the Earth's crust is already under strain, this pressure change may be sufficient to trigger an earthquake. The classic example of a cyclone-induced earthquake occurred with the Tokyo Earthquake of 1923. A typhoon swept through the Tokyo area on 1 September, and was followed by an earthquake that evening. The earthquake caused the rupture of gas lines, setting off fires that were fanned by cyclone-force winds through the city on 2 September. In all, 143 000 people lost their lives, mainly through incineration. The events of this tragedy will be described in more detail in chapter 11. There is also evidence that tropical cyclones have triggered earthquakes in other places along the western margin of the Pacific *plate* and along plate boundaries in the Caribbean Sea. In Central America the coincidence of earthquakes and cyclones has a higher probability of occurrence than the joint probability of each event separately.

The Tokyo earthquake is also unusual for one other reason. The fires were blown out of control by the typhoon winds. This dichotomous occurrence of torrential rain and uncontrolled fire is quite common in Japan. Typhoons moving northeast in the Sea of Japan produce *föhn* winds in the lee of mountains. These winds, discussed in more detail in the following chapter, can be very strong. Outside of main areas of precipitation they can also be very dry or desiccating. In September 1954 and, again, in October 1955 similar cyclone-generated winds spread fires that destroyed over 3300 buildings in Hokkaido and 1100 buildings in Niigata respectively. Nor is the phenomenon of cyclone-fire unique to Japan. The great hurricane of 1938 that devastated New England, in the United States, started fires in New London, Connecticut, that raged for six hours, and would have destroyed the city of 30 000 people if it was not for the fact that the fires were turned back on themselves as the winds reversed direction with the passage of the cyclone.

Mechanics of cyclone generation (Linacre and Hobbs, 1977; Anthes, 1982; Nalivkin, 1983; Holthouse, 1986)

Tropical cyclones derive their energy from the evaporation of water over oceans, particularly the western parts, where surface temperatures are warmest. As the sun's apparent motion reaches the summer *equinox* in each hemisphere, surface ocean waters begin to warm over a period of two to three months to temperatures in excess of 26°C. Tropical cyclones, thus, develop from December to May in the southern hemisphere and from June to October in the northern hemisphere. At the same time, easterly trades on the equatorial side of the Hadley cell intensify and blow these warm surface waters to the westward side of oceans. Hence, the warmest ocean waters accumulate to a great depth in the Coral Sea off the east coast of Australia, in the Caribbean and in the west Pacific Ocean, southeast of China. A deep layer of warming reduces the upward mixing of cold sub-surface water that can truncate the cyclone before it fully develops. Warm water also develops in shallow seas because there is a smaller volume of water to heat up. Cyclonic depressions begin to develop when ocean temperatures exceed 24°C; however, the *eye* structure will not develop until sea surface temperatures reach 26–27°C. Cyclones rarely develop poleward of 20° latitude because ocean temperatures never reach these latter crucial values. Figure 2.7 presents the origin, worldwide, of all recorded cyclones for the 20-year period 1952–71. The majority of tropical cyclones during this period originated either at the western sides of oceans, or over shallow seas, where temperatures were enhanced by one of the above processes.

Warm ocean temperatures, by themselves, are not sufficient to initiate cyclones. Additionally, there must be convergence of air taking place over these sites. This convergence is frequently induced by easterly wave development in the trade winds, and is similar to that caused by wave motion along the polar front that initiates *extra-tropical depressions.* In both situations, resulting convergence causes upward movement of air and creates instability in the upper atmosphere. However, in the case of tropical cyclones, this instability is enhanced by the release of *latent heat of evaporation.* Because of the warm ocean temperatures and intense solar heating, evaporation at the sea surface is optimum. Each kilogram of water evaporated at 26°C consumes 2425 kilojoules of heat energy. As air rises, it cools, and this energy is released as latent heat of evaporation. Vast amounts of heat energy are thus transferred to the upper atmosphere causing *convective instability* that forces air upwards even higher. Convective instability will persist as long as this heating remains over a warm ocean. Figure 2.8

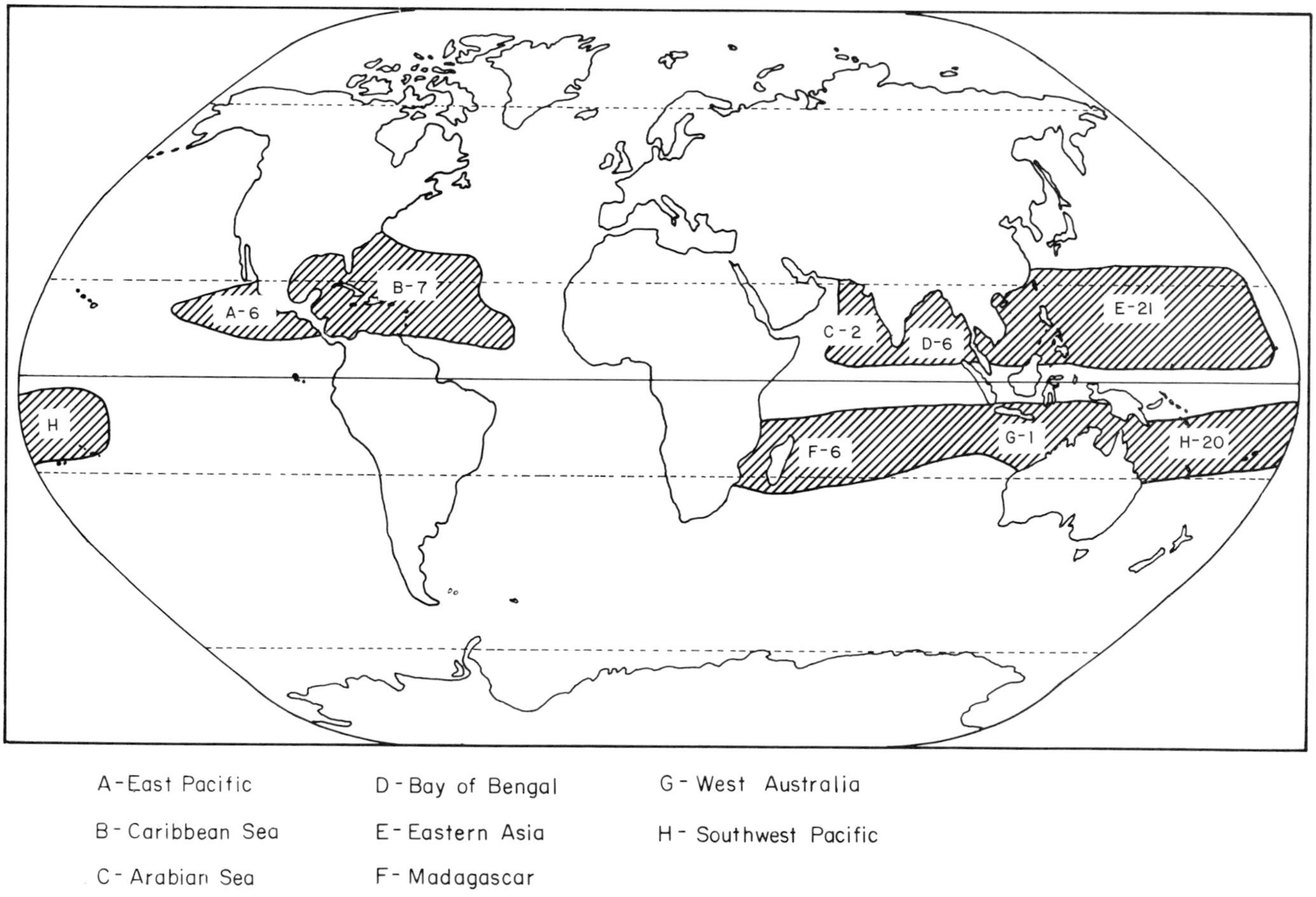

FIG. 2.7 Origin and number of tropical cyclones per year: 1952–71 (after Gray, 1975). The number in each zone refers to the average number per year.

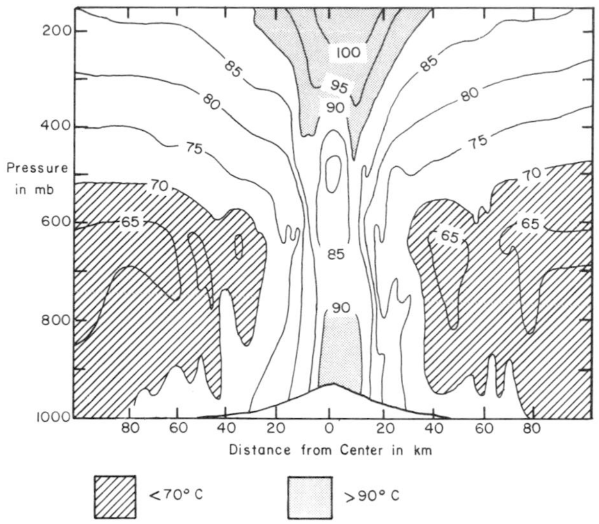

Fig. 2.8 Potential temperature structure due to latent heat of evaporation in Hurricane Inez, 28 September 1966 (adapted from Hawkins and Imbembo, 1976)

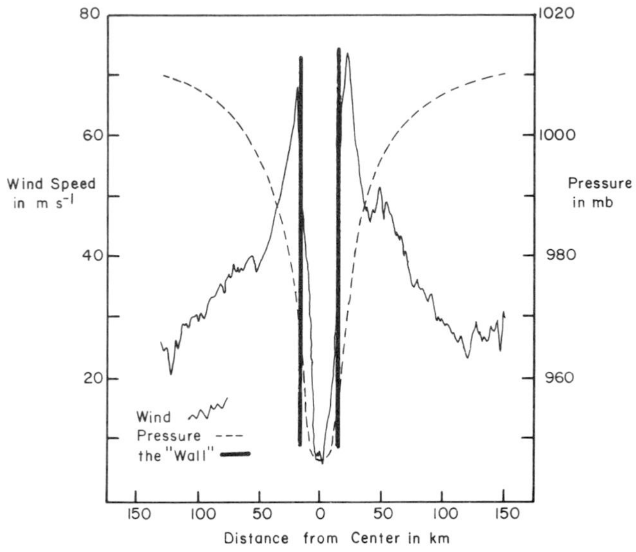

Fig. 2.9 Wind and pressure cross-sections through Hurricane Anita, 2 September 1977 (based upon Sheets, 1980)

illustrates the type of temperature structure that can develop in a tropical cyclone because of this process. The temperatures shown schematically represent the maximum possible temperature, in degrees Celsius, if all evaporated moisture was condensed. Note that there is a core of heating at the center of the cyclone that can maintain air temperatures up to 20° C warmer than the adjacent environmental level at any particular height throughout the atmosphere. This core area forms the eye of the cyclone and is a necessary prerequisite before any tropical depression can be labelled a cyclone. The intense convection that develops as a result of the formation of the eye causes more moist air to be sucked into the area of instability at the surface. As this air moves into the zone of convergence, it begins to rotate under Coriolis force and imparts to a tropical cyclone a second feature, namely a strong wind-vortex over a large area. Note that this rotation or vorticity is usually initiated by some prior feature such as an *easterly wave*. This second condition for formation excludes cyclone development within 5° of the equator because Coriolis force is too ineffective to generate rotation wind flow.

If the convective zone increases to a distance of around 10–100 km, then subsidence of air takes place in the center of the zone of convection as well as to the sides. This inner subsidence abruptly terminates convection forming a wall, towards which upward spiralling convection intensifies and peaks. Once the convective wall forms with subsidence in the core, the cyclone develops its characteristic eye structure. This is the third prerequisite for tropical cyclone formation. Subsidence causes stability in the eye, cloud evaporates and calm winds result. The eye is also the area of lowest pressure. Figure 2.9 illustrates these effects on pressure and wind-speed through the eye of a typical tropical cyclone. Usually the eye structure will not fully develop unless the central pressure is below 990 hPa. The lower the central pressure, the more intense the cyclone. Some of the lowest pressures recorded for tropical cyclones are presented in table 2.1. Note that cyclones with the lowest pressure are not always the most destructive in terms of the loss of human lives. This is mainly owing to the fact that not all cyclones occur over populated areas.

Finally, the eye will fail to develop to maturity if horizontal wind speed aloft increases by more than 10 m s^{-1} over near surface values. At this value, the upward convective center gets displaced downwind, and a vertical eye and wall structure cannot form. Cyclones tend to develop equatorward of, rather than under the direct influence of, strong westerly winds.

Magnitude and frequency (Bureau of Meteorology, 1977; Lourensz, 1981; Gray, 1984)

Tropical cyclones occur frequently. Figure 2.7 also summarizes the average number of cyclones per year worldwide in each of seven main cyclone zones.

Table 2.1 Lowest Central Pressures recorded in Tropical Cyclones

Event	*Location*	*Date*		*Pressure (hPa)*
Typhoon Tip	NE of Philippines	October	1979	870
Typhoon June	Guam	November	1975	876
Typhoon Nora	NE of Philippines	October	1973	877
Typhoon Ida	NE of Philippines	September	1958	877
Typhoon Rita	NE of Philippines	October	1978	878
Hurricane Gilbert	Caribbean	September	1988	885
[a]	Philippines	August	1927	887
Labour Day storm	Florida, United States	September	1935	892
Typhoon Marge	NE of Philippines	August	1951	895
Hurricane Allen	Caribbean	August	1980	899
Hurricane Camille	Gulf of Mexico	August	1969	905
Typhoon Babe	NE of Philippines	September	1977	906
[a]	Philippines	September	1905	909
Typhoon Viola	NE of Philippines	November	1978	911
the Cossack Cyclone	Australia	June	1881	914 (?)
Hurricane Janet	Mexico	September	1955	914
Cyclone Mahina	Australia	March	1899	914 (?)

[a] Not named.

While the numbers in any one zone appear low, when summed over several decades, cyclone occurrence becomes very probable for any locality in these zones. For example, consider the path of Hurricane Hazel in 1954, which I witnessed as a child living at the western end of Lake Ontario, Canada, in a non-cyclone area 1000 km from the Atlantic Ocean. Hurricane Hazel was a rogue event and is still considered to be the worst storm witnessed in southern Ontario. This cyclone not only had to travel far inland, but it also had to cross the Appalachian mountains, which effectively should have diminished its winds and rainfall. In actual fact, Hurricane Hazel's path over the Great Lakes was not that rare an event. Over a 60-year period, 14 cyclones have crossed the Appalachians into the Great Lakes area. Not all were of the magnitude of Hurricane Hazel, but all brought heavy rain and winds. In Australia at least one cyclone every 20 years has managed to penetrate to Alice Springs in the center of the continent, and it is possible for west-coast cyclones to cross the whole continent before completely dissipating their winds and rain.

The actual number of cyclones within the zones in figure 2.7 is highly variable. Figures 2.10 and 2.11 summarize, for this century, the incidence of cyclones in the Australian region and the number of cyclone days in the north Atlantic-Caribbean respectively. In Australia between 1909 and 1980, there have been on average ten cyclones per year.

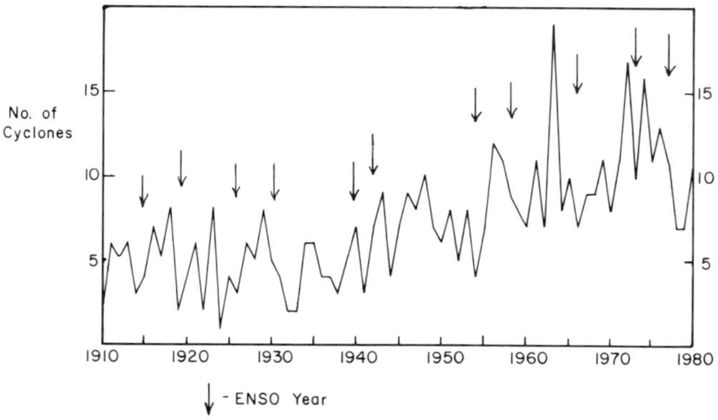

Fig. 2.10 Incidence of tropical cyclones in the Australian region: 1910–80 (based upon Lourensz, 1981)

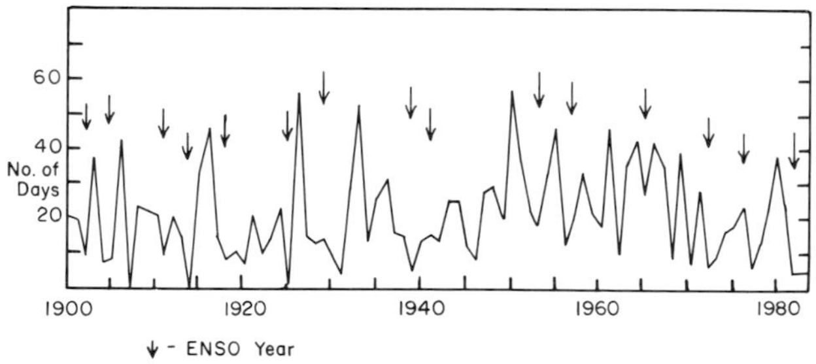

Fig. 2.11 Number of hurricane days in the north Atlantic-Caribbean Sea: 1900–83 (based upon Gray, 1984)

The number has been as low as 1 per year and as high as 19. In the Caribbean region, the presence of tropical cyclones has fluctuated just as much between 1910 and 1983. In 1950 the aggregated number of days of hurricanes totalled 57, while no hurricanes were recorded at all in 1914. These patterns for both areas are not random, and to a certain extent, are interlinked because cyclone generation in both areas is controlled by the state of trade wind circulation across the Pacific Ocean. The timing of one particular state, termed an *El Niño-Southern Oscillation* (*ENSO*) event is marked on both figures 2.10 and 2.11. The nature of these ENSO associations and the reasons for their fluctuations will be discussed in greater detail in chapter 5.

The zones depicted in figure 2.7 are also restrictive. Tropical cyclones have appeared in the 1970s and 1980s outside these limits, and outside the usually defined cyclone seasons for each hemisphere. The cyclone season around Australia usually lasts from 1 December to 1 May. In 1972 Cyclone Hannah clipped the northern coast of Australia on 10 May, while Cyclone Ida was reported on the 1 June. As late as 21 May 1986 Cyclone Namu devastated the Solomon Islands, outside the area of influence of South Pacific cyclones shown in figure 2.7. The 1987 Australian season ended with Cyclone Blanche on 26 May, while the 1988 season ended with Cyclone Harvey, which crossed the continent on 21 May. The occurrence of these storms has effectively meant that the cyclone season for Australia must now be extended to include the month of May. There also appears to be a general increase in the frequency of Australian cyclones, but this could simply be due to better monitoring methods. The worst season to date occurred in 1989, which witnessed more than 14 cyclones in the Coral Sea. Certainly, the number of cyclones affecting central Pacific islands such as Fiji and Tonga has increased significantly in the 1980s coincident with more frequent ENSO events.

World cyclone disasters (Cornell, 1976; Anthes, 1982; Whipple, 1982; Nalivkin, 1983; Holthouse, 1986)

Tropical cyclones are by far the worst and most widespread naturally occurring hazard on Earth. Tropical cyclones in the period 1960–1987 were responsible for 25 per cent more deaths than earthquakes (table 2.2). Historically, there have been some devastating if not enormous cyclones. There have been at least four cyclones in recorded history killing over 300 000 people: all but one occurred over the Indian subcontinent. In 1935 one hurricane in Florida obtained wind speeds of 400–500 km and technically became a tornado. The most destructive cyclone this century was the one that struck what was then Eastern Pakistan (now Bangladesh) in November 1970, killing over 500 000 people. It is memorable for two reasons. Firstly, most people were killed by the storm surge rather than strong winds or floodwaters. Secondly, although viewed by inhabitants as a rare event, the disaster was repeated in May 1985, resulting in another 100 000 deaths. The earlier disaster was also graphically documented in the press. The ensuing, ineffectual rescue and reconstruction effort by the Pakistan government led to political upheaval resulting in independence for Bangladesh. Ironically, while Bangladesh obtained its independence because of a cyclone, ultimately it became the poorest country in the world mainly because it never recovered economically, from the disaster. That 1970 disaster will be examined in depth when the impact of, and response to cyclone disasters is discussed later in this chapter.

Historically, the greatest number of deaths resulting from geological or climatic hazards can be attributed to cyclones. Much of this deathtoll

TABLE 2.2 Deathtolls from Natural Hazards 1960–87

Hazard type	*Deaths*	*Largest Event and Date*	*Deathtoll*
Tropical Cyclones	622 360	Eastern Pakistan (Bangladesh) 1970	500 000
Earthquakes	497 600	Tangshan, China 1976	250 000
Floods	36 350	Vietnam 1964	8 000
Avalanches, Mudslides	30 029	Peru 1970	25 000
Volcanic Eruptions	27 459	Columbia 1985	23 000
Tornadoes	4 500	Eastern Pakistan (Bangladesh) 1969	540
Snow, Hail, Wind Storms	3 078	Bangladesh 1986 (hail)	300
Heat Waves	1 000	Greece 1987	400

Note: Statistics are drawn from Encyclopaedia yearbooks 1960–87

can be attributed to flooding caused by heavy rainfall over swathes of 400–500 km. On average, a cyclone can dump 100 mm per day of rain within 200 km of the eye, and 30–40 mm per day over an area of 200–400 km. These rates can vary tremendously depending upon local topography, cyclone motion, and the availability of moisture. In 1952 a tropical cyclone at Réunion Island dropped 3240 mm of rain in three days. The most horrific cyclone-induced flooding recorded occurred in China along the Yangtze River between 1851 and 1866. A total of 40–50 million people drowned on the Yangtze River floodplain because of tropical cyclone-related rainfall during this period. Another 1.5 million died along the Yangtze River in 1887, and in 1931, 3–4 million Chinese drowned on the Hwang Ho River floodplain as a result of cyclone flooding.

Tropical cyclones have even changed the course of history in the Far East. In AD 1281 Kublai Khan invaded Japan, and after seven weeks of fierce fighting seemed set to break through Samurai defenses until a typhoon swept through the battlefield. The storm destroyed most of the 1000 invading ships and trapped 100 000 attacking soldiers on the beaches, where they either drowned in the storm surge or were slaughtered by the Japanese. The Japanese subsequently believed their homeland was invincible, protected by the *kamikaze* (or Divine Wind) that had saved them from that invasion.

Tropical cyclones can even affect non-tropical areas with disaster. As mentioned above, Hurricane Hazel swept through southern Ontario, Canada in October 1954. This storm occurred during a period of remarkable super-storms that afflicted Japan and North America between 1953 and 1957. The 1954 American season began with Hurricane Carol, a super-hurricane whose main force fortunately remained at sea. However, it caused the deaths of 60 people in New England and over $US500 million damage, the greatest damage caused by any hurricane to date in the United States. It was followed a fortnight later by Hurricane Edna and then, on 5 October, by Hurricane Hazel. Hazel swept across the Caribbean with winds exceeding 300 km hr^{-1} and torrential rains over a radius of 700 km. As it drifted along the eastern coast of the United States, it killed 95 people, and brought with it a storm surge of 5 m, together with the strongest winds ever recorded at Washington and New York. As it crossed North Carolina it dropped coconuts, shells and a cup marked 'Made in Haiti', which it had picked up three days earlier. It then intensified, turned inland over the Appalachians and headed across the Great Lakes to Toronto, where 78 people lost their lives. In the Great Lakes region it is considered the worst storm for 200 years and a benchmark against which to measure disaster. Two weeks after its birth in the Caribbean, it swept out into the Atlantic and degenerated into an intense extra-tropical depression, producing strong winds and heavy rains across the Atlantic to Scandinavia.

In Australia Cyclone Mahina (or the Bathurst Bay Cyclone) on 4–5 March 1899 crossed the northern Great Barrier Reef and killed 407 people, including about 100 Aborigines. The height of the storm surge is based on the fact that a police camp, which was situated on a 12-meter high bluff, was overwashed by waves during the storm. Small cyclones, called *howling terrors* or *kooinar* by the Aborigines, with eyes less than 20 km in diameter, appear the most destructive. For instance, the town of Mackay, Queensland with 1000 inhabitants was totally destroyed in 1918 by one of these small cyclones. Cyclone Ada in 1970 was so small that it slipped unnoticed through the meteorological observation network, and then surprised and wrecked holiday resorts along the Queensland coast. The Yongala Cyclone of 1911, where it reached landfall south of Cape Bowling Green, cleanly mowed down mature trees 2 m above the ground in a 30-kilometer swathe. In recent times Cyclone Tracy has been the most notable cyclone. Tracy with an eye diameter of 12 km obliterated Darwin on Christmas day 1974 (figure 2.12), killing over 60 people and obtaining wind speeds exceeding 217 km hr^{-1}. In world terms, its wind speeds were small and the loss of life meager. There was also no surge damage because the tidal range in Darwin is about 7 m and the cyclone occurred at low tide, with a storm surge of 4 m. However, it is one of the best documented Australian cyclones, with its graphic evidence of wind damage forming a benchmark for engineering research into the effects of strong winds. It also resulted in one of the largest non-military air evacuations in history, as the majority of the population of 25 000 was airlifted south to avoid the outbreak of disease and to provide shelter.

Geological significance

About 100–120 tropical cyclones develop every year. In a thousand years, which is not long by geological time scales, 100 000–120 000 tropical storm events occur worldwide. Thus, these seemingly rare or isolated events take on major geological significance when it is considered that they also

Fig. 2.12 Damage done to a typical suburb in Darwin by Cyclone Tracy, December 1974 (photograph courtesy John Fairfax and Sons Limited, Sydney, FFX Ref: Stevens 750106/27 #4)

intrude into continental interiors outside of the tropics. This fact can seriously confuse the environmental interpretation derived from the geological rock record. For instance, cyclones can easily carry and deposit wind-blown marine *diatoms* or *foraminifera* far inland, thus contaminating non-marine sediments with a false marine signature. Tropical cyclones not only carry salt water inland but also stir up marine sediment, which can be carried extensively inland in low-lying areas by storm surges. Such flooding could be misinterpreted as a marine transgression when, in fact, it represents storm deposition.

Geological work on the continental shelf now supports the view that the catastrophic event, the tropical cyclone, is responsible for significant sediment movement. Analysis of the geological record indicates that storm deposits in the rock record very much define the areas of reconstructed palaeo-environments for tropical cyclones and extra-tropical storms. Cyclone-generated swell can effectively travel 1000 km or more away from tropical areas into temperate latitudes. Along the south coast of New South Wales, Australia, waves 4 m in height with periods in excess of 15 seconds have travelled these distances, and swept across the continental shelf. Here, they are capable of stirring up sediment to depths of 80–100 m and moving marine sand landwards onto the inner shelf. Such waves, thus, represent a prime mechanism for replenishing shoreline beach systems with shelf sand. Along the Canadian and United States east coast, tropical cyclones are one of the major agents generating change in beach morphology. Along the quiescent coastline on the east coast of Canada in the Gulf of St Lawrence, *barrier island* gross morphology is characterized by storm effects such as barrier breaching and tidal inlet migration. These events only occur every 20 or more years because of tropical cyclone penetration into the area.

Impact and response (Burton et al., 1978; Simpson and Riehl, 1981)

The human impact of cyclones can be assessed mainly in terms of deaths or the financial damage. The response to cyclones depends mainly on these two factors. Figure 2.13 presents by five-year periods the number of deaths and amount of property damage due to tropical cyclones this century in the United States. Deaths have decreased dramatically this century mainly because the United States has adopted a policy of total evacuation together with earlier warning systems. Costs, on the other hand, have increased mainly because of

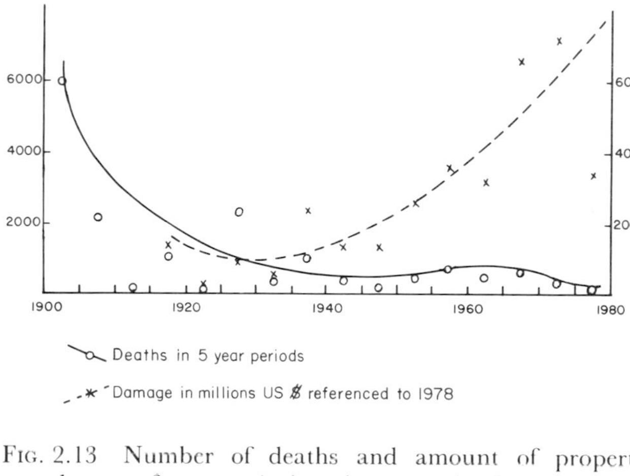

Fig. 2.13 Number of deaths and amount of property damage from tropical cyclones in the United States since 1900 (based upon Sheets, 1980)

the escalating development in cyclone-prone areas after the Second World War. The rise in damage costs in the 1960s also reflects the frequency of more intense hurricanes at this time. Of course, the damage bill from a tropical cyclone also depends upon the type of building construction. Figure 2.14 illustrates this effect for Australia: the damage to buildings as a percentage of the original cost of construction is assessed, at various wind speeds, for various cities and towns across northern Australia most likely to be affected by cyclones. Figure 2.14 shows that Cyclone Tracy, the most destructive cyclone experienced in Australia, only generated a repair bill representing 10 per cent of the initial cost of buildings in Darwin. While most houses were damaged or destroyed, larger commercial and public buildings with a high capital investment were not badly damaged. Other non-residential buildings were mildly affected because they had to comply with strict building codes formulated to minimize the effect of cyclone winds. Had Tracy occurred in Carnarvon, along the west coast of Western Australia, the damage bill would have exceeded 25 per cent of initial cost simply because buildings here are not built with the expectation that they will be subject to strong winds. In Port Hedland, which is a company town designed to withstand cyclone winds, the damage bill would have been less than 10 per cent of initial capital outlay.

The impact of cyclones goes far beyond just deaths and building damage. Figure 2.6 indicates that cyclones can cause increased soil salinity through storm-surge inundation, an effect that can have long-term economic consequences upon agricultural production which are virtually unmeasurable in monetary terms. Contamination of water supplies and destruction of crops can also lead to disease and starvation. In fact, some of the large deathtolls caused by cyclones on the Indian subcontinent in the nineteenth century were due to starvation afterwards. In certain cases, the destruction of crops can affect international commodity markets. For example, Hurricane Flora's passage over Cuba in 1963 wiped out the sugar cane crop and sent world sugar prices soaring. Of all hazard events, the effects of tropical cyclones differ substantially from each other depending upon the economic development of a country and its transport and communication infrastructure. This can be illustrated most effectively by comparing the response and impact of tropical cyclones within three countries—Australia, United States and Eastern Pakistan/Bangladesh.

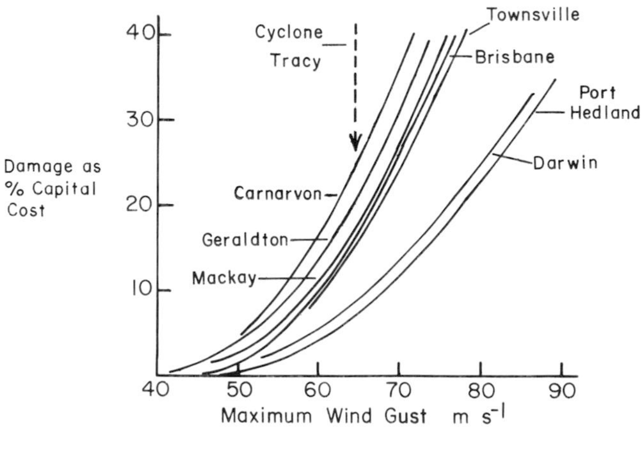

FIG. 2.14 Cost of repairs as a percentage of original capital investment due to tropical cyclones in Australian towns (based upon Leicester and Beresford, 1978)

Australia (Western and Milne, 1979)

One of the effects of tropical cyclones not covered in figure 2.6 deals with long-term human suffering caused by dislocation. Studies of hazards where people are dispossessed indicate that the incidence of psychosomatic illness (illness induced by changes in mental well-being, such as by anxiety) increases afterwards. More importantly, people who have lost a relative recover better than people who have lost the family home. A death can be grieved by a funeral within several days, but a family home has to be rebuilt over months or years. Australia has experienced this phenomenon on a large scale. In northern coastal Australia, it is accepted that cyclones will inevitably occur. Evacuation is normally not encouraged except for the most hazardous areas next to the ocean. Strict building codes have been established to minimize damage and loss of life. Such codes are policed and adhered to. A network of remote-controlled, satellite and staffed stations has been established to detect and track cyclones. These observations are reported over radio and television continually if a cyclone is imminent. Some criticism has been raised that the Australian Bureau of Meteorology is unable to predict the landfall of all cyclones, and there is some fear that the staffed stations are being run down as an economy measure. However, it should be realized that tropical cyclones in Australia as elsewhere often travel unpredictably. Residents have been instructed on how to prepare and weather a cyclone. For example, most people upon hearing about the forecast arrival of a cyclone will remove objects in the house and yard that are loose and that can become missiles in strong winds, will tape windows, and will store at least three days' supply of water and food.

Cyclone Tracy, on 25 December 1974, revealed that apathy can negate all these preventative procedures. The cyclone also led to large-scale psychological damage that cannot be measured in economic terms. Despite three days of warnings, and because of the approaching Christmas holiday, only 8 per cent of the population were aware of the cyclone. About 30 per cent of Cyclone Tracy victims in Darwin took no precautions to prepare for the storm's arrival; less than 20 per cent believed that people took the cyclone warnings seriously, and 80 per cent of residents later complained that they had not been given enough warning. Fortunately, Australia had established a disaster coordinating center which, depending on the severity of the disaster, could respond in varying degrees to any event. The director of this center, once requested to take charge, could adopt dictatorial power and even override federal government or judicial decisions. This organization responded immediately to Cyclone Tracy. Within 24 hours communications had been re-established with the city even though it was the height of the Christmas vacation season. Major-General Stretton, the head of the National Disasters Organization, reached Darwin within 24 hours on a military aircraft with emergency supplies loaded at Sydney and Mt Isa. Stretton took total control of Darwin and made the decision to evacuate non-essential residents because of the serious threat to health. The cyclone completely destroyed 37 per cent of homes and severely damaged most of the remainder (figure 2.15). There was no shelter against tropical downpours, no water, power or sanitation. Within a few days, most of the food in fully stocked deep-freezers began to decay, so that a stench of rotten meat hung over the city. Stray cats and dogs were shot and special army teams were organized to clean up rotten food. In addition, the decision was made to facilitate evacuation in order to permit orderly and manageable rebuilding. Commercial planes were commandeered for the evacuation and all vehicles leaving Darwin by road were stopped and processed. Volunteer emergency relief organizations ensured that all evacuees had food, adequate transport, accommodation and money. The response in the first week was efficiently organized and carried out.

FIG. 2.15 Effects of wind damage caused by Cyclone Tracy striking Darwin Australia, 25 December 1974 (photographs courtesy of Dr Geoff Boughton, Department of Structural Engineering, Curtin University of Technology, Perth). *Above*, high winds in excess of 217 km hr^{-1} tended to detach outside walls from the floor, leaving only the inner framework surrounding the bathroom. *Below*, in extreme cases, the high winds progressively peeled off the complete building, leaving only the iron floor beams and support pillars.

In the medium term, the evacuation caused more problems than it solved. A study carried out by Western and Milne (1979) examined the personal and social adjustment of victims within the first year. Factors studied included standards of accommodation, recreation, entertainment, community involvement, employment, income, emotional and physical health of families, family integration and marital adjustment, extended family and social networking, children's schooling and behavioural problems. One major problem that arose was the fact that, while residents were efficiently and safely evacuated, no register of their whereabouts was established. Over 25 000 people were flown out, mainly to southern cities of their choice 3000–4000 km away. Communication with, and between evacuees became impossible. Hence, there was no way to contact people personally after the cyclone to aid them with adjustment to new circumstances.

Western and Milne found that residents who stayed in Darwin suffered the least, while those who did not return suffered the most. Non-returned evacuees reported greater negative changes in lifestyle. These people also tended to be disadvantaged by the storm. They were more likely not to have insured houses for cyclone damage and more likely to have suffered total property loss. They ended up with less income after the event and suffered the greatest stress. While the stress may have been due to the fact that they bore the greatest property loss, to the evacuation or to the fact that they were people more inclined to stress in the beginning (although research indicated that they were no more prone to emotional and physical problems before the cyclone than any other group), the difference between this group and the others was striking. Both psychosomatic illness and family problems increased significantly for non-returnees. There was a high positive correlation between the occurrence of stress and the degree of impact of the cyclone upon people. It appears that the evacuation reinforced and increased the levels of stress and anxiety residents were experiencing as a result of their confrontation with Cyclone Tracy. Some of the stress was a consequence of the dictatorial powers assumed by a few after Tracy. Unintentionally, a victim-helper relationship became established to the point that little encouragement was given to victims to play an active role in their own recovery. Disaster victims in Australia are far from resourceless. To treat them as such impedes the social recovery from disaster.

The long-term response to Cyclone Tracy was also negative. The decision to rebuild and the method of allowing rebuilding did not take into consideration the longer term effects on human mental and physical health. Large sums of money were poured into rebuilding lavish government offices and facilities. The rebuilding effort was taken over by outsiders, and the growth of Darwin beyond 1974 population levels was encouraged. The rebuilding became a sacred cow, and the visible commitment to the rebuilding by the federal government led to unlimited growth. Locals who still lived in Darwin objected to redevelopment plans, stating that they had no say in the range of plans put forward. Residents evacuated to the south had no say in the redevelopment at all. Only men were allowed to return to the city to rebuild. The men, isolated from their families for up to one year, grouped together socially. When their families finally returned home, these men could not break back into family groupings, many of which had completely changed (that is, children had physically and mentally grown without a male role figure). Even though, physically and emotionally, returnees suffered less than non-returnees within the first year, returnees in the long term suffered the same consequences as non-returnees. As of 1985, over 50 per cent of the families who experienced Tracy had broken up. The personal problems resulting from Cyclone Tracy are now being imparted to the second generation, the children, to be passed on in future years to another generation unless rectified. Today, Darwin is a new city physically and in terms of its population. It is estimated that 80 per cent of the people now living in Darwin, never lived through Cyclone Tracy.

United States (American Meteorological Society, 1986; Carter, 1987)

The United States government believes that early warning and monitoring is the best method to reduce loss of life due to tropical cyclones. Weather satellites were originally deployed to predict and track cyclones, thus giving enough warning to permit the total evacuation of the population, either horizontally or vertically. Horizontal evacuation involves moving large numbers of people inland from flood-prone coastal areas. In August 1985 millions of people were evacuated, some more than once, as Hurricane Danny unpredictably wandered around the Florida coast and then headed into the Gulf states. This type of evacuation can be especially difficult where traffic tends to jam on escape routes. The alternative is vertical evacuation into high-rise, reinforced buildings. This approach is taken in Miami Beach, where limited causeways to the mainland restrict the number of people that can flee, where the population is elderly and hence physically immovable, and where safe structures can easily accommodate the majority of the population. There is a concerted effort to build coastal protection works and even try to modify the intensity of hurricanes through cloud seeding. A high priority is also put on research activities that model, predict or quantify hurricanes and related hazards. Some east coast states, such as North Carolina and Maine, have development plans that exclude development from low-lying coastal zones subject to storm surge and wave erosion due to hurricanes. These set-back lines are equivalent to the shoreline recession distances caused by 30–100 years of erosion.

As evidenced in figure 2.13, the above measures have effectively reduced the loss of life from tropical cyclones in the United States. However, there are serious flaws with the evacuation procedures. As recently as May 1986, the American Meteorological Society pointed out that most evacuations actually take up to 30 hours to carry out, whereas the general populace believes that it would require less than one day to flee inland to safety. The evacuation system also relies heavily upon warnings of hurricane movement put out by the National Hurricane Center, which attempts to give 12 hours warning of hurricane movements. In actual fact, hurricanes move unpredictably and can shift dramatically within six hours. Hurricane Hazel moved at speeds of 50 km hr^{-1} at one point in time. In the United States, the time required for evacuation currently falls well short of the monitoring and forecast lead times for hurricanes. There is also the problem that 80 per cent of the 40 million people living today in hurricane-affected areas have never experienced an evacuation. These same people, together with developers, have become complacent about the hurricane threat because the frequency and magnitude of hurricanes have decreased in recent decades. The 1970s, in fact, experienced the lowest frequency of hurricanes ever recorded. At present along the whole eastern seaboard and much of the Gulf coast of United States, record construction is taking place with dwellings being built either at the back of the beach, or just landward of ephemeral dunes. Studies have shown that there is a general lack of awareness of cyclone danger and evacuation procedures. This is despite wide publicity of the dangers of cyclones as the hurricane season approaches. Publicity includes television specials, distribution of free literature and warnings on supermarket bags and parking tickets. These warnings are meeting with resistance because, in the absence of any storm, municipalities are viewing the publicity as bad for the tourist trade and potential property investment. The actual realities of evacuation, and the poor public perception of the ferocity of hurricanes in the United States, implies that many people are sitting on a time bomb so far as the cyclone hazard is concerned. The American Meteorological Society has warned that when a large tropical cyclone strikes the United States east coast, evacuation procedures may fail and the resulting loss of life may reach unprecedented levels.

Eastern Pakistan (Bangladesh) (Burton et al., 1978; Blong and Johnson, 1986; Milne, 1986)

In Bangladesh the response is totally different. The Bay of Bengal has a history of destructive storm surges. The 300 000 deathtoll from the October 1737 cyclone in Calcutta was caused by a surge, 13 m high racing up the Hooghly River through the city. In 1876 a storm surge near the mouth of the Meghna River immediately killed 100 000 people, while another 100 000 died in the subsequent famine. Small cyclone-related surges drowned 3000 people on the Ganges River in October 1960 and another 15 000 people in May 1965. The deltaic islands, the Chars, of the mouth of the Ganges were first viewed as potentially arable in the nineteenth century (figure 2.16). In the twentieth century Dutch engineers designed embankments and dams to protect low-lying salt-water marshland from flooding, with The World Bank financing the scheme. The reclaimed land, however, was not designed to be protected from storm surge. Settlement, mainly by illiterate farmers, was permitted and encouraged to relieve overcrowding in the country. Many itinerant workers searching for temporary employment flocked into the Chars at harvest time.

The 13 November 1970 cyclone was detected by satellite three days in advance, but a warning was not issued by the Pakistani Meteorological Bureau until the evening the cyclone struck. Even then, the message was given to the sole radio station after it had shut down for the day at 11:00 pm. Even if the warning had been broadcast, the majority of the population, being asleep, would not have heard it. Much of the reclaimed land had inadequate transport links across channels to higher ground, such that effective evacuation was impossible. Recent migrants had no knowledge of the storm surge hazard and were not prepared, or organized, for evacuation. In the middle of the night a bore, 15 m high, struck the southern Chars obliterating 25 islands, swamping a further 2000 and drowning approximately 500 000 people. Over 400 000 hectares of rice paddies were inundated with salt water and 1 000 000 head of livestock killed. Over 50 million people were affected by the storm surge, flooding and winds. The pressure for land in Bangladesh is so great that the deltaic land was recolonized rapidly after the 1970 event. Nothing had really changed by 1985 and in May of that year a similar cyclone caused the same sort of destruction. Again, while there was adequate satellite warning of the approaching cyclone, evacuation was impossible. Many illiterate itinerant workers knew nothing of the hazard, the method of escape or the event. In the 1985 event, 100 000 people lost their lives.

The response in Bangladesh to tropical cyclones appears to be one of ignorance and acceptance of the disaster as the 'will of god'. Cyclones cannot be avoided and one must simply wait out the storm

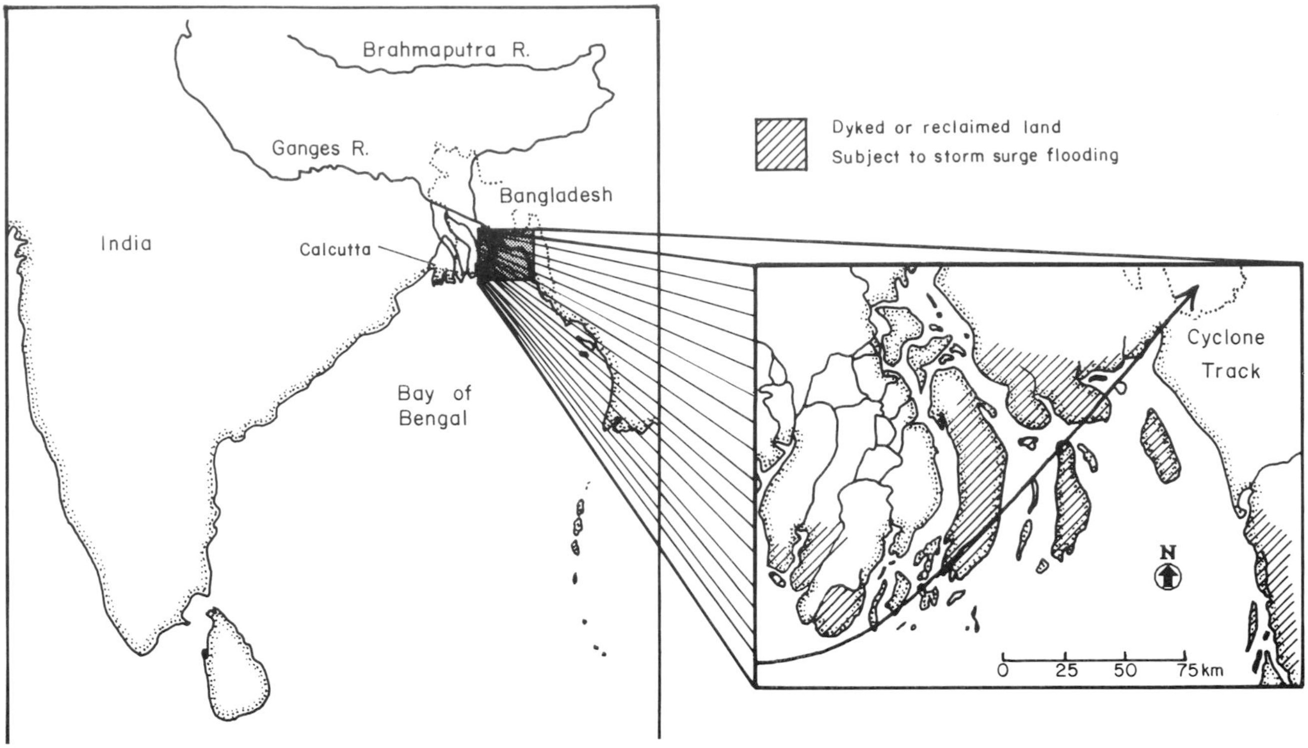

FIG. 2.16 Path of 13 November 1970 cyclone in the Bay of Bengal and location of embankments protecting low-lying islands in Bangladesh

and see what happens. With a bit of luck a family that has experienced one cyclone storm surge in its lifetime will not experience another. In 1970 only 5 per cent of the people who survived had evacuated to specified shelters while 38 per cent had climbed trees to survive. Many viewed evacuation as unwise because homes would be subject to looting. Even if evacuation had been possible, religious taboos in 1970 would have prevented Moslem women from leaving their houses because the cyclone struck during a month when women were forbidden by established religious convention from going outside. The immediate response for Bangladesh after any cyclone is to request international aid to rebuild the farming infrastructure. There is no real attempt to establish evacution procedures and routes of escape, because flooding is too widespread, and there are too many people to evacuate.

Extra-tropical cyclones

Formation

Extra-tropical cyclones or depressions are usually considered to be low-pressure cells that develop along the polar front. They also can develop over warm bodies of water outside the tropics usually off the east coast of Australia, and possibly Japan and the United States. In the case of these latter areas, the lows gain latent heat from the water but there are additional mechanisms driving the circulation. This section will describe the formation of *east-coast lows* that develop or intensify over bodies of warm water along limited sections of the world's coastline, and examine the formation of low pressure due to upward forcing of warm air along the polar front.

East-coast cyclones (Holland et al., 1988)

Research into east-coast cyclones or lows is at the pioneering stage. For some regions, where easterly moving pressure systems pass over a mountainous coastline and then a warm poleward flowing current, intense cyclonic depressions can develop which have structural features and intensities similar to tropical cyclones. These regions include the east coasts of Japan characterized by the Japanese Alps and the Kuroshio current, the United States characterized by the Appalachians and the Gulf Stream, and southeast Australia characterized by the Great Dividing Range and the east Australian current. Systems may also develop off the east coast of South Africa dominated by the Drakensburg range and the Agulhas current. The lows are not associated with any frontal structure but tend to develop in the upper atmosphere under, or downstream of a cold-core depression that probably has formed in westerly air movement at these altitudes. The depressions are also associated on the ground with a *blocking high*, a strong high-pressure system that may stall or block over the adjacent ocean. The lows do not appear initially on surface pressure charts but appear as a slight dip in isobars at mid-altitudes.

The systems intensify and reach the ground near the coastline as a result of convection that is enhanced by steep topography near the coast, cold inshore upwelled water (leading to a steep sea-surface temperature gradient between the coastline and the offshore warm current) and rapid input of latent heat of evaporation obtained from winds blowing over the warm offshore current. Topography appears to be a crucial factor as it deflects low-level easterlies towards the equator causing convergence and convection that can explode the formation of the cyclone. In this respect some lows become bombs developing rapidly within a few hours. Convection is also enhanced by cold inshore water, which forces onshore air movement upwards and enhances condensation of moisture-laden air. East-coast lows develop preferentially at night, at times when the maritime boundary layer is most unstable. In extreme cases the lows can develop the structure of a tropical cyclone. They may become warm-cored, obtain central pressures below 990 hPa, develop an eye that appears on satellite images, and generate wind speeds in excess of 200 km hr^{-1}. Once formed, they tend to travel poleward along a coastline; but if they move too far offshore over the warm current, they diminish in intensity rapidly. In some circumstances if the high pressure that spawned their formation remains in position over the ocean, or if the warm offshore current is trapped between two tongues of cold water, the systems can persist off the coastline for up to one week, directing continual heavy rain onto the coast, producing a high storm surge and generating waves in excess of 4 m.

Polar-front lows (Linacre and Hobbs, 1977; Whipple, 1982)

Low-pressure cells develop along the polar front, have cold cores and an effective diameter that is much larger than that of a tropical cyclone (2000 km versus 400 km respectively). Their intensity depends upon the difference in temperature between colliding cold and warm air masses. They always travel in an easterly direction entrapped within the westerlies. Their location depends upon the location of *Rossby waves* in the polar front around 40° North and South of the equator. In the southern hemisphere this occurs south of landmasses, but in the northern hemisphere it coincides with heavily populated areas.

The theory for the formation of polar-front lows was formulated by Bjerknes and his colleagues at Bergen, Norway, in the 1920s. Outbreaks of cold polar air flow rapidly towards the equator and clash with much warmer air masses in mid-latitudes. Coriolis force dictates that this polar air has an easterly component while the movement of warmer air on the equatorial side of the cold front has a westerly component (figure 2.17). At the same time the warm air will be forced upwards by the advancing cold air. Winds on each side of the developed cold front travel in opposite directions and spiral upwards in a counterclockwise direction because of Coriolis force. This rotation establishes a wave or indentation along the front. Warm air begins to advance eastwards and poleward into the cold air mass while cold air begins to encircle the warm air from the west. A V-shaped frontal system develops with warm and cold fronts at the leading and trailing edges respectively. The lifting of warm, moist air generates low pressure and convective instability as latent heat of evaporation is released in the upper atmosphere. This uplift enhances the rotational circulation, increasing wind flow and producing precipitation over a wide area, as the low-pressure cell advances eastwards under the direction of the *jet stream* attached to the polar front at the top of the *tropopause*. The cold front at the back of the system pushes up warm air rapidly leading to thunderstorm formation. If the temperature difference at this location between the two air masses is great enough, tornadoes can be generated. The cold front, being backed by denser air, eventually advances faster than the warm front and begins to lift the warm air within the V off the ground, occluding the fronts. In the final stage of a depression, the polar front is re-established at the ground with decreasing rotational air-movement stranded aloft. Without continued convection, cloud formation and precipitation slowly decreases and the storm cell disappears. Under some circumstances the advancing cold front may entrap another lobe of warm air, generating a secondary low to the southeast of the original one that is now dissipating. It is even possible for the rotating air aloft to circle back to the original site of *cyclogenesis*,

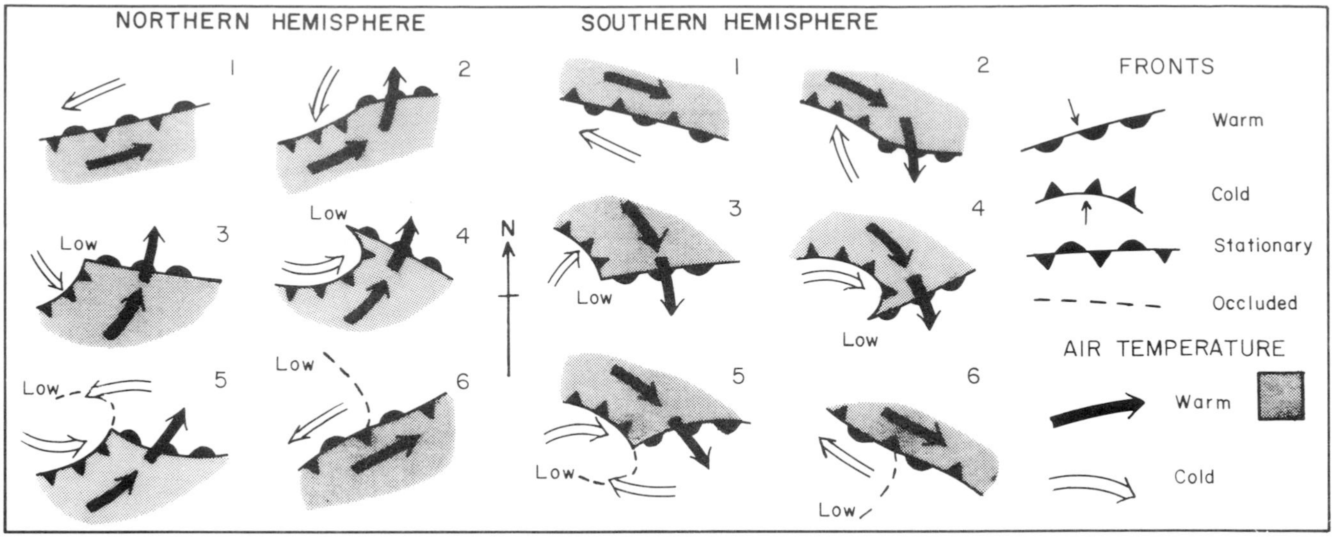

FIG. 2.17 Stages in the development of an extra-tropical depression as viewed in each hemisphere. The diagram shows (*1* and *2*) initial development of wave form; (*3* and *4*) mature front development and full extent of low; (*5* and *6*) occlusion.

and trigger the formation of a new low-pressure wave along the polar front.

These extra-tropical lows or depressions, because of their method of formation, have the potential to produce the same wind strengths and precipitation amounts as tropical cyclones; however, because they are larger and more frequent, they pose a greater hazard. Such extra-tropical storms also can generate waves the same size as those produced by tropical cyclones and storm surges of equal devastation. Geologically, there is nothing to distinguish the effects of the two hazards.

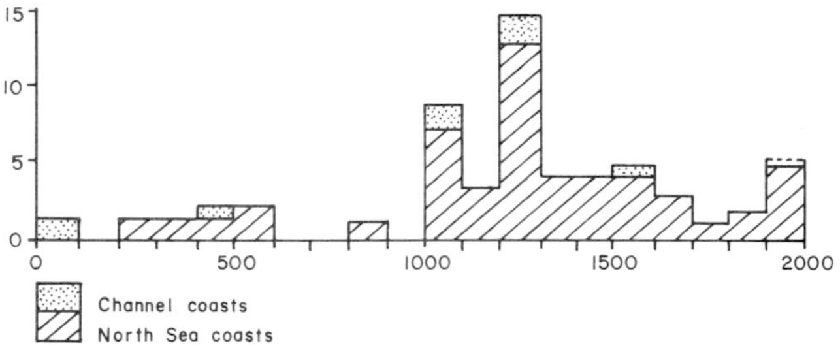

FIG. 2.18 Incidence of severe North Sea and English Channel storms per century for the last 2000 years (Lamb, ©1982; with permission from Methuen, London)

Disasters

Historical events (Davies, 1980; Anthes, 1982; Whipple, 1982; Lamb, 1982; Milne, 1986)

Mid-latitude depressions have also been as devastating as tropical cyclones. Their effects are dramatic in two areas along the eastern seaboard of the United States and in the North Sea. Historical accounts for these areas document the strongest storms recorded. Figure 2.18 plots the number of recorded cyclonic depressions in northern Europe that have produced severe flooding over the past 2000 years. The Middle Ages was an unfortunate period of cataclysmic tempests that caused great loss of life and immense erosion. Four storms along the Dutch and German coasts in the thirteenth century killed at least 100 000 people each. The worst of these is estimated to have killed 300 000 people. North Sea storms on 11 November 1099, 18 November 1421 and in 1446 also killed 100 000 people each in England and the Netherlands. By far the worst storm was the All Saints Day Flood of 1–6 November 1570. An estimated 400 000 people were killed throughout western Europe. These deathtolls rank with those produced in recent times in Eastern Pakistan/Bangladesh by storm surges from tropical cyclones. An English Channel storm on 26–27 November 1703 sank virtually all ships in the Channel with the loss of 8000 lives (figure 2.19). Other minor storms with similar deathtolls occurred in 1634, 1671, 1682, 1686 and 1717 at the height of the Little Ice Age.

Much of the modern-day coastline of northern Europe owes its origin to this period of storms. Erosion in the North Sea was exacerbated by 30 notable, destructive storm surges between AD 1000 and 1700. North Sea storms reduced the island

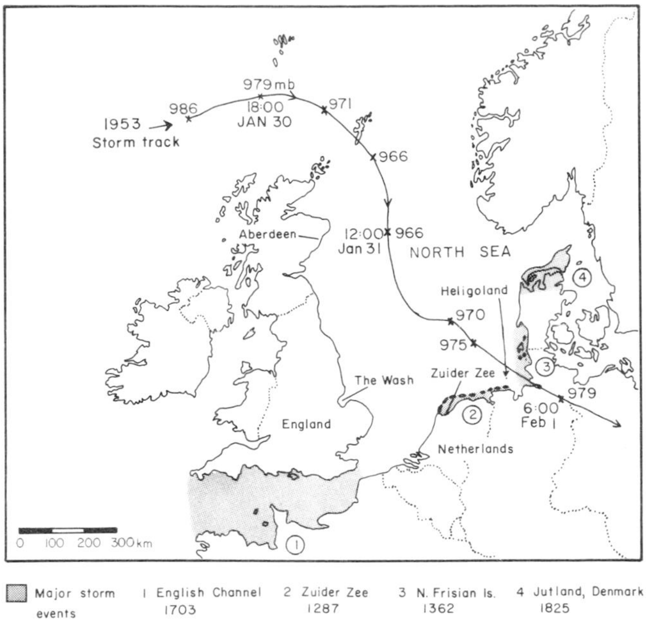

FIG. 2.19 Location of some historical, erosive North Sea storms, including the path of the 1 February 1953 storm

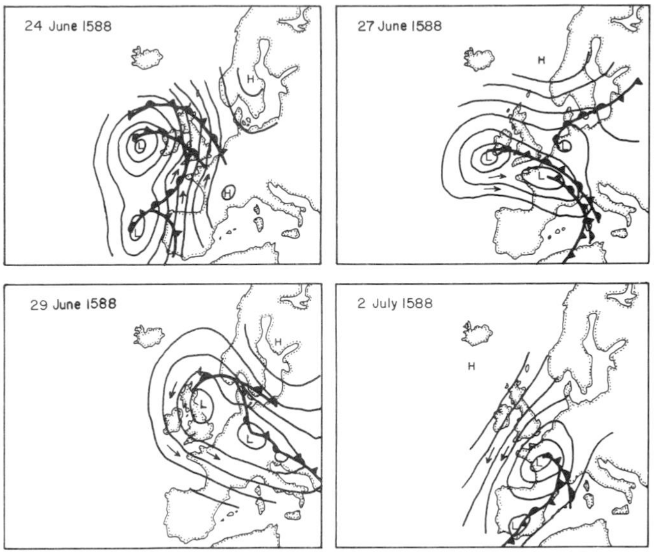

FIG. 2.20 Pressure pattern over northern Europe during the Spanish Armada's attempted invasion of Britain: 24 June–2 July 1588 (Lamb, ©1982; with permission from Methuen, London)

of Heligoland, 50 km into the German Bight, from a length of 60 km around the year 800 to 25 km by 1300 and to 1.5 km by the twentieth century. The Lucia Flood storm of 14 December 1287 in northern Europe was of immense proportions. Before that time, the Netherlands coast was bordered by a continuous barrier. This storm breached this coastal barrier and formed the Gulf of Zuider Zee (figure 2.19). It is only recently that the inlets at these breaches have been artificially dammed. Smaller embayments further east at Dollart and Jade Bay were also initiated, and the Eider estuary was widened to its present funnel shape. The same storm in northern France eroded several kilometers of marsh, making one headland a distant offshore island. In the Great Drowning Disaster storm of 16 January 1362, parts of the North Sea coast of Schleswig-Holstein eroded 15 km landward, turning a low-lying estuarine coastline into one bordered by irregular barrier islands separated from the mainland by lagoons 5–10 km wide. Over 60 parishes, accounting for half the agricultural income in this area, were destroyed. Both Danish and German coasts were severely eroded by storms in 1634, and in 1825 a North Sea storm separated part of the mainland from the northern tip of Denmark, leaving behind several islands.

These storms have also directed the course of history. In 1588 the Spanish Armada of 130 ships, after being forced into the North Sea by marauding attacks from the outnumbered British fleet, was all but rendered ineffective by a five-day storm off the east coast of Scotland (figure 2.20). What remained of the Armada was finished off by more of these storms as the fleet tried to escape west around Ireland. The summer of 1588 was characterized by an exceptional number of cyclonic depressions, corresponding to a polar jet-stream that was diverted much further south, with wind speeds much stronger than could be expected at present.

None of the extra-tropical storms mentioned above have been matched in the twentieth century for loss of life. In recent times the two worst storms have been the 1 February 1953 North Sea storm and the 7 March 1962 storm along the east coast of the United States. The worst storms recorded along the eastern coast of Australia have been the series of east-coast lows that struck between 25 May and 10 June 1974.

North Sea storm of 1 February 1953 (Wiegel, 1964; Milne, 1986)

This storm developed as a low depression north of Scotland and then proceeded to sweep directly southeast across the North Sea (figure 2.19). The lowest pressure recorded was 966 hPa. While this is not as low as pressures produced by tropical cyclones (see table 2.1), it represents an intensely low pressure for a cold-core depression. Wind speeds at Aberdeen, 200 km from the center of the storm, exceeded 200 km hr^{-1}. These velocities are similar to those experienced in tropical cyclones, but covered a much larger area. Tides in the North Sea were superimposed upon a *seiching* wave, known as a *Kelvin wave*, that moved counterclockwise along the coast because of Coriolis force. The storm moved in the same direction as this Kelvin wave. In doing so, it set up a storm surge, exceeding 4 m in height, which flooded the Wash in England and breached the dams fronting the Zuider Zee. The effect of this surge will be discussed in more detail at the end of this chapter. In Britain 307 people died, 32 000 inhabitants had to be evacuated and 24 000 homes were damaged. Over 83 000 hectares of valuable agricultural land were submerged under salt water. In the Netherlands dykes were breached in over 100 places, flooding 1 600 000 hectares or one-sixth of the country. In all 2000 lives were lost in this country and 250 000 head of livestock perished. The storm was estimated to have had a recurrence interval of 1:500 years, but does not rank as severe as some of those that occurred in the Middle Ages.

This storm became a benchmark for modern engineering design in the Netherlands and Britain. It also gave rise to renewed interest in coastal engineering in these countries. However, the lessons have not been permanently learnt. The North Sea region still has the potential to spawn some of the most intense storms witnessed by man. Economic restraint in the 1980s has seen the deterioration of the British Meteorological Service and a decline in its ability to forecast subsequent storms of similar strength. For example, on the night of 15 October 1987, very warm, humid air from the west of Africa clashed with cold Arctic air off the coast of France, triggering the formation of a very intense low-pressure cell off the Bay of Biscay. The storm, which swept up the English Channel and over the south of England, eventually developed a central pressure of 958 hPa, the lowest ever recorded in England. Wind gusts of 170–215 km hr^{-1} affected a wide area between the coast of France and England. Only 18 people were killed, but the storm virtually destroyed all large trees growing in the south of England. The French Meteorology Department, using data from the Reading-based European Centre for Medium Range Weather Forecasting, had issued a very strong wind-warning two days in advance. The same information was available to the British Meteorological Office, but was not heeded until the storm actually hit the English coastline. The storm was reported as the worst to hit England, breaking pressure and wind records, and has been followed by similar storms in January–February 1990. The above list of historical disasters for this part of Europe would indicate, however, that much larger storms than these are still feasible in the region.

United States Ash Wednesday storm of 7 March 1962

(Dolan and Hayden, 1983; Nalivkin, 1983)

North America has also experienced severe, extra-tropical storms. For instance in 1868–69, cyclonic storms on the Great Lakes sank or damaged more than 3000 ships and killed over 500 people. However, the Ash Wednesday storm of 7 March 1962 is considered to be the worst storm recorded along the east coast of the United States. At no time did the storm reach land as it paralleled the coast some 100 km offshore. Its passage along the coast took four days, with waves exceeding 4 m in height for most of this time. In the decades leading up to the storm, development had taken place on barrier islands along the coast. The destruction of the storm was immense. Whole towns were destroyed as storm waves, superimposed on a storm surge of 1–2 m, overwashed barrier islands (figure 2.21). The loss of life was minimal since most of the damage occurred to summer homes in the winter season.

Coastal retreat was in the order of 10–100 m. However, this value varied considerably over short distances. The destruction and erosion of this storm stands out along the United States east coast because no other storm has come close to its

Fig. 2.21 Storm damage produced by the Great Ash Wednesday storm of 7 March 1962 along the mid-Atlantic United States coast (UPI).

destructiveness. It must be remembered that this is a coastline also subject to the passage of tropical cyclones. The Ash Wednesday storm represented a turning point for coastal geomorphology and engineering in the United States. Afterwards, the federal government made research into coastal erosion and beach processes a research priority. Since then, the concept of a coastal *set-back line* has been developed and implemented in some states. This set-back line is based on the probability of coastal flooding or erosion derived from historical evidence. Building development is prohibited seaward of the line and the legality of such a prohibition has been tested and upheld in the courts, specifically in North Carolina, where established insurance companies now will not insure buildings against wave damage if they are built seaward of the established set-back line.

Australian east coast storms of May–June 1974

(Bryant and Kidd, 1975; Foster, et al. 1975; Bryant, 1983)

East-coast lows off the Australian coast are as intense as tropical cyclones. Notable storms afflicted the south Queensland coast in July 1935 and in June 1967. The latter storm followed a succession of six large cyclones and, after eroding 120 m of shoreline and toppling many seaside homes into the ocean, it began to threaten high-rise apartment buildings along Surfers Paradise. However, the three May–June storms of 1974, while mainly confined to New South Wales, remain the most significant and widespread series of such lows to strike the east coast of Australia in 100 years. Each storm event consisted of a classic east-coast low that developed along the western edge of the Tasman Sea over warm pools of water. In each case, blocking highs over the Tasman Sea exacerbated the duration of the storms by directing onshore SE–S winds onto the coast for several days. Figure 2.22 presents the wave heights generated by the three storms and the elevation of the associated storm surge as evidenced by the difference between actual and predicted tide heights. Maximum wave height exceeded 7 m on 25 May and again on 12 June. For most of the three weeks wave heights were above 4 m. For the whole period, tide levels exceeded predicted ones, with the greatest deviation being 0.7–0.8 m during the 25 May storm. This is the highest recorded surge along the New South

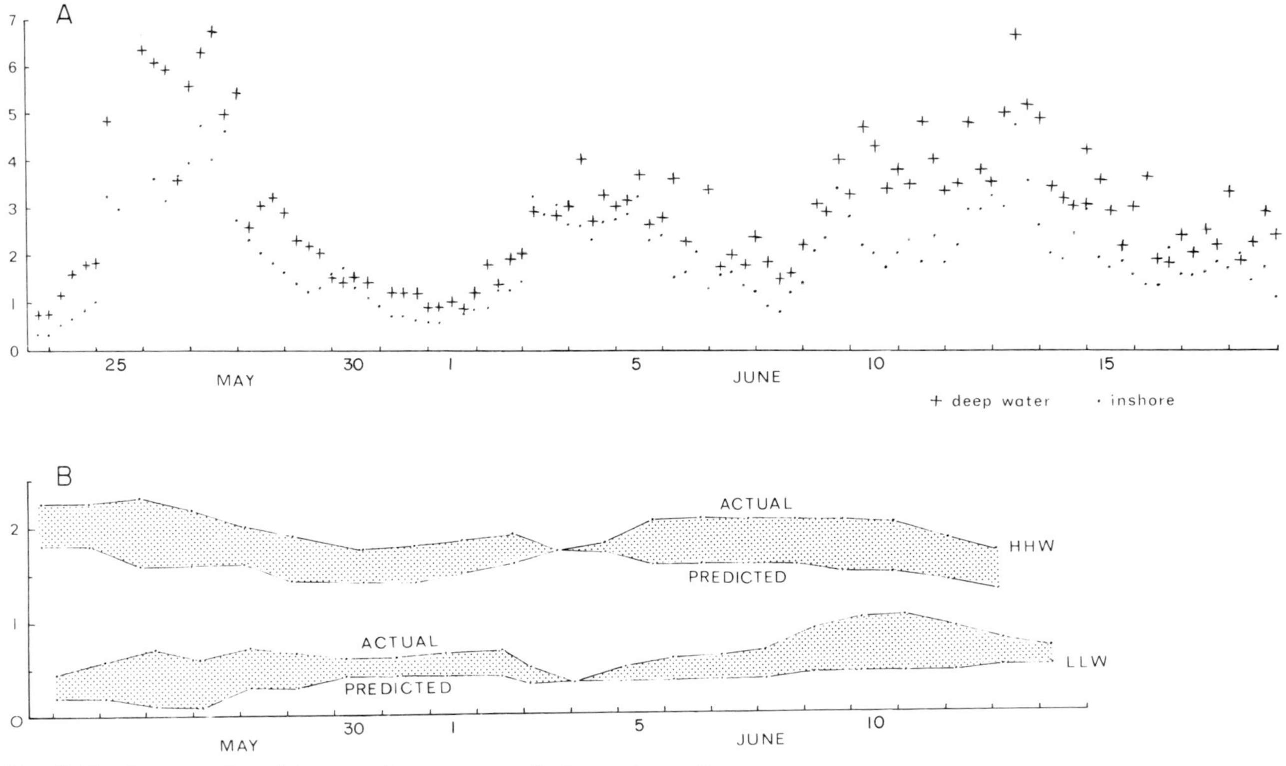

FIG. 2.22 Storms of 25 May–15 June 1974 at Sydney, Australia
A. wave heights.
B. actual and predicted tide heights (Bryant and Kidd, ©1975; with permission of the Australian and New Zealand Association for the Advancement of Science, Sydney).

Wales coast. In embayments this surge was accompanied by seiching and wave *set-up* such that water levels at the peak of each storm did not fall towards low tide but kept rising. In Sydney many homes built close to the shoreline were threatened; however, only a few homes were destroyed (figure 2.23 above). The sandy coastline suffered the greatest damage. In the three-month period from April to June, Stanwell Park beach high-tide line retreated over 100 m. At Pearl beach in Broken Bay, north of Sydney, a dune 7 m in height was overtopped by storm waves. Measured shoreline retreat amounted to 40 m on some beaches (figure 2.23 below). During the last storm, Cudmirrah beach, south of Jervis Bay, underwent dune scarping, along a face 2 m in height, at the rate of 1 m min^{-1}. Despite these large rates of retreat, coastal erosion was highly variable with some beaches escaping the storm completely unscathed. Similar results were found for the Ash Wednesday storm along the east coast of the United States.

FIG. 2.23 Beachfront damage caused at Sydney, Australia, by the storm of 25 May 1974. *Above*, undermining and collapse of beachfront houses at Bilgola beach. *Below*, this storm caused up to 40 m of erosion on other beaches. The collapsed structure shown here was a picnic table shelter situated about 5 m shoreward of the previous dune scarp at Whale beach.

Since 1974 there have been similar east-coast lows just as intense as the May 1974 storm. The autumn storms of 1978 and 1985 both produced 10-meter waves at Newcastle. Historical documents also suggest that storms rivalling the 1974 storms for intensity have been common over the past century. The fact that the frequency and magnitude of storms appears to be constant over time in New South Wales is not a unique feature in the world. A similar effect has been observed along the east coast of North America. In the long term there is no evidence that storms are solely responsible for coastal changes in New South Wales. Additional causes of beach erosion will be discussed at greater length in chapter 4.

Storm surges (Wiegel, 1964; Coastal Engineering Research Center, 1977; Anthes, 1982)

Introduction

Storm surge, as alluded to above, plays a significant role in tropical and extra-tropical cyclone damage. Storm surge was the main cause of death in Eastern Pakistan/Bangladesh in the 1970 and 1985 cyclones, the main cause of destruction in the February 1953 North Sea storm, and the main reason that waves were so effective in eroding beaches in the May–June storms of 1974 in New South Wales. In the United States storm surge is considered the main threat necessitating evacuation of residents from coastal areas preceding tropical cyclone landfall. Here, evacuation was instituted in response to the Galveston hurricane disaster of September 1900, during which 6000 people died because of storm surge inundation. The phenomenon is also a recurring hazard in the embayments along much of the Japanese and southeast Chinese coastlines. For example, approximately 50 000 people lost their lives around Shantou (Swatow), China, on 3 August 1922, and as recently as September 1959, 5500 people lost their lives in Ise Bay, Japan, because of storm surges. In this concluding section, the causes of storm surge and the concept of probability of occurrence will be discussed.

Causes

Storm surge is generated by a number of factors including:

- wind set-up,
- decreases in the atmospheric weight on a column of water,

- the direction and speed of movement of the pressure system,
- the shallowness of the continental shelf, bay or lake, and
- the shape of the coastline.

This discussion will ignore any addition effects due to river runoff, direct rainfall, wave set-up inside the surf zone or Coriolis force.

The main reason for storm surge would appear to be the piling up of water by wind. The exact amount of water piled up depends upon the speed of the wind, its duration and its location relative to the center of a cyclone. Wind set-up is difficult to calculate, but Wiegel (1964) gives an equation which shows the magnitude of wind set-up in a channel as follows:

$$h^2 \sim [(2.5\,\Pi\,\rho^{-1}\,g^{-1}) \times (x + C_1)] - d \qquad (2.2)$$

where h = height of wind set-up in meters
d = water depth of channel
$\Pi = 0.0025 U_o^2$
ρ = density of salt water
g = gravitational constant
U_o^2 = wind speed
$(x + C_1)$ = the fetch length of the surge

Equation 2.2 does not help one to easily conceptualize wind-induced surge heights, because the actual surge depends upon where you are relative to the movement of the cyclone and its center. If the wind is moving away from a coast, it is possible to get set-down. In general, for a cyclone with 200 km hr^{-1} winds, one could expect a wind set-up of at least 2 m somewhere around the cyclone.

Sea-level elevation depends more upon the weight of air positioned above it. If all air were removed from an area, then sea-level would rise proportionally. It must be remembered that some tropical cyclones have a pressure reduction of up to 13 per cent. Equation 2.3 (Coastal Engineering Research Center, 1977) expresses simply the relationship between sea-level and atmospheric pressure.

$$h_{max} = 0.0433(1023 - P_o) \qquad (2.3)$$

where h_{max} = the height of the storm surge due to atmospheric effects
P_o = the pressure at the center of the hurricane in hPa

From equation 2.3, it can be seen that the lowest central pressure of 870 hPa ever recorded for a tropical cyclone could have produced a theoretical surge-height due to atmospheric deloading of 6.6 m. The theoretical surge height for the North Sea storm of 1953 is 2.47 m, while for the 25 May 1974 storm in New South Wales, it is 1.2 m. The latter value is higher than the recorded tide-level difference and reflects cyclone movement.

If a storm moves in the direction of its wind speed, then it will tend to drive a wall of water ahead of it. This wall behaves as a wave and travels with a speed similar to that of the storm. The height of the wave is related to the size of the disturbance. Along the United States east coast, this long wave can have a height of several meters, becoming highest where land juts out into the path of the tropical cyclone. Figure 2.24 shows the track of Hurricane Carol in 1954 along the United States east coast together with records of sea-level elevation at various tide gauges. Note, that as the cyclone approaches land, the surge height increases; but it increases most where land at Long Island and Cape Cod intercepts the storm path. At these locations the long wave, which has been moving with the storm, piles up against the shoreline. The 1938 east coast hurricane, moving along a similar path, sent a wall of water 7 m in height ploughing into the Long Island coastline, where it made landfall. Along the Cape Hatteras barrier island chain that sticks out into the Atlantic Ocean, and along the Gulf of Mexico coastline of Florida, planning dictates that houses must be

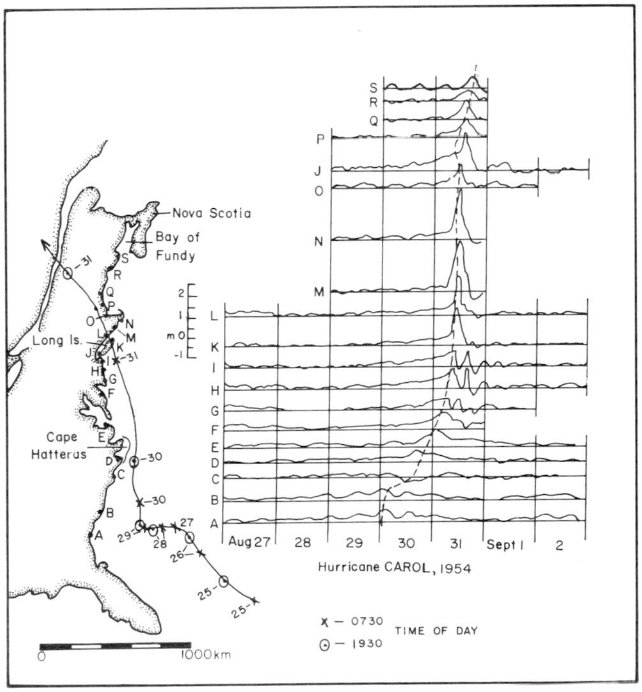

Fig. 2.24 Track of Hurricane Carol, 1954, and tide records along the United States east coast (after Harris, 1956)

constructed above the 6.5 m storm-surge flood limit, located where land juts into the preferred path of cyclones moving through these areas. In New South Wales, Australia, most storms move offshore; so, the long wave is in fact travelling away from the coast, and there is minimal storm surge felt at the shoreline. This is the reason storm surge elevations during the May 1974 storm only reached 0.7–0.8 m instead of the theoretically possible value of 1.2 m due to atmospheric deloading.

Figure 2.24 also reflects two other factors affecting surge. As Hurricane Carol moved along the coast, it began to cross a shallower shelf. The wave of water moving shoreward underwent shoaling (shallowing), which raised the wave height. Figure 2.25 illustrates the effect of shoaling on a wave. The movement of the storm surge wave is dictated by the speed of the storm and affects the complete water column. As this wave moves through shallower water, its speed decreases and, in order to conserve the energy flux through a decreasing water depth, wave height must increase. The faster the wave travels, then the greater will be the shoaling when it reaches shallow water. Any shallow body of water can generate large surges for this reason. The Great Lakes in North America are very susceptible to surges of the order of 1–2 m. In December 1985 a major storm struck Lake Erie, producing a surge of 2.5 m at Long Point, Ontario, coincident with record high lake-levels. Lakeshore cottages were floated like corks half a kilometer to a kilometer inland. In the shallow Gulf of Finland towards Leningrad, surges between 2–4 m in height can be generated. Since 1703 there have been at least 50 occasions when surges greater than 2 m have flooded Leningrad.

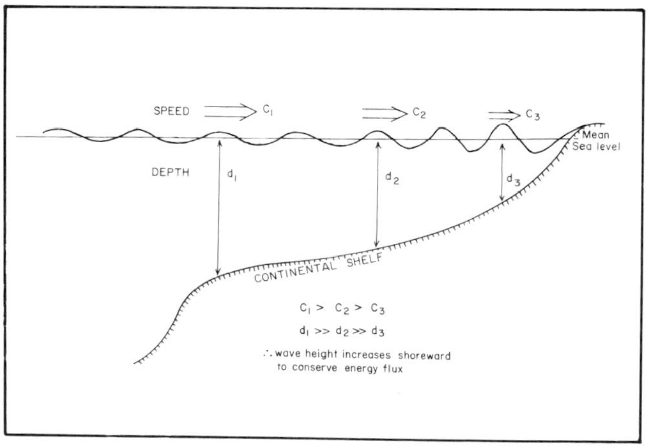

Fig. 2.25 Effect of wave shoaling on the height of long waves crossing the continental shelf

Finally, the shape of the coastline accounts for the largest storm surges produced. Examine figure 2.16 showing the Bay of Bengal, where storm surges have drowned so many people. The basin is funnel-shaped, such that any surge moving up the funnel, as the 1970 cyclone storm surge did, will be laterally compressed. In addition, the size of the basin can lead to *resonance* if it has a shape which matches the period of any wave entering it. A similar enhancement effect happened in 1770 to the tidal bore on the Qiantang River, south of Shanghai, China. A storm surge lifted the tidal bore on the funnel-shaped estuary to heights in excess of 4 m. The protecting dykes on each side of the river were overwashed and over 10 000 people drowned within minutes. The highest tides in the world are recorded in the Bay of Fundy in Canada (see figure 2.24), because the basin shape matches within six minutes the diurnal tidal period of 12.42 hours. On 4 October 1869 a cyclone, called the 'Saxby Gale', moved up the United States coast in a similar fashion to Hurricane Carol in 1954, but slightly eastwards. The storm travelled up the Bay of Fundy, moving a mass of water at about the resonance frequency (13.3 hours) of the Bay of Fundy-Gulf of Maine system. The resulting storm surge of 16 m was superimposed on a tide height of 14 m. In all, water levels were raised 30 m above low-tide level, almost *overwashing* the isthmus, 10 km in length, joining Nova Scotia to the mainland.

Probability of occurrence (Wiegel, 1964; Leopold et al., 1964)

The probability of occurrence of a surge height is highly dependent upon the physical characteristics of a coastal site. In order to define this probability, knowledge of the size of past events and how often they have occurred over a long period of time (the *magnitude-frequency*) are also required. This information is usually obtained from tide gauges. Wiegel (1964) has calculated the maximum storm-surge values for many tide stations around the United States Gulf and east Atlantic coasts. This type of information can be used for planning; however, it is fraught with the danger that one may not have records of the most extreme events. For instance, take the example of the 1953 storm surge in the Netherlands. The engineers correctly designed dykes to withstand the 100–200-year storm surge event; but the 1953 event exceeded those limits. The probability of occurrence and magnitude of this event, but not its timing, could have been foreseen from a simple analysis of past historical events.

Recurrence intervals

The probability of rare events of high magnitude can be ascertained in one of two ways: by determining the *recurrence interval* of that event or by constructing a *probability of exceedence* diagram. In both methods, it is assumed that the magnitude of an event can be measured over discrete time intervals—for example, daily in the case of storm waves, or yearly in the case of storm surges. All the events in a time series at such intervals are then ranked in magnitude from largest to smallest. The recurrence interval for a particular ranked event is calculated using the following equation:

$$\text{Recurrence interval} = (N + 1)M^{-1} \quad (2.4)$$

where N = the number of ranks
M = the rank of the individual event (highest = 1)

The resulting values are then plotted on special logarithmic graph paper. This type of plot is termed a *Gumbel* (frequency) *distribution*. An example of such a plot is shown in figure 2.26 for 70 years of maximum fortnightly tide heights in the Netherlands, before the 1953 storm surge event. Note that the recurrence interval is plotted along the x-axis, which is logarithmic, while the magnitude of the event is plotted along the y-axis. Often the points plotting any natural hazard time series will closely fit a straight line. The extrapolation of this line beyond the upper boundary of the data permits the recurrence interval of unmeasured extreme events to be determined. For example, in the century prior to the 1953 storm surge in the Netherlands, the greatest recorded surge had a height of 3.3 m. This event occurred in 1894 and had a recurrence interval of one in 70 (1:70) years. The ranked storm-surge data for the Netherlands fit a straight line which, when extended beyond the 70-year timespan of data, permits one to predict that a 4 m surge event will occur once in 800 years. The 1953 surge event fits the straight line drawn through the existing data and had a recurrence interval of once in 500 (1:500) years. If the dykes had been built to this elevation, they would be expected to be overtopped only once in any 500-year period. Note that the exact timing of this event is not predicted, but just its elevation. The engineers who designed the dykes in the Netherlands before the 1953 North Sea storm cannot be blamed for the extensive flooding that occurred because they had built the dykes to withstand only the 1:200 year event. At present, dykes in the Netherlands have now been elevated to withstand the 1:10 000 year event with a predicted surge height of 5 m.

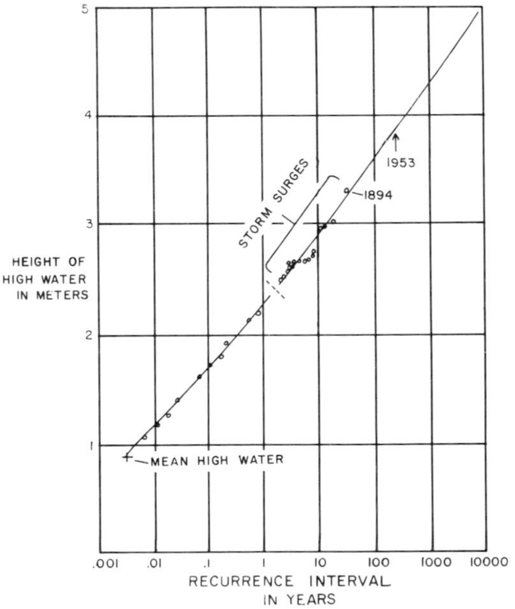

Fig. 2.26 Recurrence intervals of storm surges in the North Sea along the Netherlands coast for the 70 years before the 1953 February surge event (after Wemelsfelder, 1961)

Probability of exceedence diagrams

Recurrence intervals can be awkward to interpret, especially if the period of measurement is limited. In this case, a probability of exceedence diagram is favoured because it expresses the rarity of an event in terms of the percentage of time that such an event will be exceeded. Probability of exceedence diagrams are constructed by inverting equation 2.4 as follows:

$$\text{Exceedence probability} = M(N + 1)^{-1}\ 100\% \quad (2.5)$$

Data in this form cannot be plotted on logarithmic paper, but instead are plotted on normal probability paper. This paper has the same advantage as logarithmic paper, in that the probability of occurrence of rare events beyond the sampling period can be readily determined, because the data tend to plot along one or two straight lines. This is shown in figure 2.27 for the Netherlands storm surge data plotted in figure 2.26. For these data, two lines fit the data well. Where the two lines intersect at a high-water height of 2.4 m, coincidentally is the point separating storm surges from astronomical tides. The advantage of a probability plot is that the frequency of an event can be expressed as a percentage. For example a storm surge

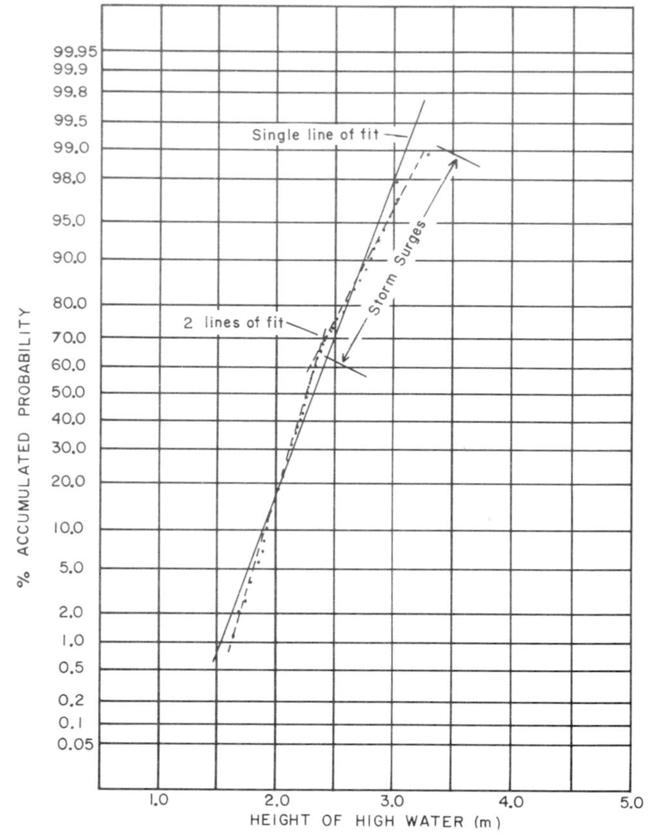

FIG. 2.27 Probability of exceedence diagram for the same data set as depicted in figure 2.26 (data from Wemelsfelder, 1961)

of 3 m, which in figure 2.26 plots with a recurrence interval of 1:30 years, in figure 2.27 has a probability of being exceeded only 3.8 per cent of the time, or once every 26 (1:26) years. Engineers find this type of diagram convenient and tend to define an event as rare if it is exceeded 1 per cent of the time. Probability plots have an additional advantage in that the probability of low-magnitude events can also be determined. For instance, annual rainfalls when ranked and plotted as recurrence intervals will not give one any information about the occurrence of drought. If the same data are plotted on probability paper, both the probability of extreme rainfall as well as deficient rainfall can be determined from the same graph.

There are a few limitations in using both types of plots. Firstly, an extreme event such as a 1:10 000 year storm surge, while appearing rare, can occur at any time. Secondly, once a rare event does happen, there is nothing to preclude that event recurring the next day. This constitutes a major problem in people's perception of hazards. Once high magnitude, low frequency events have been experienced, people tend to regard them as beyond their life experience again. The farmers of the Chars, having lived through the 1970 storm-surge event that killed 500 000 people, came back and re-established their flooded farms with the notion that they were now safe—because such an event was considered so infrequent that it would not recur in their lifetimes—but only 15 years later a storm surge of similar magnitude happened again. The 1973 flooding of the Mississippi River was the greatest on record in 200 years. Within two years that flood had been exceeded. High-magnitude events in nature tend to cluster over time regardless of their frequency. Such *clustering* is related to the persistence over time of the forcing mechanism causing the hazard. In the case of storm surges in the Bay of Bengal, the climatic patterns responsible for the cyclones may be semi-permanent over the span of a couple of decades. A third drawback about probability diagrams is that the base line for measurements may change over time. For example, deforestation of a major drainage basin will cause rarer flood events to become more common. In the case of the Netherlands, storm surges in the Middle Ages were certainly more frequent and higher than at present. The 1:10 000 storm-surge elevation planned for today could also be exceeded by an event smaller than 5 m if mean sea-level were to rise. This latter possibility will be examined later in chapter 4. Fourthly, plots of probability of exceedence and recurrence intervals need not give the same results for extreme events. In the case of the Netherlands storm surge data, the probability plot is deficient in giving a realistic probability of the 1953 event. The logarithmic plot indicated that this surge had a recurrence interval of 1:500 years. This same event cannot even be plotted on the probability plot given the straight line that has been drawn through the data. Finally, neither diagram may be the most appropriate method in engineering design for determining the frequency of rare events. In the case where high-magnitude events tend to cluster over time, engineers will resort to more complex frequency distributions in order to permit the recurrence of clustered events to be evaluated more realistically.

References

Bryant, E. A. and Kidd, R. W. 1975. 'Beach erosion, May–June, 1974, Central and South Coast, NSW'. *Search* v. 6 No. 11–12 pp. 511–13.

Burton, I., Kates, R. W., and White, G. F. 1978. *The environment as hazard*. Oxford University Press, NY.

Coastal Engineering Research Center. 1977. *Shore Protection Manual* (3 vols). United States Army.

Davies, J. L. 1980. *Geographical Variation in Coastal Development* (2nd edn). Longman, London.

Gray, W. M. 1975. 'Tropical cyclone genesis'. *Department of Atmospheric Science, Colorado State University, Fort Collins, Colorado, Atmospheric Science Paper* No. 234.

——. 1984. 'Atlantic seasonal hurricane frequency Part 1: El Niño and 30 mb Quasi-biennial oscillation influences'. *Monthly Weather Review* v. 112 pp. 1649–67.

Harris, D. L. 1956. 'Some problems involved in the study of storm surges'. *United States Weather Bureau, National Hurricane Research Project Report.* No. 4.

Hawkins, H. F. and Imbembo, S. M. 1976. 'The structure of a small, intense hurricane, Inez 1966'. *Monthly Weather Review* v. 104 pp. 418–42.

Lamb, H. H. 1982. *Climate, history and the modern world.* Methuen, London.

Leicester, R. H. and Beresford, F. D. 1978. *The resistance of Australian housing to wind forces: estimating insurance risk in tropical cyclone areas, Pt II.* Australian Department of Housing and Construction, AGPS, Canberra.

Lourensz, R. S. 1981. *Tropical Cyclones in the Australian Region July 1909 to June 1980.* Bureau of Meteorology, AGPS, Canberra.

Sheets, R. C. 1980. 'Some aspects of tropical cyclone modification'. *Australian Meteorological Magazine* v. 27 pp. 259–86.

Wemelsfelder, P. J. 1961. On the use of frequency curves of stormfloods. *Proceedings 7th Conference Coastal Engineering*, The Engineering Foundation Council on Wave Research, A.S.C.E., NY, pp. 617–32.

Western, J. S. and Milne, G. 1979. 'Some social effects of a natural hazard: Darwin residents and cyclone "Tracy"'. In Heathcote, R. L. and Thom, B. G. (eds) *Natural hazards in Australia.* Australian Academy of Science, Canberra, pp. 488–502.

Wiegel, R. L. 1964. *Oceanographical Engineering.* Prentice-Hall, Englewood Cliffs, NJ.

Further reading

American Meteorological Society. 1986. 'Is the United States headed for hurricane disaster?'. *Bulletin American Meteorological Society* v. 67 No. 5 pp. 537–38.

Anthes, R. A. 1982. 'Tropical cyclones: their evolution, structure and effects'. *American Meteorological Society Meteorological Monograph* v. 19 No. 41.

Blong, R. J. and Johnson, R. W. 1986. 'Geological hazards in the southwest Pacific and southeast Asian region; identification, assessment, and impact'. *Bureau Mineral Resources Journal Australian Geology and Geophysics* v. 10 pp. 1–15.

Bryant, E. A. 1983. 'Coastal erosion and beach accretion Stanwell Park beach, N.S.W., 1890–1980. *Australian Geographer* v. 15 pp. 382–90.

Bureau of Meteorology, Australia. 1977. *Report on Cyclone Tracy December 1974.* AGPS, Canberra.

Carter, R. W. G. 1987. 'Man's response to sea-level change'. In Devoy, R. J. N. (ed.) *Sea surface studies: a global review.* Croom Helm, London, pp. 464–98.

Cornell, J. 1976. *The great international disaster book.* Scribner's, NY.

Dolan, R. and Hayden, B. 1983. 'Patterns and Prediction of Shoreline Change'. In Komar, P. D. (ed.) *CRC Handbook of Coastal Processes and Erosion.* CRC Press, Boca Raton, Florida, pp. 123–50.

Foster, D. N., Gordon, A. D. and Lawson, N. V. 1975. 'The Storms of May–June, 1974, Sydney, NSW'. *Proceedings 2nd Australian Conference Coastal and Ocean Engineering*, Institute Engineers of Australia, pp. 1–11.

Holland, G. J., Lynch, A. H. and Leslie, L. M. 1988. 'Australian east-coast cyclones Part 1: overview and case study'. *Monthly Weather Review* v. 115 pp. 3024–36.

Holthouse, H. 1986. *Cyclone: a century of cyclonic destruction.* Angus and Robertson, Sydney.

Leopold, L. B., Wolman, M. G. and Miller, J. P. 1964. *Fluvial processes in geomorphology.* Freeman, San Francisco.

Linacre, E. and Hobbs, J. 1977. *The Australian Climatic Environment.* Wiley, Brisbane.

Milne, A. 1986. *Floodshock: the drowning of planet Earth.* Sutton, Gloucester.

Nalivkin, D. V. 1983. *Hurricanes, storms and tornadoes.* Balkema, Rotterdam.

Simpson, R. H. and Riehl, H. 1981. *The Hurricane and its Impact.* Blackwell, Oxford.

Whipple, A. B. C. 1982. *Storm.* Time-Life Books, Amsterdam.

3

Strong Wind as a Hazard

Introduction

The previous chapter was mainly concerned with the effects of strong winds generated by secondary features of general air circulation. These winds were associated basically with the development of low-pressure cells spanning areas of 10 000–100 000 km^2. While tropical cyclones and extra-tropical depressions produce some of the strongest winds over the widest area, they are by no means the only source of high wind-velocities. Highest wind speeds are generated by tornadoes, while topographic enhancement of regional wind-fields can produce localized wind damage just as severe as that produced by a cyclone. These localised wind-fields represent features of *tertiary air circulation*, and are localized in extent by either topography or the spatial dynamics of the wind-generating system. In addition, moderate-to-strong winds on a regional scale have the capacity to suspend large quantities of dust. This material can be transported thousands of kilometers, a process that represents a significant mechanism for the transport of sediment across the surface of the globe. Dust storms are exacerbated by drought in low rainfall, sparsely vegetated regions of the world; for that reason, they pose a slowly developing but significant hazard in sparsely settled semi-arid landscapes. Unfortunately, many of these landscapes have been marginalized by the degradatory agricultural practices of humans over thousands of years. However, in some regions where settlement has only occurred in the last 200 years, we are presently witnessing this marginalization process. Two of the most significant of these areas are located on the Great Plains of the United States and in the semi-arid regions of eastern Australia. Here, dust storms are a visual indication that the *degradation* process is active. This chapter will firstly examine the development and structure of tornadoes and associated vortices, together with a description of major disasters associated with these phenomena. Secondly, the response to tornadoes and methods used to mitigate their effects in the central United States will be discussed. Thirdly, the worldwide occurrence of dust storms will be described. Here, the cyclicity of dust storms on the Great Plains of the United States will be used to show how such events may be predicted, and how the process of land degradation is being dealt with. Finally, the mechanisms for topographically enhancing regional winds generating strong local winds will be described.

Tornadoes

Introduction (Miller, 1971; Nalivkin, 1983; Eagleman, 1983)

A tornado is a rapidly rotating vortex of air protruding as a funnel groundward from a cumulonimbus cloud. Most of the time, these vortices remain suspended in the atmosphere, and it is only when they connect to the ground or ocean surface that they become destructive. Tornadoes are related to larger vortex formation in clouds. Thus, they often form in convective cells such as thunderstorms, or in the right forward quadrant of a hurricane at large distances (>200 km) from the area of maximum winds. In the latter case, tornadoes herald the approach of the hurricane. Often, the weakest hurricanes produce the most tornadoes. Tornadoes are a secondary phenomenon, in which the primary process is the development of a vortex cloud. Given the large number of vortices that form in the atmosphere, tornadoes are generally rare. On average, each year, 850 tornadoes are reported, of which 600 originate in the United States. In Australia the number amounts to only one or two per year. The width of a single tornado's destructive path is generally less than 1 km. They rarely last more than half an hour and have a path length extending from a few hundreds of meters to tens of kilometers. In contrast, the destructive path of a tropical cyclone frequently can exceed 100 km, travel over thousands of kilometers and last for up to two weeks. However, tornadoes move faster than cyclones and generate higher wind velocities. They move at speeds of 50–200 km hr^{-1} and generate internal winds at speeds exceeding 400–500 km hr^{-1}. Similarly to tropical cyclones, tornadoes are almost always accompanied by heavy precipitation.

Form and formation (Nalivkin, 1983; Eagleman, 1983)

The role of parent clouds

The vertical funnel of a tornado is a secondary feature of a much larger vortex formed within a *parent cloud*. This parent cloud may be 15–20 km in diameter (some are as large as 150 km) and up to 4–5 km thick. Sometimes it is this cloud which descends to the surface without producing a funnel. The vortex motion in the cloud can be horizontal or vertical. Nalivkin considers the former feature to be dominant, and describes the vortex as consisting of several horizontal cylinders rotating at different speeds. Where these meet, a new, more intense vortex can form, which drops downward to form the tornado funnel. Usually, the horizontal vortex is up to ten times longer than the vertical funnel forming the tornado and in some cases this vortex or parent cloud can produce more than one funnel. The development of a horizontal vortex follows several stages, culminating in the formation of the tornado, followed by its demise (figure 3.1). Firstly, a *cumulo-nimbus cell* develops with small internal vortices. These grow upward as a result of convection, with parallel development of rain. The horizontal vortex then forms at the base of main uplift. A vertical vortex may be formed separately from the parent cloud, and to one side of this horizontal rotation. When the vertical and horizontal vortices link, a distinct funnel forms as material on the ground is lifted into the upper parts of the parent cloud. The tornado thus acts as a suction pump, lifting objects into the main maelstrom of the parent vortex. This material can then be transported by the storm tens of kilometers, whereupon it falls back to the Earth's surface as the vortex disintegrates.

The vortex in the parent cloud can take on three different forms. The first, and probably most common, originates as a straight or curved tube surrounding the area of horizontally rotating air at

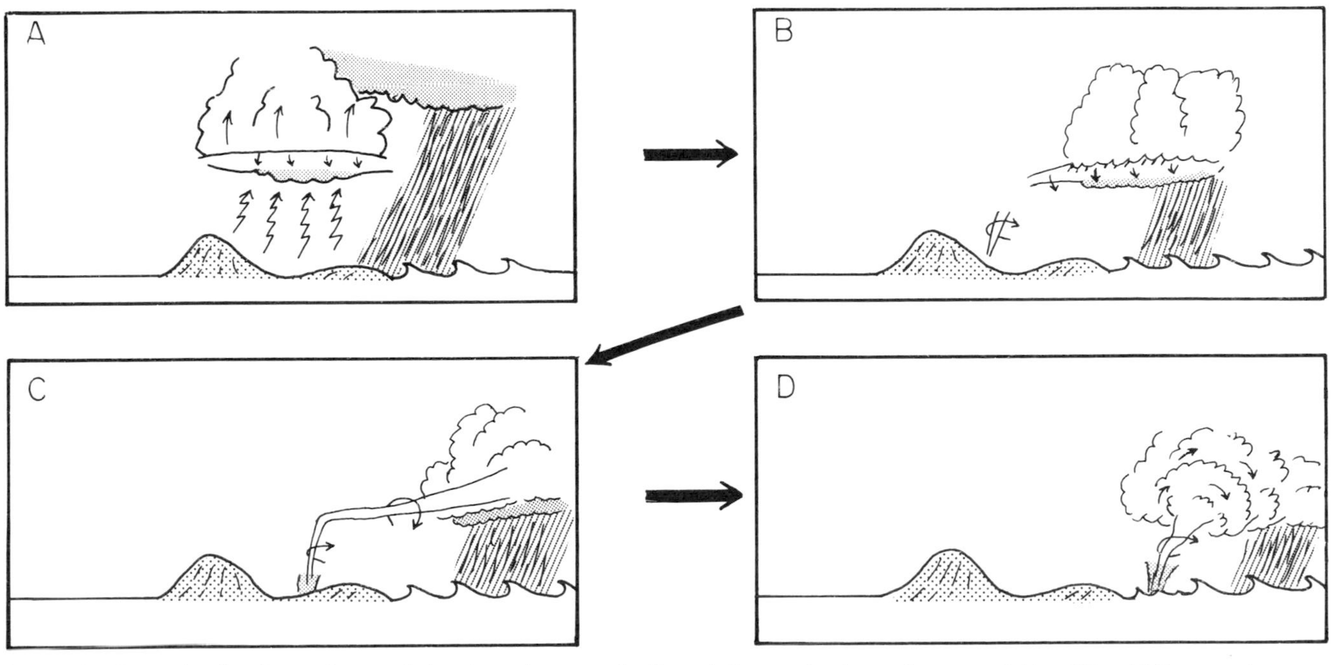

Fig. 3.1 Steps in the formation and demise of a tornado funnel from a horizontal vortex (after Dinwiddie, 1959)

the base of the parent cloud. The second type forms a distinctly horizontal funnel as one end of this tube pinches closed. The third type displays a vertical axis of rotation and is referred to as a 'tower vortex'. The latter commonly develops within thunderstorms or super-convective cells in the United States mid-west. Here, tornadoes are associated with the development of convection in thunderstorms, which generate updrafts linking the ground to the jet stream at the top of the troposphere, some 8–10 km up in the atmosphere (figure 3.2). The updraft forms two vortices, one at the leading edge of the storm and the other at the trailing end. If the trailing vortex breaches the jet stream, the rotating upward movement of air can intensify until it reaches down to the ground as a tornado funnel. In this case, thunderstorm development, giving rise to tornadoes, is linked to the path of the polar jet stream which is closely attached to the polar front. In spring and summer, temperature differences of 20°C between warm humid Gulf air and outbreaks of cold, polar air-masses over the mid-west of the United States can cause intense convection. This convection is responsible for the formation of super cells which have the capacity to tap the jet stream above. For this reason, the United States mid-west is the prime location of tornado formation in the world.

Dust devils, mountainados, fire tornadoes

There are also a number of small vortex phenomena that do not originate under the above conditions. Horizontal vortices tend to form over flat surfaces because air at elevation moves faster and overturns without the influence of frictional drag at the Earth's surface. If the horizontal vortex strikes an obstacle, it can be split in two and turned upright to form a dry tornado-like vortex (figure 3.3). Because of its small size, these vortices are not affected by Coriolis force, so they can rotate either clockwise or counterclockwise. On very flat ground, where solar heating is strong, thermal convection can become intense, especially in low latitudes. If the rate of cooling with altitude of air over this surface exceeds the usual rate of −1°C per 100 m (the normal *adiabatic lapse rate*), then the air can become very unstable, forming intense updrafts which can begin to rotate. These features are called 'dust devils', or in Australia, 'willy-willys'. While most dust devils are not hazardous, some can intensify downwind to take on tornado-like proportions. Tornado-like vortices can also be generated by strong winds blowing over topography. Strong wind shearing away from the top of an obstacle can suck air up rapidly from the ground. Such rising air will undergo rotation, forming small vortices. Around Boulder, Colorado (see figure 3.4 for the location of major place names mentioned in this chapter), strong winds are a common feature in winter, blowing across the Rocky Mountains. Vortices have often been produced with sufficient strength to destroy buildings. Such vortices behave like small tornadoes and are called 'mountainados'.

Strong updrafts can also form because of intense heating under the influence of fires. Most commonly this involves the intense burning of a forest; however, the earthquake-induced fires that destroyed Tokyo on 2 September 1923, and the fires induced by the dropping of the atomic bomb on Hiroshima

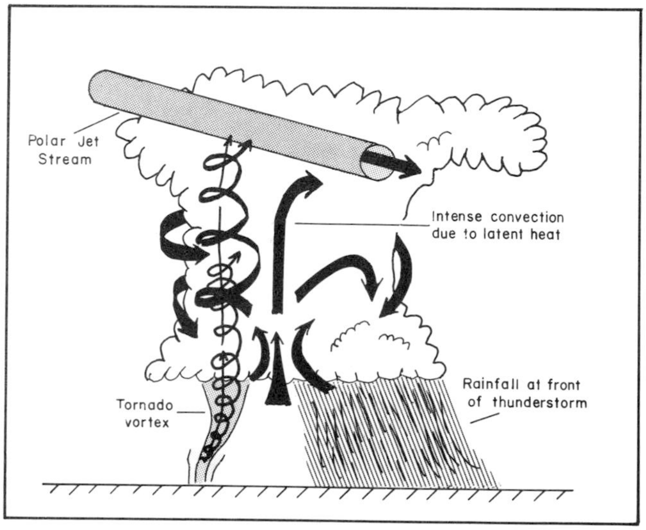

FIG. 3.2 Generation of a tornado funnel with a thunderstorm that taps the jet stream (after Eagleman, 1983)

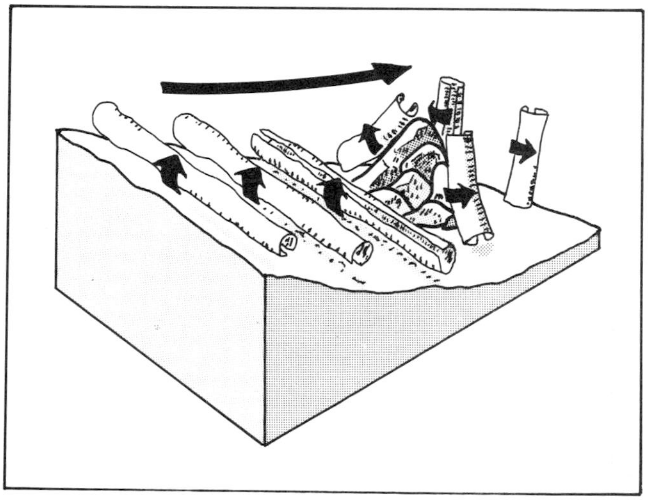

FIG. 3.3 Development of a mountainado from the splitting of a horizontal vortex rolling along the ground (after Idso, 1976)

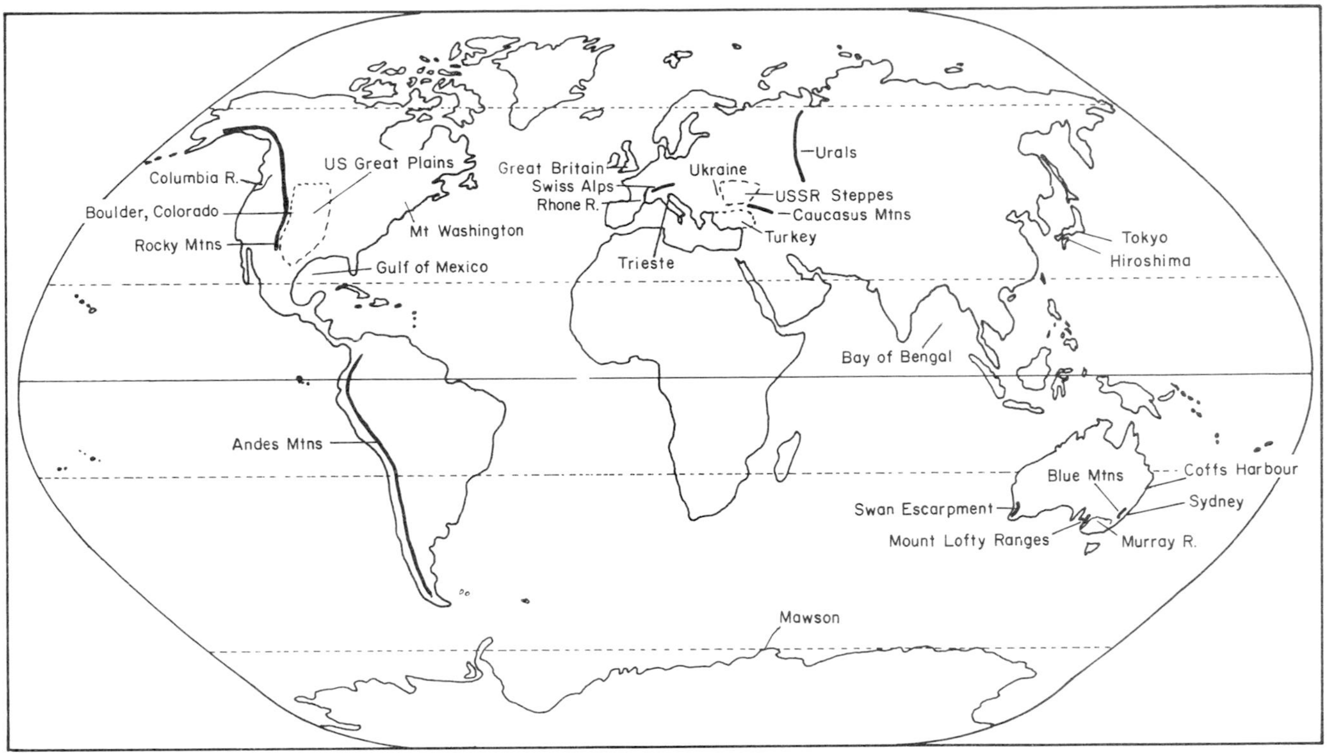

FIG. 3.4 Location map

on 6 August 1945 also produced tornadoes. Experiments, where the atmosphere has been heated from below using gas burners, commonly have generated twin vortices on either side of the rapidly rising columns of air. These fire whirlwinds can vary from a few meters in height and diameter to fire tornadoes hundreds of meters high. The most intense and hazardous fire tornadoes are generated by forest fires, and usually form on the lee side of hills or ridges during extreme bushfires, or forest fires such as that of the Ash Wednesday disaster in Southern Australia in February 1983. Rising wind velocities can exceed $250\,km\,hr^{-1}$, with winds inflowing into the maelstrom reaching velocities of $100\,km\,hr^{-1}$. Such firestorms pose a very life-threatening hazard for firefighters on the ground. Known fire whirlwinds have lifted flaming debris high into the sky and have snapped off fully grown trees like matchsticks. In many cases, they are linked to the convective column above the fire, and are fuelled by massive gaseous explosions within this rising thermal.

Structure of a tornado

Any tornado has three parts to its structure. The first is the horizontal vortex in the parent cloud, or the convective vortex in the lee of the thunderstorm, discussed above. The second feature is the funnel; the third involves the formation of auxiliary vortices, giving rise to cascades or envelopes. The funnel is virtually invisible until it begins to pick up debris. It has horizontal pressure and wind profiles similar to a tropical cyclone (figure 2.9), except that the spatial dimensions are reduced to a few hundreds of meters. In addition, tornadoes develop greater depressed central pressure and more extreme wind velocities at the funnel boundary. The sides of the funnel enclose the eye of the tornado, which is characterized by lightning and smaller, short-lived mini-tornadoes. Air movement in the eye is downwards at great speed. Vertical air motion occurs close to the wall with upward speeds of $100-200\,m\,s^{-1}$. The wall itself can be diffusely or sharply defined, and it is this latter feature that controls many of the unusual destructive features of a tornado. While diffuse wall boundaries occur at lower wind speeds, the high rotational wind velocities produce a knife-like boundary to the funnel. These velocities have not been measured, but have been determined theoretically using such evidence as eggs or windows cleanly pierced by a piece of debris without shattering, or a wooden pole pierced by a piece of straw. Wind speeds in excess of $1300\,km\,hr^{-1}$, which is close to the speed of sound, have been inferred. These wind speeds can pluck the feathers from a chicken, sometimes leaving half the chicken untouched where wind speeds sharply declined. Tornadoes are noted for their haphazardous and indiscriminate damage. People driving a horse and cart have witnessed a tornado taking their horse and leaving the cart untouched. One tornado that swept through a barn took a cow, and left the girl milking it and her milk bucket untouched. Numerous photographs exist showing houses and large buildings cleanly cut in half with the remaining contents undisturbed. Much of this rapid variability does not result from the sharp funnel boundary, but from the presence of mini-tornadoes rotating inside the vortex.

Many tornadoes are associated with multiple funnels. In June 1955 a single tornado cloud in the United States produced 13 funnels that reached the ground and left still more suspended in mid-air. Generally, the larger the parent cloud, the greater the number of tornadoes that can be produced. Parent clouds can often be separated beneath a thunderstorm, and it is possible for funnels to rise upwards linking the two cloud masses. Funnels have even been observed rising upwards into the anvil head of the thunderstorm. The most visible part of a tornado, the funnel, is actually enclosed within a slower spinning vortex. The downmoving air through the center of the funnel can kick up a curtain of turbulent air and debris at the ground in a similar process to that of a hydrofoil moving over water. This feature, termed a 'cascade', can also form above the ground before the main funnel actually develops, linking the ground to the parent cloud. If the cascade spreads laterally far enough, its debris can be entrained by the outer vortex. And if wind velocities in this outer vortex are strong enough, the main funnel may slowly be enshrouded by debris from the ground upwards to form what is called an 'envelope'.

Occurrence (Miller, 1971; Nalivkin, 1983; Oliver, 1979)

About 80 per cent of all tornadoes occur in the United States particularly on the Great Plains. Figure 3.5 illustrates the density of tornadoes and

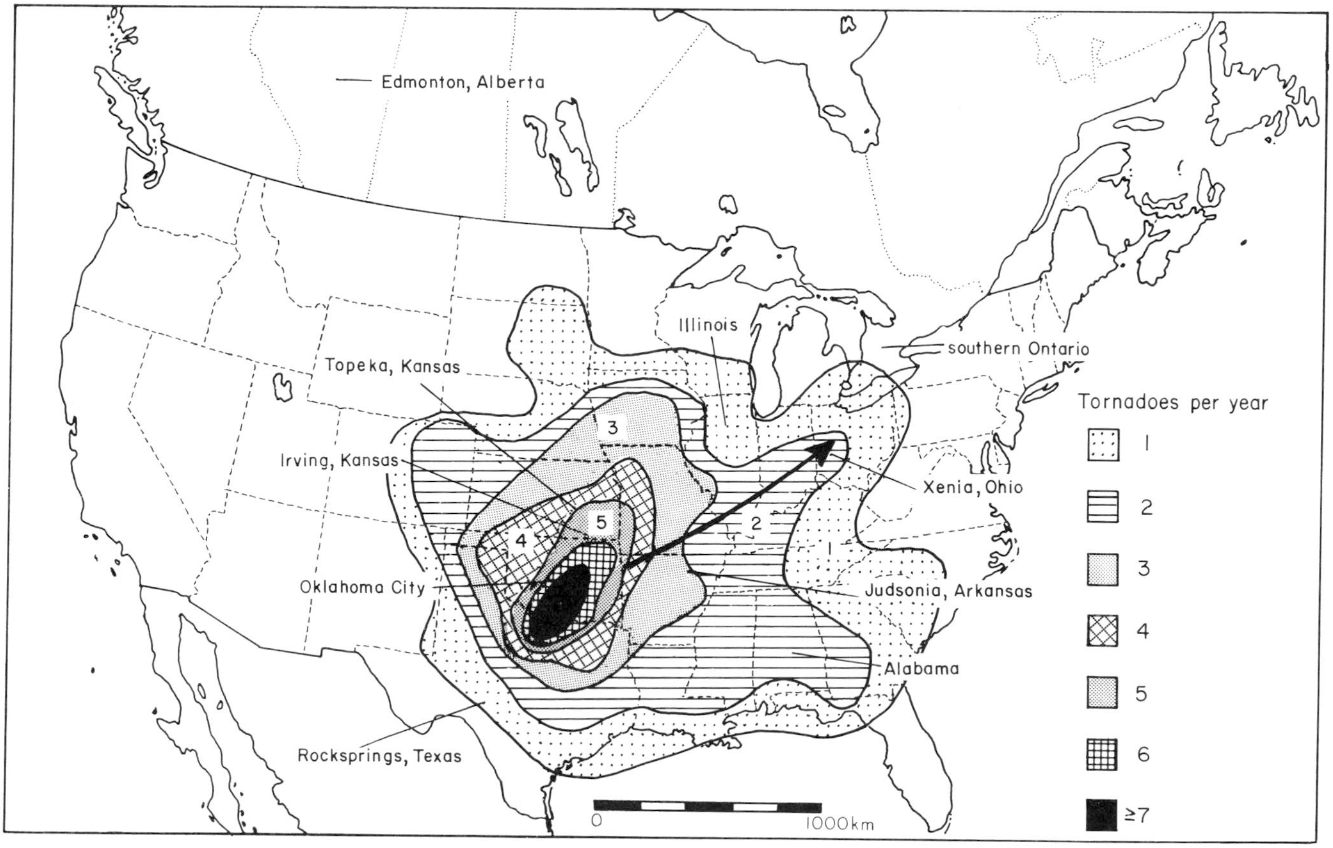

FIG. 3.5 Average occurrence of tornadoes per year and preferred path of movement in the continental United States (adapted from Miller, 1971)

the preferred path of travel in this region. The area with five or more tornadoes per year, which is referred to as 'tornado alley', takes in Kansas, Oklahoma, Texas, Missouri and Arkansas. Tornadoes occur commonly in this region, because it is the preferred location where cold Arctic air meets air from the warm Gulf of Mexico, with maximum temperature differences in the order of 20–30°C. The flat nature of the plains enhances the rapid movement of cold air, while the high humidity of Gulf air creates optimum conditions of instability in the atmosphere. Most tornadoes in the United States tend to occur in late spring, when this temperature difference between air masses is at its maximum. At this time of year, the polar jet stream still can loop far south over the Great Plains.

In the 35-year period between 1916 and 1950, 5204 tornadoes were observed in the United States. This figure probably under-represents the actual number by a factor of two to four. Since 1950 the frequency of tornadoes has been increasing. Much of this increase undoubtedly reflects the fact that monitoring has improved; however, there is disturbing evidence that tornadoes are becoming more frequent outside the main areas of influence shown in figure 3.5. For example, before 1970 tornadoes were rare events in southern Ontario, Canada (figure 3.5 shows the location of place names for tornadoes in North America). Now, between one and two funnels are reported each year from this area. Totally unexpected was the Edmonton, Alberta, tornado of 1987. This city lies at a latitude of 53.5°N, too far north usually for the occurrence of sharp temperature differences between air masses favouring tornado development.

There are other areas of the world susceptible to tornadoes. The Russian steppes are similar in relief to the Great Plains; here, however, there is no strong looping of the jet stream nor any nearby warm body of water supplying the high humidities. Great Britain averages 60 tornado-like vortices each year, although it is doubtful if many ever reach full development. In a similar manner, Italy averages ten per year. Surprisingly, tornadoes are rare on the African continent and Indian subcontinent. Extreme heating or atmospheric instability cannot singly account for tornado formation. In the case of the Indian subcontinent, tornadoes are rarely reported outside the northern part of the Bay of Bengal; there, like other similar locations in the world, they are associated with tropical cyclones. As with tropical cyclones, tornadoes do not form near the equator because Coriolis force is lacking. Nor do tornadoes form in polar or sub-polar regions, mainly because convective instability is suppressed in these regions by stable, cold air-masses. This is the reason why the Edmonton tornado mentioned above was so unusual.

As with Africa and India, in Australia tornadoes are rare, although more probably occur than are recorded. Generally, tornadoes are reported most often along the east coast of the continent during the passage of cold fronts. They can also occur because of instability in the lee of the Great Divide associated with the development of low-pressure cells, or in northern Australia under extreme convergence associated with the *Inter-tropical Convergence Zone*. In southern Australia, most tornadoes occur beneath the subtropical jet stream. Tornadoes are most prevalent from early spring to mid summer. In spring the difference between air masses is greatest, while in summer convective thunderstorms are best developed. It has been suggested that tornadoes occur as often in Australia as in the United States; however, wind speeds in many of the Australian cases has rarely exceeded 190 km hr^{-1}, a figure that is much less than the wind speed typically developing in a tornado. At these wind speeds, the term willy-willy or dust-devil seems more appropriate. Meteorological records indicate about one reported occurrence per year each in New South Wales and Western Australia. Most reported tornadoes probably represent the presence of the incipient cascade preceding full funnel development. Compared to United States, where the number of people killed by tornadoes is equal to the combined total of lives lost from floods and tropical cyclones, few if any deaths have been caused by this hazard in Australia.

Tornado destruction (Whipple, 1982; Nalivkin, 1983)

Tornadoes are one of the most intense and destructive winds found on the Earth's surface. The fact that they most frequently form along cold fronts implies that they can occur in groups over wide areas. Their destructive effects are due to the lifting force at the funnel wall, and to the sudden change in pressure across this boundary. The lifting force of tornadoes is considerable, both in terms of the weight of objects that can be moved, and the volume of material that can be lifted. Large objects weighing 200–300 tonnes have been shifted tens of meters, while houses and train carriages weighing 10–20 tonnes have been carried hundreds of meters. Tornadoes are also able to suck up millions of tonnes of water, which can be carried in the parent cloud. Tornadoes have been known to drain the water from rivers, thus temporarily exposing the bed.

While one would be unfortunate to be struck by some of the debris moved by these winds, the most effective mechanism of destruction in a tornado is the dip in barometric pressure across the wall of the funnel. It is not how low the pressure in the eye gets (about 800 hPa), but how rapidly this change in pressure is effected. The fact that the tornado is so compact and has a sharp wall means that a pressure reduction of tens of $hPa\,s^{-1}$ could occur, although this dip has been difficult to substantiate by measurement. Air in buildings affected by a tornado ends up being much higher than the outside air pressure, and is unable to escape from the structure quickly enough without causing an explosion. This exploding effect is the main reason for the total devastation of buildings, whereas adjacent structures or remaining remnants appear untouched. The pressure reduction can be so rapid that live objects appear burnt because of dehydration. One of the major causes of death from a tornado is not due to the fact that people have been struck by flying debris or tossed into objects, but to the fact that they have received severe burns. It is even possible that the de-feathered chickens described above could be cooked because of this phenomenon.

The 5204 tornadoes recorded in the period 1916–50, killed 7961 people and caused $US500 million worth of damage in the United States. In the last 50 years the deathtoll has amounted to over 9000 people. In April 1965, 257 people lost their lives as 47 tornadoes developed along a cold front in the central United States. The 'Three States' tornado of March 1925 created a path of destruction 350 km long. Over 700 people were killed by the tornadoes spawned from the movement of this single parent cloud. One tornado in March 1952 destroyed 945 buildings in Judsonia, Arkansas. Another tornado in Rocksprings, Texas, in April 1927, killed 72 people, injured 240 others, and wiped out most of the town of 1200 people in 90 seconds: in all, 26 per cent of the population was either killed or injured. Cities are not immune, although high-rise buildings tend to survive better than single-unit dwellings. For example, in June 1966, a tornado went through Topeka, Kansas, destroying most homes and shops in its path but leaving multistoried flats and office buildings undamaged. The worst tornado disaster in the United States occurred during a 16-hour period on 3–4 April 1974. A total of 148 tornadoes were recorded from Alabama to Canada, killing 315 people and causing damage amounting to $US500 million. Six of the tornadoes were amongst the largest ever recorded. Complete towns were obliterated, the worst being Xenia, Ohio, where 3000 homes were destroyed or damaged in a town with only 27 000 residents.

Geological and ecological significance

(Nalivkin, 1983)

The fact that about 600 tornadoes per year occur in the Great Plains implies that over a 1000-year timespan, which is a small period in geological terms, as many as 600 000 tornadoes could have occurred over a relatively small area in the United States. Apart from dust storms, tornadoes are a major mechanism for transporting large amounts of clay and silt-sized particles. More importantly, they are a major dispersive mechanism for *biota*. Almost all reports of rain accompanied by debris involve tornadoes as the major entrainment process. Pollen, seeds, micro-organisms, fish, and invertebrates such as frogs, toads and turtles can be transported up to 100–200 km and deposited live (or in some rare cases encased in ice). Manna from heaven, mentioned in the Bible, is simply an edible lichen transported in all likelihood by a tornado-like vortex. It is, thus, possible for small animal and plant species to be dispersed rapidly over large areas by this mechanism. The problem of species migration between unconnected lakes in North America after the last glaciation can also be solved by such a mechanism. The process of species diffusion has always been a major area of study for biologists. Tornadoes offer a simple and effective mechanism by which many species can be spread rapidly over large areas between ecosystems, or dispersed from remnant island pockets following deglaciation in the northern hemisphere.

Response (Sims and Baumann, 1972; Oliver, 1981)

For most of the world tornadoes can be put into the act of God category. For instance, in Australia, it is just bad luck if a tornado strikes you or your property. In the United States, especially in 'tornado alley', tornadoes are a frequent hazard. While it is difficult to avoid their occurrence, there are many methods for lessening their impact. For instance, children in the United States mid-west schools are taught at an early age how to take evasive action if they see, or are warned about, the approach of a tornado. I can remember, as a school child living in southern Ontario, being sent home from school one day when tornadoes were forecast. Southern Ontario, then, was not considered highly vulnerable; but the school authorities were aware that a tornado hitting a school full of children would generate a major disaster. We were sent home to minimize the risk. We were also drilled in procedures to take if a tornado did strike during school hours—the same procedures we would have taken if a nuclear attack was imminent! We knew that tornadoes travelled mainly northeast; so, if caught in the open, we were instructed to run to the northwest. In the United States mid-west, most farms or homes have a basement or cellar separate from the house, which can be used as shelter. All public buildings have shelters and workers are instructed in procedures for safe evacuation. Safety points in buildings, and in the open, are well sign-posted. In the worst threatened areas such as 'tornado alley', tornado alerts are broadcast on the radio as part of weather forecasts. The actual presence of a tornado can be detected using doppler radar or tornado watches maintained by the Weather Service. The weather watch consists of a volunteer network of ham radio operators known as the Radio Emergency Associated Citizen Team (REACT), who will move to vantage points with their radio equipment on high-risk days. Immediately a tornado is detected, local weather centers, civil defense and law enforcement agencies are alerted and warnings then broadcast to the public using radio, television and alarm sirens.

For many rural communities in the mid-west, a tornado is the worst disaster that can strike the area. State governments will declare a state of emergency, and if outbreaks of tornadoes occur over several states, federal assistance may be sought. In rural areas, community response is automatic to a tornado disaster and there are government agencies, in addition to police and fire services, that are specially formed for disaster relief. Generally, however, tornadoes are accepted as a risk which must be lived with. The above observations are not universal, but may be affected by the cultural attitudes of people living in various threatened areas. Sims and Baumann (1972) analyzed the attitudes of people to the tornado risk between Illinois and Alabama. The death rate from tornadoes apparently was higher in the latter state. They concluded, based upon interviews, that there was a substantial attitude difference between the two states. People in Illinois believed that they were in control of their own future, took measures to minimize the tornado threat, and participated in community efforts to help those who became victims. In Alabama, however, people were more likely to believe that God controlled their lives, did not heed tornado warnings, took few precautions and were less sympathetic towards victims. Oliver (1981) has analyzed this study and pointed out some of its fallacies. For instance, the deathtoll from tornadoes is higher in Alabama because the tornadoes there are more massive than in Illinois. However, the Sims and Baumann study does highlight the fact that not all people in the United States have the same perception about the tornado hazard or response for dealing with its aftermath. These differences will be elaborated further in chapter 14.

It may make little difference what your attitudes to tornadoes are, but rather where you actually live in the mid-west that dictates your chances of survival. European settlement in the mid-west inadvertently may have chosen the most susceptible sites for towns. For example, Oklahoma City on the Canadian River has been struck by more than 25 tornadoes since 1895, even though Oklahoma State ranks seventh in the United States for deaths per unit area due to tornadoes—7.3 deaths per 10 000 km^2 compared to 20.5 per 10 000 km^2 for Mississippi, which ranks first. Indian tribes in the mid-west rarely camped in low-lying areas such as the bottom of ravines, or along rivercourses, because they were aware of the fact that tornadoes tend to move towards the lowest elevation once they touch ground. This makes sense since the tornado vortex is free-moving and will tend to follow the path of least resistance. Figure 3.6 illustrates this tendency dramatically. The first tornado

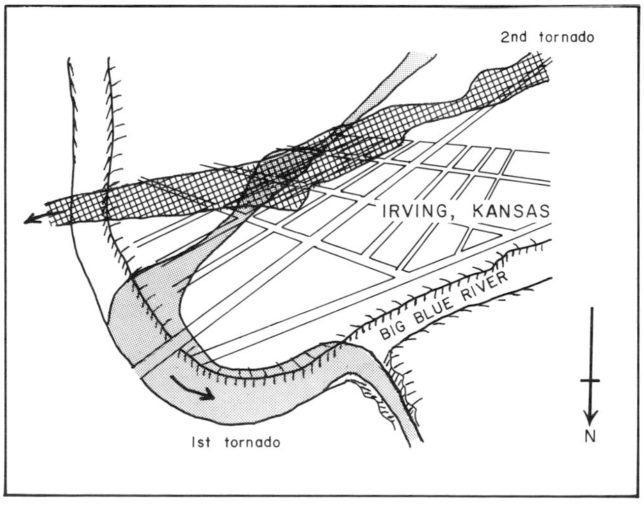

FIG. 3.6 The effect of topography on the passage of a tornado at Irving, Kansas, 30 May 1879 (after Finley, 1881)

to pass through Irving, Kansas, on 30 May 1879, ended up being trapped in the valley of the Big Blue River, and eventually in a small gully off to the side. Tornadoes do not always have to adhere to this pattern, as illustrated in the same figure by the second tornado that came through this town minutes later. Many towns in the mid-west of the United States are built near rivers or in low-lying areas for historical reasons such as access to transport and water. The fact that tornadoes prefer to travel along these paths may be one of the reasons why they have been, and will continue to be, a major hazard in the mid-west of the United States.

FIG. 3.7 Dust storm over Melbourne, Australia, 8 February 1983 (photograph courtesy of the Australian Bureau of Meteorology). This was only the second such storm in a century to afflict Melbourne; it originated in the mallee country of western Victoria.

Dust storms (Lockeretz, 1978; Nalivkin, 1983; Goudie, 1983; Middleton, 1984; Middleton et al., 1986)

Introduction

Dust storms are wind storms accompanied by suspended soil material without precipitation. Although snow blizzards can be considered within this categorization, they will be described separately in chapter 7. Dust storms are one of the main mechanisms on the surface of the Earth for the transport of fine sediment. Between 130 and 800 million tonnes of dust, with extremes as high as 5000 million tonnes, are entrained by winds each year. The average annual amount of clay-sized particles moved is about 500 million tonnes. Dust storms account for most of the terrigenous material found in ocean basins, contributing over 75 million tonnes of material per year to the Atlantic Ocean alone. At distances 5000 km out into the Atlantic Ocean, fallout from the Sahara is still deposited at a rate of 3000 tonnes $km^{-1} yr^{-1}$. Dust from the Sahara has often been deposited on Bermuda and the Bahamas, a distance of over 7000 km, while dust from China has been recorded in Hawaii and as far away as Alaska, a distance in excess of 10 000 km. In some locations dust storms are even a major source of supply of material to rivers.

While the shock stories of the United States *dust bowl* are considered by most as typical of dust-storm events, dust storms have a worldwide occurrence and are more common in Africa and central Asia than they are in North America. Dust storms often have a *cyclic* occurrence, which, in the United States, parallels drought and is linked to the 18.6-year lunar cycle. In Australia a dust storm hit Melbourne on 8 February 1983, a week before the Ash Wednesday bushfires (figure 3.7). This storm was instigated by a cold front passing over paddocks devoid of vegetation in western Victoria, following the long drought of 1982–83. The winds picked up tonnes of topsoil from farms and swept this across Melbourne, eastern Victoria and out to sea. While the storm was only the second dust storm to hit Melbourne this century, it illustrates well the vulnerability of landscapes, both in Australia and elsewhere in the world, to soil *deflation* following poor farming management practices, which are then exacerbated by droughts. In the short term, dust storms only represent to the majority of the population in an affected area, a temporary physical discomfort; however, a single storm in actual reality represents the insidious deterioration of arable topsoil, and the slow process of *marginalization* or *desertification* of semi-arid land. The long-term social consequences usually involve the dislocation of communities, famine and, ultimately, the destruction of civilizations dependent upon such areas for their existence. This has been the fate of early agriculturally dependent civilizations in Mesopotamia, which were reliant upon the Euphrates-Tigris basin, and those in Pakistan reliant upon the Indus River.

Formation

Dust storms commonly form as the result of the passage of cold fronts across mid-latitude plains, or on the westerly side of Hadley pressure cells. In the former case, they are tied into the intensity of the westerlies; in the latter case, they are associated with the strength of the Hadley cell, the subtropical jet stream and the formation of monsoonal lows. In all cases, circulation can give rise to dust storms lasting for several days. In the northern

Sahara region, dust storms are mainly produced by complex depressions associated with the westerlies. The depressions originate in the eastern Mediterranean, or the Atlantic in winter. In the southern Sahara, low pressure fronts are responsible for *harmattan* winds. These depressions track south during the northern hemisphere winter and easterly during the summer. The seasonal movement of the *monsoon* into the Sudan region of the Sahara is responsible for dust storms in eastern Africa. A similar process, tied into the occurrence of the Indian monsoon, is responsible for the majority of dust storms on the Arabian peninsula. A particularly common feature of dust storms in these arid landscapes is the *haboob*, generated by cold, downburst winds associated with large convective cells, and linked to the advance of the inter-tropical convergence. These winds also account for many of the dust storms in the Gobi and Thar deserts in central Asia, in the Sudan, and in Arizona. East of the Mediterranean, depressions moving across Turkey and northern Iraq generate most dust storms. In Australia dust storms are generated by the passage of intense, cold fronts across the continent, following the desiccating effects of hot, dry winds produced by the subtropical jet stream in the lee of high-pressure cells. In the southeast part of the continent, these dry winds blow from the centre of the continent in a southeasterly direction. The Victorian dust storm of 8 February 1983, and the devastating bushfires of Ash Wednesday exactly one week later, both originated via this mechanism. Finally, in mountainous regions such as western North America, *katabatic* or gravity induced winds can generate dust storms in places such as California and along the Colorado Front ranges. Similar, cold density winds flowing off the mountains in the Himalayas and the Hindu Kush generate dust storms southward over the Arabian Sea. Such localized winds also produce dust storms in the valleys leading from the Argentine foothills.

There is a correlation worldwide between dust storms and rainfall. Where rainfall exceeds 1000 mm yr^{-1}, dust storms occur less than 1 day yr^{-1}, mainly because vegetation prevents silt-entrainment by winds into the atmosphere, and convective instability tends to give rise to thunderstorms rather than dust storms. Dust storm frequency is greatest where rainfall lies between 100–200 mm, and declines again in areas where annual rainfall is less than 100 mm. The latter fact is due to the scarcity of dust, either because it has already been removed by wind, or because the extremely arid conditions do not permit substantial clay and silt production. Except under unusual conditions, dust storms tend to have their greatest frequency of occurrence in both hemispheres in late spring and early summer. This pattern corresponds to the changeover from minimum rainfall in early spring to intense evaporation in summer. In Australia, however, the pattern also represents the most likely period of drought exacerbated by the *aperiodic* collapse of easterly trade winds in the tropics associated with the Southern Oscillation.

Where dust storms are formed by cold fronts, cold air forces warm air up rapidly, giving a convex lobe-appearance to the front of the dust storm. As the air is forced up, it sets up a vortex that tends to depress the top of the cold front and further enhance the lobe-like appearance of the storm. The classic image is one of a sharp wall of sediment moving across the landscape with turbulent overturning of dense clouds of dust (see figure 3.7). Wind speeds can obtain consistent velocities exceeding 60 km hr^{-1} or 30 m s^{-1}. Material the size of coarse sand can be moved via *saltation* in the first tens of centimeters above the ground surface. Fine sand can be moved within 2 m of the ground, while silt-sized particles can be carried to heights in excess of 1.5 km. Clay-sized particles can be suspended throughout the depth of the troposphere and carried thousands of kilometers. Most of the dust is suspended in the lower 1 km of the atmosphere. While concentrations are such that noonday visibility can be reduced to zero, dust storms technically exist as soon as visibility is less than 1 km. Gravel-size material cannot be transported by winds, so it is left behind to form a stony or *reg desert*. Sand-sized particles in saltation are also left behind by the removal of silt and clay. These sand deposits become non-cohesive and mobile. Once the fine material has been removed from topsoil its cohesiveness and fertility cannot easily be re-established. The removal of fines often means the removal of organics as well. This is why the initial occurrence of dust storms in an area signifies the removal of soil fertility and, over extended periods, the desertification of marginal semi-arid areas.

Location and frequency of occurrence

The frequency of dust storms cannot be overemphasized. Between 1951 and 1955, 3882 dust storms were recorded in central Asia. In the Turkmenistan region of the Soviet Union, 9270 dust storms were recorded over a 25-year period. In the United States mid-west, as many as ten dust storms or more per month occurred between

1933 and 1938. Dust storms occur, geographically, in the southern arid zones of Siberia, the European part of the Soviet Union, western Europe, western United States, northwestern China and northern Africa. Figure 3.8 illustrates the frequency, source areas and tracks of main dust storms worldwide. Satellite imagery and Space Shuttle photography have refined the source areas for most dust storms. In the Sahara region most of the alluvial basins are sources for dust, with the Libyan desert being important in northern Sudan, and the large sandy or *erg deserts* in Mauritania, Mali and Algeria supplying dust to the Atlantic. In the Middle East it is the alluvial plains of the southern Euphrates-Tigris basin, and the deserts of Syria and northern Saudia Arabia that spawn most of the dust, while further east the Indian Ocean coastline of Iran is an important source area for the Indian Ocean. On the Indian subcontinent, the Thar desert and Indus River floodplains generate dust over India. Northwards, most of Afghanistan, and a belt of Kazakhstan, Soviet Union, running from the Black Sea east around the Caspian Sea, is the source area for some of the greatest numbers of dust storms in the world. In China an area centered on the Gobi desert supplies dust for most of eastern Asia. Smaller areas in the world occur around the Simpson desert and Shark Bay in Australia, west of the Kalahari desert in South Africa, around the Atacama desert of Chile, on the Pampas of Argentina, around the Colorado delta in southwestern United States and Mexico, and on the southern United States Great Plains.

While dust storm frequencies can vary substantially for any one area, there are consistent trends over time for many locations. Frequencies in the Sahara range from 2–4 per year in Nigeria to 10–20 per year in the Sudan. Around the Arabian Gulf, the incidence of dust storms rises to over 30 per year in Iraq. Many parts of the southern, arid areas of the Soviet Union average about 20 storms per year, but the frequency rises to over 60 per year in the Turkmenistan Republic on the eastern side of the Caspian Sea. Northwest India, adjacent to the Thar desert, is affected by as many as 17 events per year, with the frequency increasing northwards into Afghanistan to over 70 storms per year. The highest frequency of dust storms

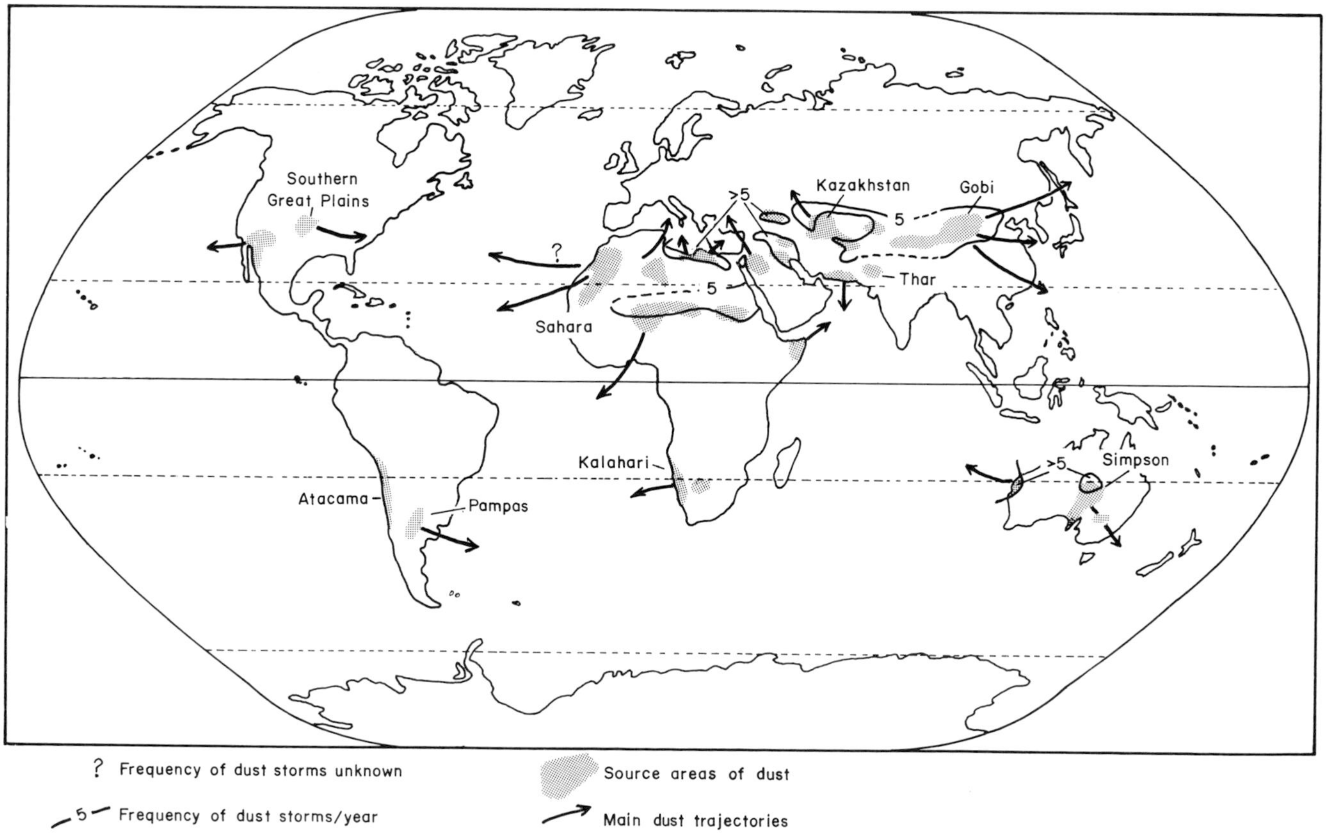

Fig. 3.8 General frequency of occurrence, source areas and tracks of dust storms (after Middleton et al., 1986; McTainish and Pitblado, 1987 and Goudie, 1983). Only frequencies for broad-scale areas shown.

occurs in the Gobi desert of China, which averages 100–174 dust storms annually under the influence of the Mongolian-Siberian high-pressure cell. Dust has been blown as far as Peking and Shanghai, where as many as four storms per year have been recorded. In Australia more than five dust storms per year are recorded around Shark Bay, in Western Australia, and Alice Springs, north of the Simpson desert.

It is debatable whether or not the incidence of dust storms is increasing as a result of growing populations. Chinese data show no such increase; but intense cultivation of the United States Great Plains in the 1920s, the opening up of the 'Virgin Lands' in the Soviet Union in the 1950s, and the *Sahel* drought of the 1960s–70s, all led to an increase in the frequency of dust storms in these locations. While the latter two regions in recent times have produced the most dramatic change in dust storm activity because of human activities, the best documented effect has been on the United States Great Plains during the dust bowl years of the great depression in the 1930s. Here, it can be shown that dust storms have a cyclic occurrence. In the 1870s dust storms during drought drove out the first group of farmers. While no major dust storm period was recorded until the 1930s, drought has tended to occur synchronously with maxima in the 18.6-year lunar cycle. This timing has global implications which will be discussed later in chapter 5. This cyclicity is shown in figure 3.9, which plots the number of acres damaged by dust storms in the Great Plains between 1936 and 1977. Dust-storm degradation of arable land in the mid-1950s and mid-1970s can be seen to follow an 18–19-year pattern. It is surprising that, while the dust bowl years of the 1930s have become part of American folklore, the drought years of the mid-1950s and 1970s actually produced more soil damage. This is despite the fact that reclamation and rehabilitation programs had been initiated by the federal government in response to the 1930s dust bowl. These programs included the return of marginal land to grassland, and the use of *strip farming*, crop rotation and *mulching* practices. These techniques build up soil moisture and nutrients while protecting the surface soil from wind deflation. In most cases, the dust storms followed a period of rapid crop expansion during favourable times of rainfall and commodity prices. In fact, there is a strong correlation in the United States between high wheat prices and dust storms several years afterwards.

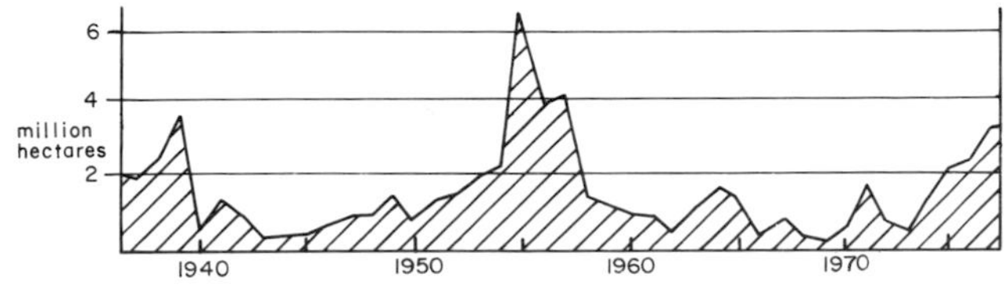

FIG. 3.9 Number of hectares of arable land in the United States mid-west subject to wind damage: 1936–77 (after Lockeretz, 1978)

A similar process of land degradation by dust storms is occurring in Australia. Despite Australia's aridity, dust storms are not as common as in other parts of the world (see figure 3.8). However, a minor area of activity in the the Mildura region along the Murray River, in the southwest corner of New South Wales, lies outside recognized centers of dust storm activity. Here, irrigation has become extensive. There appears to be a weak association between soil type and dust storm activity. Non-calcareous, self-mulching clayey or red duplex soils are more susceptible to wind deflation, while loamy, *calcareous* or crusted soils are more resistant. These latter soil characteristics are fragile and can be easily altered by human mismanagement practices. Such modification in the irrigation lands around Mildura accounts for the increasing incidence of dust storms here.

Major storm events

There have been many notable dust storms. For instance, in April 1928 a dust storm affected the whole of the Ukrainian steppe, an area in excess of 1 000 000 km^2. Up to 15 million tonnes of black *chernozem* soil was removed and deposited over an area of 6 million km^2 in Romania and Poland. In the affected area, soil was eroded to a depth of 12–25 cm in some places. Particles ranging from 0.02–0.5 mm in size were entrained by the wind. In March 1901, 150 million tonnes of red dust from the Sahara was spread over an area from western Europe to the Ural mountains in Russia. Deposits on glaciers in the Swiss Alps were used as a marker horizon for years afterwards to study the dynamics of ice movement. A similar storm in March 1962 originated in North Africa, crossed Syria, southern Greece and Turkey and, eventually, the Caucasus mountains—a distance of 5000 km.

The dust bowl of the United States mid-west in the 1930s produced some of the most dramatic dust storms of the twentieth century. Their effect was enhanced by the fact that this area had not previously experienced such extreme events. Most of the storms occurred in late winter—early spring, just as the snow cover had melted, and the frozen ground inhibiting soil deflation had thawed. However, at this time of year, the polar westerlies were still strong. The most severe storms occurred between 1933 and 1938 on the southern Plains, and between 1933 and 1936 on the northern Plains. The number and extent of storms during the worst period in March 1936 is shown in figure 3.10. These storms covered an area from the Gulf of Mexico north into Canada. During this month, the panhandle of Texas and Oklahoma experienced dust storms on 28 out of 31 days. At one point, as the United States Congress debated the issue in Washington, dust originating 2000 km to the west drifted over the capital. By 1937, 43 per cent or 2.8 million hectares of the arable land at the center of the dust bowl had been seriously depleted. In some counties over 50 per cent of the farm population went bankrupt and onto relief.

Geological significance

Geologically, as determined from deep sea-cores and sedimentation on icecaps, dust storms have had a variable occurrence through time. Low rates of dust deposition occurred 25–50 million years ago in the early *Tertiary*, when climates appeared to be wetter and atmospheric circulation less vigorous. Dust storms increased in frequency 7–25 million years BP, and accelerated rapidly in the period 3–7 million years BP. The Late *Cenozoic* ice age appears to be a time when dust storms became more frequent than at any time in the last 70 million years. During the cold phases of the *Quaternary*, when continental glaciations were at their maximum extents and temperatures coldest, dust storms may have been depositing sediment at rates 100 times greater than at present. These *Pleistocene* glacial dust storms were responsible for substantial deposition of *loess* (a substantial wind-

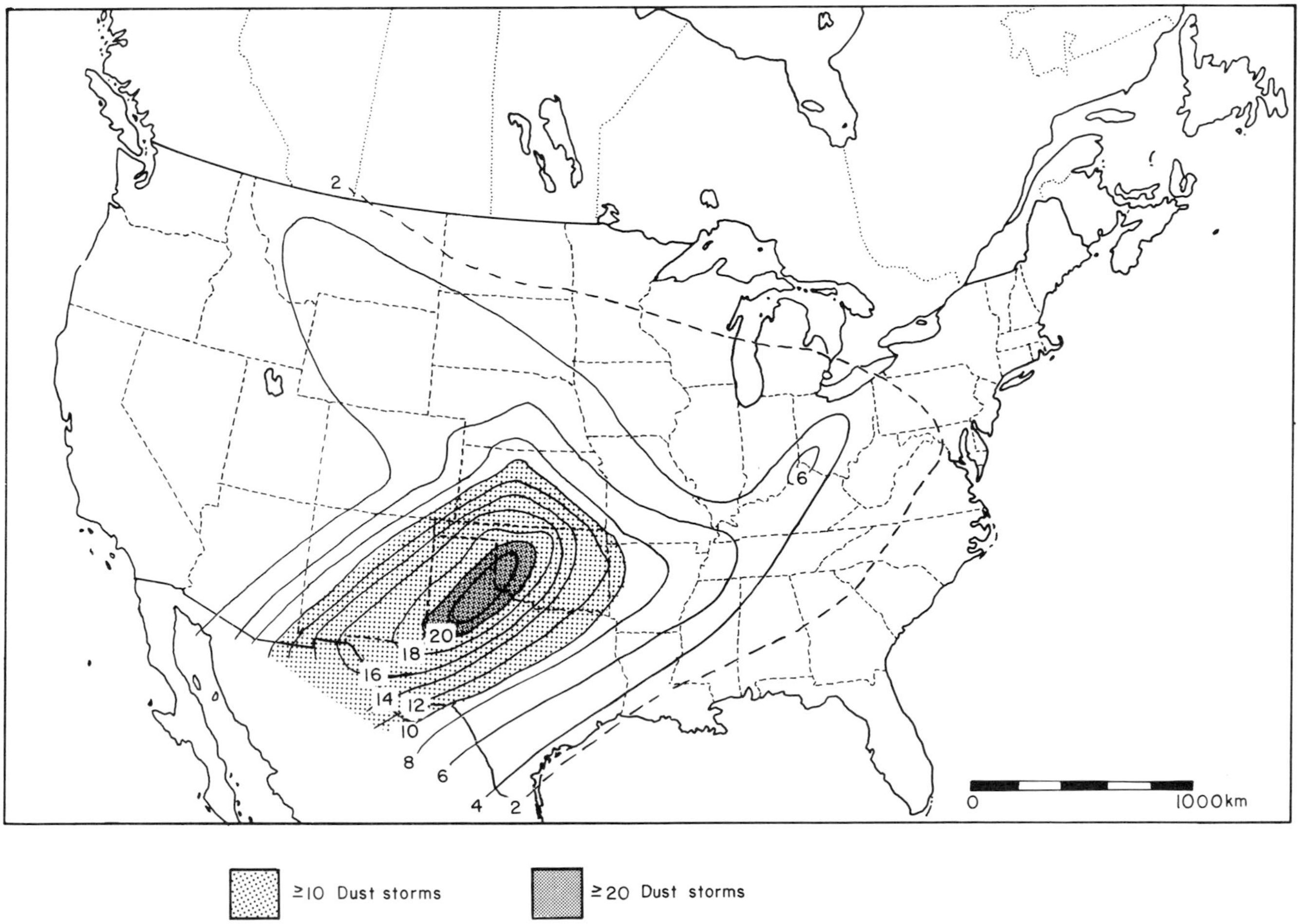

Fig. 3.10 Extent and number of dust storms in March 1936 in the United States (after Lockeretz, 1978)

blown, fine dust deposit) in China and central Europe. In the southern Ukraine region, geological rates of accumulation amounted to 0.2 m per century, while around Volgograd rates were between 0.5–0.75 m per century with sediment originating from Turkey. Loess in central Asia has accumulated at the rate of 1 m per century. High rates of silt accumulation are also encountered in ocean sediment of late glacial age off the coasts of Australia, Arabia and northwestern India. In eastern Australia red dust accumulated frequently enough in marsh deposits for it to be used to date sediment through magnetic susceptibility techniques. Rates of accumulation have always been low in the Pacific Ocean; however, Asian source sediments can still be traced in deposits that increase in thickness into the Pleistocene. The Sahara appears to be responsible for the greatest amount of dust supplied to ocean basins, principally the Atlantic Ocean. Its contribution can be traced back to the *Cretaceous*. These figures indicate that one of the principal mechanisms of sediment accumulation geologically has been via dust transport in the atmosphere. In ocean basin sedimentation, dust storms represent the dominant mode of sediment transport. Not realized, and maybe just as important, is the fact that dust storms in the subtropical latitudes over Asia and northern Africa represent the major process of land denudation. In these regions almost all weathered residues of clay and silt size are removed by wind as soon as they form. The sandy and stony deserts that remain behind attest to the degree of denudation that has occurred.

Localized strong winds (Smith, 1975; Linacre and Hobbs, 1977; Oke, 1978; Oliver, 1979, 1981)

A number of cases have now been presented above of strong winds produced by vortex storms. In the Australian context the strongest winds measured for a tropical cyclone obtained a velocity of 246 km hr^{-1} during Cyclone Trixie at Onslow, Western Australia, on 19 February 1975. The strongest wind generated by an east-coast low obtained a velocity of 209 km hr^{-1} at Coffs Harbour, New South Wales, on 29 December 1964. Strong winds can also be produced regionally within general air circulation, by small-scale convection or through topographic enhancement. These winds may occur with or without precipitation. Regionally, strong winds usually occur as a gradient wind developed between high- and low-pressure systems, especially where temperature differences are great or where isobars tend to converge. The *southerly busters* that move up the east Australian coastline represent the passage of a cold front detached from the polar front. They can bring average winds of 55 km hr^{-1}, with gusts up to 130 km hr^{-1}. Similar strong winds can be produced by cold fronts sweeping south from the Canadian Arctic across the Canadian Prairies and United States mid-west. Winds generated by extreme convection are usually associated with thunderstorms, which are the commonest source of strong winds apart from those generated rarely by associated tornadoes. Gusts during thunderstorms can reach velocities in excess of 185 km hr^{-1}, and are destructive because of the sudden acceleration in wind speed and variation in direction. Often these winds are difficult to distinguish from those generated initially by tornadoes.

Katabatic, gravity or slope winds can be generated by a process of radiative cooling of air, and subsequent movement of this chilled air downslope under the effect of gravity. These winds are very shallow, occurring in the lower 100–500 m of the atmosphere. Because they are dependent upon cooling by longwave emission, they are restricted to the night-time. A common feature of such winds is their gustiness. It appears that cold air takes time to build up before it can overcome ground friction and flow downhill. Once it does, the complete reservoir of cold air moves away like an avalanche, whereupon the cycle must be initiated again before the next gust. Gusts have typical periods of around five minutes. Katabatic winds are also subject to adiabatic warming or even compression as they descend to lower elevations. Consequently, while such winds originate because of cold-air drainage, they can be substantially warmed at the bottom of valleys. These winds can make up one component of a much larger regional wind termed a *föhn* in Europe, chinook in Canada or zonda in the Andes. The katabatic process is particularly efficient over virtually frictionless ice sheets off Alpine glaciers, and especially off the Greenland and Antarctic icecaps. On Antarctica, these winds develop best along the steep east coast, especially in winter when incoming solar radiation ceases. Mawson is affected by gale-force katabatic winds on all but 10–15 days of the year, with an average wind speed of 72 km hr^{-1}: wind speeds can average 160 km hr^{-1} on some days! Localized severe katabatic winds exist elsewhere. Cyclone-force winds can descend the Columbia River gorge, while the Santa Ana wind in southern California is known for its destructiveness. A wind known as the bora descends the Alps and afflicts the Adriatic coastline around Trieste.

Strong winds can also develop because of topographic enhancement for a number of reasons. Compression of air down valleys was mentioned as a contributing factor in the warming up of katabatic winds. This process can also generate high wind-velocities through jetting. If wind flows down a valley that constricts, flow lines will be compressed leading to high velocities at the constriction and some distance downwind (figure 3.11A). A similar process can occur as winds move over an obstacle. Flow lines are compressed vertically on the upwind side of the obstruction, and reach their maximum velocities at the crest of the hill or mountain. For this reason, the strongest winds in the United States are measured near the top of Mount Washington, New Hampshire (maximum measured speed is 371 km hr^{-1}, which is also the highest velocity measured on Earth). Finally, wherever an airstream passes over, or around an object, strong wind in the form of turbulence may be induced by flow separation. At the point where the wind flow leaves the object, a lee eddy or vortex will develop immediately behind the obstruction. At regular intervals, this vortex will detach itself and move downwind (figure 3.11B). For strong winds, which are also deflected laterally around a large obstacle such as a mountain, this vortex-shedding alternates between clockwise and counterclockwise rotation forming a *Kármán vortex street* (figure 3.11C). Such a phenomenon is of particular note as a hazard because the strong winds periodically switch direction as alternate vortices are peeled off.

In some cases, the spreading out of air near the ground surface is so violent that a phenomenon called *wind shearing* develops. Wind shearing is not restricted to topographically modified wind, but is also a feature of turbulence generated by thermal convection. This phenomenon is particularly dangerous for aircraft because planes caught in the sudden downdraft can be carried to the ground before a pilot has had time to respond to the changing wind condition.

Flow over an escarpment or mountain range will not develop these lateral vortices. Instead, a pattern of lee waves may develop downwind. Where wind, either in a stably positioned vortex or one that is moving downwind, strikes the ground, turbulence ensues as the wind spreads out along the ground surface (figure 3.11B). If the zone of turbulence moves downwind, structures are subject to periodic accelerations in wind speed. However, if the zone of turbulence oscillates in position, then buildings are subject not only to varying winds speeds but also to destructive changes in wind direction. In

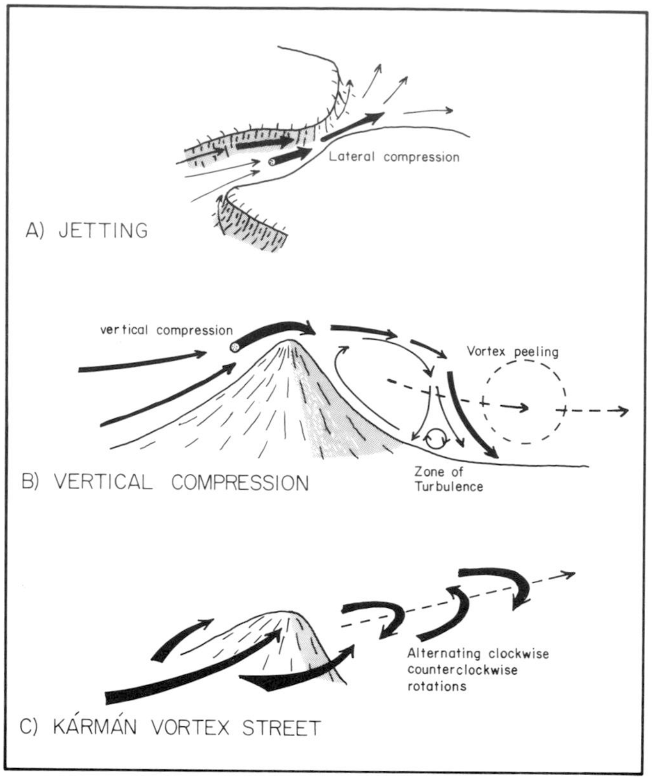

FIG. 3.11 Topographic enhancement of wind streamlines: A. by jetting, B. by vertical compression and vortex peeling, and C. by alternating clockwise and counterclockwise vortex peeling behind an obstacle to form a Kármán vortex street.

Australia coastal areas downwind of the Swan escarpment in Western Australia, the Mount Lofty ranges in South Australia and the Blue Mountains and Illawarra escarpment near Sydney in New South Wales have all experienced property destruction and loss of life because of these effects. Recorded wind gusts in all three areas have reached 140–180 km hr^{-1}.

Particularly hazardous in eastern Australia is the föhn effect superimposed upon these strong turbulent winds in the lee of the coastal escarpment south of Sydney. In summer, when the subtropical jet stream is strong on the westward side of Hadley cells, warm dry air can be drawn to the coast from the semi-arid center of the continent. Temperatures have exceeded 40°C in Sydney. Further south, adiabatic warming occurs as the winds rush over the 300–400 m high escarpment. The warming effect is enhanced by the compression of wind streamlines in the southeast corner of the continent under the jet stream. Temperatures can exceed 45°C with humidities of less than 10 per cent. These conditions pose a serious bushfire hazard, which is exacerbated by topographically enhanced wind turbulence. Fire suppression not only becomes impossible, but also life-threatening to firefighters as winds switch directions and intensity.

In the following chapters, the types of wind hazards described in this chapter will be shown to play an important role in exacerbating the damaging effects of other hazards such as coastal erosion, drought, precipitation events, and bushfires.

References

Dinwiddie, F. B. 1959. 'Waterspout tornado structure and behaviour of Nags Head, N.C., August 12, 1959'. *Monthly Weather Review* v. 89 pp. 20–23.

Eagleman, J. R 1983. *Severe and Unusual Weather.* Van Nostrand Reinhold, NY.

Finley, J. P. 1881. 'Report on the tornadoes of May 29 and 30, 1879, in Kansas, Nebraska'. *Professional Paper of the Signal Service* No. 4.

Goudie, A. S. 1983. 'Dust storms in space and time'. *Progress in Physical Geography* v. 7 pp. 503–30.

Idso, S. 1976. 'Whirlwinds, density current and topographic disturbances, a meteorological melange of intriguing interactions'. *Weatherwise* v. 29 No. 2.

Lockeretz, W. 1978. 'The lessons of the dust bowl'. In Skinner, B. J. (ed.) 1981 *Use and misuse of Earth's surface.* Kaufmann, Los Altos, California, pp. 140–49.

McTainish, G. H. and Pitblado, J. R. 1987. 'Dust storms and related phenomena measured from meteorological records in Australia'. *Earth Surface Processes and Landforms* v. 12 pp. 415–24.

Middleton, N. J., Goudie, A. S. and Wells, G. L. 1986. 'The frequency and source areas of dust storms'. In Nickling, W. G. (ed.) *Aeolian geomorphology.* Allen and Unwin, London, pp. 237–60.

Miller, A. 1971. *Meteorology* (2nd edn). Merrill, Columbus.

Further reading

Linacre, E. and Hobbs, J. 1977. *The Australian Climatic Environment.* Wiley, Brisbane.

Middleton, N. J. 1984. 'Dust storms in Australia: frequency, distribution and seasonality'. *Search* v. 15 No. 1–2 pp. 46–47.

Nalivkin, D. V. 1983. *Hurricanes, storms and tornadoes.* Balkema, Rotterdam.

Oke, T. R. 1978. *Boundary Layer Climates.* Methuen, London.

Oliver, J. 1979. 'Wind and storm hazards in Australia'. In Heathcote, R. L. and Thom, B. G. (eds) *Natural hazards in Australia.* Australian Academy of Science, Canberra, pp. 119–42.

——. 1981. *Climatology: selected applications.* Arnold, London.

Sims, J. H. and Baumann, D. D. 1972. 'The tornado threat: coping styles of the North and South'. *Science* v. 176 pp. 1386–92.

Smith, K. 1975. *Principles of Applied Climatology.* McGraw-Hill, London.

Whipple, A. B. C. 1982. *Storm.* Time-Life Books, Amsterdam.

4

Oceanographic Hazards

Introduction

In chapter 2 waves were continually mentioned as one of the main agents of erosion and destruction. In fact, destructive waves can occur without storm events, and can occur in association with other climatic and oceanographic factors such as heavy rainfall and high sea-levels, to produce coastal erosion. In this sense, waves form part of a group of sometimes interlinked hazards associated with oceans. This chapter examines these and other oceanographic hazards. Firstly, wave mechanics and the process of wave generation will be outlined followed by a description of wave-height distribution worldwide. This section will conclude with an appraisal of the hazard waves pose by themselves in the open ocean.

Wind is the prime mechanism for generating the destructive energy in waves. In cold oceans, seas or lakes, strong winds can produce another hazard —the beaching of drifting sea-ice. Sea-ice is generally advantageous along a coastline that experiences winter storms, because broken sea-ice can completely dampen the height of all but the highest waves. When frozen to the shoreline as shorefast ice, sea-ice can completely protect the shoreline from any storm erosion. However, floating ice is easily moved by winds of very low velocity. These winds might not generate appreciable waves, but they can drive sea-ice ashore and hundreds of meters inland. The force exerted by this ice can destroy almost all structures typically found adjacent to a shoreline. For this reason, sea-ice must be considered a significant hazard in any water body undergoing periodic freezing and melting. This chapter will secondly describe the problem of drifting sea-ice, its worldwide distribution, and the measures taken to negate its effects.

It was pointed out in Chapter 2 that waves may not be important as a hazard if the full intensity of a storm reaches landfall at low tide. Such was the case during Cyclone Tracy at Darwin, Australia, in 1974 (see figure 4.1 for the location of all major place names mentioned in this chapter). Even with a 4-meter storm surge, waves did not reach the high-tide mark, because the storm made landfall at low tide in an environment where the tidal range is about 7 m. More insidiously, low waves can cause considerable damage along a shoreline if sea-levels are either elevated briefly above normal levels, or rising in the long term. It is generally believed that sea-level is presently rising worldwide (*eustatic* rise) at a rate of 1.0–1.5 mm yr^{-1}. The most likely cause of this rise is the *Greenhouse effect*. Human-produced increases in CH_4, CO_2 and other gases are believed responsible for a general warming of the Earth's atmosphere. This is perceived as causing sea-level to rise because of a melting of the world's icecaps or *thermal expansion* of warming ocean waters. Accelerated melting of shelf ice, especially in the Ross Shelf area of Antarctica, may increase the rate of sea-level rise to tens of millimeters per year. Even if recent warming is not due to the Greenhouse effect, the implications are the same. The recent and threatened rise in sea-level poses a natural hazard to many coastal areas, especially where major world cities now exist close to the limits of present sea-level. The rise in sea-level is also believed to be responsible for the accelerating trend towards sandy beach erosion worldwide. This chapter will, thirdly, describe worldwide sea-level trends, and examine the causes of sea-level fluctuations.

Finally, the chapter will conclude with an evaluation of the factors responsible for beach or general coastal erosion. The processes whereby rising sea-

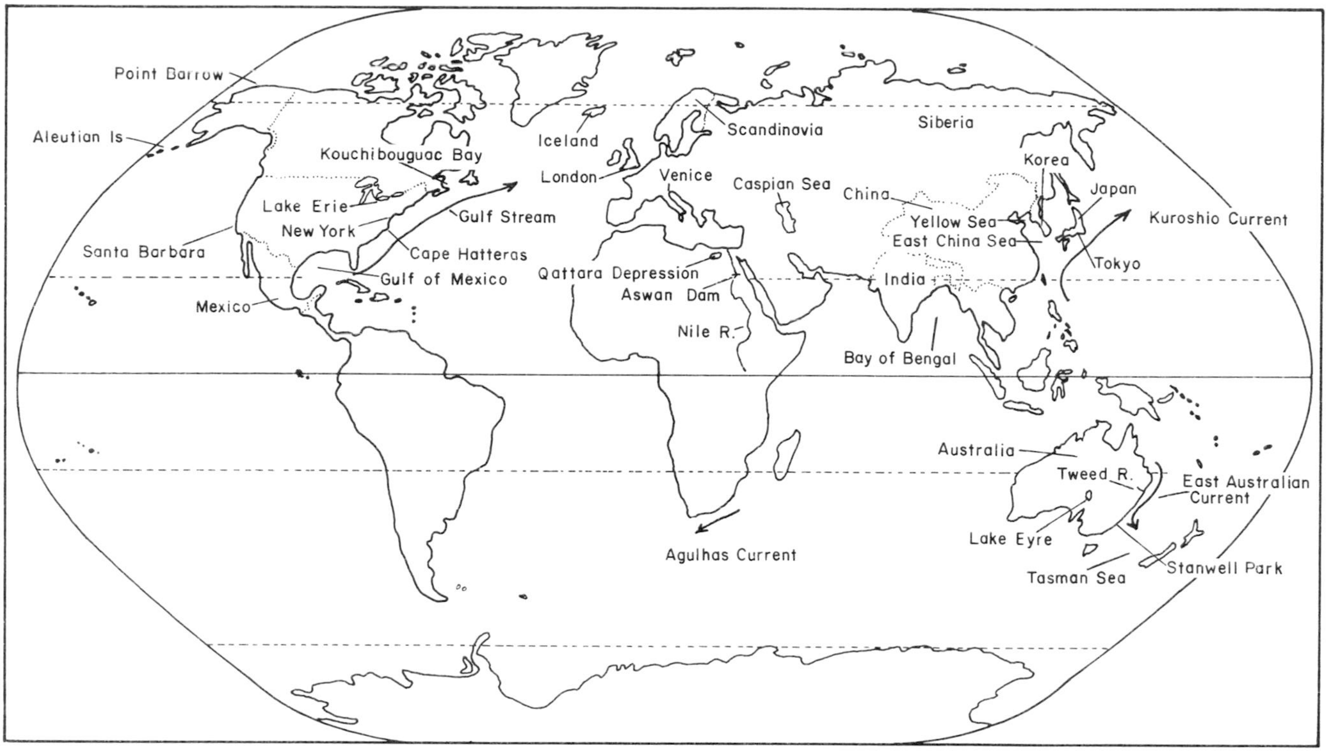

FIG. 4.1 Location map

level can induce beach erosion will be described, followed by an examination of other causes, such as increasing rainfall regimes and storminess on the east coasts of North America and Australia. The relative importance of these causes will be evaluated using as an example part of a data set compiled for Stanwell Park beach, Australia, between 1943 and 1980. This data set is one of the longest in the world and consists of accurately measured shoreline position plus ten meteorological and oceanographic variables.

Waves as a hazard

Theory (Komar, 1976; Beer, 1983)

The basic form and terminology of a linear wave is described schematically in figure 4.2. This diagram has been used to characterize general wave phenomena simply, including ocean waves, tides, storm surges and seiching. The wave form in figure 4.2 has the following simple relationships in deep water:

$$L = CT \tag{4.1}$$

$$C = gT(2\pi)^{-1} \tag{4.2}$$

$$\text{then } L = 1.56T^2 \text{ in meters} \tag{4.3}$$

where L = wave length
C = wave velocity
T = wave period
g = acceleration due to gravity
π = 3.1416

Deep water is defined as the water depth where a wave passing overhead is not discernible at the sea bed. In figure 4.2 you will note that wave motion is transmitted through the water column with water particles circumscribing elliptical paths that decrease exponentially in height and width with depth. If there is no measurable oscillation of water particles before the sea bed is reached (point A in figure 4.2), then the wave is considered to be in deep water. If oscillations are still present when the sea bed is reached (point B in figure 4.2), then the wave is in shallow water and affected by the bed. In that case, the above equations have to be modified accordingly.

The height and energy of a shallow-water wave is determined, not solely by wave form, but also by the nature of inshore topography. Under these conditions wave height can be increased: by shoaling; by *wave refraction*, where offshore *bathymetry* has focused wave energy onto a particular section of coastline such as a headland; or by *wave diffraction*

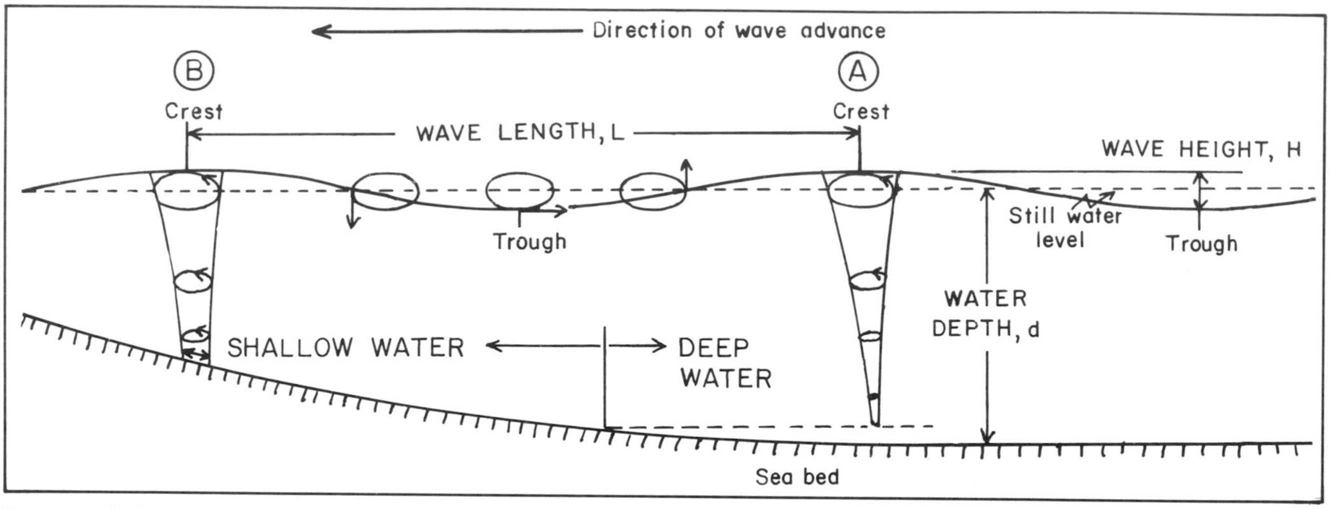

FIG. 4.2 Schematic representation of wave characteristics

around headlands, tidal inlets and harbour entrances. Figure 4.3 exemplifies each of these conditions, which are basically controlled in magnitude by wave height, period and water depth. Generally, the rate at which energy moves with the wave—the energy flux—remains virtually constant as a wave travels through increasingly shallower water towards shore. However, the speed at which the wave form is travelling decreases with decreasing water depth. In order to maintain a constant energy flux as a wave shoals into shallower water, wave height must increase. The fact that wave velocity slows down as water depth decreases also determines wave refraction. If one part of a wave crest reaches shallower water before another part, then the wave crest bends towards the shallower water segment. The process is analogous to driving one wheel of a vehicle off a paved or sealed road onto a gravel verge or shoulder. The wheel that hits the gravel first slows down dramatically, while the wheel still on the road continues to travel at the original road speed. The vehicle as a result veers rapidly towards the slower travelling wheel and careers off the road. For waves, the effect produces a bending of the wave crest towards shallow water, an effect that is particularly prominent around headlands or reefs that protrude out into the ocean. Diffraction is the process whereby energy moves laterally along a wave crest. The effect is noticeable when a wave enters a harbour between narrow breakwalls. Once the wave is inside the wider harbour, the crest width is no longer restricted by the distance between the breakwalls, but gradually increases as energy freely moves sideways along the wave crest. In some circumstances, wave height can increase sideways at the expense of the original wave height. Such an effect often accounts for the fact that beaches tucked inside harbour entrances and unaffected by wave refraction witness considerable wave heights and destruction during some storms.

The destructiveness of a wave is basically a function of the wave height and period. Height determines wave energy, the temporary elevation of sea-level, the degree of overwashing of structures, the height of water oscillations in the surf zone, and the distance of wave *run-up* on a beach. Because most waves affecting a coastline vary in period and height over a short period of time, oceanographers talk about a range of wave heights included in wave spectra. For engineering purposes, the highest one-third or one-tenth of waves, called the 'significant wave height', is used. Wave period determines wave energy and shape in shallow water. The latter aspect determines whether or not wave-generated currents at the bed move seaward or shoreward. If seaward, then the beach erodes. The following equation defines wave energy:

$$E = 0.125 \rho g H^2 L \qquad (4.4)$$

where E = wave energy
H = wave height
L = wave length
ρ = density of water

Note that wave length is much greater than wave height, so that most of the wave energy is controlled by wave length, which is related to the wave period. However, a change in wave height has a much more dramatic affect on changing wave energy than a change in wave length. Table 4.1 illustrates this effect for the most common wave periods in

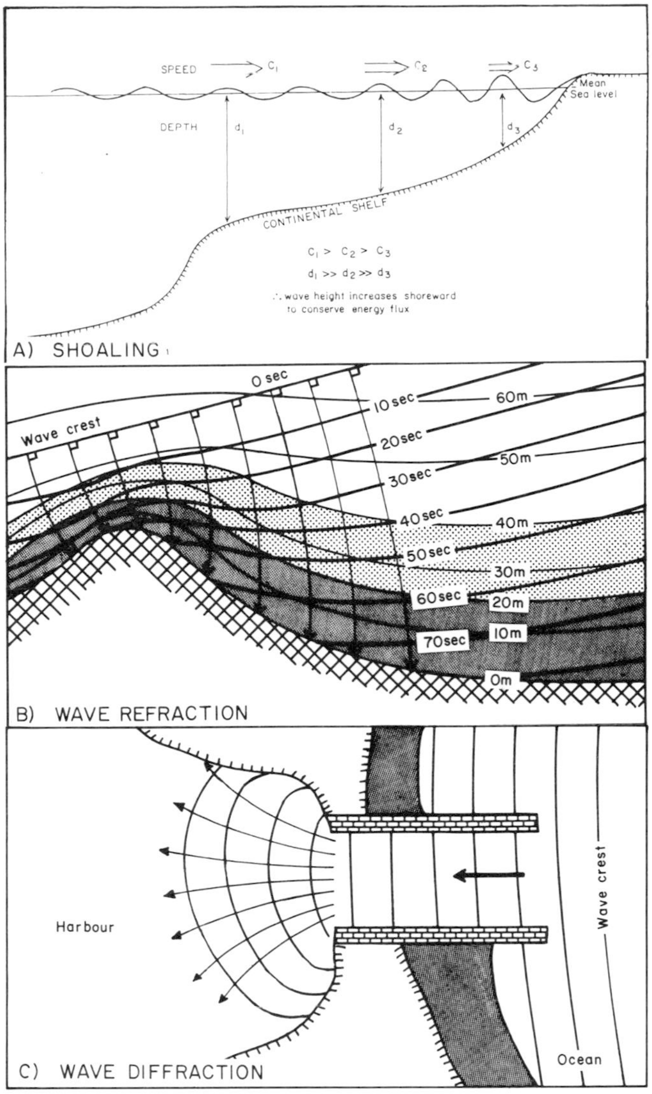

FIG. 4.3 Schematic diagrams of wave behaviour in shallow water
A. shoaling.
B. refraction.
C. diffraction.

swell environments. A doubling in wave length leads to a doubling in wave energy, whereas a doubling in wave height leads to a quadrupling in wave energy. In the coastal zone under strong winds, wave height can be changed more rapidly than wave length. The amount of wave energy in a storm wave can become abnormally large (figure 4.4). In swell environments, wave heights of 1.4 m and periods of 10 seconds are not unusual; however, if a storm wave increases in height to 7 m, a value typical of the storm waves off the eastern coasts of the United States or Australia, then wave energy will increase 25-fold. This height is probably not the largest experienced along these coastlines.

TABLE 4.1 Energy Levels for Typical Ocean Waves

Wave period	*Length*	*Height*	*Energy m^{-1} of wave crest*
10.0 seconds	156 m	1.0 m	2.39×10^5 joules
		1.4 m	4.69×10^5 joules
		2.0 m	9.57×10^5 joules
		3.0 m	21.54×10^5 joules
		7.0 m	117.28×10^5 joules
14.1 seconds	312 m	1.0 m	4.79×10^5 joules
		2.0 m	19.15×10^5 joules
		3.0 m	43.08×10^5 joules

For most coastlines of the world having average wave heights in excess of 1 meter, large storm waves might easily have 50 times their average wave energy. Note that if the above values were for fresh water, they would be 2.4 per cent lower because of lower water density. A 7-meter high storm wave in fresh water would have 2.86×10^5 joules less energy than if it had occurred in salt water. This value is roughly equivalent to the energy in a 1-meter high, 10-second period wave.

Wave generation (Komar, 1976; Wiegel, 1964; Coastal Engineering Research Center, 1977)

The generation of waves represents a transference of energy from wind to the water body. Wind itself is not steady but fluctuates. This fluctuation gives rise to air pressure differences over water, which cause water level to vary locally. Once the water surface begins to have peaks and troughs, no matter how small, a difference in wind-shear potential across the water surface rapidly develops. This

FIG. 4.4 Deep-water swell wave of 4 m breaking over the breakwall to Wollongong harbour, Australia (photograph courtesy John Telford, Fairy Meadow, New South Wales)

difference tends to transfer energy from moving air to the peaks on the water surface. The growth of these peaks or crests is exponential with time or *fetch*. Fetch is defined as the distance or length of water across which wind can blow. These two mechanisms do not totally control wave growth. There probably is some mopping-up procedure, whereby large waves absorb energy from smaller waves. Wind can generate smaller waves quite easily, while larger waves gain energy by absorbing momentum from smaller waves. There may also be a transfer of energy from wind on the upwind slope of the wave as the boundary layer of the atmosphere becomes more disturbed at this point.

The height to which a wave will grow depends upon four factors: wind speed, the duration of the wind, fetch and the processes leading to decay once the wave is established. Decay processes mainly include opposing wind drag because a wave basically retains its form and energy over long distances (half the Earth's circumference or more). Decayed waves can still cause coastal damage, but their occurrence represents only a small portion of waves at a particular site. Hence, decay processes can be ignored. Most damaging waves have only travelled short distances after forming.

Before the advent of *wave-rider buoys* to measure waves along a coastline, or *synthetic aperture side-scan radar* mounted in satellites measuring wave period and heights over large areas, data on wave characteristics depended upon ship observations and *hindcasting* procedures. Hindcasting theoretically estimates wave height and period based on wind duration, speed and fetch. The procedures for hindcasting were developed during the Second World War. Figure 4.5 is a *nomogram* representing the standard hindcasting technique based on Bretschneider's work published in the United States Army Coastal Engineering Research Center Manual (1977). At first sight, the nomogram appears quite complex. It contains three input variables: wind duration, speed, and fetch length.

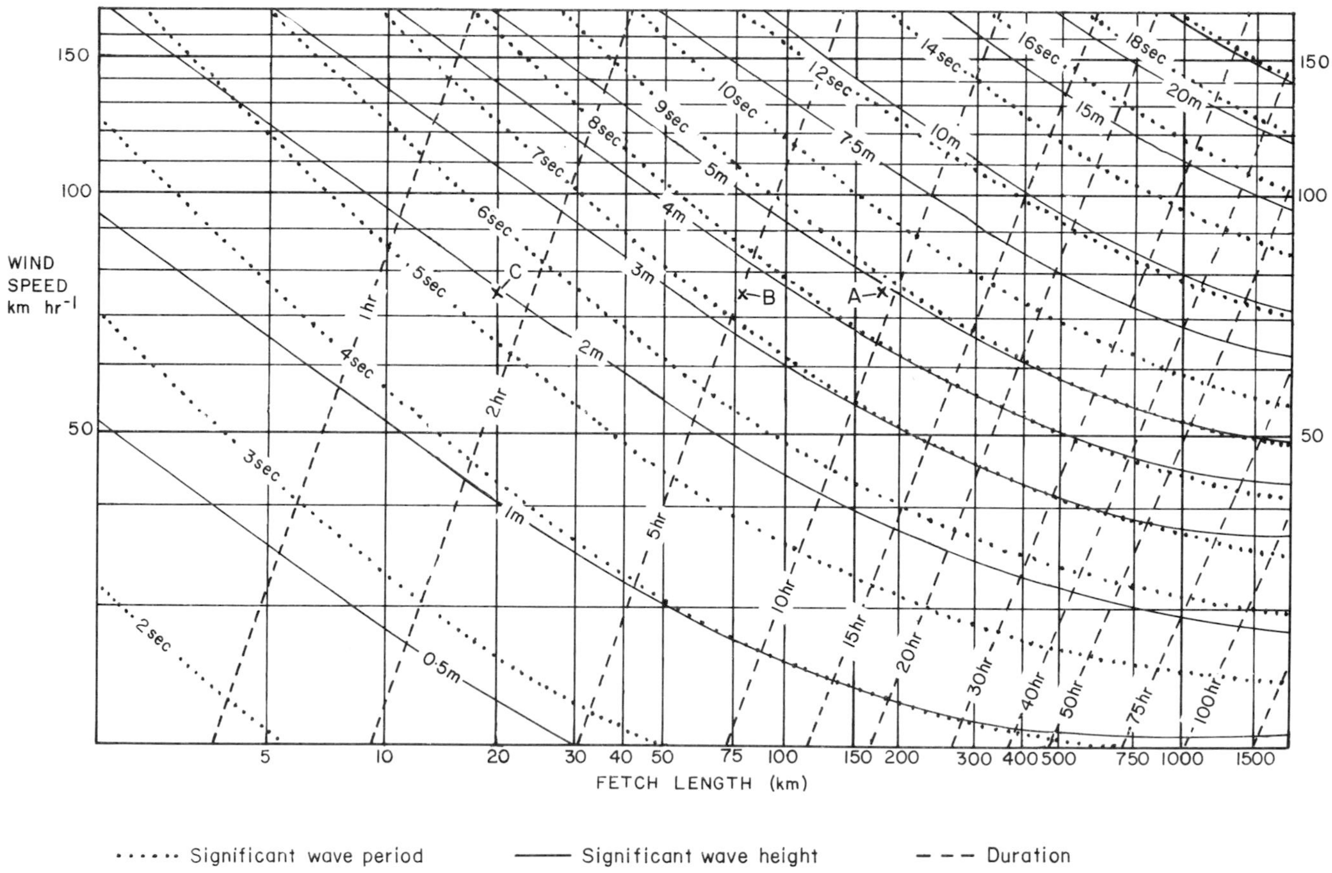

FIG. 4.5 Nomogram for hindcasting waves. Wave height and period are based upon fetch, duration and wind speed (after Coastal Engineering Research Center, 1977).

It also contains two output variables: wave period and wave height. Fetch is usually the length of water a wind has been blowing across as determined from the position of wind-generating pressure systems. Where coastal outline affects a site, then fetch becomes dependent on the length of the water body. The accuracy of the technique depends very much on the accuracy and frequency of synoptic weather charts, and the method of reading wind speed from these charts. If weather charts are only available at 12- or 24-hour intervals, errors in estimating duration severely hamper hindcasting. For instance, hindcasting of the 25 May 1974 storm in eastern Australia produced a maximum wave height of 4 m whereas the actual height was 7 m. The storm lasted no longer than nine hours, after which decay processes took over and wind-generating forces diminished.

An example using the nomogram will give an idea how it works. Suppose the fetch length is 185 km (100 nautical miles) with a wind speed of 75 km hr^{-1} (40 knots) lasting for ten hours. Locate the fetch along the x-axis and wind speed along the y-axis. Where the two lines intersect, point A in figure 4.5, the closest wave period and height can be read off from the graph. For the above example, the wave period is approximately 8.8 seconds with a height of 4.8 m (16 ft). Suppose the wind only blew for five hours. Repeating the procedure now involves a consideration of duration. For the original example, winds blowing for ten hours were sufficient to generate the interpolated wave height and period. A wind of 75 km hr^{-1} blowing for only five hours cannot generate such a large wave. In this case, the wave period and height are controlled not by fetch and speed, but by duration and speed. The 5-hour duration line intersects the 75 km hr^{-1} wind speed line at point B, giving a period of roughly 7.4 seconds and a height of approximately 3.5 m (12 ft). In this case, the wind did not blow long enough to generate the maximum possible wave, and the waves are considered duration-constrained.

Using another example, suppose a 75 km hr^{-1} wind blows for five hours over a fetch of only 20 km (10 nautical miles). Ignoring fetch, the wave period and height would be the same as the previous example. However, fetch must be considered, which is point C. This results in a wave period of only 5.4 seconds with a height of 1.9 m (6.3 ft). In this case, the wave is considered fetch-limited. The wind can blow at any speed over any period of time, but the maximum wave period and height cannot exceed these latter values. Generally, wave characteristics are determined from the nomogram using, first, fetch and speed, and then, secondly, duration and speed. The lower of the values produced gives the wave characteristics generated by the wind field.

World distribution of high waves (Davies, 1980; Allen, 1984)

The above procedures have been used quite effectively to build up wave climates for coastlines without any wave observations. On the open ocean, wave parameters are estimated by every ship at least once per day. These observations have been collected for the past century, and now amount to over 40 million observations spanning most of the world's oceans. At the beginning of chapter 2, diagrams of maximum winds over oceans were shown. It can be expected that maximum waves would occur in similar belts around the globe. Figure 4.6A summarizes wave-height data taken from ship observations and supplemented by some hindcasting. It has been difficult to calibrate or check the accuracy of these data because ships tend to avoid high seas.

In 1978 the SEASAT satellite was put into a near polar orbit for 100 days. It contained an altimeter which could measure the height of the satellite above the Earth's surface to an accuracy of 10 cm. Calibration with wave-rider data showed that the altimeter could measure wave height to an accuracy of ± 0.2 m. Figure 4.6B summarizes these wave height data globally. Realize that the satellite was recording wave heights during the southern hemisphere winter when wind conditions were calm in the northern hemisphere. Both data sets in figure 4.6A and 4.6B are very similar. The southern storm belt can be discerned in both diagrams at about 40° S. SEASAT did not measure the extreme wave conditions shown by ship observations because it only operated for 100 days. Wave heights in excess of 5 m were recorded for most of the time in the southern hemisphere storm belt during the SEASAT functioning time. Luckily, this belt is south, and seaward of, major landmasses, because waves in excess of 4 m can mobilize boulder material on beaches.

The altimeter cannot record wave period; however, this can be recorded using synthetic aperture side-scan radar (SAR). This technique gives ambivalent wave-period values if used on a satellite, because the waves are moving at the time they are

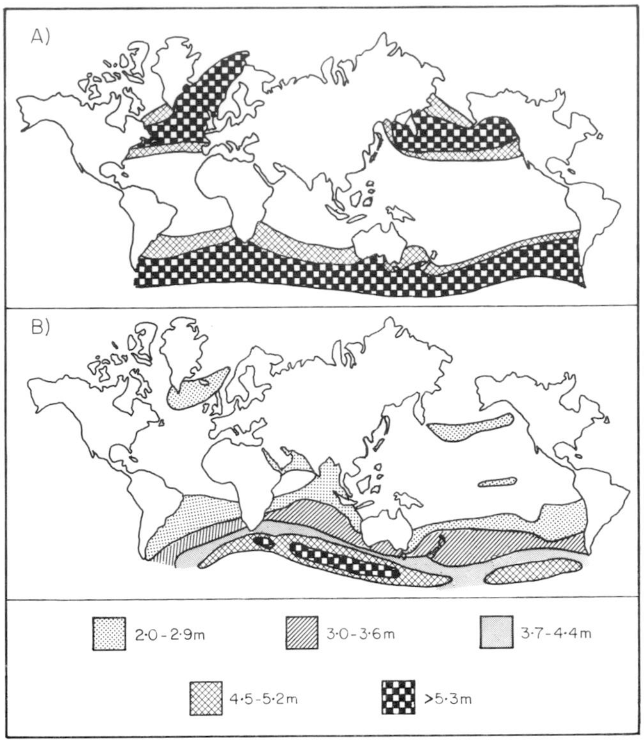

FIG. 4.6 Worldwide distribution of wave heights
A. Observed heights. Adapted from Davies (1980) and representing maximum wave heights occurring more than 5 per cent of the time, using ship observations and hindcasting.
B. Radar satellite altimeter data. Based upon SEASAT measurements July–October 1978 (after Allen, 1984).

being sensed. The fact that this technique can displace a wave crest up to 60 m in either direction can lead to two totally differently calculated wave periods. For instance, a 10-second period wave with a wavelength of 156 m (see equations 4.1–4.3) would be recorded as having a wavelength of 96 m or 216 m depending on its direction of travel. This represents periods of 7.8 or 11.8 seconds. The error is too large for most coastal hazard work. However, correction procedures can be applied to the technique and future satellite SAR imagery will be able to measure wave periods to an accuracy of ±1 second.

Unfortunately, nothing has replaced SEASAT to date. Some altimeter and side-scan radar work was performed by crew on American space flights, but these flights did not go beyond ten days. Two satellites should, however, overcome this problem in 1990. The TOPEX satellite will have a package called the Poseidon, which will contain a radar altimeter. The European Space Agency Remote Sensing Satellite System will carry microwave, synthetic aperture radar and radar altimeter instruments; these instruments will permit wave heights and sea-levels to be measured to an accuracy of 0.5 m and 0.1 m respectively. Towards the end of the 1990s, several polar orbiting platforms are also planned that will duplicate these instruments with higher accuracy. By this time, a global picture of inter-annual wave height and period variability that includes storm waves will be routinely available.

Waves as a hazard at sea (Beer, 1983; Bruun, 1985)

The effect of waves as a hazard really depends upon where the wave is experienced. In the open ocean, freak or giant waves pose a hazard to navigation. This does not involve a single wave. Instead, it involves the unfortunate meeting of several waves of different height, period and direction, which is referred to as 'wave spectrum interaction'. This concept of wave spectrum interaction is more realistic and characterizes most wave fields. Suppose there are two waves with heights of 5 m and 6 m and different periods. The waves will travel at different speeds. Rarely will all the crests overlap, but when they do, the resulting wave height is determined by the summation of the energy in the two waves, in this case resulting in a wave height of 7.8 m. This wave height is unique in time and space. If a ship happens to be at the point of confluence at that time, then it is under threat of capsizing. Such waves have always plagued shipping, but today they also pose a potentially serious hazard to oil platforms, with a consequential loss in terms of investment and potential ecological damage. This hazard has been monitored in the North Sea, where there appears to be a real increase in the overall height of waves. Freak waves with heights in excess of 15 m have been recorded on many platforms. In the Gorm field, five waves in excess of 8 m were recorded in a 13-hour period on 6 September 1983. All of these waves appeared to result from the additive interaction of 5-meter high sea waves coming from different directions. This directional variation arose because of the swift passage of cold fronts or cyclonic depressions through the affected area. Under these conditions, strong winds are able to swing nearly 180° through the compass, generating new high waves before the previous ones have passed through the oil fields.

Wave heights for a region can also increase substantially over time. For instance, waves in the

North Atlantic have increased in height by 20 per cent over the past 20 years, according to the Institute of Oceanographic Sciences in the United Kingdom. Largest wave heights offshore from Cornwall have risen from 11.9 m in the 1960s to 17.4 m in the late 1980s, while mean wave heights have increased from 2.2 m to 2.7 m over this period of time. These heights represent a 32 per cent increase in wave energy, an increase that may not have been accounted for in the engineering design of shoreline or offshore structures. This amplification in individual wave heights has resulted from an intensification of the North Atlantic wind field.

Sometimes waves will increase in height because they meet an opposing current. The region along the southeast coast of Africa is notorious for high waves due to current interaction. Here, waves generated in the roaring forties travel against the Agulhas current moving south, trapped between the submerged Agulhas plateau and the continental shelf. Waves steepen in this region as a result. One confluence of wave crests steepened by this current swamped and sunk a supertanker, the *Gigantic*, in the late 1970s. There are other regions of the world with similar oceanographic conditions that pose severe hazards to shipping.

Inshore, freak waves are less of a problem. While freak waves can pick up objects weighing several hundreds of tonnes and dislodge them, a single wave probably is not enough to completely erode a beach or cliffed shoreline. What is the ultimate sea-level elevation at shore? If a 7-meter high wave occurs at low tide, then beaches will not be damaged as much as they would be if affected by this storm wave at high tide. Large storm-waves of this magnitude often are superimposed on a storm surge that can increase in height towards semi-enclosed embayments. Even along an open coast such as the eastern coast of Australia, where storm surge is not a problem, the largest storms can elevate sea-level up to 1 m above normal. Along the eastern seaboard of the United States, the height of this storm surge can exceed 6 m. Once a 7-meter high wave breaks, it will lead to an elevation of water within the surf zone termed *set-up*. Set-up increases in height as the surf zone becomes wider. Typically, beaches with wide surf zones can have wave set-up on the order of 1–2 m. Storm-wave heights by themselves are meaningless at shore unless consideration is given to the local factors, which may elevate sea-level during storm events. Unfortunately, these factors are spatially variable between different sections of coast and even between beaches. The storm wave penetration line along the United States seaboard varied considerably in height during the Ash Wednesday storm of 1962. In Sydney, Australia, during the storm of 25 May 1974, waves overtopped a 7-meter high dune at the back of one particularly steep pocket beach, but 1 km away, left unscathed a much flatter beach with similar exposure. Waves as an erosional hazard are, thus, site specific. As will be shown later in this chapter, as erosional agents, they must be considered together with a large number of interrelated meteorological and oceanographic variables.

Sea-ice as a hazard

Ice in the ocean (Defant, 1961; Skov, 1970; Gross, 1972; Pickard, 1979; Bailey, 1982)

Sea-ice has two origins: through the freezing of sea-water and by the discharge of ice from glaciers into the ocean. The former mechanism is the more common method of forming sea-ice. Because the ocean has a salinity of 31–35‰ (parts per thousand), it freezes at temperatures around −1.9°C. New sea-ice does not contain pure water but has a salinity of 5–15‰. The faster ice forms, the more saline it is likely to become. As ice ages, the brine drains from it and the ice becomes saltless and clear. These characteristics can be used to identify the age and origin of sea-ice. At −30°C, sea-ice can form at the rate of 10 cm day^{-1}. Typically, in the Canadian Arctic, a winter's freezing cycle will produce ice 2–3 m thick. The process is similar for the freezing of ice in lakes and rivers; but, because of low salinities, ice forms faster in these latter environments. The density of sea water is 1027 $kg\,m^{-3}$. With an air content of 1 per cent, 89 per cent of floating ice is submerged. If the air content rises to 10 per cent, as it can do with some icebergs and shelf-ice, then the submerged part is reduced to 81 per cent by volume.

Sea-ice can be divided into four categories. The first consists of the polar shelf- and sea-ice forming a permanent cover 3–4 m thick over 70 per cent of the Arctic Ocean and completely surrounding the Antarctic continent (figure 4.7). This ice is very hummocky and may form open leads (polynyas) in summer. In the succeeding winter, these leads will often be squeezed shut by the shifting ice to form pressure ridges tens of meters high. The north polar icecap rotates in a clockwise direction and about one-third of it is removed by the East Greenland current each year. Because of the pressure ridges, it is impenetrable even to ice-breaking ships.

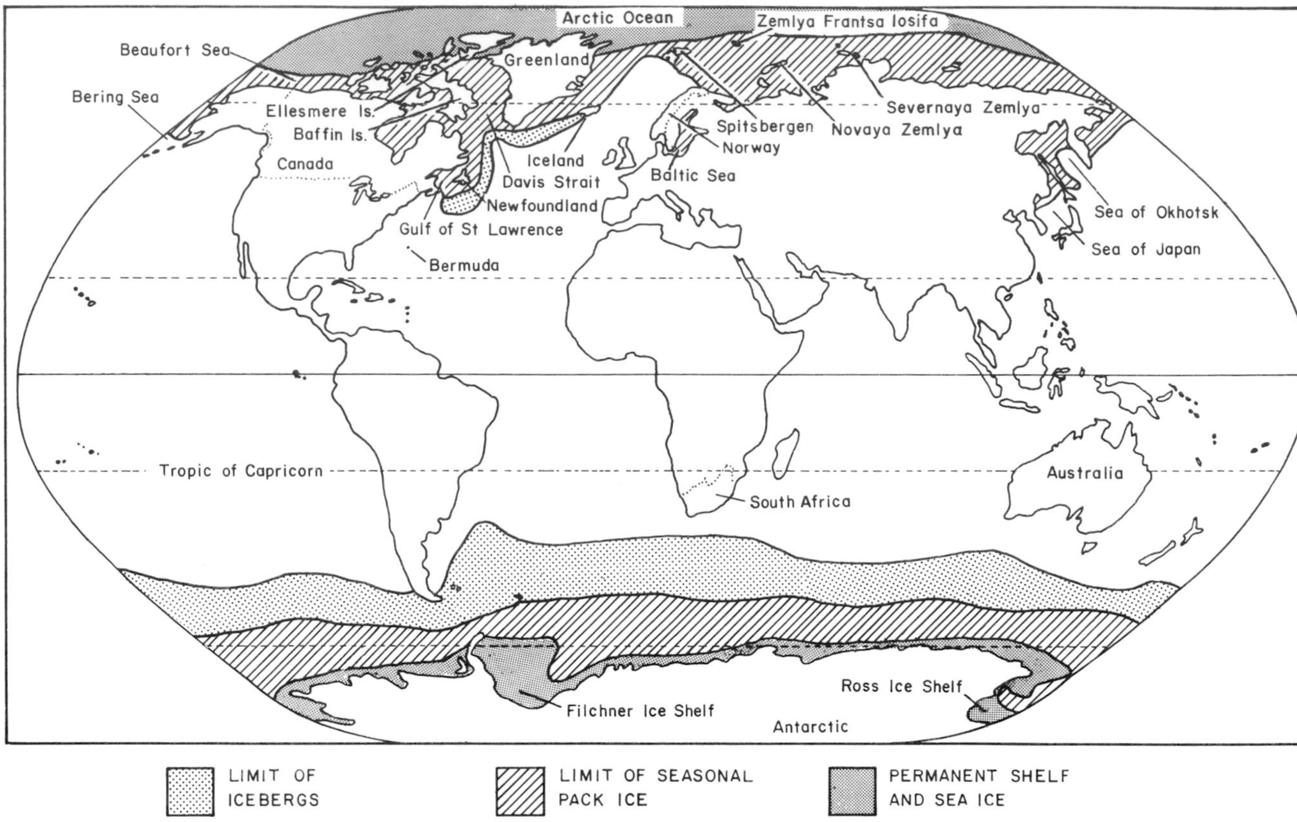

FIG. 4.7 Worldwide distribution of permanent polar shelf and sea ice, seasonal pack ice and icebergs (after Gross, 1972)

The second type of ice is the annual pack ice that mainly forms within the seas of Japan, and Okhotsk, the Bering and Baltic seas, the Gulf of St Lawrence and along the southeast coast of Canada to Newfoundland. Pack ice also forms in inland lakes where winter freezing takes place. Pack ice generally extends more southward in the northern hemisphere along the western sides of oceans because of prevailing southward-moving cold currents at these locations. In particularly severe winters, pack ice can extend towards the equator beyond 50° latitude in both the southern and northern hemispheres. If the following summers are cold, or the ice build-up has been too severe, then it is possible for large areas of pack ice not to melt seasonally.

Scenarios for non-melting pack ice have been invoked as triggering mechanisms for future ice ages. The termination of the Greenland Viking colony was brought about by excessive pack ice between Greenland and Iceland that prevented contact with Europe after the middle of the fourteenth century. Generally, pack-ice limits retreated poleward in the Arctic after 1918 during the twentieth century's warm climatic maximum. These conditions came to an abrupt halt in the early 1970s, and some scientists expressed fears that the Arctic Ocean might become ice-locked throughout the year. This has been hypothesized as one of the triggering mechanisms for initiation of glaciation in the northern hemisphere. However, the extent of pack ice varies dramatically from year to year. Since the mid-1970s, the Arctic has become much warmer and pack-ice limits have retreated northwards. Figure 4.8 plots an index of ice severity for 1820–1928 in Davis Strait between Greenland and Baffin Island. There have been dramatic shifts in the occurrence of ice in this area, with some years having little pack ice. It appears that ice severity peaks at intervals of four to five years, and, while not shown, there is a negative correlation between ice conditions in Davis Strait and Iceland. The Iceland conditions appear to depend upon the severity of temperature in northwest Siberia four and a half years previously. It takes that long for ice to drift from the Siberian coast to the Greenland Sea and Iceland. The extent of ice in Davis Strait depends upon the climate over the Canadian Arctic in the previous winter. Davis Strait ice coverage correlates directly with iceberg occurrence

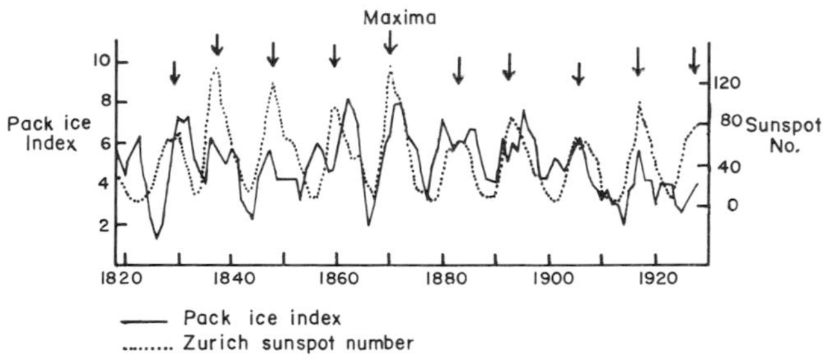

FIG. 4.8 Relative area of pack ice in Davis Strait and sunspot numbers: 1820–1928 (Defant, 1961)

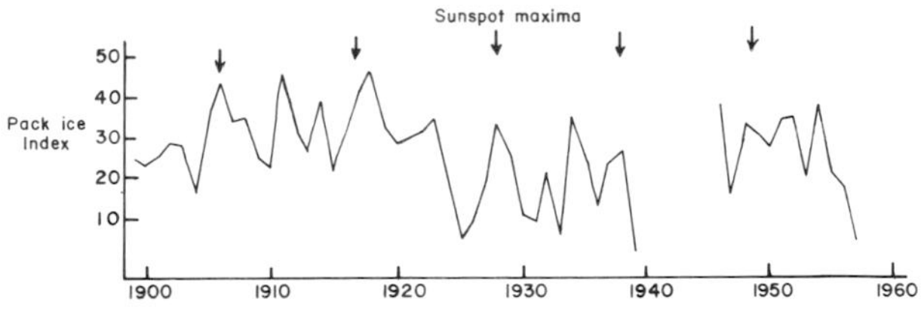

FIG. 4.9 Relative area of pack ice in the Greenland Sea: 1900–57 (after Skov, 1970)

off Newfoundland, as the Davis Strait area off Greenland is the main source area for icebergs in the North Atlantic. Interestingly, the extent of Davis Strait ice also correlates highly with the 11-year sunspot cycle. Years of maximum *sunspots* are followed two years later by increased pack ice.

A similar sunspot relationship with pack ice is also present in the Greenland Sea for this century (figure 4.9). The peaks are not as prominent as those for the Davis Strait (figure 4.8). Other peaks also appear at sunspot minima. There is also a good statistical relationship between ice-cover fluctuations in the Greenland Sea and easterly wind components along the Norwegian coastline, with stronger winds leading to an increase in the extent of ice coverage. These wind components are linked to the location of the jet stream controlling the general air circulation of the northern hemisphere in winter. They determine the magnitude of currents entering and leaving the Greenland Sea. As will be shown in the next chapter, sunspot cycles can affect parts of the general circulation, and it is through this mechanism that sunspot periodicities appear in the ice coverage data for Davis Strait and the Greenland Sea. Expansion of Antarctic pack ice is also statistically linked to increases in the easterly trade winds (*Walker circulation*) across the equatorial Pacific. Changes in sea-ice extent precede changes in the strength of the trades. Also to be shown in the next chapter, any factor which precedes a change in the trade winds is a precursor to short-term climatic change for 80 per cent of the globe in the following one to two years.

Pack ice only becomes a hazard when it drifts into shore, or when it becomes so thick that navigable leads through it have a serious chance of reclosing and forming pressure ridges. Whaling and exploration ships have been crushed in Antarctic waters and as recently as 1985–86 the Australian Antarctic research program was severely hampered by thick pack-ice which closed in, stranding supply vessels, and in one instance crushing and sinking one ship. In the Canadian Arctic, thick sea-ice historically has posed a threat to all shipping, and increases the cost of re-supply for northern settlements. Numerous expeditions to the Canadian Arctic in the nineteenth century were terminated by pack ice. The most famous was the loss of the Franklin expedition. Unsuccessful attempts at finding that expedition led to much of the mapping of the Canadian Arctic archipelago.

The third type of ice consists of shore-fast ice. This ice grows from shore and is anchored to the sea bed. It effectively protects a shoreline from wave attack during winter storms but, upon melting, can remove significant beach sediment if it floats out to sea. This type of ice poses a unique hazard at the shoreline and will be discussed as a separate topic later in the chapter.

The fourth category of ice consists of icebergs. Icebergs form a menace to shipping as exemplified by that most notorious disaster, the sinking of the *Titanic* in April 1912 with the loss of 1503 passengers and crew. Icebergs in the northern hemisphere originate in the North Atlantic, as ice calves from glaciers entering the ocean along the west coast of Greenland and the east coast of Baffin Island. Icebergs coming from the east coast of Greenland usually drift south, round the southern tip of Greenland, and join this concentration of ice in the Davis Strait. Along the southern half of the west coast of Greenland, icebergs collect in fjords and discharge into the Strait at fortnightly intervals on high tides. Along the northern half of the Greenland coast, glaciers protrude for several kilometers into the sea and icebergs calve directly into Davis Strait. About 90 per cent of all icebergs in the northern hemisphere collect as swarms in Davis Strait during severe periods of pack ice, and move south into the Labrador current as the pack ice breaks up in spring. These icebergs then pass into the Atlantic Ocean, off Newfoundland, with widely varying numbers from year to year. Less than 10 per cent of icebergs originate in the Barents Sea from glaciers on the islands of Spitsbergen, Zemlya Frantsa

Iosifa (Franz Josef Land), Novaya Zemlya and Severnaya Zemlya.

The movement of icebergs is mainly determined by ocean currents, whereas the movement of pack ice is determined by wind direction. Ice drifts to the right of the wind in the northern hemisphere and to the left in the southern hemisphere because of Coriolis force. Icebergs in the North Atlantic usually have heights of tens of meters but have been measured up to 80 m in height and over 500 m in length. The most southerly recorded sighting in the northern hemisphere occurred near Bermuda at 30° 20′ N. In the Antarctic, icebergs usually originate as calved ice from the Ross and Filchner ice shelves. Here, glaciers feed into a shallow sea forming an ice bank 35–90 m above and below sea-level. The production of icebergs in the Antarctic is prolific, probably numbering 50 000 at any one time. Icebergs breaking off from the Ross Shelf may reach lengths of 100 km. The largest yet measured was 334 km by 96 km in size with a height above water of 30 m. This one tabular ice-block had a volume that exceeded two times the annual ice discharge of the entire Antarctic continent. Tabular icebergs have been found as far north as 40° S, with the most northerly recorded sighting being 26° 30′ S, 350 km from the Tropic of Capricorn. This is unusual because the Antarctic Circumpolar current between 50° and 65° S forms a warm barrier to ice movement northward. About 100 smaller tabular icebergs, ranging up to 30 km in size, also exist in the Arctic Ocean, drifting mainly in a clockwise direction with the currents in the Beaufort Sea. These ice rafts have calved from five small ice shelves on Ellesmere Island. Because they have little chance of escaping south to warmer waters, individual blocks can persist for 50 years or more.

In the southern hemisphere, icebergs pose little threat to navigation because they do not frequently enter major shipping lanes. However, it is not unknown for shipping between South America and Australia, and between Australia and South Africa to be diverted north because of numerous sightings of icebergs 40–50° S. North Atlantic icebergs do enter the major shipping lanes between North America and Europe, especially off the Grand Banks, east of the Newfoundland coast. In severe years, between 30 and 50 larger icebergs may drift into this region. Historically, they have resulted in an appallingly high deathtoll. Between 1882 and 1890, 14 passenger liners were sunk and 40 damaged by ice around the Grand Banks. The iceberg hazard climaxed on 14 April 1912, when the 'indestructible', 46 000 tonne *Titanic* scraped an iceberg five times its size. The encounter sheared the rivets off plates below the waterline over two-fifths the length of the ship. So innocuous was the scrape that most passengers and crew members did not realize what had happened. Within three hours the *Titanic* sunk dramatically, bow first. Only one-third of the passengers survived in lifeboats, while over 1500 people died as the ship sunk. Immediately, the United States Navy was ordered to begin patrols for icebergs in the area. In the following year, 13 nations met in London to establish and fund the International Ice Patrol, administered by the United States Coast Guard. The Coast Guard issues advisories twice daily, and all ships in the North Atlantic are obliged to notify authorities of the location of any sighted icebergs. Since the Second World War, regular air patrols have also been flown. With the advent of satellites large icebergs can be spotted visually on satellite images, while the more threatening ones can be monitored by using radio transmitters implanted on the surface. Individual ships can also spot icebergs using very sophisticated radar systems.

While the iceberg hazard to shipping has all but disappeared, the hazard to oil exploration and drilling facilities has increased in recent years, especially in the Beaufort Sea and along the Labrador coastline. In the Arctic icebergs can be entrapped in pack ice. Because over 80 per cent of their mass lies below sea-level, large icebergs can gouge the sea bed to depths of 0.5–1.0 m, forming long, linear grooves on the continental shelf. This process could easily gouge open a pipeline carrying oil or gas across the sea bed. Drifting icebergs also have enough mass and momentum to crumple an oil rig. Not only could there be substantial financial loss, with the possibility of deaths, but the resulting oil spill could pose a significant ecological disaster. Attempts have been made to divert icebergs by towing them away using ocean tugboats; however, the large mass of icebergs gives them enormous inertia, and the fact that they are mostly submerged means that they are easily moved by any ocean current. At present, waves pose the most serious threat to oil and gas activities in ice-prone waters; but it is only a matter of time before an iceberg collides with a drill rig or platform. Present disaster mitigation procedures for this hazard simply consist of evacuating personnel from threatened facilities. Such procedures have been used on several occasions over the last few years.

Ice at shore (Taylor, 1977; Kovacs and Sodhi, 1978; Bruun, 1985)

Nowhere is the destructive effect of ice evidenced more than at the shoreline. During winter, the ice-foot, which is frozen to shore at the water line, protects the beach from storm-wave attack. Pack ice, which develops in the open ocean or lake, also attenuates the fetch length for wave development. The middle of winter sees the whole shoreline protected by ice. In spring the pack ice melts and breaks up, whereupon it may drift offshore under the effect of wind. At this time of year, shorelines are most vulnerable to ice damage if the wind veers onshore and drive the pack ice shoreward. At shore, pack ice has little problem overriding the coastline especially where it is flat. This stranding ice can pile up to heights of 15 m at shore, overriding wharfs, harbour buildings and communication links. Ice can be actively deriven inland distances fo 250 m, moving sediment as big as boulders 3–4 min diameter. In low-lying islands in the Canadian Arctic, shallow pack-ice, 1–2 m thick, can push over the shoreline along tens of kilometers of coastline (figure 4.10). Large ridfes of gravel or boulders — between 0.5 m and 7 m in heights, 1–30 m in width and 15–200 m in length — can be ploughed up at the shoreline. The grounding of ice and subsequent stacking does not require abnormally strong winds. Measured wind velocities during ice-push events have rarely exceeded 15 m s^{-1} (54 km hr^{-1}). Nor is time important. Ice-push events have frequently occured in as little as 15 minutes. All that is required is for loosely packed floating ice to be present near shore, and for a slight breeze to drive it ashore.

FIG. 4.10 Pack ice 1 m thick pushing up gravel and muds along the south shore of Cornwall Island, Arctic Canada. Scars in the upper part of the photograph represent ice push at slightly higher sea-levels.

Ice pushing can also occur because of thermal expansion, especially in lakes. With a rapid fall in temperature, lake ice can contract and crack dramatically. If lake water enters the cracks, or subsequent thawing infills the crack, warming of the whole ice surface can begin to expand the ice and push it slowly shorewards. If intense freezing alternates with thawing, then overriding of the shore can be significant for lakes as small as 5 km^2 in area. Larger lakes tend not to suffer from ice expansion because ice will simply override itself, closing cracks which have opened up in the lake.

The main effect of grounded sea-ice or ice push is the destruction of shore structures. In holiday lake resorts in northeastern United States and southern Canada, docks, breakwalls and shoreline buildings are almost impossible to maintain on small lakes because of thermal ice expansion. On larger lakes, and along the Arctic Ocean coastline in Canada, northern Europe and the Soviet Union, wind-driven ice-push has destroyed coastal structures. In the Beaufort Sea area of North America oil exploration and drilling programs have been severely curtailed by drifting pack ice. Sites have had to be surrounded by steep breakwalls to prevent ice overriding. In the Canadian High Arctic, oil removal was delayed for years while the government tried to determine the cheapest and safest means of exporting oil. Pipelines laid across the sea bed between islands are exposed to drifting icebergs, or to ice-push damage near the shoreline. Alternatively, harbour facilities for oil tankers are threatened by ice push. Even where ice push is not a major problem, drifting sea-ice can push into a harbour and prevent its being used for most of the short (two to three months), ice-free navigation season.

Sea-level rise as a hazard

Introduction

Sea-level rise also poses a significant long-term hazard. Over the past century there has generally been a worldwide (eustatic) rise in sea-level of 15.1±1.5 cm. In the early 1980s Emery proposed that this rate had accelerated to 3 mm yr^{-1}, fuelling speculation that the Earth was about to enter a period whereby sea-levels could rise 50–100 cm eustatically in the next 100 years. The Villach Conference in Austria in 1985 fanned the flames of speculation further, predicting a rise of 20–140 cm in the next 50 years, mainly because of thermal

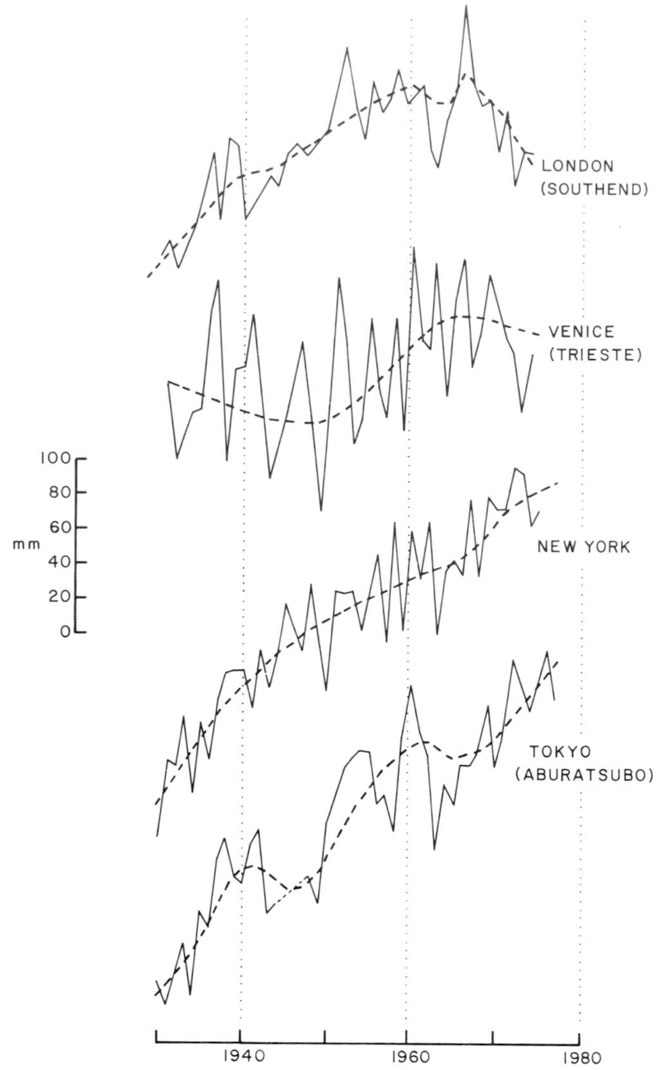

FIG. 4.11 Annual changes in sea-level 1935–78. This is based on data taken at four diverse stations, which are perceived as threatened by rising sea-level (Bryant, ©1987; reproduced with permission from the Geographic Society of New South Wales).

expansion of the oceans due to Greenhouse climate-warming. A sea-level rise of this magnitude has destructive implications for the world's coastline. The beach erosion hazard would be accelerated, low-lying areas would be permanently flooded or subject to more frequent inundation during storms, and the base line for water-tables raised. Figure 4.11 shows sea-level curves, 1935–78, for four major cities—London, New York, Venice and Tokyo—generally perceived as being threatened by a rise in ocean levels. The cities are not representative necessarily of eustatic change in sea-level because local (*isostatic*) effects operate in some areas. London is in an area where the North Sea basin is tectonically sinking, Venice suffers from subsidence due to groundwater extraction; Tokyo has had tectonic subsidence due to earthquake activity. However, the diagrams illustrate the point that some of the world's major cities would require extensive engineering works to prevent flooding, given the predicted rates of sea-level rise. This section will examine the evidence for sea-level rise, its causes, and the implications for beach erosion.

Role of the Greenhouse effect (Gribbin, 1983; Barnett, 1983; Gornitz and Lebedeff, 1987; Bryant, 1987)

The culprit for much of the rise in sea-level this century has been perceived as the melting of icecaps, concomitantly with increasing temperatures because of increasing concentrations of manufactured gases, mainly CO_2, from the burning of fossil fuels. This is known as the Greenhouse effect, whereby certain gases are more transparent to incoming short-wave radiation than they are to outgoing longwave radiation. If an atmospheric Greenhouse gas increases in concentration, then surface temperature must increase to compensate for the increased downward emitted long-wave radiation. There is no doubt that the CO_2 content of the atmosphere has been rising as shown by the record of measurement at Mauna Loa in Hawaii between 1957 and 1987. Over this period CO_2 increased from 311 to 348 parts per million (ppm) at a rate of 0.2 per cent per year, a rate that may be accelerating.

The warming of the atmosphere can also be induced by the substantial increase in other Greenhouse gases, including both stable and unstable molecules such as carbon monoxide (CO), nitrogen oxides (N_2O, NO_x), methane (CH_4), ozone (O_3), and chlorofluorocarbons (CCl_2F_2, CCl_3F). Methane has increased because of the growth in the number of grazing animals since the Second World War and increased coal-mining activities; nitrogen oxides, because of increased fertilizer use and urban pollution from motor vehicles; and chlorofluorocarbons, because of increased aerosol and coolant manufacture. The three gases N_2O, O_3 and CH_4 are producing a Greenhouse effect that is 50–100 per cent of the effect produced by CO_2 alone.

The Greenhouse scenario is a hypothesis, not a substantiated prediction. Computer-generated general circulation models have been used to model global temperature increases (about 1.5–4.5°C over the next 50 years), but such models are far from perfect. Additionally, there is no statistical basis for associating warming global air temperatures with the effects of increasing atmospheric CO_2. Statistical, linear regression models have been used to unravel the relative roles of volcanic dust, variations in *solar luminosity*, sunspot cycles and CO_2 in determining the temperature fluctuations of the northern hemisphere over the past century. One of the significant facets of this research is the control volcanic dust and solar luminosity, rather than increases of CO_2, have had on temperature fluctuations. The former two variables can account for most of the warming in the last century, including the cooling which took place between 1960 and 1980. If temperature records are examined before the Industrial Revolution, an additional question arises regarding a hypothesis based on atmospheric warming due to CO_2. The summer temperatures of the temperate latitudes of the northern hemisphere in the 1980s are matched only by those in the thirteenth and fourteenth centuries. What caused the warming in those centuries? There was no human-induced CO_2, CH_4 or other gas increase then. The role of Greenhouse gases in determining future air temperature increases—and, by inference, future sea-level rises—must be set within the context of these past temperature fluctuations.

The scenario of a globe-warming due to CO_2 has as one of its prime consequences, a worldwide rise in sea-level caused by thermal expansion of ocean waters because of heating from a warming atmosphere. It was on this basis that the Villach Conference in 1985 predicted an increase in sea-level of 20–140 cm by 2030. However, the evidence that the current eustatic rise in sea-level (1.5 mm yr^{-1}) is due to thermal expansion is not conclusive. Other factors including deceleration of large-scale ocean *gyres*, or melting of alpine glaciers may be responsible. For instance, the postulated rise in sea-level this century requires only a 20 per cent reduction in alpine ice-volume. While this reduction could be generated by global warming, it may simply reflect regional climatic changes in a few key areas. The eustatic rise in sea-level may not be natural, but it might simply reflect increased pumping of groundwaters by humans this past century. The United States obtains 20 per cent of its annual water consumption from groundwater. This amounts to 123 km^3, equivalent to a rise in sea-level of 0.3 mm yr^{-1}. If the rest of the world's annual water consumption of 2405 km^3 was divided into the same proportions as in the United States, then it would result in an annual rise in sea-level of 1.3 mm yr^{-1}, a figure that is remarkably close to the present estimated rate of worldwide change. The latter point raises the possibility of an economical solution for negating future sea-level rises. Sea water could be diverted, with minimal effort, into depressions such as the Dead Sea, The Qattara depression in Egypt, Lake Eyre in Australia and the Caspian Sea. These basins have very large areas below sea-level that could hold increased ocean volumes easily. At the same time electricity could be generated because seawater would have to flow downhill to get into the basins.

Current rates of change in sea-level worldwide (Bryant, 1985, 1987)

The concept of sea-level rising uniformly across the globe over time is not tenable. A detailed examination of figure 4.11 indicates that sea-levels, even in London and Venice, can drop at times. More importantly, sea-levels are affected significantly by local factors. These aspects have been evaluated globally using 219 records of monthly mean sea-level between 1960 and 1979: the results are shown in figure 4.12. In this diagram, areas with annual changes in sea-level >3 mm yr^{-1} have been differentiated from those with changes of 0–3 mm yr^{-1}. Contrary to the general belief that sea-level is presently rising everywhere, sea-level has fallen since 1960 along significant stretches of coastline, mainly in western Europe, western North America and eastern Asia. Some areas of falling sea-level are outside zones affected by glacial deloading, while others are in the zone of collapsing-forebulge submergence, where sea-level should still be rising. Increasing sea-levels are restricted mainly to the Gulf of Mexico and the eastern coastlines of North America and Japan. Large areas of consistent sea-level change are difficult to delineate because rates can change from positive to negative over distances of a few hundred kilometers or less. This is even apparent where rates of change are >3 mm yr^{-1}, as evidenced by the patterns of variation in northern Europe and eastern Asia.

There are a range of factors producing the sea-level trends depicted in figure 4.12. In the Yellow Sea and East China Sea between China and Japan, sea-level motions simply reflect long-term tectonic and isostatic behaviour that is documented in a series of extensive raised or submerged terraces along the coast. Land is rising by as much as 5 mm yr^{-1} in areas of massifs and ancient foldbelts in the Korea-north China region, and subsiding by as much as 9 mm yr^{-1} in areas of Cenozoic basins and foldbelts in east China. Over periods of less

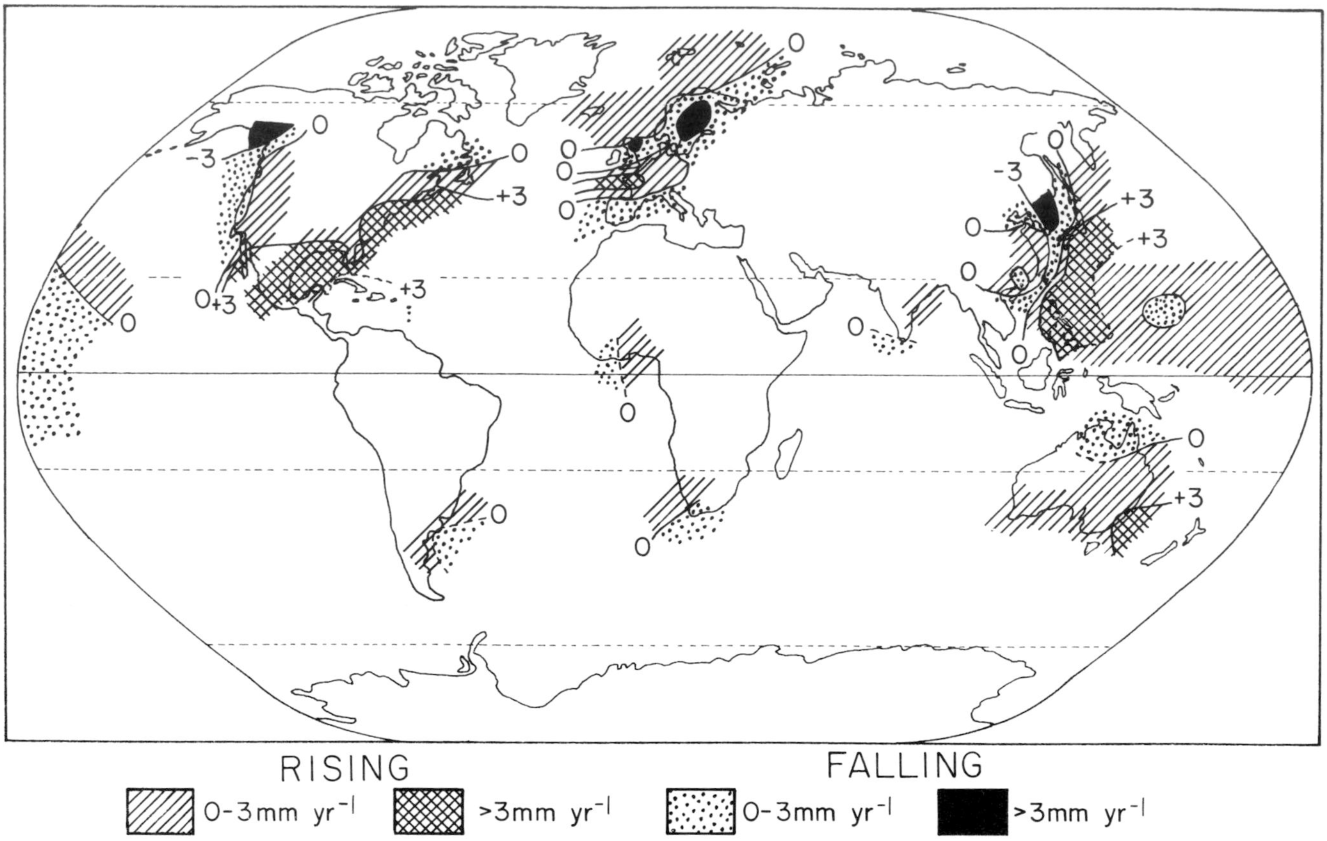

FIG. 4.12 Average annual rates of change in worldwide sea-level mainly between 1960 and 1979 (Bryant, ©1988; reproduced with permission from CSIRO Editorial and Publishing Unit, Melbourne)

than 25 years, fluctuations relate to the shifting behaviour of the Kuroshio current that dominates oceanographic effects in eastern Asia, or to the large volume of freshwater runoff mainly from China. Sea-level trends around Japan can be ascribed purely to tectonic factors. Rates range from 24 mm yr^{-1} of submergence along the southeast coast, to 6.8 mm yr^{-1} of emergence along the northwest coast. Shorter term fluctuations again correlate well with the behaviour of the Kuroshio current. Along the east coast of North America, differing trends for three independent coastal segments can be delineated between 1940 and 1980. These trends do not parallel depths in the shelf-break or any other obvious topographic or structural feature. The eastern North American coastline has some of the most complex factors affecting sea-level. It can be speculated that mean sea-level here is increasing at a steady rate approximating 1.7 mm yr^{-1} because of upwelling and current effects over the shelf, changes in longshore wind speeds piling up water at shore, runoff, increasing rainfall, or substantial groundwater extraction leading to land subsidence. In Australia sea-level is rising at the rate of 1.8 mm yr^{-1} in the south and falling at the rate of 0.5 mm yr^{-1} in the extreme tropical north. These patterns are not tectonically induced but appear to respond strongly to meteorological and oceanographic factors that include storminess, rainfall and sea-surface temperature.

More importantly, it is evident from these results that there is insufficient information to state emphatically, whether or not sea-levels are rising worldwide. More data are required from the centers of oceans and systematically from the coastlines surrounding these oceans to take into account the speeding up, or slowing down of ocean gyres, which could affect sea-level in the short term. UNESCO is trying to establish the Global Sea Level Observing System (GLOSS) to obtain these data. Over 250 sea-level stations will comprise this network worldwide. The program requires the establishment of 100 new coastal and mid-ocean stations, mainly in the southern hemisphere. As a prelude to this program, 22 countries in the Pacific Ocean now send data from 55 stations to Honolulu, where monthly mean sea-level anomaly charts are prepared and distributed. However, it should be realized that land-based stations are severely affected by the boundary conditions that affect sea-levels at the edges of water bodies. A comprehensive assessment of global sea-level also requires knowledge of elevations in the middle of oceans and in the polar seas. Towards the 1990s, this monitoring will be performed using accurate radar altimeters planned for the European Space Agency satellite and polar orbiting remote sensing platforms. These instruments, with an accuracy of less than 0.1 m, can provide information on the variability of seasonal sea-levels and long term trends after noise in the data has been removed using averaging techniques. Towards the year 2000, it should be possible to ascertain the true rate of any worldwide change in sea-level, as well as to describe the extent of regional anomalous variations.

Factors causing sea-level oscillations

(Bryant, 1987)

Sea-level also has great temporal and spatial variability. This variability can operate over timespans of days to years, and can greatly complicate delineation of statistical trends. Changes averaged over the short term can even exceed Emery's accelerated 3 mm yr^{-1} rate, with a variance 10–100 times greater than this value. The causes of this variability can be grouped under four main headings. Firstly, there are the daily or monthly fluctuations in sea-level of several tens of centimeters caused by inherent climatic or oceanographic variability. These include changes in atmospheric pressure, sea temperature, salinity, onshore wind stress components and storm surge, upwelling on the shelf, and mixing of surface and deep ocean waters. Also included here are shelf-waves, which in the Australian region, travel counterclockwise around the continent with a periodicity of one to seven days and amplitudes exceeding 50 cm in some instances.

Secondly, there are seasonal fluctuations caused by the same factors but with a range of 15–60 cm in such diverse locations as the Bay of Bengal, the west coast of Mexico, northeastern Siberia and Australia. In the northern hemisphere seasonal river discharge is also important, as mentioned above in the case of sea-levels in the East China Sea. Along the eastern United States coastline, 7–21 per cent of the annual variance in sea-level is due to this factor.

Thirdly, and of particular note, there are the persistent inter-annual fluctuations in sea-level associated with the strength of trade winds in the equatorial Pacific region. Generally, tropical air movement in the Pacific is dominated by strong easterly trade winds, the Walker circulation. Warm surface-water is blown towards the western side of the Pacific, where it piles up to heights of 20 cm or more. The Walker circulation oscillates (the Southern Oscillation) in strength every three-to-five years and has teleconnections with meteorological change worldwide. The Southern Oscillation has wide-ranging implications in natural hazards research that will be discussed in more depth in following chapters. If the trades fail completely, the water piled up in the west Pacific moves as a Kelvin wave eastwards across the Pacific in the space of two months. When this happened in 1982–83, sea-levels changed by 30–40 cm across the equatorial Pacific. These sea-level fluctuations are not restricted to the tropics. When the Kelvin wave reaches South America, it then propagates north and south along the coastline over several months, with minimum loss of elevation. The 1982–83 failure of the trades raised sea-levels 35 cm above average as far north as the Oregon coast. Sea-level increases in the Pacific region may not indicate an eustatic rise, but may simply reflect the regional consequences of more frequent failings of the easterly trades.

Finally, there are the long-term fluctuations reflecting tectonic, climatic or oceanographic change. The varying rates in the East China Sea region are undoubtedly due to differential movements in the Earth's crust; however, climatic change cannot be ruled out for some of the variation. The low-latitude asymmetry in rates of sea-level change in the Atlantic and Pacific evidenced in figure 4.12, with a western boundary rise and an eastern boundary decline, is consistent with either a change in intensity of westerly winds, or a shift in the *orographically controlled* and oceanographically fixed pressure cells of the northern hemisphere. Even a long-term increase in rainfall over the coastal sector of an ocean can cause an increase in sea-level measured at a tide gauge. Rainfall regimes globally have altered significantly in the last three decades, particularly along coastlines in the Arctic, eastern South America and northwestern Australia.

Global associations between sea-level and climate (Bryant, 1988a)

Regional studies in particular demonstrate that the causes of sea-level change along a coast, if known at all, can be dominated by local factors. General atmospheric circulation traits may also have a major

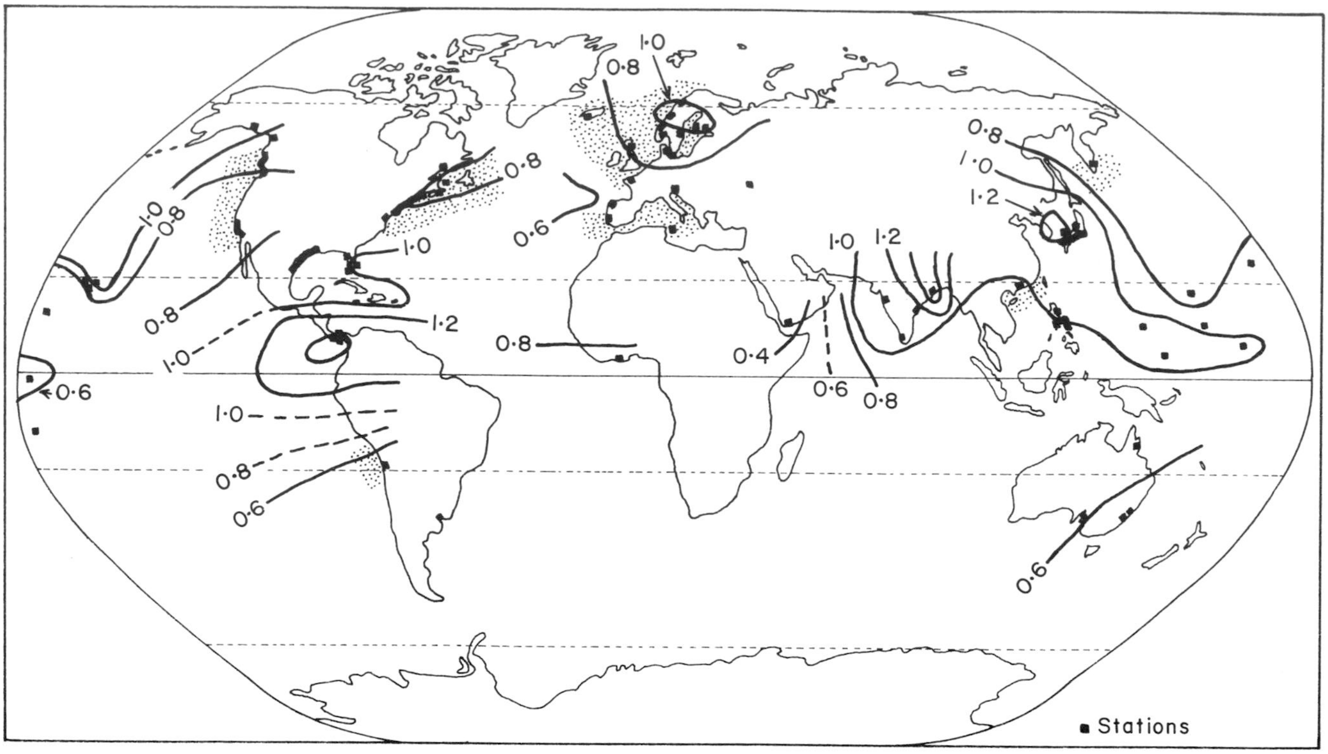

FIG. 4.13 Global distribution of summarized principal component scores for seasonal relationships amongst sea-level, precipitation and temperature (Bryant, ©1988; reproduced with permission from CSIRO Editorial and Publishing Unit, Melbourne)

impact upon regional sea-level behaviour, both seasonally and over the long term, to the extent that global increases in sea-level attributed to eustatic factors may only reflect changes in these traits at crucial locations. This aspect has been evaluated by combining the monthly values of sea-level, used above, with monthly values of atmospheric pressure, surface air-temperature and precipitation. Atmospheric pressure influences sea-level directly via the *inverted barometer* effect, or indirectly via wind stress components, established by pressure gradients as part of general atmospheric circulation. Changes in precipitation regimes affect coastal sea-levels directly or via runoff, as discussed above. Air temperature at monthly, seasonal and annual time scales, especially in the northern hemisphere, reflects sea surface temperatures, which control water density and, hence, *steric* sea-level elevations.

The analysis was performed using principal component analysis for the period 1933–80 at 91 locations around the globe, each with an average record length of 301 months. Principal component analysis is beyond the scope of this book; basically, however, it is a technique for defining relationships amongst variables when there are many data points or variables, and the relationships are not very obvious. The technique makes few assumptions about the data and has the ability to cluster together variables that have strong inter-relationships. The interested reader is referred to Davis (1973) or Johnston (1980) for a more elaborate discussion of the technique. The strength of a relationship can be assessed at any sampling point by calculating what is termed a 'principal component score'. The greater the score, the stronger the relationship.

Our analysis found that sea-level was strongly linked to temperature and precipitation seasonally. This is indicated in figure 4.13, which plots a summary of principal component scores for this relationship. The strongest link amongst sea-level, temperature and precipitation was found around southeast Asia, the Indian subcontinent, Scandinavia, Central America and the northeast Pacific. These regions correspond to areas of intense atmospheric instability. In the northern hemisphere the strongest relationships lie underneath the preferred path of the jet stream from the north Pacific, across North America to northern Europe. This path also corresponds to areas where the 18.6-year lunar tide has a strong influence upon atmospheric forcing and sea-level fluctuations. Two areas, namely Central America and the western Pacific Ocean, where summary scores exceed 1.0, also match very closely areas where mean oceanic rainfall seasonally exceeds 2500 mm yr^{-1}. All areas, except for the Bay of Bengal, correspond to coastlines where major currents influence coastlines. While this relationship groups areas such as western North America, northern Europe and the Mediterranean, where sea-level has fallen recently, it also includes areas of substantial rising sea-level in eastern North America. Note, that large areas where sea-level trends are very variable in figure 4.12 are remarkably coherent in figure 4.13. This is most evident again around the East China Sea region. While rates of sea-level change may be affected by tectonics, the year-to-year variation is dominated by climatic factors.

A strong, inverted barometer relationship between sea-level and air pressure can also be found in other groupings. These data are not presented here; however, they can be shown to relate to areas of cyclonic generation associated with the Aleutian and Icelandic lows. This response is strongest in winter, when these low-pressure areas are most active. Changes in the intensity of low-pressure areas over oceans in the northern hemisphere have a major impact upon nearby sea-levels either directly through the inverted barometer effect, or indirectly through the enhancement of major ocean gyres. This is why the whole question of rising sea-level must involve consideration of, not only the Greenhouse effect, but also of regional climatic changes evident through measurements of barometric pressure, precipitation and air temperature. Where sea-level is rising in the northern hemisphere, it may simply be reflecting changes in jet stream paths and the intensity of low-pressure cells.

Beach erosion hazard

Introduction

There is little argument that 70 per cent of the world's sandy beaches are eroding at average rates of 0.5–1.0 m per year. Some of the reasons for this erosion include increased storminess, coastal submergence, decreased sediment movement shoreward from the *continental shelf* associated with leakage out of beach compartments, and shifts in global pressure-belts and, hence, changes in the directional components of wave climates. As well, humans have had a large impact on coastal zone erosion this century. No single explanation has worldwide applicability because all factors vary in importance regionally. Evaluation of factors is complicated by a lack of accurate, continuous, long-term erosional data. Historical map evidence spanning 100–1000 years has been used in a few isolated areas, for instance on Scolt Head Island in the United Kingdom, Chesapeake Bay in the United States, and on the Danish coast; however, temporal resolution has not been sufficient to evaluate the effect of climatic variables. Aerial photographic evidence is restricted to the past 50 years and often suffers from insufficient ground control for accurate mapping over time. Ground surveying of beaches was rarely carried out before 1960 and is often discontinuous in time and space. Examples of long time-series of beach change data include: the 16-year record for Scarborough beach, Western Australia, that also goes back discontinuously into the 1930s; the fortnightly record for Moruya beach, New South Wales, from 1973 to the present; and the periodic profiling of the Florida coastline at 100 m intervals since 1972. These examples are isolated occurrences. To date coastal geomorphologists are still not certain exactly why so much of the world's coastline is eroding. This section will examine some of the more popular means of explaining the causes of this erosion.

Sea-level rise and the Bruun rule (Bryant, 1985)

The general emphasis on rising sea-level—an assumption made without firm evidence— has led to this factor as being perceived as the main mechanism for eroding beaches. Beach retreat operates via the *Bruun rule*, first conceived by Per Bruun, a Danish researcher, and formulated by Maurice Schwartz of the United States. The Bruun rule states that, on beaches where offshore profiles are in equilibrium and net longshore transport minimal, a rise in sea-level must lead to retreat of the shoreline. The amount of retreat is not simply a geometrical effect because of the slope angle of the beach (figure 4.14). Instead, Bruun postulated that excess sediment must move offshore from the beach, to be spread over a large distance in order to raise the sea bed, and maintain the same depth of water as before. The Bruun rule implies that beaches do not steepen over time as they erode, but maintain their configuration, be it barred, planar or whatever. Generally, a 1-centimeter rise in sea-level will cause the shoreline to retreat 0.5 m. This rate

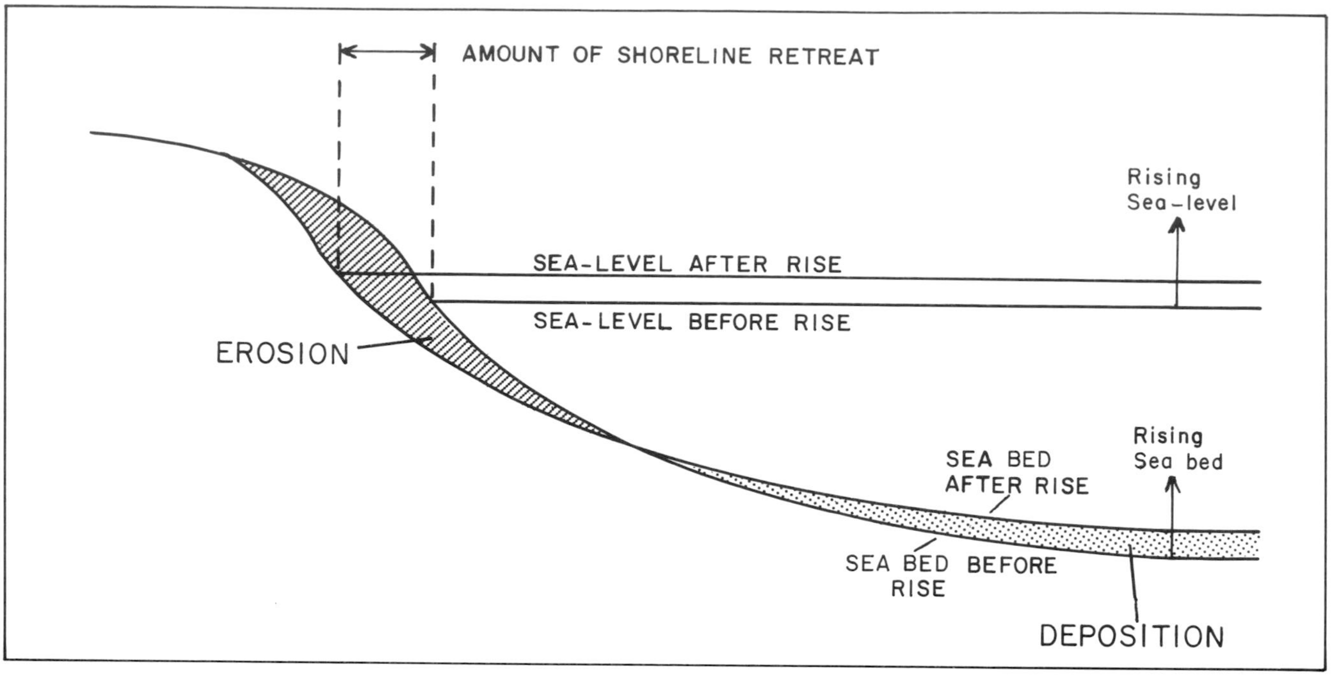

FIG. 4.14 Bruun's rule for the retreat of a shoreline with sea-level rise

has been substantiated by numerous studies from varying coastal locations, ranging from high energy, exposed tidal environments, to lakes and wave tanks. The Bruun rule operates by elevating the zone of wave attack higher up the beach. This is exacerbated by the fact that the base line of the beach water-table is also raised. Water draining from the lower beach-face decreases sediment shear resistance, and enhances the possibility of sediment *liquefaction*, causing seaward erosion of sand and shoreline retreat. The liquefaction process is described in more detail in chapters 11 and 13.

Unfortunately, while the Bruun rule has been proven universally, there are limitations to its use. The actual volume of sediment needed to raise the sea bed as sea-level rises depends upon the distance from shore and the depth from which waves can move that material. The Bruun rule was formulated when it was believed that this depth was less than 20 m. The distance offshore to the 20 m bathymetric contour was a simple calculation to perform for most coastlines. Individual waves can, in actual fact, move sediment offshore to depths exceeding 40 m, and well-defined groups or sets can initiate sediment movement out to the edge of the continental shelf. Coastlines, where Bruun's hypothesis has been applied most successfully, include the United States east coast, and Japan, where sea-level is rising rapidly. There are many areas of the world where sea-level is rising at slower rates, or declining, and coastal erosion is still occurring. It would appear that sea-level change as a major, long-term cause of beach erosion has evolved from evidence collected on, and applied with certainty to, only a small proportion of the world's coastline. A dichotomy also exists between measured rates of beach retreat and the present postulated eustatic rise in sea-level. Common magnitudes of beach retreat in the United States, and elsewhere in the world, range from 0.5–1.0 m yr^{-1}. According to the Bruun rule, shoreline retreat at this rate would require sea-level rises three to seven times larger than Emery's projected rise of 3 mm yr^{-1} that presently may be occurring eustatically. Obviously, other factors besides sea-level rise must be involved in beach erosion.

The actual rise in sea-level may be irrelevant to the application of Bruun's rule. As shown above, sea-level oscillations at all timescales are one of the most common aspects about sea-level behaviour. More sections of the world's coastline have significant oscillations than simply have rising sea-level. The emphasis upon Bruun's rule has been upon rising levels, but the rule also can predict accreting coastlines where sea-level is falling. If Bruun's rule is to account for long-term beach erosion in these locations, then what has to be taken into account is the disequilibrium between the amount of erosion during high sea-level fluctuations and the amount of accretion when sea-level drops. It takes a shorter time to remove material seaward during erosional, rising sea-level phases than it does to return sediment landward during accretional, falling sea-level periods. If eroded sediment that has been carried seaward cannot return to the beach before the next higher sea-level phase, then the beach will undergo permanent retreat. Bruun's hypothesis only provides the mechanism by which these shifts in sea-level operate. The more frequent the change in sea-level, the greater the rate of erosion. Bruun's hypothesis as developed to date is thus static. The hypothesis has the underlying assumption that sea-level change along a section of coastline is continuous in one direction only. As shown here, and from a simple observance of any tidal record, this is not an accurate view of reality—a fact which may limit the universal application of the hypothesis as presently conceptualized.

General causes of erosion (Bird, 1985, 1988)

Clearly, more than one factor can account for the world's eroding coastlines. Table 4.2 lists most of the commonly perceived factors leading to beach erosion. Most of the table is self-explanatory, but a few comments are necessary for some factors (including increased rainfall and storminess, which will be discussed in more detail later). Human impact on beach erosion is becoming more important, especially with the technological development in coastal engineering this century. Breakwalls have been used as a ubiquitous solution to coastal erosion in some countries such as Japan, to the extent that it has become a major reason for continued erosion. Wave energy reflects off breakwalls leading to sediment drift—either offshore drift, or *longshore drift* away from structures. Classic examples of breakwall obstruction of longshore sediment drift, leading to erosion downdrift, include the Santa Barbara harbour breakwall in California, and the Tweed River retaining walls, south of the Queensland Gold Coast in eastern Australia. Offshore mining affects refraction patterns at shore, or provides a sediment sink to which inshore sediment drifts over time. Erosion of Point Pelèe, Ontario, in Lake Erie, can be traced to gravel extraction some distance offshore. Many cases also exist along the south coast of England of initiation of beach erosion after offshore mining. Dam construction is very prevalent on rivers supplying sediment to the coastline. Californian beaches have eroded in many locations because dams, built to maintain constant water-supplies, have also trapped sediment that maintains a stable, coastal sand budget. The best example is the case of the Aswan dam in Egypt; since its construction it has resulted in the Nile delta shoreline retreating by 1 km or more.

Not realized are two factors that may be more important than any of the above. Each time someone walks across a beach, that person becomes a mini-excavator. A person who does not disrobe and endeavour to shake all clothing free of sand after swimming can remove 5–10 gm of sand on each visit to the beach. On popular bathing beaches, this can amount to several tonnes per kilometer of beach per year. Secondly, rapid urban expansion near the coastline since the Second World War has led to modification of urban drainage-channels and

TABLE 4.2 Factors recognized as Controlling Beach Erosion

Exclusively Human-induced

- Reduction in longshore sand supply because of construction of breakwalls.
- Increased longshore drift because of wave reflection off breakwalls behind, or updrift of, the beach.
- Quarrying of beach sediment.
- Offshore mining.
- Intense recreational use.

Exclusively Natural

- Sea-level rise.
- Increased storminess.
- Reduction in sand moving shoreward from the shelf because it has been exhausted, or because the beach profile is too steep.
- Increased loss of sand shoreward from the beach by increased wind drifting.
- Shift in the angle of wave incidence due to shifts in the location of average pressure cells.
- Reduction in sediment volume because of sediment attrition, weathering or solution.
- Longshore migration of large beach lobes or forelands.
- Climatic warming in cold climates, leading to melting of permanently frozen ground and ice lens at the coastline.
- Reduction in sea-ice season, leading to increased exposure to wave attack.

Either Human-induced or Natural

- Decrease in sand supply from rivers because of reduced runoff from decreased rainfall, or dam construction.
- Reduction in sand supply from eroding cliffs, either naturally, or because they are protected by seawalls.
- Reduction in sand supply from dunes because of migration inland, or stabilization.
- Increase in the beach water-table because of increased precipitation, or drainage modification by humans.

Source: Based on Bird, 1988.

water tables. Water pumped from coastal aquifers, to supply urban water-needs, not only may cause local subsidence—as is the case for most cities along the United States east coast—but could also lead to accelerated beach erosion as that water is eventually discharged either into streams or into the water table through septic systems, thus raising local coastal water-tables.

Natural factors are perhaps still the major cause of beach erosion. Long-term changes have been documented best on beach forelands along the channel and east coast of England at Dungeness, Orford Ness and Scolt Head Island. These forelands are tens of kilometers in width, and appear to have developed since the *Holocene* rise in sea-level, following the last glaciation. However, some are slowly migrating alongshore. Material is being eroded from the updrift-ends only to be re-deposited as a beach ridge on the downdrift-end. Shorter term beach erosion has manifested itself this century, mainly because of the change of global climate that has occurred since 1948. This change has not been one of magnitude as much as one of variation: the storms have become worse followed by longer periods of quiescence; severe droughts are broken by record rain and floods; the coldest winter on record is followed by the warmest summer. Unfortunately, the emphasis of this climate shift has focused on temperature change, whereas the shift also involves dramatic changes in coastal rainfall regimes, and changes in pressure-cell location leading to an upsurge in storminess. These latter aspects are examined in the remaining part of this chapter.

Erosion resulting from increased rainfall

(Bryant, 1985)

Although sea-level rise enhances beach erosion by elevating beach water-tables, logically any factor affecting water-tables can control beach erosion. For instance, a sandy backshore absorbing rainfall acts as a time-delaying filter for subsequent discharge of this water to the sea through the lower foreshore. The process can raise the water-table substantially for periods up to two months after heavy rainfalls. Many coastlines, especially those where seasonal *orographic* rainfall is pronounced, can have over 1000 mm of precipitation within a month. Under these intense rainfalls, sand suspension by groundwater flow is even possible. Rainfall may be better related to beach change than sea-level. Figure 4.15 plots the relationship defined, using linear regression analysis, between average

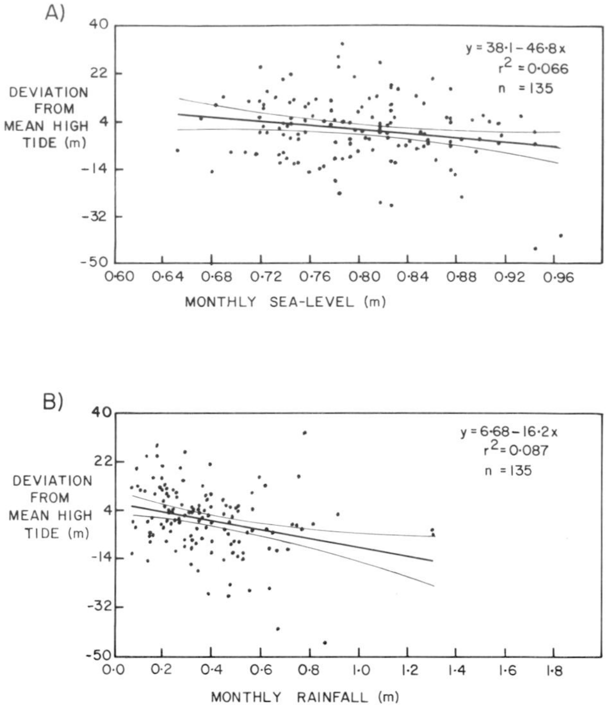

FIG. 4.15 Linear regression lines (with 95 per cent confidence limits) between average deviation from the mean high tide position and A. monthly sea-level, and B. monthly rainfall: at Stanwell Park beach, Australia, 1895–1980.

deviations from the mean high-tide position and (A) monthly sea-level and (B) monthly rainfall for Stanwell Park beach, Australia, using 135 points in time between 1895 and 1980. Stanwell Park beach forms a *compartmentalized*, exposed, ocean beach with no permanent longshore leakage of sediment, and little human interference (figure 4.16). The beach lies on the southeast coast of Australia, and faces the main southeast swell originating in the Tasman Sea. This swell averages 10 seconds in period and 1.2 m in wave height. Inshore topography varies rapidly, from alternating shore-tied shoals and rip channels, to a single, shore-parallel bar-trough in response to storm waves, which often exceed a deep-water height of 4 m. Deviations of the high-tide beach position, averaged for all of Stanwell Park, were determined from oblique photographs accurately dated to the nearest month.

Linear regression analysis is a statistical technique for relating two variables to each other, assuming that this relationship is linear or forms a straight line. Generally, one of the variables depends upon the other. For instance, we have stated above that ice cover in the North Atlantic is related statistically to easterly winds along the Norwegian

FIG. 4.16 Differences in beach volume, Stanwell Park, Australia. *Above*, the beach after sustained accretion in February 1982. *Below*, the same beach when eroded severely by a storm in August 1986 (photograph by courtesy of Dr Ann Young, Department of Geography, University of Wollongong).

coast. In this case, ice cover is the dependent variable and the strength of easterly winds is the independent variable. The regression line is a straight line that is fitted to a scattering of data points so that the distance between all the points and the line itself is minimized. In other words, the fitted line accounts for the maximum variation or noise in the scattering of points. The amount of variation that is accounted for by the line is measured using the regression coefficient, and is often expressed in terms of a percentage of the variation in the data. The regression coefficient can be tested for its statistical significance or departure from randomness. This technique is very good when the data points have been accurately measured, and a researcher wants to define a trend amongst a scattering of points that may number in the hundreds or thousands. That is why this method was chosen to characterize the relationships shown by the scattering of points in figure 4.15. It should also be pointed out that there are many assumptions involved in satisfactorily using linear regression analysis, and readers should be willing to consult relevant literature if they wish to pursue the subject. Linear regression analysis shows that rainfall is more efficient than sea-level in explaining the variation in beach change (8.7 per cent versus 6.6 per cent).

Erosion attributed to storminess

In chapter 2, the mid-latitude extra-tropical depression was viewed as one of the most destructive weather phenomena observed. Some of the worst events to occur in the world were described in that chapter, along with a summary of the types of mid-latitude depressions that can develop along coasts. Since the classic observations both of the effects of Hurricane Carla in 1961 upon the normally inactive barrier islands of the Central Texas coast, and of a late autumn storm upon Point Barrow, Alaska, large magnitude storms have been viewed as the major agent of sediment transport, and large-scale geomorphic change in the coastal zone. Storms supposedly shape in a few hours what would take decades to achieve under normal conditions. Research has shown that storms dominate not only barrier washovers and inlet breaches but also sedimentation in estuaries, bays and lagoons. Nowhere in the coastal zone have the effects of storms been so preponderant, as on the continental shelf. Here, grading of sediment is controlled by storms through landward sediment displacement over thousands of years, sorting of sediment under high energy and then waning energy conditions, migration of mega-ripples and bottom liquefaction. The idea that storms dominate the formation and preservation potential of coastal morphology has led to the view that recent worldwide coastal erosion can be attributed partially to increased storminess, and that the rates and preferred locations of coastal erosion by storms is somehow different from that attributable to everyday processes. In this concluding section, the role of storms as agents responsible for increased beach erosion is evaluated for two coastlines with varying morphologies, the barrier island coastline of eastern North America, and the compartmentalized coastline of south-eastern Australia.

The North American east coast (Bryant, 1978; Dolan and Hayden, 1983; Dolan et al., 1987)

Evidence from the barrier island coastline of eastern North America not only indicates large spatial variation in long-term rates of erosion of barrier islands induced by storms but also shows that particularly large storms tend to erode, or overwash the shoreline worst in the those places evidencing the highest historical rates of erosion produced by lesser events. For the 35-kilometer long, barrier island complex of Kouchibouguac Bay in the Gulf of St Lawrence along the east coast of Canada, gross island morphology for the past 150 years can be linked to wave-power patterns characteristic of the overall wave climate. Wave refraction patterns were simulated for the complete wave spectrum, consisting of seven directions, periods ranging from 4 to 10 seconds and heights of 0–2 m. Wave-power values at shore were then integrated along the whole barrier-island chain, and matched with historically determined areas of morphological change. Stable inlet channels occupy positions with the lowest onshore components of wave power while areas of historical major inlet-breaching occupy positions with higher wave power. Even areas of present overwashing and substantial lagoon infilling, some of which are unrelated to any known inlet position, lie within high wave-power sections of coastline. Catastrophic storms add to, rather than detract from this average wave-power pattern.

These results are repeated on the 1300 km, higher wave-energy coastline between New Jersey and South Carolina. Here, the high-water and storm-penetration limits have been mapped at 100-meter intervals every five years between 1930 and 1980. Beaches along the Atlantic coastline are changing at rates ranging from +10 to −30 m yr^{-1} and average 1.5 m yr^{-1} of erosion (figure 4.17). Islands with more southerly exposure have slower rates of erosion, implying that the direction of storm tracking and wave approach are important. Comparison of storm waves hindcasted between 1885 and 1985, with actual measurements since 1945, indicates that storms are becoming more frequent along this coast. The accumulated storm-wave record, excluding tropical cyclones, between 1942 and 1984 for this mid-Atlantic coastline is shown in figure 4.18. This diagram presents the annual duration in hours of waves greater than 1.6 m elevation, and the number of hours of waves in excess of 3.4 m. On average, there are 64 hr yr^{-1} when waves exceed 3.4 m along this coastline. East-coast lows appear incapable of generating large waves and account for only 13 per cent of waves over 3.4 m. Onshore winds resulting from anti-cyclones account for 24 per cent of all storms. The most severe storms along this coast are related to extra-tropical cyclones tracking along the polar front eastwards across the continent, especially south of Cape Hatteras. Both severe storm waves

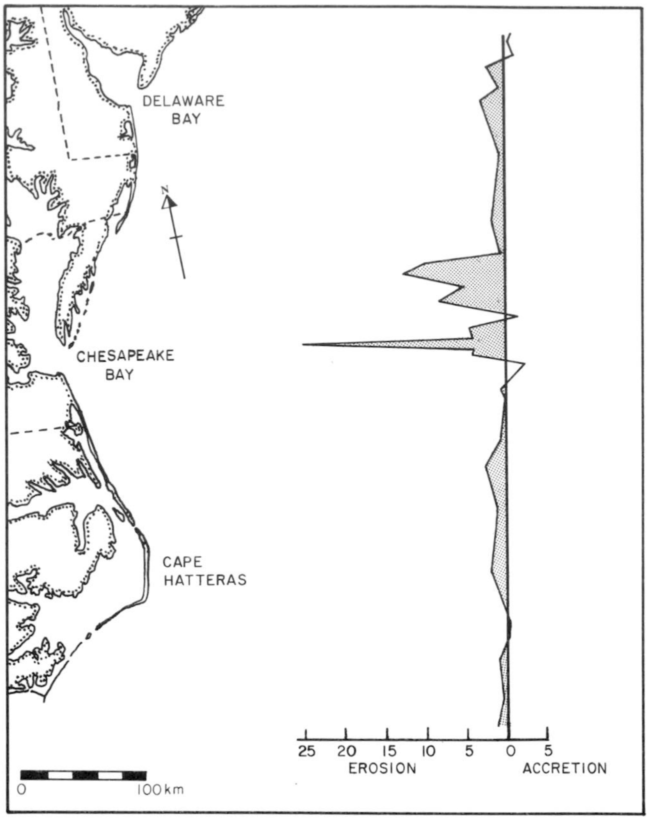

Fig. 4.17 Mean rate of erosion along the United States Atlantic coast (reprinted from Komar, P. D. [ed.] ©1983 *CRC Handbook of Coastal Processes and Erosion*, with permission CRC Press Inc., Boca Raton, Florida pp. 123–50)

and the frequency of occurrence of all storm waves cluster over a number of years, as shown for the seven-year period around 1970. In contrast, years such as the late 1970s that were deficient in storm waves, also lacked excessive wave heights. The correlation between the two sets of data is highly significant even allowing for serial *correlation* effects (r = 0.73, significant at the 0.01 level). Storm over-washing of barriers is greatest in those areas that historically have had substantial overwashing, and/or where long-term average shoreline erosion rates are greatest. Ironically, while storms produce significant overwash, their overall effect generates subsequent accretion at the shoreline mainly because small storms tend to move sand shoreward on the inner shelf. Even the 1:100-year, Great Ash Wednesday storm of 1962, had no effect on long-term erosion determined for a 42-year period. While this storm produced dramatic erosion, it is in fact embedded in the overall storm climatology shown in figure 4.18. This storm occurred in a year when storms were near average in number and intensity.

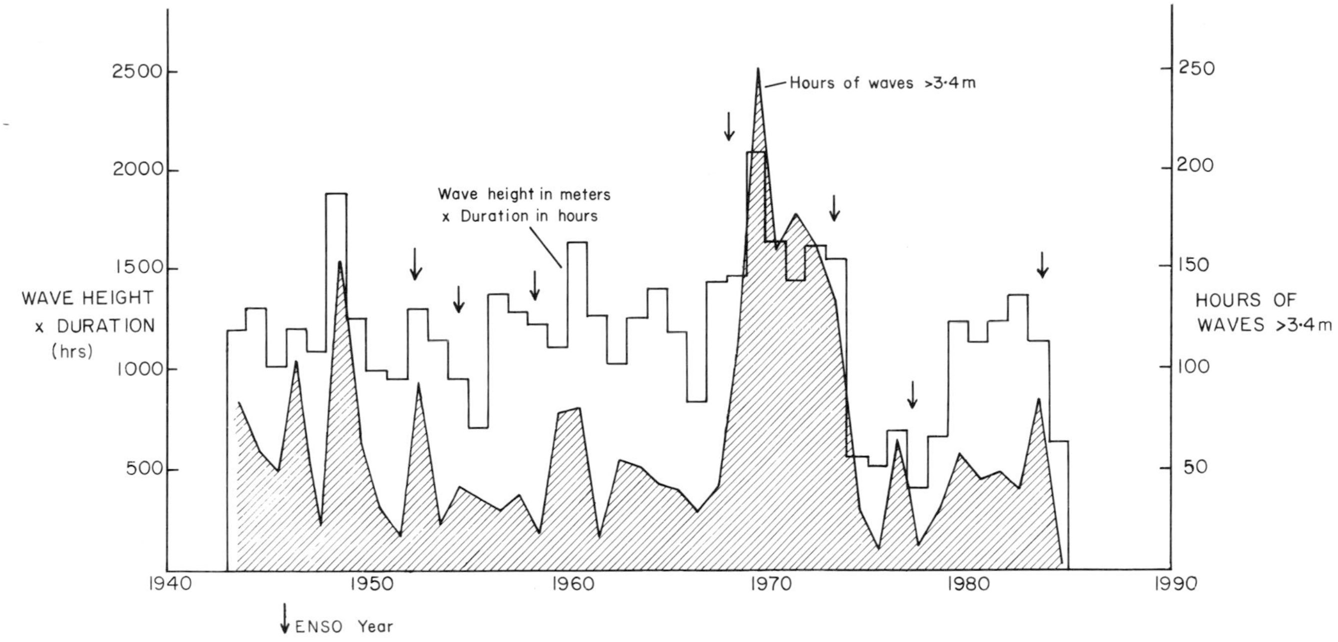

FIG. 4.18 Annual frequency of waves over 3.4 m and annual total hours of wave height above 1.6 m: United States mid-Atlantic coastline, 1942–84 (based on Dolan et al., 1987)

The Australian east coast (Blain et al., 1985; Bryant, 1988b)

The effect of storms along the Australian east coast was evaluated using the beach erosion data set for Stanwell Park described above, and the frequency and magnitude of storms along this coast complied by Blain et al. (1985). Both of these data sets are amongst the best of their kind in the world. The Blain et al. report classifies storms affecting the New South Wales coast into six types: tropical cyclones, easterly trough lows, inland trough lows, continental lows, secondary lows and anticyclonic intensification. Tropical cyclones originate in the Coral Sea and rarely travel south of 30° S. Only the Queensland coast and the north coast of New South Wales are affected directly by these cyclones; however, the rest of the New South Wales coast does receive long period attenuating waves originally generated by these systems. The four types of lows are all variations of the east-coast low described in chapter 2. Such storms develop because of sea-surface temperature gradients that steeply decrease towards shore, because of easterly wind deflection northwards by the Great Dividing Range, and because of abundant latent heat obtained through evaporation over a warm east Australian current. These east-coast lows develop preferentially in autumn or winter and can generate a strong wind-field over distances of 1000 km. They have a profound potential for beach erosion on New South Wales beaches in their area of influence. Anticyclonic intensification refers to the strengthening, and/or stalling, of easterly moving high-pressure systems in the south Tasman Sea, and subsequent directing of easterly or southeasterly winds onto the coast. Such systems usually develop pressures above 1030 mb, and may be associated with the development of east-coast lows. Such systems may be devoid of rain, but give rise to wave heights in excess of 4 m along long stretches of coastline over several days.

Using this data base, indices of storm waves were constructed for Stanwell Park beach for the period 1943–78, by accumulating the effective storm-wave height for each quarter and year to form seasonal and annual indices. These indices then take into account the additive effect of storms over these periods. Beach change and accumulative storm data are plotted in figure 4.19, and show that major erosional events occur in those years when storms have been more frequent, and/or larger. The May–June 1974 storms described earlier (in chapter 2) were responsible for over 60 m of cutback on Stanwell Park beach. Recovery following storm events is remarkably quick.

Linear regression analysis was performed using these time series. There is no trend for increased storminess along this section of coastline over the

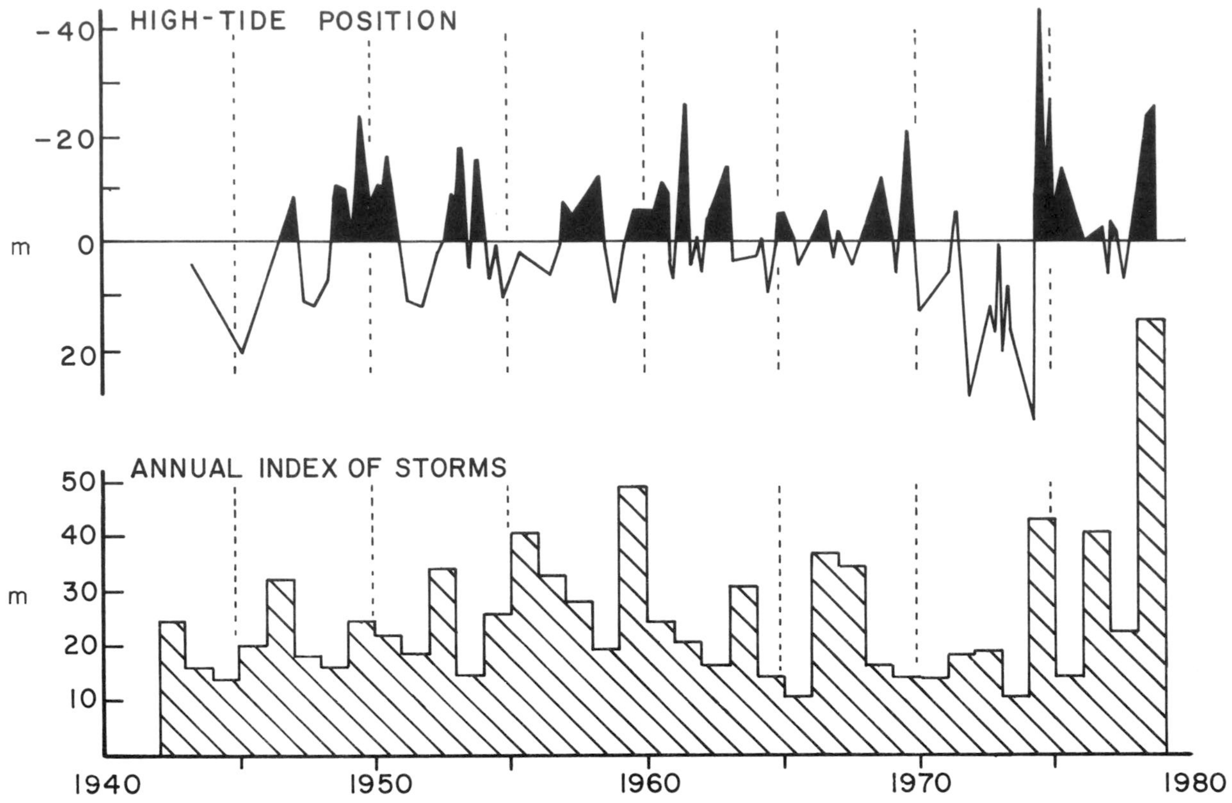

FIG. 4.19 Deviations from mean high-tide position and annual accumulated storm wave heights: Stanwell Park beach, Australia, 1943–78 (Bryant, ©1988 reprinted with permission from Elsevier Science Publishers, Physical Sciences and Engineering Division, Amsterdam, The Netherlands)

period of record. However, both storm indices account for a higher degree of the variation in the beach change record (9.9 per cent and 16.6 per cent for seasonal and annual indices respectively) than either rainfall or sea-level. A 1-meter increase in the accumulated, storm-wave height over the previous year will shift the high-tide line shoreward 0.47 m. While this value is equivalent to the effect generated by a 1-centimeter rise in mean sea-level, it represents a very minor increase in storm-wave height. It also implies that beach erosion may not be due to a single storm event. Erosion can be triggered by storms up to a year previously. The erosion caused by the 1974 May–June storms culminated retreat that had actually begun four or five months earlier. Clearly, the effect of storms on beach retreat accumulates over time, leading to subsequent beach erosion irrespective of the behaviour of other factors like sea-level or rainfall. The reverse situation is also true. If storms have been absent or are diminishing in intensity over the previous year, then a beach is more likely to undergo seaward progradation. It should also be noted that many storms that affect this coastline are also accompanied by heavy rainfall and higher sea-levels related to storm surges of about 0.5–0.7 m.

Concluding remarks

While it is generally believed that sea-level rise is responsible for recent sandy beach erosion (the Bruun rule), this factor may be subservient to the role played by rainfall and storms. Rainfall controls beach position through its influence on the water-table as does rising sea-level. Storms operate on the beach through wave action. It should be pointed out that coastal storms are not catastrophic events. Over the timespan of decades or longer, storms should be viewed as ordinary events, playing a significant role in beach erosion along with many other environmental variables such as rainfall, sea-level and atmospheric circulation factors. Even the 1:100-year storms affecting the Australian and American east coasts fit within this overall picture. The presence of fair weather is not a separate entity either, but part of a fair weather–storm continuum.

Three variables—rainfall, sea-level and storms—together form part of a suite of changing variables which are interrelated. Analysis of data for Stanwell Park beach illustrates this aspect. Groups of environmental parameters that individually can be linked to erosion, collectively and logically operate together. Furthermore, changes in these environmental parameters reflect the large-scale climatic change which the Earth is presently undergoing. The more dramatic and measurable increases in temperature (supposedly leading to icecap melting and higher sea-levels) have unfortunately detracted from the more significant nature of that change, namely the increasing variability of climate since 1948. This increase accelerated in the 1970s and shows no sign of abating. The world's beaches will continue to respond to this change; unfortunately, it will be in an erosive manner.

Finally, a note of caution must be raised. This chapter has cast doubt on the relevance of emphasizing a uniform, worldwide increased sea-level, and shown that this factor is not solely responsible for eroding beaches. Erosion of the world's beaches may not be as common, or as permanent, as generally believed. For example, the beach high-tide record for Stanwell Park shown in figure 4.19, when averaged over the long term, is stable even though sea-level off this coast is thought to be recently rising by 1.6 mm yr^{-1}, and even though the rainfall regime has increased about 15 per cent since 1948, and there are no presently active sediment sources. In Florida detailed surveys at 100 m intervals around the coast have shown that beaches have not eroded between 1972 and 1986, even though both sea-level and rainfall have increased, and sand supplies have remained meager. Here, the absence of storm activity over this timespan may be the reason for stability; researchers still cannot agree, however, on the main reason for the lack of erosion. These examples emphasize the point that research into worldwide, long-term beach change is still clouded by some uncertainty.

References

Allen, T. D. 1984. 'An eye on ocean waves and currents: satellite-borne radars scan the sea surface'. *Physics Bulletin* v. 35 no. 2 pp. 239–41.

Bryant, E. A. 1985. 'Rainfall and beach erosion relationships, Stanwell Park, Australia, 1895–1980: worldwide implications for coastal erosion'. *Zeitschrift für Geomorphologie Supplementband* v. 57 pp. 51–66.

——. 1987. 'CO_2-warming, rising sea-level and retreating coasts: review and critique'. *Australian Geographer* v. 18 No. 2 pp. 101–113.

——. 1988a. 'Sea-level variability and its impact within the greenhouse scenario'. In Pearman, G. I. (ed.) *Greenhouse: planning for climate change*. Brill, Leiden, pp. 135–47.

——. 1988b. 'The effect of storms on Stanwell Park, N.S.W. beach position, 1943–80'. *Marine Geology* v. 79 pp. 171–87.

Coastal Engineering Research Center. 1977. *Shore Protection Manual*. (3 vols). United States Army.

Davies, J. L. 1980. *Geographical Variation in Coastal Development*. (2nd edn). Longman, London.

Davis, J. C. 1973. *Statistics and Data analysis in Geology*. Wiley, NY.

Defant, A. 1961. *Physical Oceanography* v. 1 Pergamon, NY, pp. 243–84.

Dolan, R. and Hayden, B. 1983. 'Patterns and Prediction of Shoreline Change'. In Komar, P. D. (ed.) *CRC Handbook of Coastal Processes and Erosion*. CRC Press, Boca Raton, Florida, pp. 123–50.

Dolan, R., Hayden, B., Bosserman, K. and Isle, L. 1987. 'Frequency and magnitude data on coastal storms'. *Journal Coastal Research* v. 3 no. 2 pp. 245–47.

Gross, M. G. 1972. *Oceanography: a view of the Earth*. Prentice-Hall, Englewood Cliffs, NJ, pp. 201–206.

Johnston, R. J. 1980. *Multivariate Statistical Analysis in Geography*. Longman, London.

Komar, P. D. (ed.). 1983. *CRC Handbook of coastal processes and erosion*. CRC Press, Boca Raton.

Skov, N. A. 1970. 'The ice cover of the Greenland Sea'. *Meddelelser om Grønland* v. 188 No. 2.

Further reading

Bailey, R. H. 1982. *Glacier*. Time-Life Books, Amsterdam.

Barnett, T. P. 1983. 'Recent changes in sea level and their possible causes'. *Climatic Change* v. 5 pp. 15–38.

Beer, T. 1983. *Environmental Oceanography*. Pergamon, Oxford.

Bird, E. C. F. 1985. *Coastline changes: a global review*. Wiley, Chichester.

——. 1988. 'Physiographic implications of a sea-level rise'. In Pearman, G. I. (ed.) *Greenhouse: planning for climate change*. Brill, Leiden, pp. 60–73.

Blain Bremner & Williams Pty Ltd & Weatherex Meteorological Services Pty Ltd. 1985. 'Elevated ocean levels: storms affecting N.S.W. coast 1880–1980'. *New South Wales Public Works Division Coastal Branch Report*. No. 85041.

Bruun, P. (ed.). 1985. *Design and construction of mounds for breakwalls and coastal protection*. Elsevier, Amsterdam.

Bryant, E. A. 1978. 'Wave climate effects upon changing barrier island morphology Kouchibouguac Bay, New Brunswick'. *Maritime Sediments* v. 15 pp. 49–62.

Emery, K. O. 1980. 'Relative sea levels from tidegauge records'. *National Academy of Science Proceedings* v. 77 pp. 6968–72.

Gornitz, V. and Lebedeff, S. 1987. 'Global sea-level changes during the past century'. In Nummedal, D., Pilkey, O. H. and Howard, J. D. (eds) *Sea-level change and coastal evolution.* Society of Economic Paleontologists and Mineralogists, Tulsa, Oklahoma, Special Publication No. 41 pp. 3–16.

Gribbin, J. 1983. *Future Weather: the causes and effects of climatic change.* Penguin, Harmondsworth.

Komar, P. D. 1976. *Beach processes and sedimentation.* Prentice-Hall, Englewood Cliffs, New Jersey.

Kovacs, A. and Sodhi, D. S. 1978. *Shore ice pile-up and ride-up: field observations, models, theoretical analyses.* United States Army Corps Eng. Cold Regions Research and Engineering Laboratory Report.

Pickard, G. L. 1979. *Descriptive Physical Oceanography* (3rd edn). Pergamon, Oxford.

Taylor, R. B. 1977. 'The occurrence of grounded ice ridges and shore ice piling along the northern coast of Somerset Island, N.W.T.' *Arctic* v. 31 No. 2 pp. 133–49.

Wiegel, R. L. 1964. *Oceanographic Engineering.* Prentice-Hall, Englewood Cliffs, NJ.

5

Causes and Prediction of Drought and Flood

Introduction

Since civilizations first developed urban-agricultural systems, they have been plagued by drought and famine. However, there is strong evidence to suggest that *flood* is intricately linked to these two natural calamities. Often droughts are broken, or followed within a year by heavy rainfall and floods. This is more apparent in semi-arid and arid parts of the world dominated by the Hadley cell. This association is very prevalent in those parts of the world where drought and flood are linked to the Southern Oscillation, which has been alluded to in previous chapters, but will be discussed in detail here. In fact, floods were also a major hazard to early civilizations and so important that flooding features prominently in many of the creation stories from different cultures. For instance, the story of Noah and the Ark probably reflects an actual event. The Noah story had its origins in the basin of the Euphrates-Tigris rivers (see figure 5.1 for the location of major place-names mentioned in this chapter), where there is strong geomorphic evidence of a major flood about the time agricultural-urban civilization was first developing there. This flood was truly catastrophic. The sediment record depicts sediment transport and deposition on the order of several meters over a wide area of the *floodplain*. For the purpose of this discussion, we cannot separate the occurrence of drought from that of heavy rain and flood. However, it should be pointed out that, given the high population densities in semi-arid parts of the world, the human effect of drought in the 1970s and 1980s has been much greater than that of flood. No more poignant examples can be given than that of the Ethiopian-Sudanese famines in these decades.

While many definitions of drought exist, it can be defined simply as an extended period of rainfall-deficit during which agricultural *biomass* is severely curtailed. In some parts of the world, such as northeast United States and southern England, a drought may have more of an effect on urban water supplies than on agriculture. The actual period of rainfall-deficit before a drought exists also varies worldwide. In southern Canada a drought is any period where no rain has fallen in 30 days. Lack of rain for this length of time can severely reduce crop yields in an area where crops are sown, grown and harvested in a period of three to four months. In Australia such a definition is meaningless as most of the country receives no rainfall for at least one 30-day period per year. Indeed, in tropical areas subject to monsoons, drought conditions appear each dry season. Most of tropical Australia, even in coastal regions, endures a dry season lasting several months with no rain. In Australia drought usually is defined as a calendar year in which rainfall amounts are in the lower 10 per cent of the records. Unfortunately, a calendar year splits the summer growing season into two in the southern hemisphere. A more effective criteria for drought declaration should consider abnormally low rainfall in the summer growing season.

There is one other aspect of drought and rainfall that should be emphasized. We tend to think that our climate is constant. Nothing could be further from the truth. If temperature and rainfall are averaged at a station for 30 years, then these values represent the norms for that center. The World Meteorological Organization (WMO) first defined the 'normal climate' as the average of observations for the years 1901–30 and then

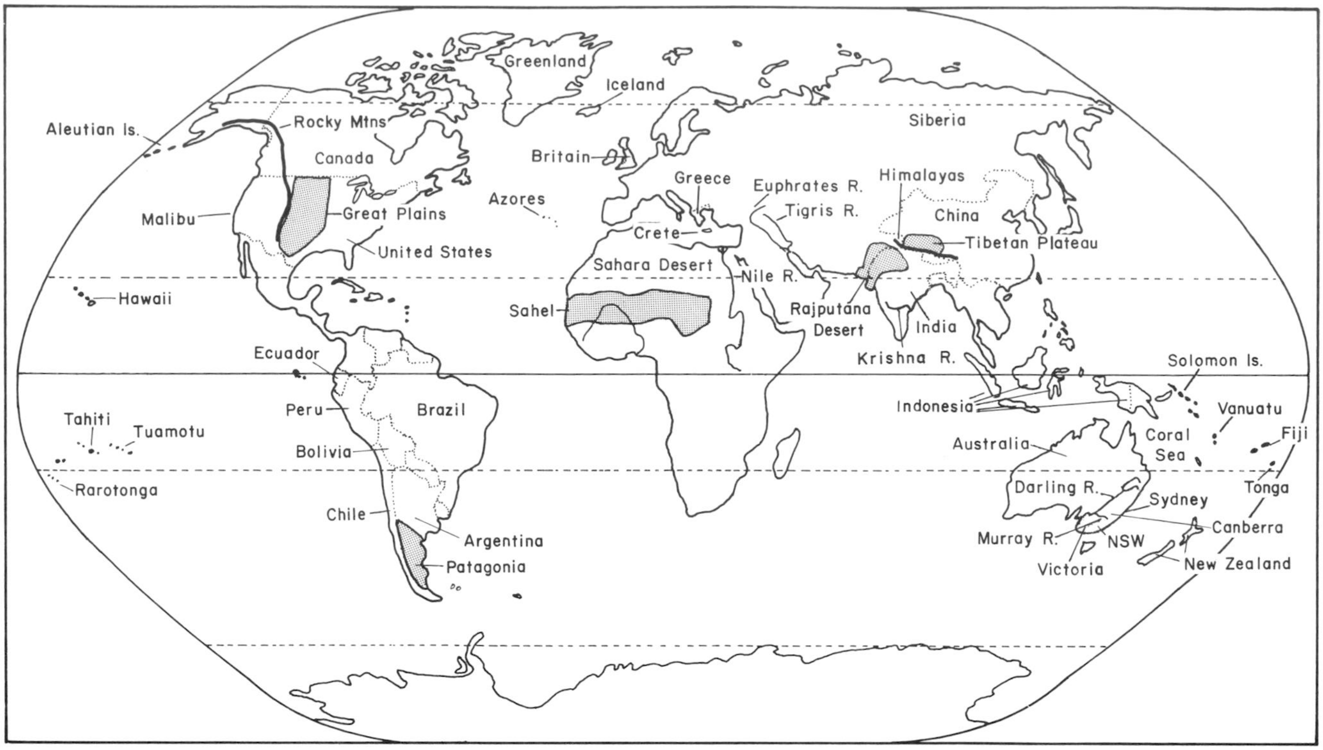

Fig. 5.1 Location map

1931–60. Many countries still follow this practice. Firstly, climate during these latter decades was very different from the preceding 30-year period or century. Secondly, climate within these decades has varied greatly. Figure 5.2 plots the mean change in rainfall for the period 1960–79 over the preceding decade 1950–59, using over 700 stations. Areas of decreased and increased rainfall are respectively cross-hatched and stippled in this diagram and coastlines having a change in rainfall regime >20 per cent are completely shaded. Changes depicted in this figure parallel trends for the northern hemisphere traceable as far back as 1931. There have been some dramatic shifts in rainfall since then. Most of Africa has become drier. In some places around the fringes of the Sahara desert rainfall has decreased by 50 per cent. Northwestern Europe, Japan and the east coast of the United States have become drier as well. Precipitation has increased throughout most of the Soviet Union, Australia, the eastern half of South America, the United States Great Plains and the southeastern portion of Africa (although in more recent years, 1980–85, this latter area has been struck by drought). Increases in the north and western parts of Australia exceed 20 per cent, while in Siberia they exceed 100 per cent. The point to be made is that societies and their agricultural systems may be adjusted to higher rainfall regimes, which are now dramatically changing. The Sahel region of Africa is one such region. Any attempt to redevelop pre-existing agricultural systems here following recent droughts may be totally fruitless. What may be required is a shift to different, more arid, agricultural practices.

This chapter will examine changes in precipitation regimes in terms of changes in general air circulation patterns and human land-use. These two factors mainly emphasize drought. The chapter will also evaluate the role of two additional factors, the Southern Oscillation and astronomical terms such as sunspot and the 18.6-year lunar cycles, which not only control the timing and intensity of drought and rainfall but also have predictive attributes.

Changes in circulation patterns

(Bryson and Murray, 1977; Gribbin 1978; Currie 1981; Lamb, 1982)

Regional climatic changes responsible for drought or flood must reflect changes in general air circulation patterns. Air movement throughout the

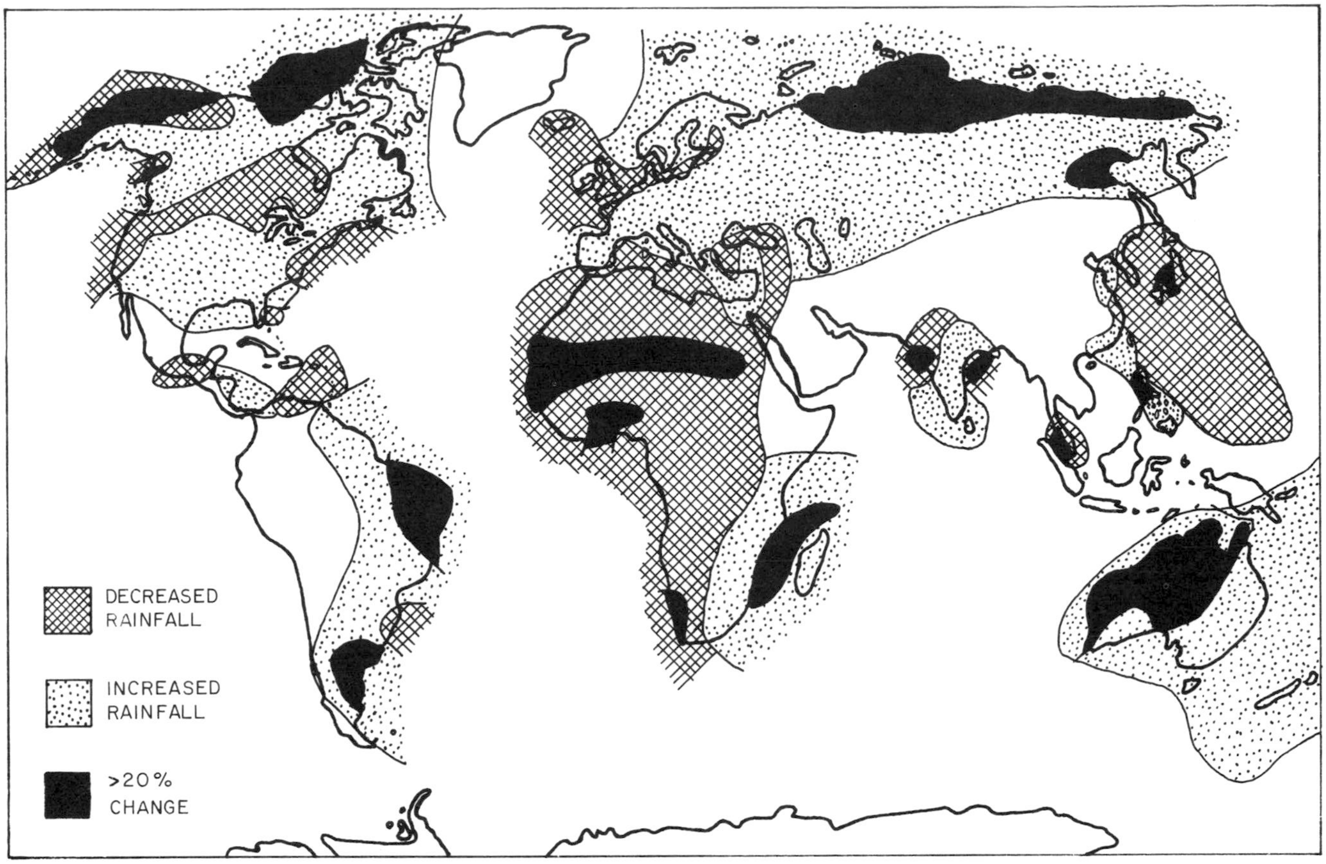

FIG. 5.2 Changes in mean annual world rainfall for the period 1960–79 compared to 1950–59 (Bryant, ©1985 reproduced with permission from Gebrüder Borntraeger, Stuttgart, West Germany)

atmosphere between the equator and poles was illustrated schematically in figure 2.3. Intense heating by the sun at the equator causes air to rise and spread out polewards in the upper troposphere. As this air moves toward the poles, it cools and begins to descend back to the Earth's surface at 20–30° North and South of the equator. Upon reaching the Earth's surface, this air either returns to the equator or moves polewards. Where air rises in this circulation, low air-pressure forms and there is intense instability and condensation of moisture, with the possibility of subsequent heavy rainfall. Where air descends, high air-pressure forms with intense evaporation, clear skies and stability. This tropical circulation forms Hadley cells that encircle the globe coincident with the great subtropical deserts.

At the pole, air cools and spreads towards the equator along the Earth's surface. Where this air meets warmer air from the tropics, a cold polar front develops with strong uplift and generation of extra-tropical depressions. Between the polar front and the Hadley cell strong westerly winds develop in both hemispheres. Controlling the position of the polar front is the polar jet stream in the upper troposphere. The path of this jet is not constant around the Earth. Instead, the jet stream and, with it, the westerlies tend to follow a looping path as they encircle the globe, especially in the northern hemisphere (figure 5.3). This looping, termed 'Rossby waves', is quasi-stationary and orographically fixed in place by the position of the Tibetan plateau and the Rocky Mountains. Usually the jet stream will originate downwind of the Tibetan plateau, cross China, move up over Japan and the north Pacific, and either through the center, or to the north of the Rocky Mountains. If a line plotting atmospheric pressure was drawn from Siberia, across the north Pacific to the center of North America, it would form a wave-like phenomenon of alternating high and low pressure that is stable over time (figure 5.4). This wave is defined in amplitude by the difference in pressure between the highs and low, and in position by the

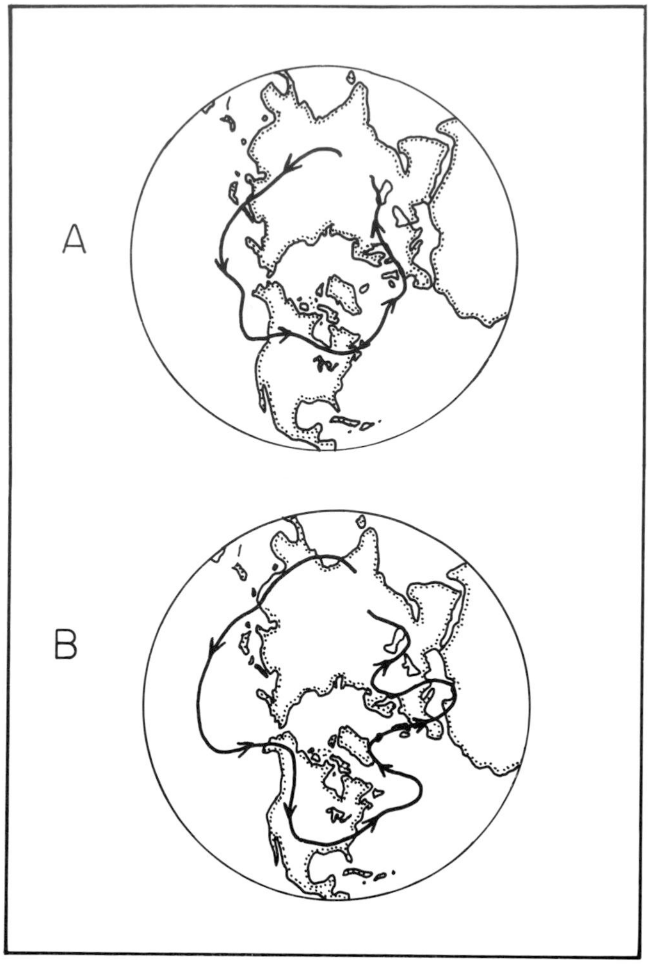

FIG. 5.3 Path of jet stream in the northern hemisphere showing likely Rossby wave patterns: A. for summer, B. for extreme winter (Bryson and Murray, ©1977, with the permission of The University of Wisconsin Press).

degree of looping. If the intensity of the pressure cells increases, then the degree of looping will also increase. As mentioned in the previous chapter, the position of the jet stream in figure 5.4 (and intensity of pressure cells) matches coastlines where sea-level has recently been changing the most in response to climatic forcing (figure 4.13). As will be shown later, the intensity of Rossby waves can be influenced by external factors.

An added feature of this jet-stream tracking is the fact that extra-tropical depressions tend to develop downwind of mountain systems along its path. For instance, China and the Great Plains of the United States receive significant amounts of precipitation as a result of mid-latitude cyclogenesis along the path of the jet stream. If the path of the jet stream moves too far south in either of these localities, then adjacent areas will come under the influence of high pressure, and will not receive cyclonic rainfall. Figure 5.3 also illustrates such a pattern globally, and is typical of extreme winter conditions wherever the jet stream loops too far south. Many researchers believe that changes in this looping are responsible not only for short-term drought (or rainfall) but also for semi-permanent climatic change in a region extending from China to Europe. Indeed, Bryson and Murray believe that the pattern can tie in with droughts in the Mediterranean and Africa. They also show that on the United States Great Plains, expansion of the westerlies or intensified looping correlates with the collapse of mid-western Mill Creek Indian culture about AD 1215–1380, and movement of the *boreal forest-tundra* line southwards in North America. At the same time, other Indian cultures in the American southwest also disappeared, the Greenland Viking colony died out, and the advancement of industrialization in Europe came to an abrupt halt as cooler weather set in. Similar movement of the westerlies could have led to collapse of the Mycenaean civilization in Greece and Crete around 1200 BC because of drought. The main point that Bryson and Murray make is that changing climatic patterns can persist for long periods, and that such patterns can mean drought conditions severe enough to destroy civilizations in certain areas. More recently, Bryson and Murray have also tried to link expansion of the westerlies with failure of monsoon rains, and with drought in the Sahara region in the early 1970s. In the Sahara region rain normally develops after intrusion of moist tropical air, in conjunction with seasonal movement of the inter-tropical convergence northwards over regions dominated by the Hadley cell. If the westerlies expand, then such movement of the inter-tropical convergence is impeded and drought conditions prevail.

The 1976 drought in Britain can also be linked to displacement of the jet stream and polar-front lows northwards, into the Arctic instead of across Scotland and Norway. This displacement carried westerly airstreams and associated precipitation northwards, leaving Britain under the influence of an extension of the Azores high-pressure system that was stable and rainfall-deficient. The Azores high was characterized by one other aspect that exacerbated the drought. It remained in the same location for several months, refusing to budge. It was such a process, termed 'blocking', that kept the jet stream locked into a northerly, drought-prolonging path over England. Blocking highs have

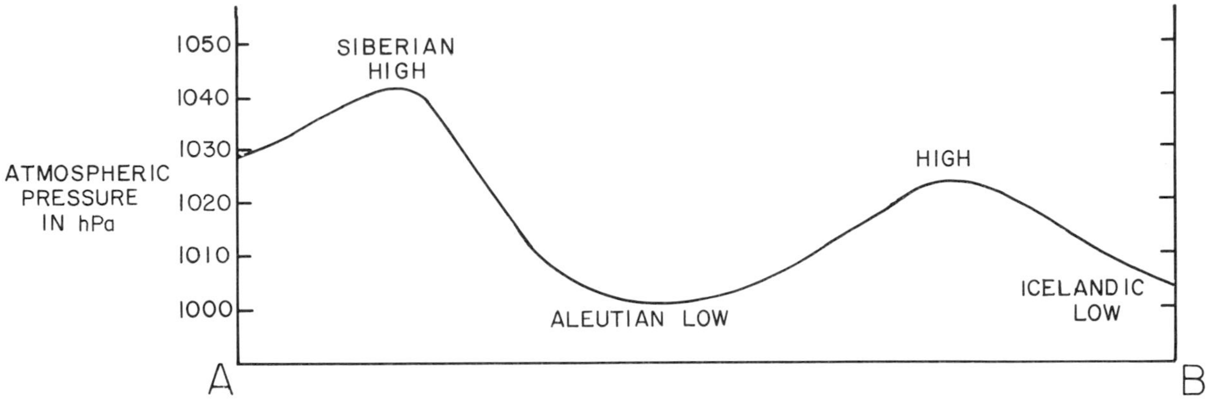

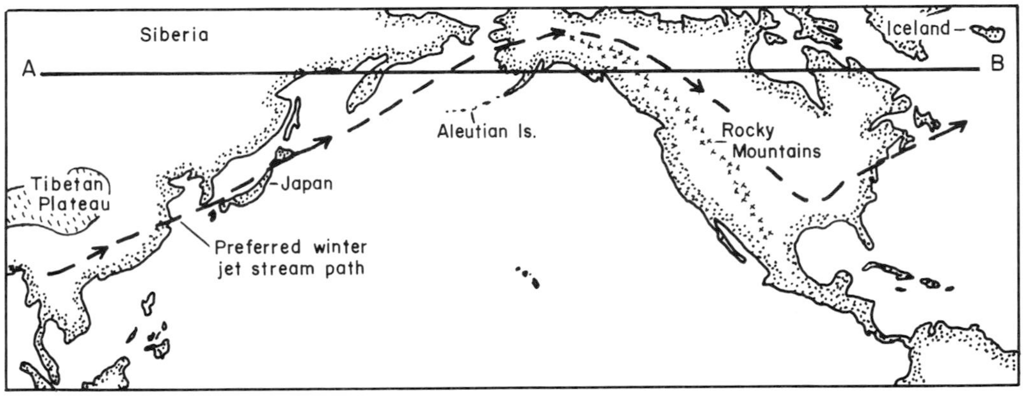

FIG. 5.4 Northern hemisphere winter pressure profile between Siberia and North America

also been responsible for drought elsewhere in the world. The 1977 drought in the mid-United States was exacerbated by a blocking high off the Californian coast that prevented rain-bearing westerlies from penetrating the continent. A similar type of blocking occurred in the 1982–83 drought in Australia. Here, the development of the high-pressure cell over the continent in winter deflected rain-bearing westerlies south of the continent. The resulting drought was extended into the summer by collapse of an important, tropical, *secondary air circulation* phenomenon—the Southern Oscillation.

Drought conditions exacerbated by humans

Rajputana desert (Indian subcontinent)

(Bryson and Murray, 1976)

In Chapter 3 dust storms on the Great Plains of the United States were viewed as more enhanced in the mid-1930s, 1950s and 1970s, because people had over-cropped marginal farmland as a response to increasing crop prices before drought periods. Such a relationship raises the prospect that the intensification of a drought, but not necessarily the timing, may be due to people's activities. Nowhere is this exemplified more than in the Rajputana desert in the Indian subcontinent, which appears to be an artificially made desert simply because of poor agricultural practices. Historically, the Rajputana area was one of the cradles of civilization, with a well-developed agrarian society based on irrigation around the Indus River. There have been successive cultures occupying the desert but, each time, they have collapsed. At present the area is still densely populated. The Rajputana area is affected by monsoons and much of the air is humid; however, rainfall generally amounts to less than 400 mm yr^{-1}. At the time of the development of civilization by the Harappan culture (4500 years BP) rainfall exceeded 600 mm yr^{-1}.

The atmosphere over the desert is presently very dusty. The dust is put into the air by light winds blowing across a land surface denuded of most vegetation through overgrazing. This dust rises throughout the atmosphere, and in the upper troposphere absorbs heat during the day. As a result, in daytime, the ground surface is slightly cooler than normal, resulting in less convection. At night, the dust gives off longwave radiation in the upper atmosphere, cools and sinks. At the ground, however, the dust entraps longwave radiation that would normally escape from the Earth's surface, causing the air above the surface to remain warmer than normal, thus preventing dew formation. The sinking air aloft leads to conditions of stability, while the lack of dew keeps the ground surface dry and friable, conditions favouring dust transport. The desert is self-maintaining as long as dust can get into the atmosphere.

If agricultural practices were changed and grasses were permitted to grow, then it is possible that the Rajputana desert would revert to a semi-arid landscape with a higher moisture regime. Undoubtedly droughts would still occur, but the effect of a drought on an area with a rainfall regime of 400 mm yr^{-1} is much more severe, than the effect on an area with a rainfall regime exceeding 600 mm yr^{-1}.

Sahel region (northern Africa) (Bryson and Murray, 1977; Glantz, 1977; Lockwood, 1986)

A similar process of desertification—whereby semi-arid areas are converted to arid environments—is occurring at present in the Sahel region of Africa. In fact, it is almost certain that human response to drought in the Sahel prolongs and intensifies the effects of drought. In wet years, such as the 1950s, the equatorial trough of low pressure associated with the inter-tropical convergence was displaced northwards into the Sahel region, concomitantly with a warming of tropical north Atlantic surface waters. These warm waters increased the water content of air masses travelling over the Sahara, and increased rainfall. During recent droughts, the inter-tropical convergence failed to move northward, and the adjacent ocean temperatures were cooler than normal. Globally, cooling or warming of ocean temperatures occurred synchronously in both hemispheres between 1900 and 1960. The coolest ocean temperatures occurred between 1905 and 1910, while the warmest ones existed in the 1940s. Beginning in 1960, fluctuations between hemispheres grew out of phase with the result that northern oceans, especially adjacent to western Africa, cooled relative to southern oceans. Air masses, travelling over the Sahel from the Atlantic Ocean, have not been able to evaporate as much moisture from these cool surface waters; consequently, rainfall amounts have decreased over land.

This decreasing precipitation generates a negative, bio-geophysical *feedback* mechanism: decreased annual rainfall reduces biomass growth, which leads to reduced evapotranspiration, decreased moisture content in the atmosphere, and a further decrease in rainfall. Over time, soil moisture slowly diminishes, adding to the reduction in evaporation and cloud cover. As the soil surface dries out and vegetation dies off, the surface *albedo*—the degree to which shortwave solar radiation is reflected from the surface of plants—is reduced, leading to an increase in ground heating and a rise in near-ground air temperatures. This process also negates precipitation. The destruction of vegetation exposes the ground to wind, permitting more dust in the atmosphere, a process that causes heating of air higher up in the troposphere. This has the effect of increasing the stability of the atmosphere, and reducing the possibility of dew formation at the ground surface at night. Once initiated, these mechanisms simply enhance droughts in the Sahel until rain-bearing, tropical air circulation patterns re-establish and terminate these negative feedback processes. This persistence in dry climatic conditions in the Sahel region is unusual compared to other semi-arid areas, where droughts rarely last longer than two to three years.

Rainfall has also decreased concomitantly with an increase in population. Certainly, human mismanagement of marginally arable land, under stress during drought, has exacerbated the process of desertification at the edges of the Sahara desert. In the Sahel agriculture is marginal to nomadic in character and highly dependent on the monsoon. At the beginning of the 1970s, when climatic conditions over the Sahel switched to drought, world fuel prices soared and people switched from kerosene to wood for cooking and heating purposes. Such a change led to rapid chopping down of shrubs and trees. Increased population also meant that there was increased pressure put on land to supply food. Traditionally, 35 per cent of land was left fallow; but in the 1970s and 1980s almost all arable land was continuously cultivated. Tall shrubs and trees, which increase the roughness layer of the atmosphere near the ground, have decreased in extent by 50 per cent in the 1980s in these regions. During increasingly harsher droughts, trees and shrubs have also been used for animal fodder. The atmosphere has become dustier, as winds near the ground have become more effective at removing fine sediment from grazed and cultivated land denuded of vegetation. Western techniques of ploughing, introduced to increase farming efficiency, have proven inappropriate in these regions. Such practices have destroyed soil structure, leading to the formation of surface crusts which increase runoff and prevent rain infiltration. All of these practices have enhanced the negative feedback mechanisms, favouring continued drought and desertification.

The spread of the Sahara today into once semi-arid, but arable land is on the order of 5–6 km yr^{-1}, although this rate has been recently challenged as exaggerated. Along the Niger River, that once flowed through a semi-arid landscape much like the Murray River in Australia, the desert has spread far to the south. Towns such as Tombouctou and Niamey now lie within the desert. Nomads and marginal farmers who used to seek refuge in the cities only during the most severe drought, now permanently occupy the cities. A similar process is occurring in the Sudan. At present, the feedback mechanisms continuing drought in the Sahel have not abated nor has tropical air circulation switched into a heavier rainfall cycle.

Australia (Holmes, 1976; Allison and Peck, 1987)

Western developed countries with technologically superior farming practices need not feel complacent about the processes of desertification in their own countries. In Australia, for example, over half the continent is semi-arid with a marginal, but extensive and efficient, agriculture industry. Native vegetation consists of shrubs and grasses specially adapted to an arid, moisture-stress regime. Here, plants do not store moisture, instead they conserve it. On a world basis, plants such as mallee, acacia, eucalyptus and saltbush have a high biomass for their environment. During extreme drought, most plants go into a dormancy period, dropping their leaves to provide a mulch around their root system, and to reduce soil evaporation. Such vegetation is also palatable to a native kangaroo population that responds quickly to droughts by reducing birth rates. This population is also kept in check through natural attrition by dingoes.

Humans have interfered with this ecosystem to a great extent. The reduction of the dingo population preserves the kangaroo population at pre-drought numbers, thus increasing competition with domestic stock for food during drier conditions. The extensive distribution of dams and ponds for stock watering permits native animals to survive into a drought. Introduction of feral animals—such as rabbits, pigs, goats, cats, cattle, water buffalo and donkeys—has replaced passive native species with more aggressive species. Many of these introduced animals have a hoof structure which compacts soil. Others, mainly goats, when pressed for food will eat the roots and branches of a plant, thus killing off vegetation that normally would regenerate after drought. Water buffalo have increased the siltation of watercourses and water holes, and reduced the drainage of saline groundwaters. The lifestyle of such animals means that in some areas it is impossible for new tree seedlings to germinate, let alone survive.

Western farming methods have also altered land use. The concept of a 99-year lease on most large properties has meant that graziers do not own their land, but use it for maximum profit. Many graziers faced with bankruptcy in the increasingly more difficult economic times of the 1970s and 1980s, retirement or the end of leases have overstocked properties for maximum return. Such policies have led to overgrazing and depletion of native saltbush and bluebush (*Kochia*), which provide the main source of food for livestock during droughts. Along with these grazing policies, wholesale ringbarking of trees has occurred, to such an extent that in many areas there are not enough parent trees to re-seed an area naturally. Older trees that remain are nearing the end of their lifespan with no replacement seedlings growing. The spread of dieback has severely reduced tree cover, which is necessary as a windbreak. More importantly, trees, through evapotranspiration, maintain a depressed water-table. While surface runoff increases after trees are removed, more water can also percolate from the surface to the water-table, which can lie up to 30 m below the surface. It only takes a small amount of water to raise the water-table substantially, because excess water is only filling the *voids* that exist in the soil matrix. The proportion of voids may be as low as 1 or 2 per cent the volume of the soil. In some places under a rainfall regime of 500–1200 mm yr^{-1}, water-tables are rising between 0.2–1.0 m yr^{-1}. Once raised, these water-tables do not fall nearly as much during subsequent drought. Since Australia possesses deeply weathered soil profiles several millions of years old, rising waters are also dissolving the chemically weathered residuals that have accreted over these timespans, and are thus becoming saline. When saline water-tables reach the surface, insoluble salts are precipitated, rendering the soil irreversibly infertile. The salt produces a scalding which is devoid of vegetation and susceptible to wind erosion. Soil structure is altered and surface crusting takes place, with the result that wetting during rain becomes more difficult. Throughout the Murray-Darling basin and Western Australia, groundwaters are already reaching the surface. In Victoria salinity affects about 4500 km^2 and is increasing by 2–5 per cent year, while in Western Australia 2500 km^2 of arable land has been abandoned. Large sections of the Murray River drainage basin are threatened in the next decade with this increasing problem, which is exacerbated by irrigation. Tree-planting programs are being carried out in the hopes of stemming rising water-tables, but in some places such as the New South Wales side of the Murray River, tree felling is still continuing on a large scale.

As a result, arable land in some agricultural river basins within Australia is being ruined at a rate ten times faster than it took Babylonian civilization to degrade the Euphrates-Tigris river valley, and five times faster than it took to change the Indus River floodplain into the Rajputana desert. Figure 5.5 schematically illustrates human effects in a small catchment area near Canberra, typical for Australia. Since the first settlement around 1820, people have cleared the catchment, imposing a different vegetation and introducing foreign practices such as fertilizers and ploughing. As a result of clearing, sediment movement into streams rapidly accelerated within 20 years, and has been maintained at rates five times larger than pre-European levels. Some 150 years after initial land disturbance, groundwater salinity has become evident. While existing land-use practices within Australia do not cause drought, they certainly exacerbate its effects because of the added stress placed on land and vegetation. At present, within westernized and underdeveloped countries, periodic droughts are often dramatized. In actual fact, it is the accumulating deterioration of land quality that needs much more publicizing.

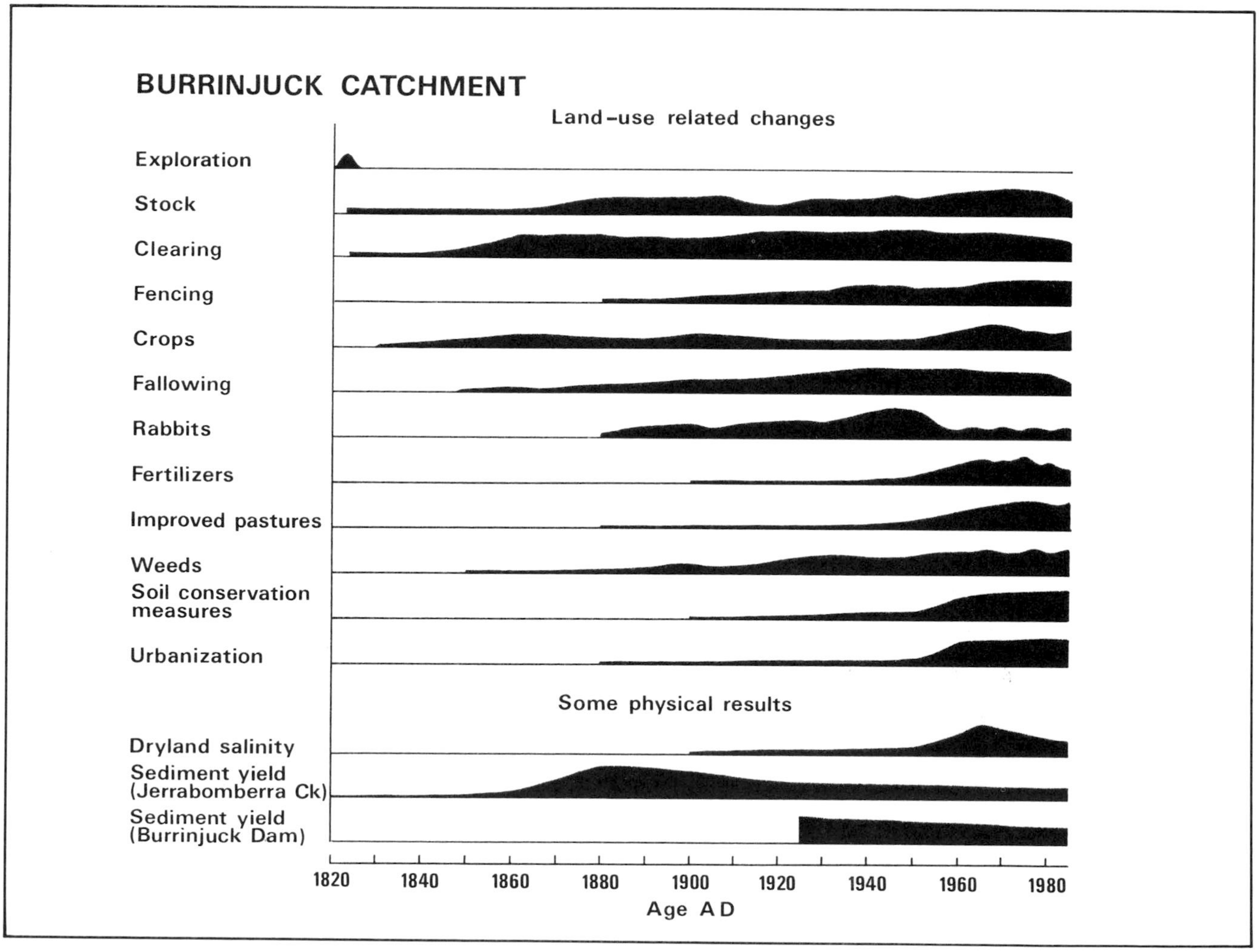

FIG. 5.5 Schematic representation of land-use change and natural response on a small catchment near Canberra, Australia, following human settlement (courtesy Dr Bob Wasson and Ms Wendy Gallagher, CSIRO, Canberra)

The Southern Oscillation

Introduction

The Earth's general atmospheric circulation in the tropics and subtropics, as shown above, is simply described. However, the actual scene is slightly more complex. The Hadley cells migrate annually with the apparent movement of the sun north and south of the equator. While the heating at the equator also tends to migrate about the equator with the seasons, there is a subtle, but important, shift in the most intense location of that heating. In the northern hemisphere summer, heating shifts from equatorial regions to the Indian mainland with the onset of the Indian monsoon. Air is sucked into the Indian subcontinent from adjacent oceans and landmasses, to return via upper air movement, either to southern Africa or the central Pacific. In the northern hemisphere winter this intense heating area shifts to the Indonesian-northern Australian 'maritime' continent (figure 5.6A), with air then moving in the upper troposphere, to the east Pacific. In both cases, an area of high pressure develops over the equatorial ocean west of South America as part of the Hadley cell. The high pressure off the Peruvian coast is intense, because cold water cools the air, causing it to descend. Air flows from this region, across the Pacific, as an easterly trade wind back to the the region of heating in the western Pacific Ocean. In doing so, the easterlies blow warm surface-water across the Pacific Ocean, where it piles up in the Coral Sea offshore from Australia. As a result, sea-level is slightly depressed off the South American coast and raised in the Coral Sea. The easterlies also cause upwelling of cold water along the South American coast. The pools of distinctly different temperature water form a positive feedback system between the atmosphere and

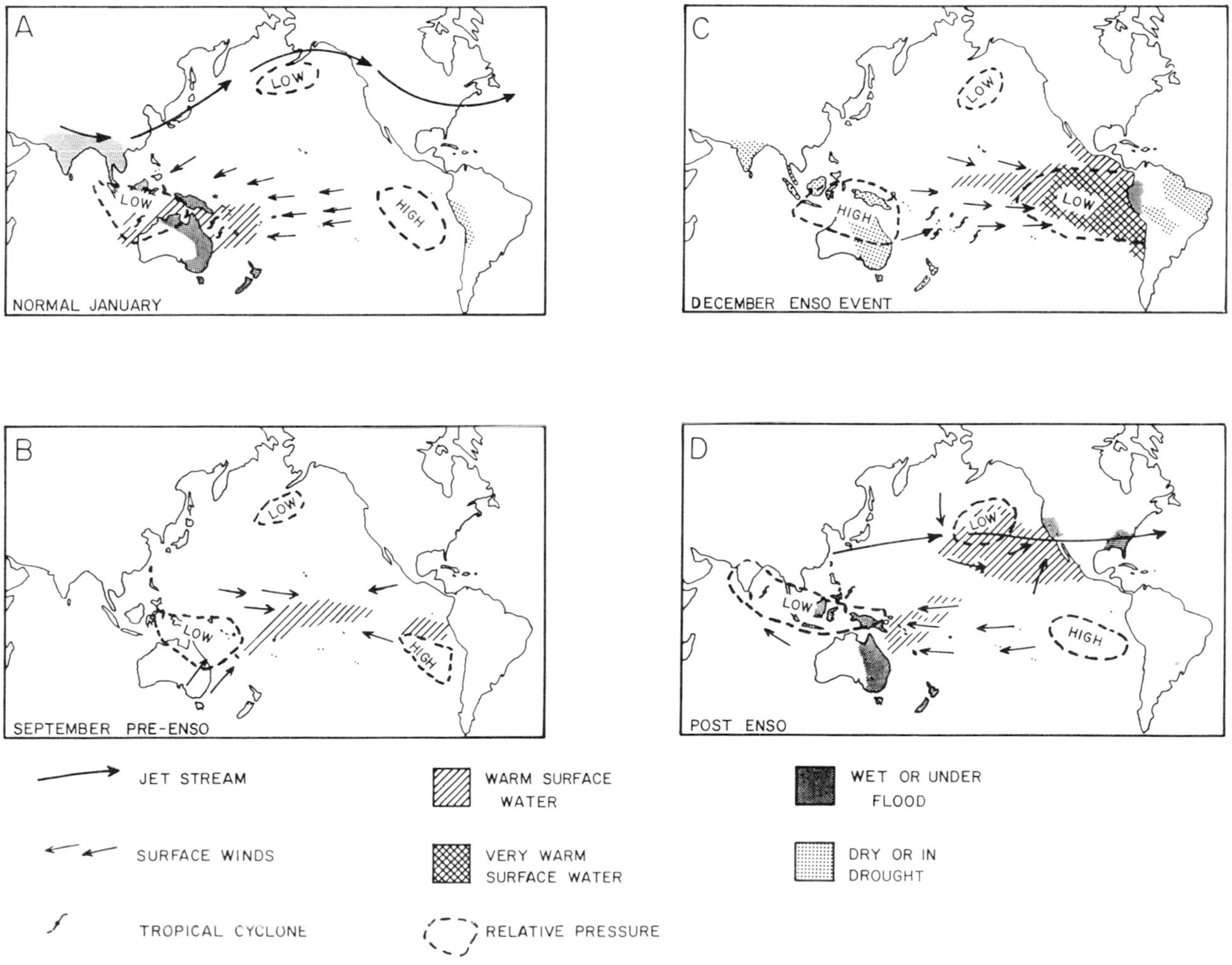

FIG. 5.6 Idealized representation of evolution and effects of an El Niño-Southern Oscillation event

ocean, reinforcing the east-west circulation. Cold water in the eastern Pacific Ocean creates high pressure that induces easterly air flow; this process causes upwelling of cold water along the coast, which cools the air. On the western Pacific side, easterlies pile up warm water, thus enhancing convective instability, causing air to rise, and perpetuating low pressure.

The system is very stable, existing beyond the annual climatic cycle. The Hadley cell off the Peruvian coast migrates only 5° seasonally, while the two centres of heating, and their associated low pressures, appear to be one and the same cell undergoing a small seasonal shift. In December–January the cell is narrow extending only across the Pacific; but in June–July it extends half-way around the globe from the east Pacific to the Indian subcontinent. The circulation cell is thus *zonal*, rather than *longitudinal* as with general air circulation. If the cell was imaged from space using a thermal sensor, it would appear red; in fact, some planetary scientists have suggested that the rising end of the circulation represents Earth's 'great red spot', analogous to the one on Jupiter. The easterly trade wind flow is called 'Walker circulation', after the Indian meteorologist who first identified it at the beginning of the twentieth century.

For some inexplicable reason, this quasi-stationary heating process weakens in intensity, or breaks down completely every two to seven years, frequencies very suggestive of a chaotic system. High pressure can become established over the Indonesian-Australian area, while low pressure develops over warm water off the South American coast. The easterly winds abate and are replaced by westerly winds in the tropics. The rainfall belt shifts to the central Pacific and drought replaces normal or heavy rainfall in Australia. Such conditions persisted in the Great World Drought of 1982–83, which affected most of Australia, Indonesia, India and South Africa. It is this tendency to fluctuate that is called the 'Southern Oscillation'. What is more important, extreme climate changes can occur worldwide as a result of these reversals. These climatic changes can be responsible for such minor occurrences as increased frequencies of snake bites in Montana, United States, and funnel-web spider bites in Sydney, Australia. More significantly, these changes can generate extreme economic and social repercussions, including the damaging of national economies, the fall of governments, and the deaths of thousands through starvation, storms and floods.

Southern Oscillation phenomena

El Niño-Southern Oscillation (ENSO) events (Horel and Wallace, 1981; Cane, 1983; Rasmusson and Wallace, 1983; Canby, 1984; Caviedes, 1984; Lockwood, 1984; Glantz, 1984; MacKenzie, 1987)

Zonal, oceanic-atmospheric feedback systems in the southern hemisphere are powered by three tropical heat-sources located over Africa, South America and the Indonesian-Australian maritime continent (figure 5.7A). The latter site initiates Walker circulation, by far the strongest of the three systems. While it is relatively stable, Walker circulation is the least sedentary, because it is not anchored directly to a landmass. It is highly migratory during transition periods of the Indian monsoon, in either March–April, or August–September. For instance, in the annual changeover from northern to southern hemisphere circulation, the rising part of the Walker circulation may be displaced eastwards towards the central Pacific. The reason for this displacement is not clear, but part of the answer may lie with the behaviour of southern hemisphere Hadley cells. It appears that strong westerlies between 35° S and 55° S precede stronger Walker circulation. These westerlies are generated by more intense, east-west orientated, Hadley cell development. Intensified Hadley cells appear to lock heating over the Indonesian-Australian maritime landmass. Weak or north-south orientated Hadley cells, especially over eastern Australia, permit the Walker circulation heat source to wander into the western Pacific. This situation causes southwest displacement of the South Pacific Convergence Zone as the Indonesian-Australian heating center moves eastwards. At this point, westerly air flow moves across the western edge of the Pacific ocean, such that there is no easterly wind-holding water at the western boundary of the Pacific. This mass of water begins to move eastwards, triggering the collapse of Walker circulation (figure 5.6B).

As warm water shifts eastwards, it shifts with it a zone of convective instability from the western Pacific, to normally drier central Pacific islands. Cyclones that usually form in the Coral Sea now ravage Tahiti, Tonga and Hawaii, which are mostly unaffected by such intense storms. The 1982–83 event caused widespread destruction in these areas, with Tonga receiving four tropical cyclones, each as severe as any single event this century. The Tuamotu archipelago, east of Tahiti, was devastated by five cyclones between January and April

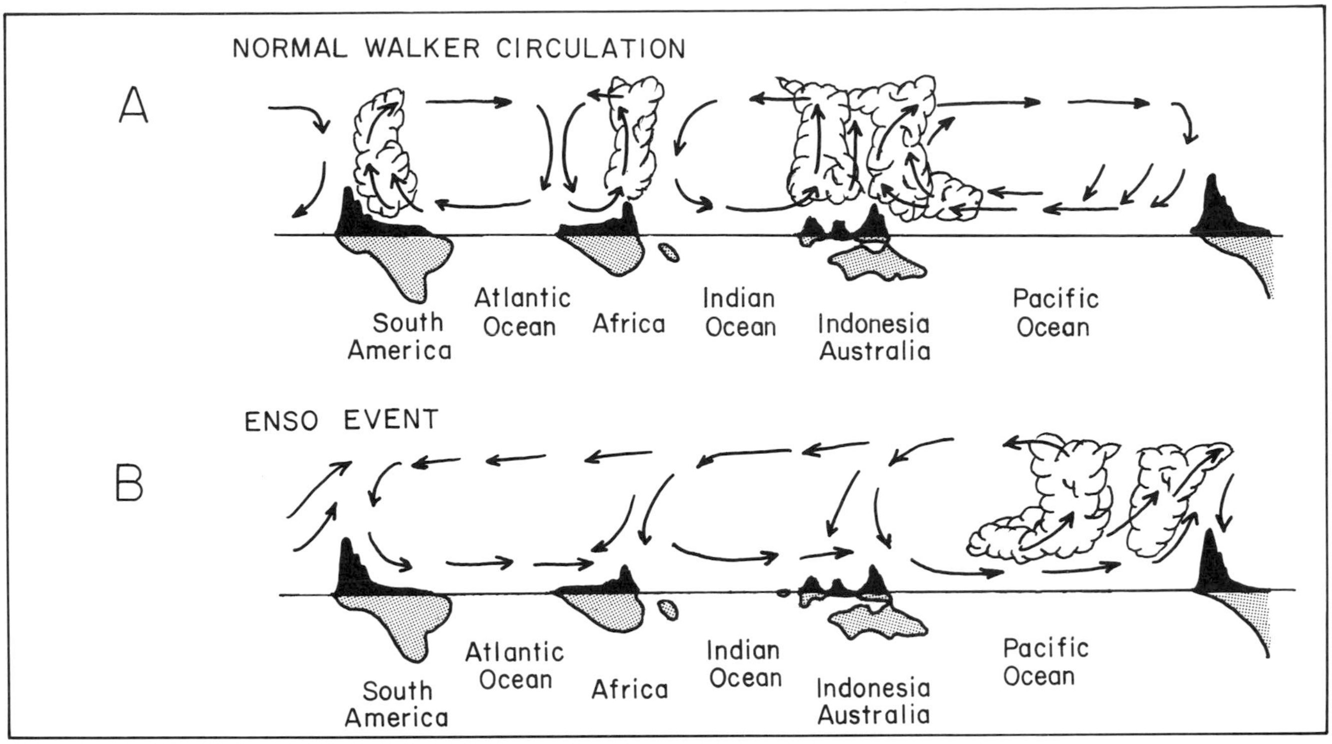

FIG. 5.7 Schematic representation of zonal air circulation mainly in the southern hemisphere (based on Mackenzie, 1987)
A. when the Southern Oscillation is 'turned on'—an exaggerated Walker circulation.
B. when the Southern Oscillation is 'turned off'—an ENSO event.

1983. Not since 1906 had a tropical cyclone occurred this far east in the tropics. The beginning of the 1986–87 event was also unusual. The breakdown of warm water began in March of 1986 at the end of the cyclone season. As the pool of warm water migrated eastwards, the last remnants of cyclone development followed the pool of water, but with no abatement in cyclone intensity. First, Fiji was struck by a cyclone and ten days of torrential rain that caused widespread damage to crops about to be harvested. The worst effects were reserved for the Solomon Islands, which lie to the northeast of cyclone development in the Coral Sea and, hence, rarely witness the effects of cyclones. Cyclone Namu, one of the latest cyclones ever recorded in the Australian region struck on 21 May. The storm surge swept all fishing boats inland and massive landslides occurred in mountain regions unaccustomed to such heavy rainfall. The slides swept down rivers and onto the coastal plains, obliterating villages and covering prime agricultural land and crops in mud and logs. The deathtoll was less than 200; but, of the 200 000 inhabitants on the islands 50 per cent were left homeless, and every building suffered major structural damage. With the shift of solar heating back to the southern hemisphere in September, this pool of warm water was reactivated, spawning a succession of tropical cyclones of devastating power. First Fiji, then Rarotonga in the southern Cook Islands, and finally Vanuatu—all experienced cyclones of unprecedented ferocity. In the latter case, Cyclone Uma paralleled the effects that Namu had wrought on the Solomon Islands: almost every building was destroyed in Vanuatu, over 100 people were killed and it left a damage bill of \$A250 million, \$A2000 for every person on the islands.

Accompanying this shift of warm water eastwards is the strengthening of westerly winds in the tropics, and the weakening of easterlies. This has two effects. Firstly, because the warm water which has piled up in the west Pacific Ocean has nothing to hold it there, it now continues to surge as a Kelvin wave across the Pacific to the eastern side within a single month. Secondly, sea-level increases in the east Pacific, and the cold water at the surface maintaining the high-pressure cell off the South American coast is swamped by a layer of warm water extending 50–100 m below the ocean surface (figure 5.6c). As low pressure replaces the high pressure over this warm water in the east Pacific, the easterlies fail altogether because there is no pressure difference across the Pacific to maintain them. Globally, the three heating cells of figure 5.7A give way to single cell that has only rising air motion over the east Pacific (figure 5.7B). Elsewhere, air flow is either stable or subsiding.

The appearance of warm water along the Peruvian coast usually occurs by Christmas. This warming occurs in a weakened form each year at this time, and is termed the *El Niño* (which, in Spanish, refers to the 'Christ Child'). However, when Walker circulation collapses, this annual warming becomes exaggerated, with sea-surface temperatures increasing 4–6° C above normal and remaining that way for several months. Until a few years ago, the El Niño appearance was thought to be the triggering mechanism for the climatic oscillation; however, it is now clear that most El Niño events appear several months after warming in the central Pacific has been initiated. Because extreme El Niño events are so well documented back to the 1700s, these climatic changes are now termed El Niño-Southern Oscillation or, for brevity, ENSO events. The warm water of the El Niño along the South American coastline forces fish feeding off plankton to migrate to greater depths, or causes vast fish kills. Guano-producing birds feeding off anchovy die in their thousands, and the anchovy and fertilizer industries in Peru collapse. The warm water, along what is normally an arid coastline, leads to abnormally heavy rainfall, causing floods in the arid Andes rainfall region, and drought in southern Peru, Bolivia and northeastern Brazil.

In 1982–83 in the coastal desert region of Ecuador and Peru, flash floods ripped out roads, bridges and oil pipelines. Towns were buried in mud, and shallow lakes appeared in the deserts. Agriculture based on irrigation was devastated, first by the flooding, and then by swarms of insects that proliferated under the wet conditions. Mean sea-level rose by 60 cm, subjecting low-lying communities to flooding and a pounding surf. Warm water coalesced from the central Pacific to South America and slowly spread into the northern Pacific. This movement of water can cause intensification of the winter low-pressure cell off the Aleutian islands in the north Pacific Ocean (figure 5.6D). As a result, larger than normal storms can develop, affecting the United States west coast. In the 1982–83 ENSO event, coastal storms wreaked havoc on the luxurious homes built along the Malibu coastline of California while associated rainfall caused widespread flooding and landslides. The storms were made all the more destructive because sea-levels were elevated by 25–30 cm along the complete United States west coast, from California to Washington State. The intense release of heat by moist air over the central Pacific Ocean causes the westerly jet-streams in the upper atmosphere to shift towards the equator. Any shift in the jet-stream path across the Rocky Mountains of America leads to dramatic changes in climate across the continent. The 1982–83 ENSO event caused heavy rainfall in the southern United States, record-breaking mild temperatures along the American east coast, and heavy snowfalls in the southern Rocky Mountains. In contrast however, the 1976–77 event caused record-breaking low temperatures and heavy snowfall in the eastern half of North America, because that event sent jet streams looping north of the Rocky Mountains into Arctic regions.

Finally, with the spreading out of warm water in the northern part of the Pacific, easterly air circulation begins to re-establish itself in the tropics. However, the return to Walker circulation can be anything but normal. The pool of warm water moves with the easterlies slowly back across the Pacific, dragging with it exceptional atmospheric instability, and a sudden return to rainfall that breaks droughts in the west Pacific. In 1983 drought broke first in New Zealand in February, one month later in eastern Australia with rainfalls of 200–400 mm in one week, and then over the next two to three months in India and southern Africa. With the collapse of the ENSO event, the Southern Oscillation appears to 'turn on' strong Walker circulation, leading to above-average rainfall and floods for the next year at the western end of the cell and in the other two heating cells shown in figure 5.7A. There is a strong *coherence* in rainfall over five-year periods between these tropical heat sources. In fact, rainfall, or the lack of it, is strongly in phase in an arc looping up the east coast of Africa, over the Indian subcontinent and down over Indonesia and Australia.

La Niña events

Exceptionally 'turned on' Walker circulation is now being recognized as a phenomenon in its own right. The cool water that develops off the South American coastline can drift northwards, and flood a 1–2° band around the equator in the central Pacific with water that may be as cold as 20° C. This phenomenon is termed *La Niña* (meaning, in Spanish, 'the girl') and peaks between ENSO events. The 1988–90 La Niña event appears to have been one of the strongest in 40-years and, because of enhanced easterly trades, brought record-breaking flooding in 1988 to the Sudan, Bangladesh and Thailand in the northern hemisphere summer. Regional flooding also occurred in China, Brazil and Indonesia. An unusual characteristic of this event is the late onset of flooding and heavy rain. Both the Bangladesh and Thailand flooding occurred at the end of the monsoon wet season. Bangladesh was also struck by a very intense tropical cyclone in December 1988 outside the normal cyclone season.

In eastern Australia this same La Niña event produced the greatest rainfall in 25 years, with continuation of extreme rainfall events. Significant damage was caused along a stretch of the coastal road between Wollongong and Sydney by more than 250 mm of rainfall on 30 April 1988. There, numerous landslides and washouts undermined the roadway and adjacent main railway line, leading to a total road repair cost of $A1.5 million along a 2-kilometer stretch of coastline (figure 5.8). Slippage problems were still in evidence here 15 months later. The Australian summers of 1989–90 not only witnessed record deluges but also flooding of eastern rivers on an unprecedented scale. Lake Eyre has filled in both years, whereas it had only filled twice in the previous century. Towns such as Gympie, Charleville (in Queensland) and Nyngan (in New South Wales) have been virtually erased from the map.

More significantly for North America, abnormally warm water appeared in 1988 between Hawaii and the west coast. In combination with the cold equatorial water, the jet stream over North America was shifted northwards, deflecting rain-bearing cyclonic depressions from the United States Great Plains to the Canadian Prairies. A stable, high-pressure cell developed over the Great Plains leading to severe, summer drought in the mid-west. This was followed by unusually cold winter temperatures. La Niña, thus, may be as important as ENSO events in controlling extremes in flood and drought over large sections of the globe.

FIG. 5.8 Damage to the Wollongong-Sydney coastal road caused by more than 250 mm of rainfall on 30 April 1988 (photographs by Bob Webb, courtesy of Mal Bilaniwskyj and Charles Owen, New South Wales Department of Main Roads, Wollongong and Bellambi offices). *Above*, fans deposited on the road by this rainfall event. *Below*, one of the many landslides and washouts caused by this same event.

Global long-term links to drought and floods (Quinn, et al., 1978; Pant and Parthasarathy, 1981; Bhalme et al., 1983; Pittock, 1984; Adamson et al., 1987)

Where the Southern Oscillation is effective in controlling climate, it is responsible for about 30 per cent of the variance in rainfall records. The effects of an ENSO event are widespread. Southern Peru, western Bolivia, Venezuela and northeastern Brazil are affected most in South America by the aseasonal lack of rainfall; however, greatest devastation occurs in the western part of the cell in southern Africa, India, Indonesia and Australia. Failure of the Indian monsoon in the late nineteenth century at times of ENSO events was responsible for the deaths of millions as a result of famine. Such events may even contribute to drought in the African Sahel region, especially in Ethiopia and Sudan. There is a strong coherence in the timing of both flood and drought on the Nile in Africa, the Krishna in central India and the Darling River in Australia. As well, rainfall coincides for parts of Australia, Argentina and Chile, New Zealand and southern Africa. An index of the intensity of fluctuations in the Southern

Oscillation can be simply derived by subtracting the barometric pressure over Darwin from that over Tahiti. Darwin is situated in the western part of the cell, while Tahiti lies basically at the edge of the eastern Hadley cell. Figure 5.9 presents this index as well as indices of drought and above-normal monsoons in India. Superimposed on these latter indices are the times when flows on the Nile, Krishna and Darling rivers were coincidentally high or low. When the Southern Oscillation index is positive (solid shading in figure 5.9), then the Walker circulation is 'turned on'. Widespread flooding in India has occurred at these times, especially in 1894, 1916–17, 1933–38, 1961, 1970–78. There is also a strong link between high Southern Oscillation index values and high discharges on the Nile, Krishna and Darling rivers. Most notable were the years 1894, 1910, 1917, 1961 and 1970. There is even a better correspondence between droughts in India and 'turned off' Walker circulation, or ENSO events. The severe droughts in India of 1899, 1905, 1918, 1951, 1966 and 1971 all occurred before, or at times when, El Niño events were at their peak around Christmas or when the Southern Oscillation index was negative. All of these years, except for 1951, were also times when discharges on the Nile, Krishna and Darling river systems were either very low, or non-existent.

A similar relationship can be traced back even further in Indonesia. This is surprising because most climatic classifications label Indonesia as a tropical country with consistent rainfall from year to year. Table 5.1 summarizes the relationship between east monsoon droughts and ENSO events in the period 1844–1983. In Indonesia, over 93 per cent of monsoon droughts are associated with El Niño events and 78 per cent of ENSO events coincide with failure of the monsoon. The relationships are so strong that in Java the advent of an ENSO event can be confidently used as a prognostic indicator of subsequent drought. In recent years, both India and Indonesia have established distribution infrastructures, which prevent major droughts from causing starvation on as large a scale as they once did; however, the 1982–83 event caused the Indonesian economy to stall as the country shifted foreign reserves towards the purchase of rice and wheat.

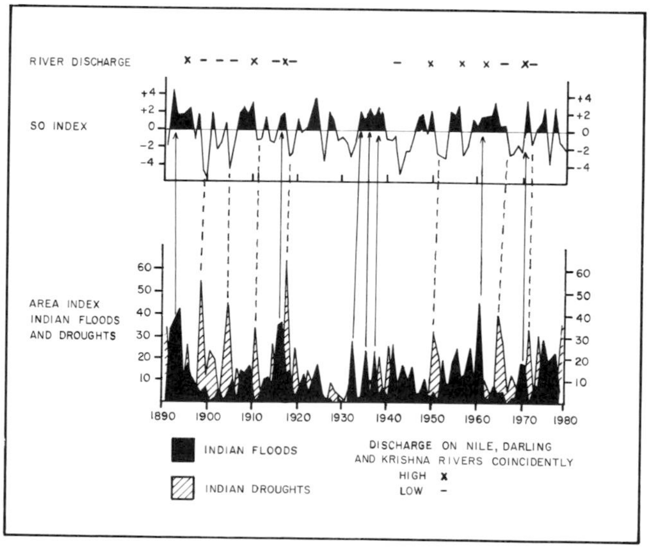

Fig. 5.9 Southern Oscillation index, drought and rainfall indices for India, and timing of coincident low or high discharges on the Nile, Darling and Krishna rivers: 1890–1980 (based on Bhalme et al., 1983; Adamson et al., 1987)

Links to other hazards (Gray, 1984; Marko et al., 1988)

The effect of ENSO events and the Southern Oscillation go far beyond just drought and flood. They are responsible for the timing of many other natural hazards that can influence the heart and fabric of society. As an example, not many countries witnessed such natural and sociological turmoil as Australia did during the 1982–83 ENSO event,

Table 5.1 Comparison of Droughts in Indonesia with El Niño-Southern Oscillation Events, 1844–1983

1844–96		*1902–83*	
Drought	*El Niño event*	*Drought*	*El Niño event*
1844	1844	1902	1902
1845	1845–46	1905	1905
1850	1850	1913–14	1914
1853	none	1918–19	1918–19
1855	1855	1923	1923
1857	1857	1925–26	1925–26
1864	1864	1929	1929–30
1873	1873	1932	1932
1875	1875	1935	none
1877	1877–78	1940	1939–40
1881	1880	1941	1941
1883	none	1944	1943–44
1884–85	1884–85	1945–46	1946
1888	1887–89	1953	1953
1891	1891	no data	1954–75
1896	1896	1976	1976
		1982–83	1982

Source: Quinn et al., 1978

which was the strongest in a century. This event developed between May 1982 and March 1983, coinciding with the worst drought Australia has experienced. Approximately $A2000 million was erased from the rural economic sector as sheep and cattle stock were decimated by the extreme aridity. Wheat harvests in New South Wales and Victoria were, respectively, 29 per cent and 16 per cent the average for the previous five years. The resulting lack of capital expenditure in the rural sector deepened what was already a severe recession, and drove farm machinery companies such as International Harvester and Massey-Ferguson into near-bankruptcy. By January, dust storms had blown thousands of tonnes of topsoil away in Victoria and New South Wales. Finally, in the last stroke, the Ash Wednesday bushfires of 16 February 1983, now viewed as one of the greatest natural conflagrations observed by humans, destroyed another $A1000 million worth of property in South Australia and Victoria. Just at the peak of the event, Malcolm Fraser's Liberal-Country Party coalition government called a national election. The timing could not have been worse, and the government of the day was swept convincingly from power.

The occurrence of other natural hazards at peaks in a fluctuating Southern Oscillation characterize hazards in Australia. Figure 5.10 presents the magnitude of the Southern Oscillation between 1851 and 1980. The index is an extension of the one used in figure 5.9. Shaded along this index are the times when eastern Australia was either in drought or experiencing abnormal rainfall. Sixty-eight per cent of the strong or moderate ENSO events between 1851 and 1974 have produced major droughts in eastern Australia. Sixty per cent of all recoveries after an ENSO occurrence have led to abnormal rainfall and floods. In more recent memory, the floods of January 1974 in eastern Australia were associated with the return to normal circulation after the 1972–73 ENSO event, while exceptional localized rainfalls in the Sydney-Wollongong area in 1984 followed the 1982–83 ENSO event.

Some indication of the geomorphic importance of the Southern Oscillation can be gained by looking at time series of several natural hazards in Australia. Figure 5.10 also includes the number of tropical cyclones in the Australian region between 1910 and 1980, described in chapter 2; the discharge of the Campaspe River between 1887 and 1964 flowing into the Murray River in the central north of Victoria; beach change at Stanwell Park, south of Sydney, between 1930 and 1980, described in chapter 4; rainfall this century at Helensburgh, south of Sydney; and the incidence of landslip in the Kiama area, south of Sydney, between 1864 and 1978. Drawn on the diagram are lines passing through all data at the times when the Southern Oscillation was strongly 'turned on'. At this time, tropical cyclones should be more frequent, the Campaspe River discharge high, Sydney rainfall heavy, beach erosion at Stanwell Park more severe, and Kiama landslip more frequent. The diagram indicates that 'turned on' Walker circulation results in this predictable geomorphic response. The 1933, 1951 and 1961 periods of strong Walker circulation show up in all five time series at both regional and localized levels. Much of the Kiama landslip occurred during positive phases of the Southern Oscillation, and it appears that about 33 per cent of the variance in the Stanwell Park beach-change record can be accounted for by the Southern Oscillation index, in combination with the previous season's rainfall and sea-levels. The reverse conditions apply during ENSO events, except for the frequency of tropical cyclones in the Australian region, which historically is not well correlated to ENSO events (see figure 2.10). This is owing to the fact that many cyclones are still counted within the Australian region even though they have shifted eastwards during these events. More importantly, the Southern Oscillation appears to switch months in advance of the arrival of the meteorological conditions driving these responses. For instance, the heavy rainfall in eastern Australia, beginning in the autumn of 1987, was predicted 12 months in advance. This rainfall appears to be substantiating the associations described above. Unfortunately, while the timing of rain can be forecast, the exact whereabouts or amounts which may fall still remain unpredictable because of the spatial variability of precipitation along the coast. Even during the worst of droughts, there have been areas of eastern Australia receiving normal rainfall.

Links to other hazards are also appearing globally as researchers rapidly investigate teleconnections between hazards and the Southern Oscillation. For instance, the duration and incidence of tropical cyclones in the equatorial Atlantic is profoundly depressed during ENSO years (see figure 2.11). For hurricane prevalence, upper tropospheric winds between 0° and 15° N must be easterly, while those between 20° N and 30° N must be westerly. Under these conditions, easterly wave depressions and disturbances are more likely to develop into tropical cyclones. During ENSO events, upper westerlies

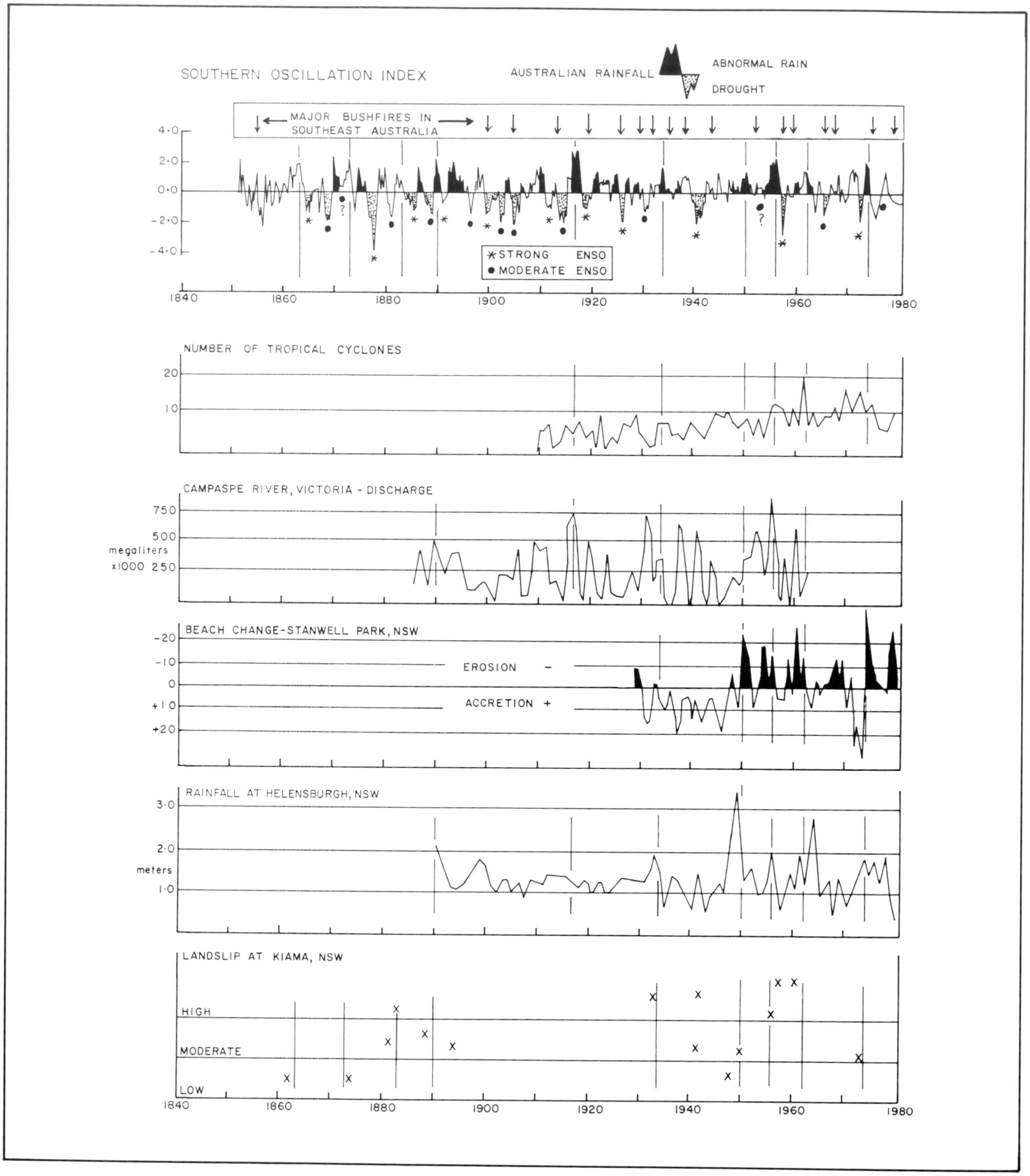

FIG. 5.10 Time series. These time series provide a comparison between the Southern Oscillation index, major bushfires (from Luke and McArthur, 1978), Australian cyclone frequency (from Lourensz, 1981), Campaspe River discharge (from Reichl, 1976), beach change at Stanwell Park (from Bryant, 1985), rainfall at Helensburgh near Sydney, and landslip incidence at Kiama (courtesy Jane Cook, Faculty of Education, University of Wollongong).

tend to dominate over the Caribbean and the western tropical Atlantic—conditions that suppress cyclone formation. Years leading up to an ENSO event have tended to produce the lowest number of tropical cyclone days in the Atlantic over the past century. Additionally since 1955, the number of icebergs passing south of 48° N has been highly correlated to the occurrence of ENSO events. ENSO events may trigger a concomitant response in the much weaker north Atlantic Oscillation involving the Icelandic Low and the Azores Hadley cell. A strengthened Icelandic Low enhances the number of icebergs. Alternatively, ENSO events may change wind-stress components along the North American east coast, thus exacerbating the production and southward movement of icebergs. Both the 1972–73 and 1982–83 ENSO events produced over 1500 icebergs south of 48° N, a number that dramatically contrasts with non-ENSO years, when less than 100 icebergs per year were recorded this far south.

Early detection of the onset of ENSO events can thus provide advance warning not only of drought and rainfall over a significantly large part of the globe but also of the occurrence of numerous other associated hazards. More significantly, the probability of the occurrence of ENSO events no longer appears to be as random as generally believed. Large sections of the globe, affected by highly variable precipitation, undergo periods of drought and rainfall that coincide to two very stable, astronomical phenomena—sunspot cycles and the 18.6 year M_N lunar cycles. These will now be described.

Sunspots and the 18.6-year lunar cycles

Description

Sunspots (Brandt, 1966; Giovanelli, 1984)

Sunspots are regions on the sun's surface of intense magnetic disturbance 100–1000 times the average for the sun. They are a product of disturbance in the sun's magnetic field. The spots themselves appear dark, because the magnetic field is so strong at that point that convection is inhibited and the spot cools by radiation emission. *Solar flares*, representing ejection of predominantly ionized hydrogen in the sun's atmosphere at speeds in excess of 1500 km s^{-1}, develop in sunspot regions. Flares enhance the solar wind that ordinarily consists of electrically neutral, ionized hydrogen (protons and electrons). Accompanying any solar flare is a pulse of electro-magnetic radiation that takes eight minutes to reach the Earth. This radiation, in the form of soft X-rays (0.2–1.0 nm wavelengths), interacts with the Earth's magnetic field increasing ionization in the lowest layer of the ionosphere at altitudes of 65 km. The enhanced solar wind arrives one to two days after this magnetic pulse, and distorts the magnetosphere, resulting in large, irregular, rapid worldwide disturbances in the Earth's geomagnetic field. Enhanced magnetic and ionic currents during these periods heat and expand the upper atmosphere. There are correlations between periods of solar activity called 'sunspot cycles', and upper atmospheric and climatic effects. For instance, it is known that thunderstorm and lightning activity increases worldwide during peak periods of sunspot activity (see chapter 7).

Sunspots generally occur in groups which are bipolar—that is, the first spots forming in any hemisphere of the sun are of one magnetic polarity, while subsequent spots are of the opposite polarity. This sequence has reversed polarity in the opposite hemisphere. Every 11.2 years, the polarity of a sequence within a single hemisphere reverses. For example, if sunspots developing in the northern hemisphere of the sun in the first 11-year cycle show a north-south polarity, they will evidence a south-north polarity in the following cycle. Thus, the sun has a 22-year magnetic cycle, consisting of two 11-year sunspot cycles.

Figure 5.11 plots the number of sunspots recorded since 1840 (as well as the Southern Oscillation index and the 18.6-year lunar cycle). It is evident that sunspots peak every 11 years; however every second peak is slightly higher. This is the 22-year cycle. Further inspection indicates that the peaks also become higher every 80–90 years. There is also a longer cycle of 180 years, which correlates with the 179-year alignment of planets. While the number of sunspots within this cycle may wax and wane—for instance, there was virtually no activity between 1650 to 1700 (called the *Maunder minimum*) at the peak of the *Little Ice Age*—the 11- and 22-year periodicity appears not to have changed since the Pre-Cambrian, 680 million years ago.

Climatic associations with the 90- and 180-year cycles are speculative, mainly because climatic time series have not been measured accurately for that length of time. Most researched links between climatic variables and sunspots deal with shorter 11- and 22-year cycles. How sunspot number affects climate over such long periods has not yet been established. While sunspots on a timescale of weeks can severely disturb the Earth's magnetic field,

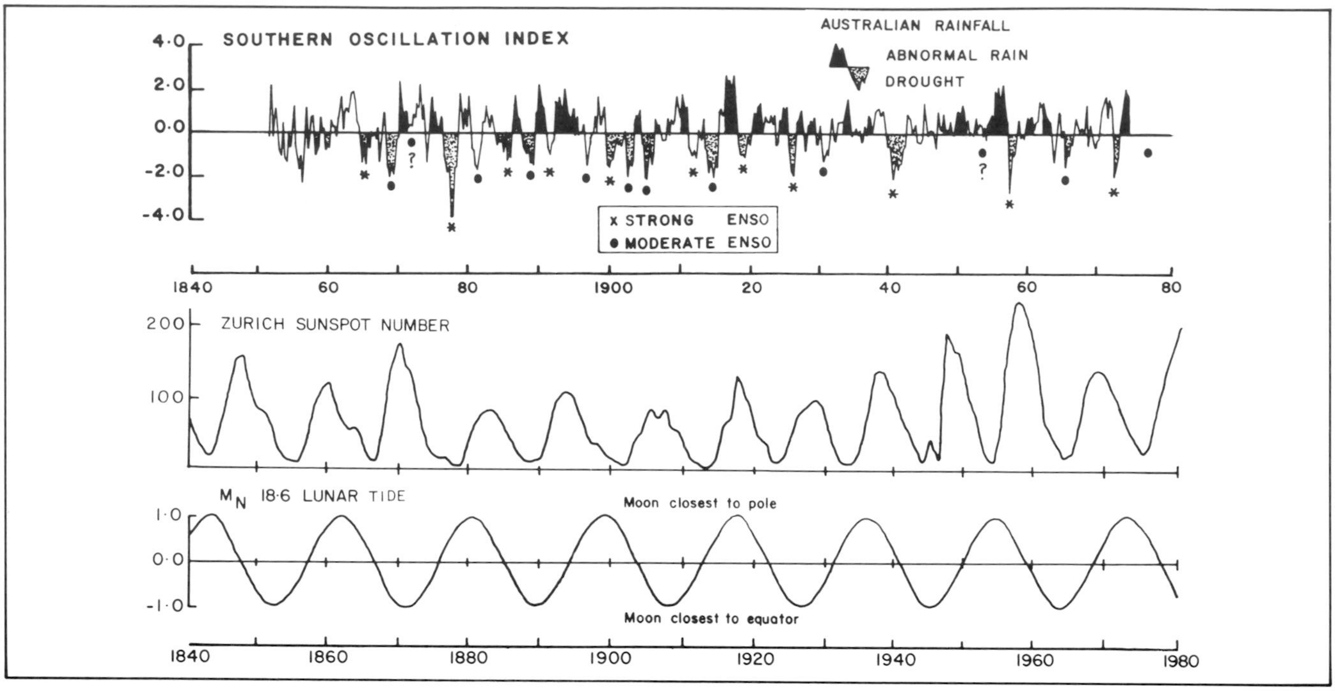

Fig. 5.11 Time series of the Southern Oscillation index, annual Zurich sunspot number and phases of the 18.6-year M_N lunar cycle: 1840–1980.

and affect the temperature structure of the *stratosphere*, there is no proven mechanism linking sunspots to climatic change at longer periodicities. This fact has tended to weaken the scientific credibility of research into the climatic effects of sunspot cycles. Unfortunately, many studies do not find exact 11- and 22-year cycles in climatic records, a fact that has discredited the research. Some of the latter problem undoubtedly rests with the fact that the 22-year cycle is close to another important astronomical cycle, the 18.6-year *M_N lunar cycle*.

The 18.6-year M_N lunar cycle (Currie, 1981, 1984)

The 18.6-year lunar cycle represents a fluctuation in the orbit of the moon. The moon's axis of orbit is related to the sun's equator, not to the Earth's. It forms a 5° angle with the sun's equator as shown in figure 5.12; however, the moon does not return to the same location relative to the sun each orbit. Instead, it moves a bit further in its orbit. The process is analogous to a plate spinning on a table. The plate may always be spinning at the same angle relative to the table, but the high point of the plate does not occur at the same point. Instead, it moves around in the direction of the spin. The moon's orbit does the same thing, such that 9.3 years later the high point of the orbit is at the opposite end of the solar equator, and 9.3 years

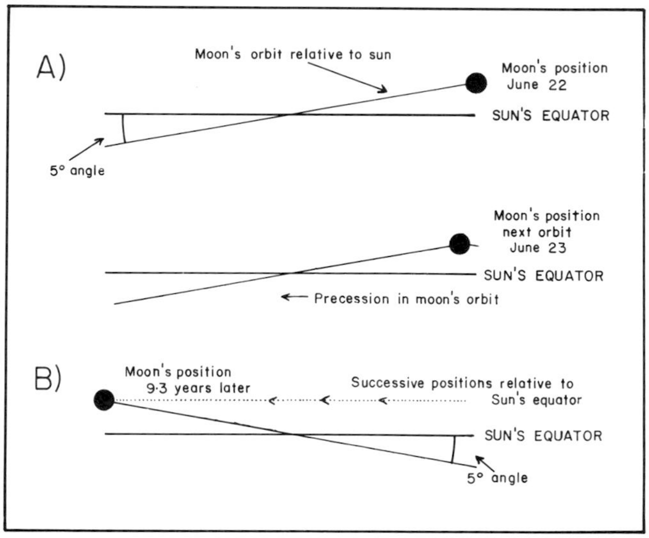

Fig. 5.12 Schematic diagram showing the precession of the Moon's orbit relative to the Sun's equator for: A. lunar maxima, and B. lunar minima.

after that it returns to its original position. Thus, there is an 18.6-year perturbation in the orbit of the moon.

This perturbation appears trivial until you consider the moon's orbit in relationship to the Earth (figure 5.13). Relative to the Earth, the solar equator moves seasonally, reaching a maximum of

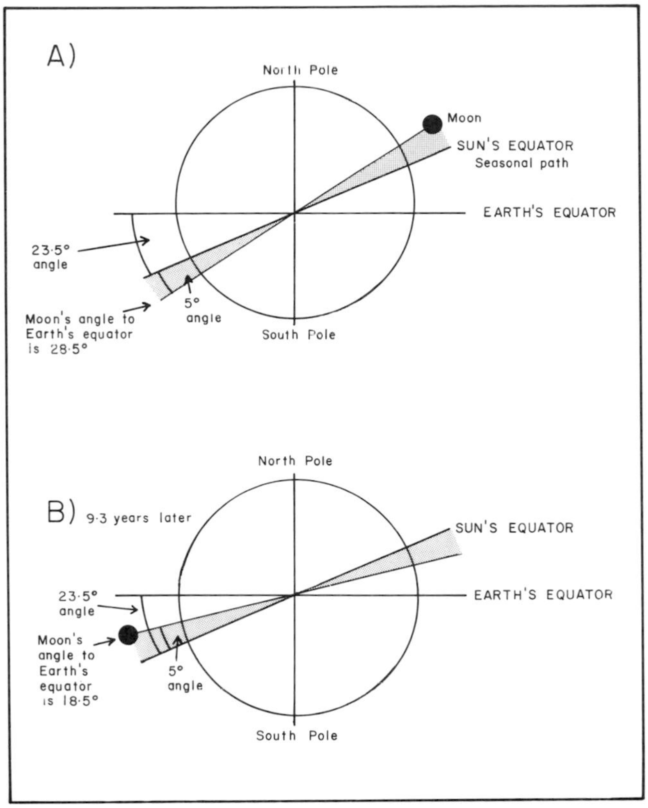

FIG. 5.13 Schematic presentation of the 18.6-year lunar orbit relative to the Earth for: A. lunar maxima, and B. lunar minima.

23.46° N of the equator on 22 June and a minimum of 23.46° S of the Earth's equator on 22 December. If the moon-sun orbital configuration in figure 5.12A is superimposed onto the Earth, then the moon is displaced in its orbit closest to the Earth's poles, 28.5° (23.5° + 5°) north and south of the equator. The moon's gravitational attraction on the Earth is slightly greater when the moon is displaced poleward. In 9.3 years' time, the moon will move to the opposite side of the sun's equator. Relative to the Earth, the moon will be closer to the equator, or in its minimum position (figure 5.12B). This fluctuation in the gravitational attraction of the moon follows an 18.6-year cycle. The fluctuation in the gravitation attraction of the moon throughout this cycle is only 3.7 per cent the daily effect; however, this is significant enough to be included in the prediction of tides for ports and harbours. Figure 5.11 plots the 18.6-year lunar cycle, which was last at its maximum phase in 1973. The next maximum will be in 1991, a crucial year for the occurrence of drought in the United States Great Plains.

Astronomical cycles and worldwide drought and rainfall (Tyson, et al., 1975; Wang and Zhao, 1981; Currie, 1981, 1984)

Detailed examination of the occurrence of drought on the United States Great Plains since 1800 has shown a strong link between drought, and either the 22-year sunspot cycle, or the 18.6-year M_N lunar cycle. Currie proposed that these droughts were induced by the influence of the Rocky Mountains on atmospheric tidal phenomena at maxima of the lunar 18.6-year tide. It would appear initially that the M_N tide is not strong enough to exert a noticeable effect on the atmosphere, which is 800 times less dense than ocean water. However, over time, the tide makes up for its lack of force because it has two to three years to act on any standing, planetary wave phenomena. Well-defined standing waves exist within the general air circulation. The most pronounced wave is the one shown earlier in this chapter (figure 5.3) locked orographically between the Tibetan plateau and Rocky Mountains. The lunar tide produces a resonance enhancement of this wave pattern that results in decreased precipitation in the lee of the Rocky Mountains. Also tied into the wave is the path of the westerly jet stream, which loops initially from the Tibetan plateau, across Japan and the northern United States, to Scandinavia. Throughout a band influenced by this path, and only within this band, M_N lunar and 11-year sunspot signals have been found consistently in sea level, air pressure and temperature data.

Astronomical cycles are not unique to United States mid-west rainfall records. A 20-year cycle is evident in the amount of rainfall in the rainy season of the Yangtze River of China that coincides with the timing of American mid-western drought. Statistically, significant (0.05 level) 22- and 11-year cycles in the frequency of droughts and floods have also been found for the whole of China since 1440. A 20–25-year cycle in drought is also present in the Dnieper basin from 1650 BC onwards while a 10–15-year cycle appears from AD 750 onwards. The Indian flood record presented in figure 5.8 evidences a 22-year cycle with peaks at sunspot maxima, while the drought record supports the 11-year sunspot cycle. In the southern hemisphere, 10- and 20-year periodicities appear in the occurrence of drought in, respectively, the yearly uniform and summer rainfall areas of southern Africa. Rainfall peaks in this 20-year cycle parallel the Indian situation, corresponding to the 18.6-year lunar maximum, while peaks in the 10-year cycle correspond to 11-year sunspot cycle minima. Astronomical cycles, where they appear in the above locations, account for about 15 per cent of the variance in rainfall records.

While the above results support a sunspot cycle association with drought and rainfall worldwide, Currie in fact shows conclusively, using a thorough and sensitive analysis of rainfall records that only the 18.6 M_N lunar tide and 11-year sunspot cycle are in fact dominant. Table 5.2 summarizes his results for the M_N tidal effect on rainfall in the western United States and Canada, northern China, India, the Nile region of Africa and mid-latitudes of South America. The table shows an excellent coincidence between maxima in the 18.6-year lunar tide, and either flood or drought for these locations. In all cases, except for India, the mean discrepancy between the occurrence of the hazard and peaks in the lunar tide is less than one year. Not only is the 18.6-year cycle dominant, but the data also evidence what Currie labels temporal and spatial *bistable phasing*. In this process, drought may coincide with maxima in the 18.6-year cycle for one period of time, but then switch suddenly to minima subsequently. Evidence suggests that bistable 'flip-flop' occurs every 100–300 years, with the last occurrence being the turn of the century in South America, China, Africa and India. In North America no bistable 'flip-flop' has occurred since 1657. Thus drought and flood timings between the United States Great Plains and the rest of the world are presently out of phase and have been since the turn of the century. For instance, Indian flood periods tend to occur at the same time as drought in the United States. Not only can phase differences exist between continents, but intra-continental differences can also exist. Thus, when drought is occurring on the United States High Plains, heavier than normal rainfall is falling to the north in British Columbia and Alberta.

These results pose an interesting scenario for the next peak in the 18.6-year lunar cycle around 1991. At this time, part of the United States wheatbelt will be entering drought. This could lead to a significant decrease in world grain production over several years. However wheat-growing areas of Argentina, China, India and Canada will be wet, and could make up the shortfall, only if the weather

TABLE 5.2 Timing of Flood and Drought in North America, Northern China, Patagonia, the Nile Valley, India, and the Coincidence with the 18.6-year Lunar Tide

Lunar Maxima	*Canadian Prairies*	*United States Great Plains*	*Northern China*	*Patagonian Andes*	*Nile Valley*	*India*
1583	1581D	–	1582D	–	–	–
1601	1601D	–	1600D	1606F	–	–
1620	1622D	–	1620D	1621F	–	–
1638	1640D	–	1640D	1637F	–	–
1657	1652F	–	1659D	1656F	–	–
1676	1674F	–	1678D	1675F	–	–
1694	1694F	–	–	1693F	–	–
1713	1712F	–	–	1710F	1713D	–
1731	1729F	–	–	1727F	1733D	–
1750	1752F	–	–	1752D	1750D	–
1768	1768F	–	–	1768D	1766D	–
1787	1786F	–	–	1784D	1789D	–
1806	1805F	1805D	1806F	1802D	1806D	–
1824	1823F	1824D	1823F	1822D	1822D	–
1843	1843F	1844D	1846D	1841D	1837D	–
1861	1859F	1861D	1862D	1864D	–	–
1880	1881F	1879D	1881D	1880D	1885D	–
1899	1900F	1901D	1900D	1895D	1902D	1902D
1918	1916F	1919D	1918D	1918F	1917F	1913F
1936	1932F	1935D	1935F	1937F	1936F	1939F
1955	–	1955D	1954F	–	1953F	1958F
1973	–	1975D	–	–	–	1976F

Source: From Currie, R. G., *Journal of Geophysical Research*, v. 89, 1984, p. 7216, copyrighted by the American Geophysical Union.
Note: D = drought; F = flood.

is not so wet that harvest yields are depressed. If they are, then the world could enter a period of significant grain shortage with associated high commodity prices and recessional economic effects at the beginning of the 1990s.

Astronomical cycles and the Southern Oscillation

Orographically controlled wave phenomena are not prevalent in the southern hemisphere; however, the Southern Oscillation is an oceanographically controlled, standing wave phenomenon that could also be affected by sunspot cycles or the 18.6-year lunar cycles. The fact that the intensity of the Indian monsoon and movement of the Hadley cell precedes the El Niño effect implicates changes in global atmospheric circulation as the triggering and/or enhancing mechanism for ENSO events. Because northern hemisphere circulation is related to 11-year sunspot and the 18.6-year cycles, there should be

1 an association between the Southern Oscillation and these astronomical cycles, and
2 a link between the two hemispheres in orographically and oceanographically induced wave phenomena.

The statistical relationships between these astronomical cycles and the Southern Oscillation were assessed by lagging the time series shown in figure 5.9 relative to each other, and then performing linear regression analysis. The procedure is termed 'lagged *cross-correlation* analysis'. Results were corrected for the fact that observations in the time series tend to be correlated to previous ones. The significant correlations can be summarized as follows:

1 on a hemisphere scale, there is a weak tendency for enhanced Walker circulation as sunspots decrease in number following peaks in the 22-year Hale sunspot cycle ($r = -0.189$ significant at the 0.05 level), and
2 in the tropics, there is a tendency for enhanced Walker circulation around maxima in the 18.6-year lunar cycle ($r = +0.196$).

While these results are promising they are not nearly as strong as correlations between the 18.6-year lunar cycle and rainfall found in the northern hemisphere. It would appear that the Southern Oscillation behaves as a truly chaotic system, a fact implying that the ultimate intensity of an ENSO event, while measurable, is virtually unpredictable in advance.

Concluding comments

While droughts and floods have plagued people throughout recorded history, and are responsible probably for the greatest loss of life of any natural hazard, it is only now that it is being realized that they represent two inseparable hazards that follow each other in time like night does day. Not only are the two events linked, but their occurrence is remarkably coincident across the globe. At no time was this more evident than in 1982–83, which witnessed the worst drought recorded in southern Africa and Australia, while at the same time bringing record flooding to the normally dry, eastern Pacific islands and parts of South America.

There is now indisputable proof that numerous droughts and subsequent heavy rainfall periods are cyclic, not only in semi-arid parts of the globe, but also in the temperate latitudes. This cyclicity is not random, but correlates surprisingly well with the 18.6-year lunar tide in such diverse regions as North America, Argentina, north China, the Nile valley and India. There is also evidence that the 11-year sunspot cycle is also present in rainfall time series in some countries. Onset of El Niño-Southern Oscillation events are detectable up to nine months in advance and are responsible for dramatic fluctuations in precipitation mainly within the southern hemisphere. Other hazards—such as tropical and extra-tropical cyclones, wave erosion, and land instability—are also correlated to the Southern Oscillation, and thus subject to prediction. The effect of sunspots and other solar activity upon climatic change is being treated so seriously that the International Geosphere-Biosphere Program, which is the largest coordinated scientific search into global climatic change ever mounted, has this topic as one of its major research areas for the 1990s.

Long-term astronomical periodicities, especially in the northern hemisphere, permit us to pinpoint the most likely year that drought or abnormal rainfall will occur, while the onset of an ENSO event with its global consequences permits us to predict climatic sequences up to two years in advance. These two prognostic indicators should give countries time to modify both long- and short-term economic strategies to negate the effects of drought or heavy rainfall. For instance, on the United States Great Plains the coincidence of drought with the 18.6-year lunar tide implies that the next drought will begin in 1991 and affect that country for the following two years. At the same time, other semi-arid regions of the world will more than likely be influenced by bountiful rains. While the latter prediction bodes well for world grain supplies, the two worst hazards that people can experience, drought and flood, may be occurring at the same time in different parts of the globe. Never before has it been possible to predict such a global picture in detail. Unfortunately, however, it should be realized that very few governments in the twentieth century have lasted longer than 11 years—the length of the smallest astronomical cycle. Realistically, positive political responses, even in the United States, to dire drought warnings are probably impossible. Societies must rely upon the efficiency and power of their permanent civil services to broadcast the warnings, to prepare the programs negating the effect of the hazards, and to pressure governments of the day to make adequate preparations.

While it appears that drought and heavy rainfall over large parts of the globe can now be forecast, there are significant areas of the globe where the predictions have not been as easily defined. For instance, the Sahel region of Africa has been drying out since 1970 because of changes in general air circulation, cooling of the adjacent Atlantic Ocean and people's misuse of marginally arable land. This situation represents a negative trend in rainfall upon which may be superimposed some astronomically controlled cyclicity in drought occurrence. Additionally, the Southern Oscillation, whose climatic imprint is detectable over 80 per cent of the globe, appears little affected by any astronomical cycle.

Finally, it should be realized that the astronomical cycles and the Southern Oscillation respectively account for at most 15 per cent and 30 per cent of the variance in rainfall amounts in countries where their effects have been defined. This means that 70 per cent or more of the variance in rainfall records must be due to other climatic factors. It is possible for heavy rainfall even during a drought. Consider the situation in Sydney, Australia, in August 1986. Five months previously, the onset of the 1986–87 ENSO event had been detected, and Australia should have been slowly drifting into drought. Instead, while warm water drifted eastward into the Pacific in early 1986, some of it returned into the Tasman Sea during the southern hemisphere winter. In the first week of August, Sydney was inundated with over 400 mm of rainfall, breaking the previous 48-hour rainfall record, and generating the largest storm waves to affect the coast in eight years.

References

Adamson, D., Williams, M. A. J. and Baxter, J. T. 1987. 'Complex late Quaternary alluvial history in the Nile, Murray-Darling, and Ganges basins: three river systems presently linked to the Southern Oscillation'. In Gardiner, V. (ed.) *International Geomorphology* Pt II, Wiley, NY, pp. 875–87.

Bhalme, H. N., Mooley, D. A. and Jadhav, S. K. 1983. 'Fluctuations in the Drought/Flood area over India and relationships with the Southern Oscillation'. *Monthly Weather Review* v. 111 pp. 86–94.

Bryant, E. A. 1985. Rainfall and beach erosion relationships, Stanwell Park, Australia, 1895–1980: worldwide implications for coastal erosion. *Zeitschrift für Geomorphologie Supplementband* v. 57 pp. 51–66.

Bryson, R. and Murray, T. 1977. *Climates of hunger.* Australian National University Press, Canberra.

Currie, R. G. 1984. 'Periodic (18.6-year) and cyclic (11-year) induced drought and flood in western North America'. *Journal Geophysical Research* v. 89 No. D5 pp. 7215–30.

Lourensz, R. S. 1981. *Tropical Cyclones in the Australian Region July 1909 to June 1980.* Bureau of Meteorology, AGPS, Canberra.

Luke, R. H. and McArthur, A. G. 1978. *Bushfires in Australia.* Australian Government Publishing Service, Canberra.

MacKenzie, D. 1987. 'How the Pacific drains the Nile'. *New Scientist* 16 April, pp. 16–17.

Quinn, W. H., Zopf, D. O., Short, K. S. and Kuo Yang, R. T. W. 1978. 'Historical trends and statistics of the southern oscillation, El Niño, and Indonesian droughts'. *U.S. Fisheries Bulletin* v. 76 pp. 663–78.

Reichl, P. 1976. 'The Riverine Plains of Northern Victoria'. In Holmes, J. H. (ed.) *Man and the environment: regional perspectives.* Longman, Hawthorne, pp. 69–95.

Further reading

Allison, G. B. and Peck, A. J. 1987. 'Man-induced hydrologic change in the Australian Environment'. *Bureau of Mineral Resources Geology and Geophysics Report* No. *282* pp. 29–34.

Brandt, J. C. 1966. *The sun and stars.* McGraw-Hill, NY.

Canby, T. Y. 1984. 'El Niño's ill wind'. *National Geographic* v. 165 No. 2 p. 143–83.

Cane, M. A. 1983. 'Oceanographic events during El Niño'. *Science* v. 222 pp. 1189–95.

Caviedes, C. N. 1984. 'Geography and the lessons from El Niño'. *Professional Geographer* v. 36 No. 4 pp. 428–36.

Currie, R. G. 1981. 'Evidence of 18.6 year M_N signal in temperature and drought conditions in N. America since 1800 A.D.' *Journal Geophysical Research* v. 86 p. 11055–64.

Giovanelli, R. G. 1984. *Secrets of the sun.* Cambridge University Press, Cambridge.

Glantz, M. H. 1977. *Desertification: environmental degradation in and around arid lands.* Westview, Boulder, Colorado.

——. 1984. 'Floods, fires and famine: is El Niño to blame?'. *Oceanus* v. 27 No. 2 pp. 14–20.

Gray, W. M. 1984. 'Atlantic seasonal hurricane frequency Part 1: El Niño and 30 mb Quasibiennial oscillation influences'. *Monthly Weather Review* v. 112 p. 1649–67.

Gribbin, J. 1978. *The Climatic Threat.* Fontana, Glasgow.

Holmes, J. H. 1976. 'Extensive grazing in Australia's dry Interior'. In Holmes, J. H. (ed.) *Man and the environment: regional perspectives.* Longman, Hawthorne, pp. 24–48.

Horel, J. D. and Wallace, J. M. 1981. 'Planetary scale atmospheric phenomena associated with the Southern Oscillation'. *Monthly Weather Review* v. 109 pp. 813–29.

Lamb, H. H. 1982. *Climate, history and the modern world.* Methuen, London.

Lockwood, J. G. 1984. 'The Southern Oscillation and El Niño'. *Progress in Physical Geography* v. 8 No. 1 pp. 102–110.

——. 1986. 'The causes of drought with particular referencc to the Sahel'. *Progress in Physical Geography* v. 10 No. 1 pp. 111–19.

Marko, J. R., Fissel, D. B. and Miller, J. D. 1988. 'Iceberg movement prediction off the Canadian east coast'. In El-Sabh, M. I. and Murty, T. S. (eds) *Natural and man-made hazards.* Reidel, Dordretch, pp. 435–62.

Pant, G. B. and Parthasarathy, B. 1981. Some aspects of an association between the Southern Oscillation and Indian summer monsoon. *Arch. Meteor. Geophys. Bioklim.* v. B29 pp. 245–52.

Pittock, A. B. 1984. 'On the reality, stability and usefulness of southern hemisphere teleconnections'. *Australian Meteorological Magazine* v. 32 no. 2 pp. 75–82.

Rasmusson, E. M. and Wallace, J. M. 1983. 'Meteorological aspects of the El Niño/Southern Oscillation'. *Science* v. 222 pp. 1195–202.

Tyson, P. D., Dyer, T. G. S. and Mametse, M. N. 1975. 'Secular changes in South African rainfall 1880–1972'. *Quarterly Journal Royal Meteorological Society* v. 101 pp. 817–33.

Wang, S-W. and Zhao, A-C. 1981. 'Droughts and floods in China 1440–1979'. In Wigley, T. M. L., Ingram, M. J. and Farmer, G. (eds) *Climate and history, studies in past climates and their impact on man.* Cambridge University Press, Cambridge, pp. 271–88.

6

Response to Droughts

Introduction (Garcia, 1972; French, 1983; Morren, 1983; Watts, 1983; Scott, 1984)

The previous chapter emphasized the fact that drought onset is aperiodic, slow, and insidious. As a result, communities in drought prone areas must be constantly prepared to insulate themselves from the probability of drought and finally, when overwhelmed, to ensure survival. However, a community's response to drought varies depending upon its social and economic structure. In nomadic settings, for instance, there is a heavy reliance upon social interaction as drought develops. The !Kung Bushmen in the the Kalahari Desert of southwest Africa congregate around permanent water holes only during the dry season (see figure 6.1 for the location of major African place-names). If conditions become wetter and more favourable for food gathering and hunting, they split into progressively smaller groups with as few as seven people. During severe droughts, however, the !Kung gather in groups of over 200 at two permanent water holes, where they attempt to sit out the dry conditions until the drought breaks.

In more agrarian societies there is a heavier reliance upon agricultural diversification to mitigate all but the severest drought. This is exemplified by pre-colonial Hausa society in northern Nigeria on the border of the Sahara desert. Before 1903, Hausan farmers rarely mono-cropped, but planted two or three different crops in the same field. Each crop had different requirements and tolerances to drought. This practice minimized risk and guaranteed some food production under almost any drought circumstance. Water conservation practices were also instigated as the need arose. If replanting after an early drought was required, then the spacing of plants was widened or fast-maturing cereals planted in case the drought persisted. Hausan farmers also planted crops suited to specific micro-environments. Floodplains were used for rice and sorghum, *interfluves* for tobacco or sorghum, and marginal land for dry season irrigation and cultivation of vegetables. Using these practices, the Hausan farmer could respond immediately to any change in the rainfall regime. Failing these measures, individuals could rely upon traditional kinship ties to borrow food should famine ensue. Requests for food from neighbours customarily could not be refused, and relatives working in cities were obligated to return money back to the community or, in desperate circumstances, were required to take in relatives from rural areas as refugees.

A resilient social structure had built up over several centuries before British colonial rule took over in 1903. Hausan farmers lived in a feudal Islamic society. Emirs ruling over local areas were at the top of the social pecking order. Drought was perceived as an aperiodic event where the role of kinship and descent-grouping generally insured that the risk from drought was diffused throughout society. There was little reliance upon a central authority, with the social structure giving collective security against drought. Taxes were collected in kind rather than cash, and varied depending upon the condition of the people. The emir might recycle this wealth in a local area during drought in the form of loans of grain. This type of social structure characterized many pre-colonial societies throughout Africa and Asia. Unfortunately, colonialism enhanced these agrarian societies' susceptibility to droughts.

After colonialism arrived for the Hausa people, taxes were collected in cash, remained fixed over time and went to a centralized government rather than a local emir. The emphasis upon cash led to cash cropping, which became dependent upon world commodity markets. Production shifted from subsistence grain farming, which directly supplied everyday food needs, to cash cropping, which only provided cash to buy food. If commodity prices fell on world markets, then people had to take out loans of money to buy grain during lean times. As a result, farmers became locked into a cycle of seasonal debt that accrued over time. This indebtedness and economic re-structuring caused the breakdown of group and communal sharing practices. The process of social disruption was abetted by central government intervention in the form of food aid to avert famine. Hausan peasant producers became increasingly vulnerable to even small variations in rainfall. A light harvest could signal a crisis of famine proportions, particularly if it occurred concomitantly with declining export prices. Rather than living under the threat of aperiodic famine caused by drought, the Hausan peasantry live under the threat of constant famine.

During drought a predictable sequence of events unfolded. Firstly, after a poor harvest farmers tried to generate income through labouring or craft activity. As the famine intensified, they sought relief from kin and subsequently began to dispose of assets. Further steps included borrowing of money or grain and, if all else failed, either selling out or emigrating out of the drought-afflicted region. With outward migration, governments noticed the drought and began organizing food relief. After the drought, the farm household had few resources to re-establish pre-existing lifestyles. Rich households in the city had not only the resources (the farmer's sold assets) to re-establish but also the resources to enable self-sufficiency after poor harvests. The natural hazard of drought now affected exclusively the Hausan farmer. In this process, society became differentiated such that the wealthy could beneficially control their response to drought, while the Hausan farmer became largely powerless and subjected to the worst effects of drought disaster. This sequence of events now characterizes drought response throughout Africa and other Third World countries, even though colonialism has disappeared. Dependency upon cash cropping, export commodity prices or the vagaries of war have repetitively witnessed large segments of various societies at the crisis point during major drought throughout the 1970s and 1980s.

In contrast, westernized societies rely heavily upon technology to survive a drought. This is exemplified in Australia; there, individual farmers, who bear the brunt of most droughts, have developed one of the most efficient, extensive agricultural systems in the world. Self-reliance is emphasized over kin or community ties, and it is efficiency in land management that ultimately determines a farmer's ability to survive a drought. As a drought takes hold, these practices are subtly altered to maximize return and minimize permanent damage to the land and to the farming unit. Where land is cultivated, land management practices are directed towards moisture conservation and crop protection. Practices promoting rainfall penetration into the subsoil include mulching, tillage and construction of contour banks and furrows that minimize runoff. In certain cases, catchments will be made bare and compacted to increase runoff into farm dams that can be used for irrigation during drought. Because most crops mature at times of highest moisture stress, the time of sowing is crucial to the final yield of a crop. Early sowing in winter ensures grain flowering and maximum growth before high water-stress conditions in summer. In addition, application of phosphate fertilizers can double crop yield for each millimeter of rainfall received. Where land is used for grazing, some pasture is set aside in spring so that hay can be cut to use as feed over a dry summer period. In mixed farming areas grain must also be held over from the previous year's harvest to supplement sheep and cattle feed. When a drought persists for several years, stock numbers are either dramatically culled, or breeding stock can be shipped considerable distances and grazed at a cost (agistment) in drought-free areas.

Technological advances also play a key role in farming drought-prone, semi-arid areas. These techniques include aerial spraying of herbicides to control weed growth, and aerial sowing of seed to prevent soil compaction by heavy farm machinery. Aerial monitoring is used to track livestock and give early warning of areas that are threatened by overgrazing. Computer management models are also utilized to maximize crop yields, given existing or forecast climate, moisture and crop conditions. Finally, efficient financial management becomes cruical as much in drought survival as in recovery. Most farmers aim for three profitable years per decade and must be able to sustain the economic viability of the farm for up to seven consecutive drought years. Such a mode of operation requires either cash reserves or a line of credit with a bank to permit breeding stock to be carried for this period of time. Until the rural crisis of the 1980s, less than 15 per cent of farmers in Australia entered financial difficulty following a major drought.

The above management practices set the Australian farmer apart from those in nomadic and underdeveloped countries. As a consequence, the Australian farmer can survive droughts as severe as any that have wracked the Sahel region of Africa in recent years. However, because of its efficiency, the Australian farm does not have to support the large number of people that units in Africa do. As a result, drought-induced famine in Africa can detrimentally disturb the complete social and economic fabric of a country, whereas in Australia or the United States, much of the population may be unaware that a major drought is in fact occurring. In the 1970s and 1980s the intensity of drought worldwide not only crippled many African countries; but, for the first time since the depression years of the 1930s, it also perturbed many westernized societies. Our response to drought in the 1980s involves, not isolated communities, but whole nations and ultimately the global community. The rest of this chapter concentrates upon these responses. Firstly, the divergent reactions or inactions of various nations to drought are described. While westernized countries have generally fared better than underdeveloped countries congregated mainly in Africa, it will be shown that national perception and policy ultimately determines success in drought mitigation. The chapter concludes by comparing the success of drought relief organized by world organizations, mainly the United Nations, to that of the Band-Aid appeal headed by Bob Geldof in response to the Ethiopian drought of 1983–84.

Droughts and national governments

African policy

Those that lose (Glantz, 1977; Bryson and Murray, 1977; Whittow, 1980; Hidore, 1983; Gribbin, 1983)

The Sahel area of Africa, which occupies the semi-arid margin south of the Sahara desert, has been subjected to a hemispheric change in air circulation linked to sea surface temperatures in the adjacent Atlantic since the 1960s. The process and consequences of this change were described in chapter 5. Its impact was first felt between 1968 and 1974, when a region centered around 15° N of the equator underwent drought from Mauritania to Ethiopia (figure 6.1). It is now known that these climatic changes caused subsequent Sahel droughts, and there appears to be no indication that wetter rainfall regimes will return in the near future. Some of the possible feedback mechanisms, exacerbated by changed human land-use practices because of the droughts, may be responsible for this persistence. With drought enduring in Ethiopia and the Sudan, it now appears that a number of Sahelian countries are in a losing battle—with desertification claiming land that was suited to agriculture twenty years ago. At present, the Sahara desert is spreading southward at the rate of 1–5 km yr^{-1}. As a result, the coping responses of people in the Sahel are either out-of-date, or inadequate to ensure the survival of populations under the influence of long-term drought.

One of the reasons that Sahelian countries have suffered so dramatically is the relative inaction of national governments in responding to drought. In many of these countries local groups have traditionally played a crucial role in minimizing the effects of drought. Farmers have always been able to adjust their mode of farming to increase the probability of some crop production, to diversify, to resort to drought-tolerant crops, or to respond instantly to changing soil and weather conditions. Nomads may resort to crops under favourable con-

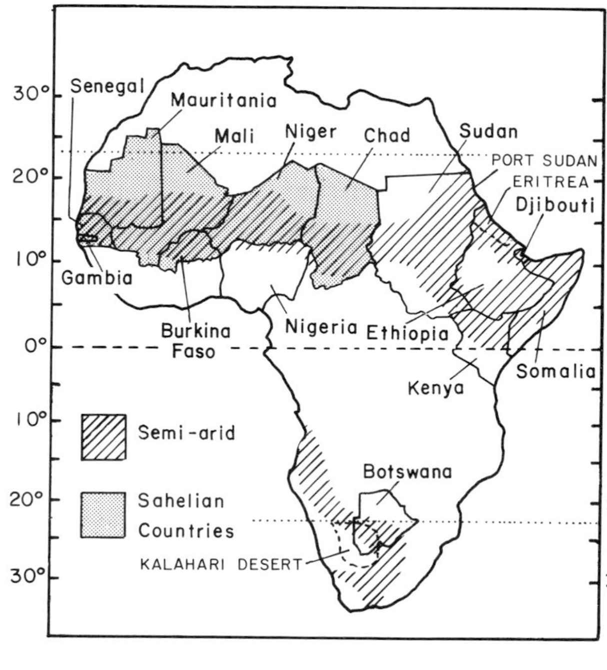

Fig. 6.1 Location map of semi-arid regions and the Sahel in Africa

ditions, follow regional migratory patterns to take advantage of spatially variable rainfall, or under extreme conditions move south to wetter climates. All groups have strong kinship ties, which permit a destitute family to fall back on friends or relatives in times of difficulty. There is very little national direction or control on this type of activity, especially where national governments are weak and ineffective. In many of these countries colonialism has weakened kin group ties and forced a reliance upon cash cropping. Westernization, especially since the Second World War, has brought improved medical care that has drastically reduced infant mortality and extended lifespans. Unfortunately, birth rates have not decreased and all countries are burdened with some of the highest growth rates in the world (3 per cent and 2.9 per cent for Sudan and Niger respectively in 1982). Technical advances in human medicine also apply to veterinary science, with the result that herd sizes grew to keep pace with population growth. Between 1924 and 1973 in the Sudan, livestock numbers increased fivefold abetted by international aid development and the reduction of cattle diseases. The government responded to these increases during past droughts by providing money for well-digging or dam construction. Otherwise, the national neglect was chronic. The primary economic activity of 80–90 per cent of people in the Sahel is subsistence farming. Almost 90 per cent of the population is illiterate, and four of the countries making up the Sahel are among the twelve poorest countries in the world, and getting poorer.

During the first series of droughts, the collapse of agriculture was so total that outward migration began immediately. By 1970, 3 million people had been displaced and needed emergency food in the west Sahel. The international response was minimal. When the drought hit Ethiopia, requests for aid to the central government were ignored. As the drought continued into 1971, large dust storms spread southward. The United States was responsible for 80 per cent of aid and Canada, a further 9 per cent, even though many of the afflicted countries had once been French colonies. The fifth year of the drought saw not only community groups but also national governments overwhelmed by the effects. Outward migration of nomads led to conflict with pastoralists over wells, dams and other watering areas. Inter-tribal conflicts increased. Migration to the cities swelled urban populations, leading to high unemployment and large refugee camps, which were often detested by locals. Migrations occurred beyond political (but not necessarily traditional) borders. Up to one-third of the population of Chad migrated across international borders. Chronic malnutrition and the accompanying diseases and complaints such as measles, cholera, smallpox, meningitis, dysentery, whooping cough, malaria, schistosomiasis, trypanosomiasis, poliomyelitis, hepatitis, pneumonia, tuberculosis and intestinal worms overwhelmed medical services. Chronic malnutrition usually results in death, but not necessarily by starvation. Lack of nutrition makes the body more prone to illness and disease and less able to recover.

The silence of aid efforts deafened. By 1972, every national government and international aid group knew of the magnitude of the disaster, but no plan for drought relief was instituted. The countries were either politically unimportant, or everyone believed the drought would end with the next rainy season. A survey of natural disasters named between 1968 and 1975, in the *New York Times*, contains not a single reference to the Sahelian drought. After 1972, national governments gave up and relied completely upon international aid to feed over 50 million people. Six west African countries were bankrupt. In the sixth year of the drought the first medical survey was carried out in the area. In Mali 70 per cent of children, mostly nomads, were found to be dying from malnutrition. In some countries food distribution had completely broken down and people were eating any vegetation available. Until 1973 only the United States was contributing food aid. The Food and Agricultural Organization (FAO) of the United Nations did not begin relief efforts until May 1973, the sixth year of the drought, when it established the Sahelian Trust Fund. The difficulties of transporting food aid were insurmountable. Transport infrastructures in Sahelian countries were either non-existent or antiquated. Seaports could not handle the unloading of such large volumes of food. The railroad inland from Dakar, Senegal, was over-committed by a factor of 5–10. Where the railways ended, road transport was often lacking or broken down. Eventually, camel caravans were pressed into service to transport grain to the most destitute areas. In some cases expensive airlifts were utilized. In 1973 only 50 per cent of the required grain got through; in 1974, the seventh year of the drought, this proportion rose to 75 per cent. In some cases, the aid was entirely ineffective. For instance, grain rotted in ports, found its way to the black market or was of low quality.

The environmental impact of the drought was dramatic. Grasslands, overgrazed by nomads moving out of the worst-hit areas, subsequently fell victim to wind deflation. The process of desertification accelerated along all margins of the Sahara. Between 1964 and 1974 the desert encroached southwards 150 km upon grazing land. And yet the lessons of this earlier drought were ignored. By the early 1980s, and despite the deathtoll of the early 1970s, the population of countries in the Sahel grew by 30–40 per cent. Five of the poorest nine countries now occupied the Sahel. Land degradation continued, with more wells being drilled to support the rebuilding and expansion of livestock herds. Governments could not control the population expansion, and relied heavily upon food imports to feed people even under favourable conditions. The 1983 drought was a repeat performance. The same script was dusted off, the same national and international responses were replayed. This time rather than beginning in the west and moving east, the drought began in Ethiopia and moved westward into the Sudan. If it had not been for a chance filming by a British television crew, the drought again would have gone unnoticed. This chance discovery evoked two international responses never seen before—the first attempt to implant drought aid in an area before a drought, and Bob Geldof. Both of these responses will be discussed at length towards the end of this chapter.

Besides the obvious point about weak, impoverished national governments being incapable or unwilling to deal with drought, there are two points to be made about the continuing drought in the Sahel region of Africa. Firstly, droughts in the Sahel have large temporal variability. It has already been emphasized, in chapter 5, that most droughts appear to coincide with the 18.6-year lunar cycle worldwide. What has not been emphasized is the fact that this lunar cycle may be superimposed upon some longer termed variation. The intense Sahelian droughts have now been present for two decades. Secondly, some developed countries may not have experienced their worst drought. For instance the 1982–83 drought in Australia, which represented the culmination of four years of below normal rainfall in the east, severely tested this nation's capacity economically to survive a drought. There is evidence from tree-ring studies in Australia that some droughts in the last two to three centuries have been as long as the present Sahelian conditions. Western countries should not be so ready to chastise the governments of Sahelian countries on their inability to deal with drought over the past two decades. Australia, and most likely other countries, have yet to be tested under similar conditions over such a long period—conditions which should be viewed as plausible under our present climatic regime.

Those that win (Charnock, 1986)

Given the continuing news from Africa about the persistence of drought, and the fact that so many people face a losing battle, the question arises whether any black African country can really cope with drought. There is one notable exception to the bleak news, namely what has been achieved in Botswana. Botswana, in 1986, was in its fifth consecutive year of drought—a record which easily matches that of any country in the Sahel. Yet, no one appeared to be dying from starvation, although two-thirds of its inhabitants were dependent upon drought relief. Botswana occupies the southern hemisphere equivalent of the Sahel region of Africa under the influence of the Hadley cell, at the edge of the Kalahari desert. Long term droughts have continually afflicted the country, with average rainfall varying erratically from 300 to 700 mm yr^{-1}. The country has become newly independent in the last two decades; but, unlike other decolonized African countries, it has an administration that is open to self-criticism, and a country which has not been wracked by inter-tribal conflict. After a single year of drought in 1980, the government commissioned an independent study on drought-relief efforts, and found measures severely wanting. Following that report, the government overhauled the storage and distribution network, to the point, that 90 per cent of the population requiring food aid during the 1986 drought year received uninterrupted supplies. The government also instituted a program of early warning that tracked the intensity of the drought in that country. Every month rainfall figures were collected and analyzed to produce a contour map that pinpointed those areas favourable for cropping and those areas in stress. In 1983 a computer model to predict soil moisture and maximum yield of crops was initiated. The model takes rainfall data and makes projections of crop yield until harvesting, based upon the premise of 'no stress' and 'continuing stress'. This model is updated with daily rainfall readings fed into a centralized collecting agency via two-way radio from all parts of the country. The model permits farmers to ascertain the risk to crops if they are planted now, the risk to crops if they are cultivated, and the risk to crops if harvesting is prolonged. Because soil moisture is calculated in the model, it is possible to determine the reserves in soil moisture from the previous crop year.

Botswana monitored the state not only of its agriculture, but also of its people. Each month, two-thirds of its children under the age of five were weighed at health clinics, to detect undernourished children with a body weight less than 80 per cent the expected weight for that age. Areas where drought effects are manifesting themselves in undernourishment can be pinpointed and reported to a central drought committee. This committee, with access to government channels, relays information about supplies back to local committees. The government firmly believes that the loss of rural income during drought, and its attendant unemployment, exacerbates the shortage of food. The government temporarily employed over 70 000 people in rural areas replacing one half of lost incomes. These programs were not in the form of food-for-work programs but operated strictly on a cash-for-work basis. Projects, selected by local committees, were usually drought mitigating. In addition, the drought program offered free seed, grants for land cultivation, subsidies for oxen hire, and mechanisms for selling drought-weakened animals at above market prices. Government-trained agricultural officers made efforts to enthuse the rural community, to provide the expertise to maintain farmers on the land during the drought, and to make certain that maximizing the farmers' potential does not result in denigrating the available arable land. The emphasis of this program was on keeping the rural population on the land, and keeping them basically self-reliant at existing living standards. As of 1986, the only factors which might jeopardise this program were growing national apathy and a 3.9 per cent annual growth rate in population.

Western societies

Societies that expect drought: the United States
(Rosenberg and Wilhite, 1983; Warrick, 1983)

The above examples of national response to drought would imply that underdeveloped countries have great difficulty in coping with drought conditions. Some countries such as Ethiopia even go so far as to ignore the plight of victims and deny to the outside world that they are undergoing a major calamity. The western world should not sit smug just because our societies are developed to a higher technical level. Often, by delving into history, it becomes evident that western countries have responded to drought in the exact same manner that some Third World countries do today. The national response to drought is well documented in the United States. In the previous chapter, it was pointed out that there was a well-defined cyclicity to drought occurrence on the Great Plains, synchronous with the 18.6-year lunar cycle. Major droughts have occurred in the 1890s, 1910s, 1930s, 1950s, and 1970s. The next drought will occur in the early 1990s. Most people have a vivid picture of the 1930s drought with mass outward migration of destitute farmers from a barren wind-swept landscape. However, the 1930s drought was not the worst to occur. Before the 1890s, the Great Plains were opened up to cultivation on a large scale. Settlement was encouraged via massive advertising campaigns sponsored by governments and private railway companies in the eastern cities and abroad. During the 1890s drought, widespread starvation and malnutrition occurred in the central and southern High Plains. Similar conditions accompanied the drought in the 1910s in the Dakotas and eastern Montana. There was no disaster relief in the form of food or money from the federal government. During the 1890s drought, many state governments refused to acknowledge either the drought, or the settler's plight, because they were trying to foster an image of prosperity in order to attract more migrants. In both the 1890s and 1910s, mass outward migration took place with settlers simply abandoning their farms. In some counties the migration was total. Those that stayed behind had to shoulder the burden individually, or with modest support from their own, or neighbouring states. The similarity between responses to these two droughts and to the 1980s Ethiopian one is striking.

The 1930s drought is notable simply because the government began to respond to drought conditions on the Great Plains. Outward migration was not as massive as some people think, but merely followed a pattern that had been established in the preceding years. Many over-capitalized farmers, as their finances collapsed, simply liquidated or signed over their farms to creditors. At a national level, the 1930s drought had little impact. It received massive federal aid but this was minor in comparison to the money being injected into the depressed American economy. There were no food shortages (in fact, there was a surplus of food production), no rising costs and few cases of starvation or malnutrition. For the first time the government assisted farmers so that by 1936, 89 per cent of farmers in some counties were receiving federal funds. The aid was not in the form of food, but cash. The government undertook a number of programs to aid farmers staying on the land. Irrigation projects were financed, government-sponsored crop insurance initiated, and a Soil Conservation Service established. All programs were designed to rescue farmers, re-establish agricultural land, and mitigate the effect of future droughts. The Botswana initiative of the 1980s is very similar to this stage begun in the 1930s in the United States.

The 1950s drought was totally different. The programs begun in the 1930s were accelerated. There was a sixfold increase in irrigated land between the 1930s and the 1980s. The amount of land under crop insurance increased, while marginal land was tucked away under the Soil Bank program. In the 1950s few farms were abandoned: they simply sold off to neighbours to form larger economies of scale. People selling their properties retired into adjacent towns. Government aid poured into the region, but at a lesser scale than in the previous drought. Again, there was little effect upon the national economy. The government responded by initiating more technical programs. Irrigation and river-basin projects were started, and weather prediction, control and modification programs researched and carried out. The Great Plains Conservation Program established mechanisms for recharging groundwater, increasing runoff, minimizing evaporation, desalinizing soil and minimizing leakage from irrigation canals. Since the 1950s, these programs have increased in complexity, to the point that drought now has little impact on the regional or national economies.

It would appear that, in the United States, the effect of drought on the Great Plains has been dramatically reduced to a stage unachievable at present in most African countries. The evolution of drought-reduction strategies indicates an increasing commitment to greater social organization and technological sophistication. Hence, the local and regional impacts of recurrent droughts has diminished over time; however, the potential for catastrophe from rarer events may have increased. The latter effect is due to the reliance placed on technology to mitigate the drought. If the social organization and technology cannot cope with a drought disaster, then the consequences to the nation may be very large indeed. Whereas droughts at the turn of the century had to be borne by individuals, and had little effect outside the region, society today in the United States is so structured that the failure of existing technology to minimize a drought will severely disrupt the social and economic fabric of that nation. A ripple effect will also permeate its way into the international community, because many Third World countries are reliant upon the Great Plains breadbasket to make up for shortfalls if their own food production should fail. This international effect will only be exacerbated if droughts occur worldwide at the same time. At present this appears unlikely because the timing of drought on the Great Plains is out of sequence with that in other continents. It is conceivable that Australia will not be affected by widespread drought in the early 1990s, when the United States enters its next mid-western drought. If this drought is severe enough to diminish American crop yields, Australia may be able to reap economic gain by taking up the slack in American grain exports.

Societies that don't expect drought: the United Kingdom (Whittow, 1980; Morren, 1983; Young, 1983)

The United Kingdom is not usually associated with drought; however, the south and southeast sections of England have annual rainfalls of less than 500 mm. In 1976 England was hit with 16 months of the worst recorded drought in its history. For the previous five years, conditions had become drier, as the rain-bearing low-pressure cells, which usually tracked across Scotland and Norway, shifted northwards into the Arctic. They were replaced with unusually persistent high-pressure cells that often blocked over the British Isles, and directed the northward shift of the lows. By the summer of 1976, many farmers were affected by poor rainfalls and temperatures which uncharacteristically soared into the 30s. Domestic water supplies were hard hit as wells dried up. By the end of July water tankers were carrying water to some towns throughout the United Kingdom, and total water restrictions were brought into force. In some cases municipal water supplies were simply turned off to conserve water. In mid-August, at the height of the drought, the river Thames ceased to flow above tide limits. By September there were calls for national programs to mitigate the worsening situation. When the government finally shifted into action, the drought had broken. By November of 1976, most of the country had returned to normal rainfall conditions.

The drought was severe, not because of a lack of rain, but because of an increase in competition for available water supplies that in some parts of the country had always been traditionally low. Since the Industrial Revolution, public priority ensured water supplies to cities and industry. Commercial water projects were common, and where they were not available, councils were empowered to supply water. The Water Act of 1945 encouraged amalgamation of local authorities and nationalization of private companies. A severe drought in 1959 resulted in the Water Resources Act of 1963, which established river authorities to administer national water resources, and a Water Resources Board to act in the national interest. By 1973 there were still 180 private companies supplying water, and the Water Act of 1973 set up ten regional water authorities to take control of all matters relating to water resources. In addition, a centralized planning unit and research centre were set up. It would appear that the United Kingdom should have been well prepared for the drought; but, despite this legislation, per capita consumption of water had increased as standards of living increased. Watermains had been extended into rural, as well as suburban areas, so that agriculture was reliant upon mains supply for irrigation and supplementary watering.

As the drought heightened in June of 1976, regional water authorities complained that their legal responses were either too weak or too extreme. Either they could ban non-essential uses of water such as lawn watering by ordinary citizens, or they could go to the other extreme and cut off domestic water supplies and supply water from street hydrants. They had no control on industrial or agricultural usage whatsoever. Except in south Wales, industry at no time had water usage curtailed drastically; on the other hand, watermains in many places were shut down, and special wardens had to be appointed to open up watermains whenever a fire broke out. In August the national government gave the water authorities more powers through the Drought Act. The authorities were free to respond to conditions as they saw fit in their areas. Publicity campaigns managed to get domestic consumers, who were the largest water users, to decrease consumption by 25 per cent. Because there were virtually no water meters in the United Kingdom, it was not possible to reduce consumption by imposing a limit on the volume of water used at the individual household level, and charging heavily for usage beyond that limit. Smaller towns had their water shut off, but at no time was there any question that London would have its water supply reduced. When canals and rivers dried up, there was concern that the clay linings would desiccate and crack, leading to expensive repairs to prevent future leakage or even flooding. Ground subsidence on clay soils became a major problem, with damage estimated at close to £50 million. This latter aspect was one of the costliest consequences of the drought.

The agriculture sector was badly affected and treated. Grain and livestock farmers were disadvantaged by the drier weather from the onset, and got little assistance. Irrigation farmers found their water supplies cut in half, just when they needed water most at the time of highest moisture stress. Britain's agriculture had become so dependent upon mains water supply, that it virtually lost all flexibility in responding to the drought. This was not just a problem with the 1976 drought but had been a growing concern throughtout the 1960s. Fortunately, because of Britain's entry into the European Common Market, the agricultural sector was insulated economically against the full consequences of the drought. The early part of the summer saw production surpluses, but at no time did agricultural production for the domestic market decrease to unbearable levels. In fact, many farmers had responded early to the drought and had increased acreages to make up for decreased yields.

The government's response to the drought was minimal. It only began to act at the beginning of August 1976, giving Water Authorities additional powers through the Drought Act and looking in detail at the plight of farmers. By October, there was considerable debate in parliament about subsidies and food pricing, but in fact the drought was over. The drought of 1976 cannot be dismissed as inconsequential; it did disrupt people's lives at all levels of society. The consequences were inevitable given the change in water usage in the United Kingdom, especially the dependence of agriculture upon mains supplies and the domestic increases in water consumption. The national government was poorly prepared for the drought and still has not taken steps to rationalize water management in the United Kingdom to prepare for the inevitability of future droughts. This case illustrates two poorly realized aspects about drought. Firstly, drought is ubiquitous: all countries can be affected by exceptional aridity. Drought is as likely in the United Kingdom (or Indonesia or Canada) as it is in the Sahel, given the right conditions. Secondly, the national response to drought in westernized countries is often similar to that of many Third World countries. The main difference is the fact that in the Third World governments are inactive, usually because they are not able economically to develop mechanisms of drought mitigation and alleviation, while governments in the Westernized countries like the United Kingdom (or other western European countries which also experienced the

1976 drought) are not prepared to accept the fact that droughts can happen.

Laissez-faire: the Australian policy (Waring, 1976; Tai, 1983)

Australia is a country plagued by drought, where an efficient, grazing and cultivation system has been developed to cope with a semi-arid environment. Droughts rarely occur at the same time across the continent. Only the 1895–1903 drought affected the whole country. Most Australian farmers—being subtly, if not overtly, financially adapted to survive on three successful years of production per decade—can experience drought conditions for seven out of ten years with no farm profit, and yet remain viable. This expectation makes it difficult to separate dry years from drought years. The 1982–83 drought, which was the worst in Australian history, was preceded by several dry years in the eastern States. Even during the severest times in 1982, the federal government was slow to instigate drought relief schemes. This inability to perceive the farmer's plight was one of the reasons the Liberal-National Party government of Malcolm Fraser lost power in the February 1983 federal election. Response to drought in Australia, thus, can be broken into two distinct categories. The first line of defense occurs with the individual farmer's response as already discussed above. Secondly, when the farmer's ability to profitably manage a property fails, then the State and federal governments attempt to mitigate a drought's worst effects, and ensure that quick economic recovery is possible when the drought breaks. Unfortunately, this response is too piecemeal and at times politically motivated.

One of the first responses to drought is the perception that areas are being affected by exceptional deficiencies in rain. The Bureau of Meteorology maintains an extensive network of rain-gauging stations, and classifies an area as drought-affected when annual rainfall is below the 10th percentile. Many States use this information together with data on dam levels, crop, pasture and range condition, and field reports through government agencies and politicians, to declare districts drought-affected. This status usually is maintained until sufficient rainfall has fallen to recharge groundwater, or regenerate vegetation to sustain agriculture at an economical level. State relief to farmers can take many forms. Initially, attempts will be made to keep stock alive. Subsidies may be given to sink deeper bores, import fodder to an area, carry out agistment of livestock to unaffected areas, or truck in water using road or rail tankers.

The 1982–83 drought exemplifies the range of measures that can be adopted to lessen the hardship of drought. Over 100 000 sheep of breeding stock were shipped from eastern States to Western Australia, a distance of 3000 km. The Australian National Railway arranged special freight concessions and resting facilities en route, while the federal government provided a 75 per cent freight subsidy. In addition, assistance was given to farmers paying interest rates over 12 per cent, and a 50 per cent subsidy was provided for imported fodder for sheep and cattle. Despite these measures, droughts can be so severe that it is only humane to destroy emaciated stock. Local councils forced to open and operate slaughtering pits were reimbursed their costs, and graziers were given a sheep and cattle slaughter bounty. In South Australia each head of sheep and cattle generated a subsidy of $A1.10c and $A11 respectively. In addition, the federal government matched this sheep subsidy up to a maximum of $A3 per head. However, the main aim of the federal government's drought-aid package, which totalled approximately $A400 million, was to prevent the decimation of breeding stock so vital in rebuilding Australia's sheep and cattle export markets.

Attempts at conserving and supplying water were also energetically pursued. The Victorian government cancelled all leave indefinitely for drilling-rig operators working in the mining industry, and redirected them to bore-drilling operations. The number of water tankers in that State proved insufficient to meet demand, and the State resorted to filling plastic bags with water and shipping them where needed. In New South Wales water allocations from dams and irrigation projects were cut in half and then totally withdrawn as the capacity of many dams dwindled to less than 25 per cent. The drying up of the Darling River threatened the water supply to Broken Hill. Water in Adelaide, which is drawn from the Murray, reached unsafe drinking limits as saline levels increased through evaporation. In order to save trees within the downtown area, trenches were dug around the trees and roots were hand-watered. Many other country towns and cities suffered a severe curtailment of water usage except for immediate domestic requirements.

While the 1982–83 drought was economically severe, the State and federal response financially was relatively minor. The drought cost farmers an estimated $A2500 million and resulted in a $A7500 million shortfall in national income. Government relief (both State and federal) probably did not exceed $A600 million, with most of this aid directed towards fodder subsidies. In Australia, in most cases, the individual farmer must bear the financial brunt of the drought and 1982–83 was no exception. Many farmers during the drought were forced to increase their debt load under personal interest rates as high as 20 per cent, to the point where their farming operations became economically un-viable even when favourable times returned. The continued high interest rates following this drought and the over-capitalization of many Australian farmers has only ensured a rural decline throughout the 1980s.

International drought response

International relief organizations

By far the largest and most pervasive international organizations are the ones associated with the United Nations or the Red Cross. The United Nations Disaster Relief Office (UNDRO) was created without a dissenting vote by the General Assembly in 1972, and financed partly by the United Nations budget and partly by the Voluntary Trust Fund for Disaster Relief Assistance. The latter channel permits countries, who are touchy about contributing to the overall running cost of the United Nations, to participate in disaster relief funding. UNDRO gathers and disseminates information in the form of situation reports to governments and potential donors about impending disaster. It conducts assessment missions and, working through 100 developing countries, uses trained, experienced personnel who can open channels of communication with government officials in disaster-prone areas, irrespective of the political beliefs of that government or area. It has a responsibility for mobilizing relief contributions, and ensuring the rapid transport of relief supplies. During times of disaster, it can call upon other agencies for personnel in the implementation of programs. It also has the duty to assess major relief programs, and to advise governments on the mitigation of future natural disasters.

There are also a number of other United Nations offices which are involved in a less direct way in drought relief. The United Nations Children's Fund (UNICEF), Food and Agriculture Organization (FAO), World Health Organization (WHO), and World Food Program (WFP), while mainly involved in long-term relief in underdeveloped countries, can provide emergency service and advice. In addition, the World Meteorological Organization (WMO) monitors climate, and the United Nations Educational Scientific and Cultural Organization (UNESCO) is involved in the construction of dams for drought mitigation. The International Bank for Reconstruction and Development (World Bank) finances these mitigation projects, as well as lends money for reconstruction to Third World countries after major disasters.

Temporary infrastructures to handle specific disasters can also be established from time-to-time. In response to the Ethiopian-Sudanese drought of the early 1980s, the United Nations Office for Emergency Operations in Africa (UNOEOA) was established to coordinate relief activities. This organization channelled $US4.5 billion dollars of aid from 35 countries, 47 non-government organizations and half a dozen other United Nations organizations, into northern Africa during this drought. Credited with saving 35 million lives, this office is one of the success stories of the United Nations. It was headed by Maurice Strong, a Canadian with a long-time involvement in the United Nations and a reputation for being able to pull strings especially with non-cooperative governments with obstructive policies. He persuaded Sudan's Gaar Nimeiny to open more refugee camps to relieve the desperate and appalling conditions in existing camps, and coerced Ethiopia's Colonel Haile Mengistu to return trucks, commandeered by the army for its war effort against Eritrean rebels, to the task of clearing the backlog of food stranded on Ethiopian docks.

The United Nations is not always the altruistic organization seemingly implied in the above descriptions. Many of its branches, most notably the Food and Agricultural Organization (FAO), are politicized and riddled with inefficiency, if not outright cronyism. FAO, which has a $US300 million annual budget to develop long-term agricultural relief programs, has over the past ten years spent 40 per cent of its budget on administration that is paternalistic and corrupt. At the height of the Ethiopian famine in 1984 FAO administrators delayed signing over $US10 million in emergency food aid for 20 days, simply because the FAO had been slighted by the Ethiopian ambassador to FAO. The most efficient FAO field worker in Ethiopia was recalled during the drought, because he inadvertently referred to himself as a UN, instead of a FAO worker in statements to the press. Much of the FAO administration operates as a Byzantine merry-go-round, where the mode of operation is more important than the alleviation of drought in member countries.

Second in effort to the United Nations is the International Red Cross (Red Crescent in Islamic countries), which constitutes the principal non-governmental network for mobilizing and distributing international assistance in times of disaster. Founded in 1861, it consists of three organizational elements as follows: the International Committee of the Red Cross (ICRC), an independent body composed of Swiss citizens concerned mainly with the victims of armed conflict; the 162 national societies, which share the principles and values of the Red Cross and conduct programs and activities suited to the needs of their own country; and the League of Red Cross Societies (LORCS), which is the federating and co-ordinating body for the national societies and for international disaster relief. The International Red Cross, because of its strict non-aligned policy, plays a unique and crucial role in ensuring disaster relief can reach refugees in areas stricken by civil war, or controlled by insurgency groups with little international recognition.

The Red Cross organization's aims are duplicated by a number of religious and non-affiliated organizations. One of the largest of these is the Caritas (International Confederation of Catholic Charities), which was set up for relief work during the Second World War, and which continued its relief efforts afterwards during natural disasters. The Protestant equivalent is the World Church service. Other denominational groups maintaining their own charities include the the Salvation Army, Lutheran Church, Adventist Development and Relief Agency (ADRA) and the Quakers (who in the United States sent aid to Cuba after it was devastated by hurricanes). Religious charities also exist in Islamic, Hindu and Buddhist countries. Finally, there are a number of non-profit organizations, which can raise money for immediate disaster relief, long-term reconstruction and disaster mitigation. These organizations include AustCare, CARE, World Vision, OXFAM, Save the Children and the '40-hour famine' program. Some have internalized administrations that pass on as much as 95 per cent of all donations, while others utilize professional fund-raising agencies, which may soak up as much as 75 per cent of donations in administrative costs. These organizations collect monies periodically using door-knock appeals, telethons, or solicitations through regular television, radio, newspaper and magazine advertisements.

International aid flops

Whenever the international community is mobilized to respond to large droughts and impending famine for millions of people, the question arises 'Does the aid get through?'. There are three difficulties with supplying aid to Third World countries. Firstly, the sudden increase in the volume of goods may not be efficiently handled because transport infrastructures existing in those countries at the time are inefficient, or non-existent. If relief is sent by boat, it may sit for weeks in inadequate harbours awaiting unloading. If it is unloaded, it may sit for months on docks biding time for transhipment. Such was the case with grain supplies sent to Ethiopia during the drought at the beginning of the 1980s. If a drought has occurred in isolated areas, there simply may be no transport network to get relief to the disaster site. This problem may be exacerbated by the destruction of transport networks because of ongoing civil war. Secondly, Third World countries may have an element of corruption or inefficiency to their administration. The corruption, if it exists, may siphon off relief aid for sale on the black market, for export or simple ransom. Finally, inefficient administrations may simply keep the relief aid bottled up in paper work or be unaware that relief aid is even necessary.

By far one of the biggest flops to occur for these reasons was the international aid effort for the Sudanese drought of 1984–85. This drought was the first to be predicted and prepared for before its occurrence. In the previous two years Ethiopia had been wracked by drought which virtually went unnoticed for one year, and then was exposed on British television. The Ethiopian drought generated considerable attention as to why it had occurred, and how relief was getting through to the people. For this reason, it became a foregone conclusion that the drought was in fact spreading westwards, and would sooner or later enter the Sudan, a country that had magnanimously permitted over 1 million Ethiopian refugees to cross its borders for relief aid. The management of the Ethiopian famine was a disaster in itself. As it had in 1975, the Ethiopian government at first failed to admit that it had a famine problem, mainly because the drought was centered in rebellious northern provinces. It continued to ship garden produce to other countries for sale, instead of diverting food and produce to the relief camps that had been set up or had spontaneously developed. It deliberately strafed columns of refugees fleeing the drought-affected areas, and it permitted relief aid to pile up in its eastern ports. At the same time an enormous sum of money was spent to dress up the capital for a meeting of African nations.

The United States government recognized these problems, and decided that a similar situation would not hinder relief aid for the impending drought in the Sudan. It pledged over $US400 million worth of food aid, and set about overcoming the distribution problem, which consisted in getting the food from Port Sudan on the Red Sea to the western part of the Sudan, which was isolated at the edge of the Saharan desert. The United States hired consultants to evaluate this problem of distribution, and it was decided to use the railway between Port Sudan and Kosti, or El-Obeid, to ship grain west before the onset of the rainy season. The grain would then be distributed westwards using trucks. The United States went so far as to warn the residents of the western part of Sudan that a drought was coming, and that food aid would be brought to their doorstep. The inhabitants of that part of the country were not encouraged to migrate eastwards. Thus, there was going to be no repeat of the large and deadly migrations which occurred in Ethiopia.

However, no one bothered to evaluate the condition of the Sudanese railway system to handle the transportation of such a large volume of grain. In fact, trains on the scrap heap in western countries were in better shape than the Sudanese railway system. A decision was made to repair the railway, but inefficiencies delayed this for months. Finally, it was decided to move the food for most of the distance by truck. Needless to say, no one had evaluated the trucking industry in the Sudan. The operation absorbed most of the trucks in the country. The local truck drivers, realizing they now had a monopoly, went on strike for higher wages. By the time the railway and trucking difficulties were overcome, the drought had well and truly set in, and few if any people had migrated to the east. They were now starving. When the transport situation was ironed out, the rains set in, and many trucks found themselves wandering down mud tracks, attempting to ford impassable rivers, or hopelessly broken down. Meanwhile, grain supplies were rotting in the open in staging areas in the east, because no one had ever considered the possibility that they would still be sitting undistributed when the rainy season broke the drought.

The decision was finally made to airlift food supplies into remote areas using Hercules aircraft, a decision that immediately increased transport costs fourfold. That was not the end to the story. The rains made it impossible for the planes to land, and there was no provision to parachute supplies to the ground. Like sheep in an Australian flood awaiting the airdrop of bales of hay, the starving inhabitants of the west Sudan waited for airdrops of bags of grain. The Hercules would fly as low and as slow as possible, as bags of grain were shoved out the back without any parachutes to slow them down. The bags tumbled to the ground, bounced and broke scattering their contents. The Sudanese then raced on foot to the scene to salvage whatever grain they could before it was completely spoilt in the mud.

Bob Geldof (Geldof, 1986)

And this is where Bob Geldof comes in, although he was there well and truly before the Sudanese crisis. There is no other way to entitle this section except to use the man's name, for without Bob Geldof there would not have been Band-Aid or Live-Aid. His name is now synonymous with spontaneous, altruistic, international disaster relief efforts. Bob Geldof was a musician from Ireland who worked with a band called the Boomtown Rats, which had one international hit single in 1979 called 'I don't like Mondays'. They were the number one band in England that year. Because he was a musician, Geldof had contacts with most of the rock stars at the time and also was aware of the sporadic efforts by musicians to help the underprivileged in the Third World. For example, following the 1970 storm surge in Eastern Pakistan (Bangladesh) and that country's independence, musicians had organized a concert to raise money to help overcome this disaster—the 'Concert for Bangladesh'. It was staged free by several famous rock groups, and raised several millions of dollars through ticket sales and royalties on records.

Bob Geldof, like many other people in 1984, was appalled by the plight of the Ethiopians and the fact that the famine remained virtually unknown to the outside world until a British television crew led by reporter Michael Buerk tripped over it. Buerk's film showed the enormity of the disaster, and painted a dismal, hell-like picture. Children were being selectively fed. Only those who had a hope of surviving were permitted to join food queues. People were dying throughout the film. In the midst of this anarchy, the Ethiopian people stoically went on trying to live in the camps, which were over-crowded and without medical aid. Their honesty formed a striking contrast to the greed of most people viewing the show. Though starving, Ethiopian refugees would not touch food in open stores unless it was given to them. No one appeared to be doing anything to help the refugees. The Ethiopian government had denied the existence of the drought, and the international monitoring agencies had failed to bring it to world attention. UNDRO in particular appeared hopelessly inefficient.

The emotional impact of the BBC film stunned all viewers, including Bob Geldof; however, a rock musician was hardly the most obvious candidate to organize and lead the largest disaster relief effort in history. The day after the television documentary, Geldof began to organize, coerce, and lie to put a group of musicians together to perform a record for the Christmas of 1984, the profits of which were to be donated to Ethiopian relief. The group was made up of the most popular rock musicians in Britain at the time—Sting, Duran Duran, U2, Style Council, Kool and the Gang, Boy George and Culture Club, Spandau Ballet, Bananarama, Boomtown Rats, Wham, Ultravox, Heaven 17, Status Quo, David Bowie, Holly and Paul McCartney. The band was called Band-Aid, an obvious play on words, but chosen in reality because the name would make people aware of the futility of this one act of charity in overcoming the famine. In the end, Bob Geldof managed to convince all people associated with the record to donate their services free. Even the Value Added Tax was waived by the British government after protracted lobbying. The proceeds of the record were to be put into a special fund to be distributed for relief aid as a committee saw fit. The record contained the song 'Do they know it's Christmas?' on side one, and some messages from rock stars on side two. It cost £1 30p and reached number one the day it was released, six weeks before Christmas. Within weeks it was selling 320 000 copies per day and utilizing the printing facilities of every record factory in Britain, Ireland and Europe to meet demand. People bought 50 copies, kept one and gave the rest back for resale; they were used as Christmas cards, sold in restaurants, and replaced meat displays in butchery windows. Musicians in the United States (USA for Africa) and Canada (Northern Lights) subsequently got together to produce a similar type of album, which had the same success in North America. Over twenty-five Band-Aid groups ('Austria für Afrika', 'Chanteurs Sans Frontiers', and three German groups amongst others) were formed worldwide raising $US15 million dollars for famine relief.

Following the success of the Christmas album, Bob Geldof then organized in 1985 a live concert. It was to become the first global concert and would run for 17 hours on television, using satellite hookups. It would start in Wembley stadium in London, and then end in Philadelphia with five hours of overlap. Famous bands would alternate on a rotating stage. The concert would also draw in television coverage of smaller concerts held in Australia and Yugoslavia. Countries would pay for the satellite link-up, and at the same time be responsible for organizing telethons to raise money. Paraphernalia would also be sold with all profits going to the Band-Aid fund. As much as possible, overheads were to be cost-free.

Critics said that it could not be done, that the organization would be too horrendous, the musicians too egotistical, that no one would watch it or give donations. A group of middle-aged consultants hired to evaluate the income potential of such a concert, said it would raise less than half a million dollars (US). The whole organization was done in three months and it was not known until the day of the show, on 13 July 1985, exactly what groups would be participating. Many countries only telecast the concert at the last minute. The chaos, especially in Philadephia, spelt disaster; but through the efforts of talented organizational people in television and concert promotion, the show was viewed by 85 per cent of the world's television audience.

The show blatantly, but sincerely, played upon the television audience's emotions. At one point, when telephone donations were lagging, a special video clip of a dying Ethiopian child was played to the words of the song 'Drive' by the group Cars. As the child tried repetitively to stand up only to fall down each time, the song asked who was going to pick him up when he fell down. The clip brought a response that jammed the telephone lines worldwide. Donations were received from both communist and non-communist countries. One private donation from Kuwait totalled $US1.7 million. Ireland gave the most money per capita of any country, in total $US10 million. People there even cashed in their wedding rings and houses. In 24 hours Live-Aid, as the concert became known, raised $US83 million dollars.

Live-Aid was to ad hoc disaster relief, what Woodstock was to rock concerts. It became just one of many unique fund-raising efforts. Geldof convinced people in many industries and recreational pursuits to run their own types of money-raising programs, spawning an alphabet-soup list of organizations beginning with Actor Aid, Air Aid, Art Aid, Asian Live Aid, Bear Aid, Bush Aid, Cardiff Valley Aid, Fashion Aid, Food Aid, and so on. In France, which had not successfully linked in with the fund-raising activity for Live Aid, students raised money in School Aid, whose success was ensured by the refusal of teachers to participate. In May of 1986, Sport-Aid saw joggers worldwide paying for the privilege of participating in a 10-kilometer run. Over 20 million joggers participated worldwide. Sport-Aid raised nearly $US25 million with donations being split between the Band-Aid organization and UNICEF.

Bob Geldof personally oversaw the distribution of relief aid in Ethiopia and the Sudan, and was responsible for ensuring, in a last desperate measure, that Hercules planes were used to salvage the Sudan drought relief operation. A committee, Live-Aid Foundation, was established to distribute monies for relief efforts. It was run by volunteers, which included professional business people, lawyers and accountants. Bureaucratic administration was shunned as much as possible. A board of trustees (Band-Aid Trust) was established to oversee the relief distribution. This board acted on the advice of a team of eminent academics, with regular working experience in Africa, who were from various institutions that included departments in the universities of Sussex and Reading; the School of Hygiene and Tropical Medicine, and the School of Oriental and African Studies in London; and Georgetown University in Washington. Of the money raised, 20 per cent went for immediate relief, 25 per cent for shipping and transport, and the the remainder for long-term development projects vetted by these experts. In addition, the Band-Aid Trust approached countries to undertake joint aid. For instance, the United States government was ideologically opposed to the Ethiopian government, but not to the Sudanese government. Band-Aid teamed up with United States aid agencies to ensure that relief efforts were directed equally to these two countries. Band-Aid handled the Ethiopian relief, while the United States tackled the Sudan. Other countries were approached for transport and monies. The Australian government was asked to refurbish ten Hercules transport planes for food drops; but in the end donated only one plane without support funding. Other countries faced with similar requests, donated all the support needed to keep their planes in the air. President Mitterand of France, in one afternoon of talks with Geldof, pledged $US7 million in emergency relief aid for areas and projects which had been recognized by Band-Aid officials in the field.

The Band-Aid organization also decided that money should go for drought reconstruction and mitigation. Proposals for financing were requested, and then checked for need by a four-person field staff paid for by private sponsorship. Over 700 proposals were evaluated, with many being rejected because of their lack of relevancy or grassroots participation. By the end of 1986, $US12 million had been allocated for immediate relief, $US20 million for long-term projects, and $US19 million for freight. Project monies went mainly to Burkina Faso, Ethiopia and the province of Eritrea, and Sudan. The largest sum was for $US2.25 million for a UNICEF program to immunize children, and the smallest was for $US3000 to maintain water pumps in refugee camps in eastern Sudan.

All these efforts had their critics. Many argued that Geldof's fund-raising efforts were siphoning money that would have gone to legitimate aid organizations. Geldof always contended that the efforts were raising the level of consciousness amongst ordinary people. In fact, following the Live-Aid Concert, donations to Oxfam and Save the Children increased 200 per cent and 300 per cent respectively. Geldof foresaw criticism that he might personally profit from his fund-raising activities, and refused to use one cent of the money for administration or personal gain. All work had to be voluntary, using equipment, materials, labour and fares which had been donated free and with no strings attached.

Of all the international relief programs, Bob Geldof's has been the most successful at pricking the consciences of individuals worldwide regardless of their political ideology. He initially tapped the music industry for support, but has shown that the methods for raising aid, with involvement of people from all walks of life, are almost endless. Geldof's techniques for money-raising and distributing funds without any significant administration costs is one of the most successful and efficient programs mounted, putting to shame the efforts of individual countries and the United Nations. The euphoria of Band-Aid and Live-Aid have not worn off; however, Geldof saw them as being temporary measures. At the end of 1986 the Band-Aid umbrella organization was wound up and converted into a standing committee. Future donations are being channelled through this committee to permanently established aid organizations. At the time of writing, drought again is gripping Ethiopia: it remains to be seen if Band-Aid resurfaces in the same enthusiastic, altruistic manner.

Long-term consequences of drought

The effects of drought go far beyond the immediate crisis years when communities and nations are trying to survive physically, socially and economically. Of all hazards, drought has the greatest impact beyond the period of its occurrence. This does not necessarily affect population growth. For instance, while many Sahelian countries suffered appalling deathtolls in the 1968–74 drought, high birth-rates of 3 per cent ensured that populations had grown to pre-drought levels within 10–13 years. However, this growth was disproportionate, with the number of children under the age of 18 making up 50 per cent of the population in some countries. Such a young population taxes medical and school facilities and overwhelms economic development. Unless exploited, such numbers of young people do not contribute directly to the labour force or to a nation's economic growth. In addition, survivors of droughts can weigh down family units and community groups with an excessive numbers of invalids, especially amongst the young. Many of the diseases which afflict malnourished children can lead to permanent intellectual and physical impairment. Malnourished children who survive a drought may never overcome the condition of malnutrition especially if they are orphaned, have many siblings, exist within a family unit that cannot economically recover from the drought or are displaced from home areas, where kinship ties can traditionally provide support.

Additionally, droughts pose difficult problems in recovery. If seed stock is consumed, it must be imported quickly to ensure that a country can re-establish the ability to feed itself. While many people migrate out of drought areas to towns where food can be more easily obtained, the transport infrastructure is not adequate to ship seed into isolated rural areas when these people return home. A similar situation applies to livestock. During major droughts, even in developed countries such as Australia, breeding stock may either perish or be used for food. Livestock herds require several years to be rebuilt, a process that can put a heavy financial strain upon a farm's resources because this period of rebuilding generates little income. In the longer term, droughts are often accompanied by human-induced or natural land degradation. Rangeland may be overgrazed during a drought to keep livestock herds alive, and in some cases foliage stripped from shrubs and trees to feed stock. Landscapes can be denuded of any protecting vegetation, resulting in wholesale loss of topsoil during dust storms. In some places in New South Wales, Australia, up to 12 centimeters of topsoil blew away during 1982–83 drought. This soil can never be replaced; while vegetation regrowth, required to protect the remaining topsoil, can be a slow process taking years to decades to complete. This problem is especially acute in semi-arid areas. Even in the United States, where a major federal government program was initiated on the Great Plains following the mid-1930s drought, land rehabilitation was still continuing when the next major drought struck in the mid-1950s. For developing countries strapped for foreign aid and burdened by debt, such long-term recovery and mitigation programs are receiving such a low priority that complete land degradation, caused by successive droughts, may become a permanent feature of the landscape.

References

Further reading

Bryson, R. and Murray, T. 1977. *Climates of hunger.* Australian National University Press, Canberra.

Charnock, A. 1986. 'An African survivor'. *New Scientist* No. 1515 pp. 41–43.

French, R. J. 1983. 'Managing environmental aspects of agricultural droughts—Australian experience'. In Yevjevich, V., da Cunha, L. and Vlachos, E. (eds) *Coping with droughts.* Water Resources Publications, Littleton, pp. 170–87.

Garcia, R. 1972. *Drought and man.* v. 1 Pergamon, Oxford.

Geldof, B. 1986. *Is that it?.* Penguin, Harmondsworth.

Glantz, M. H. 1977. *Desertification: environmental degradation in and around arid lands.* Westview, Boulder, Colorado.

Gribbin, J. 1983. *Future Weather: the causes and effects of climatic change.* Penguin Harmondsworth.

Hidore, J. J. 1983. 'Coping with drought'. In Yevjevich, da Cunha and Vlachos (eds) *Coping with droughts.* pp. 290–309.

Morren, G. E. B. 1983. 'The bushmen and the British: problems of the identification of drought and responses to drought'. In Hewitt, K. (ed.) *Interpretations of Calamity.* Allen and Unwin, Sydney, pp. 44–65.

Rosenberg, N. and Wilhite, D. A. 1983. 'Drought in the U.S. Great Plains'. In Yevjevich, da Cunha and Vlachos (eds) *Coping with droughts*. pp. 355–68.

Scott, E. P. (ed.). 1984. *Life before the drought*. Allen and Unwin, London.

Tai, K. C. 1983. 'Coping with the devastating drought in Australia'. In Yevjevich, da Cunha and Vlachos (eds) *Coping with droughts*. pp. 346–54.

Waring, E. J. 1976. 'Local and Regional Effects of Drought in Australia'. In Chapman, T. G.(ed.) *Drought*. AGPS, Canberra, pp. 243–56.

Warrick, R. A. 1983. 'Drought in the U.S. Great Plains: shifting social consequences?'. In Hewitt (ed) *Interpretations of Calamity*. pp. 67–82.

Watts, M. 1983. 'On the poverty of theory: natural hazards research in context'. In Hewitt (ed.) *Interpretations of Calamity*. pp. 231–62.

Whittow, J. 1980. *Disasters: the anatomy of environmental hazards*. Pelican, Harmondsworth.

Young, J. A. 1983. 'Drought control practices in England and Wales in 1975–76'. In Yevjevich, da Cunha and Vlachos (eds) *Coping with droughts*. pp. 310–25.

7

Associated Precipitation Hazards

Introduction

Associated precipitation hazards can be discussed under four separate headings: thunderstorms and associated phenomena; frozen precipitation; high-magnitude, short-period rainfall leading to flash flooding; large-scale floods. While thunderstorms are generally associated with heavy rains leading to flash flooding, their associated hazards are common enough to warrant separate discussion. Tornadoes derived from thunderstorms have already been covered in detail in chapter 3. Hail and lightning also represent significant and ubiquitous hazards and will be discussed in this chapter under thunderstorms. Apart from hail, frozen precipitation also includes snow, *blizzards*, and *freezing rain*. Snow and freezing rain are most often produced by extra-tropical depressions when air temperatures approach freezing. Blizzards are unique wind-generated storms similar to dust storms; however, instead of clay and silt, snow and ice crystals are transported. Snowfall becomes a hazard in those areas receiving moderate to heavy precipitation over a period of days or months. Whereas the total precipitation as liquid may not pose a problem, snow can represent a tenfold increase in volume. When it is considered that most landscapes can absorb 200–300 mm of rain over a week's period without much physical effect, the same amount of precipitation converted to snow would cover the ground to a depth of 2–3 m. It is this fact which makes snow a winter hazard in mid-latitudes. Snow also poses a hazard in that it represents storage of precipitation, which can lead to flooding in the following melt season. Finally, freezing rain occurs when the air temperature is not low enough to form snow, but ground temperatures and those of adjacent objects such as trees and power lines are below freezing. In these circumstances, falling rain adheres and freezes quickly to these objects and can quickly accumulate beyond their bearing capacity.

Thunderstorms, lightning and hail (Linacre and Hobbs, 1976; Whipple, 1982; Eagleman, 1983; Geophysics Study Committee, 1987)

Thunderstorms

Thunderstorms are a common feature of the Earth's environment. There are about 1800 storms per hour or 44 000 per day. In tropical regions they can occur daily in the wet season. However, in these regions thunderstorms may not represent a hazard because they do not intensify. As thunderstorms represent localized areas of instability, their intensity is dependent upon factors which increase this instability. On a world scale, instability is usually defined by the rate at which the base of the atmosphere is heated by incoming solar radiation, especially where evaporation at the ground and condensation in the atmosphere occurs. In this case, the saturated adiabatic lapse rate prevails. Under these conditions, large quantities of heat energy (2400 joules gm^{-1} of liquid water that condenses) are released into the atmosphere. This process causes convective instability, which only terminates when the source of moisture is removed.

On a localized scale, the degree of instability is also dependent upon topography and atmospheric conditions such as convergence. If air is forced over a hill, then this may be the impetus required to initiate convective instability. Convergence of air masses by topography, or by the spatial arrangement of pressure patterns, can also initiate uplift. The most likely occurrence of instability takes place along cold fronts, mainly the polar front where it intrudes into moist tropical air. Such intrusions are common over the United States mid-west or along the southeastern part of Australia (major place names in this chapter are shown on individual locality maps).

Figure 7.1 illustrates all of the above processes over the United States. The highest incidence of thunderstorms, greater than 60 per year, occurs in Florida, which is dominated by tropical air masses, especially in summer. A smaller area of high thunderstorm frequency over the Great Plains corresponds to an area affected by topographic uplift. From here across to the east coast, thunderstorm activity occurs because of the interaction of polar and tropical air. In Australia the distribution of thunderstorms follows a similar pattern (figure 7.2). The greatest intensity of storms (greater than 30 per year) occurs in the tropics and along the eastern Divide, where topographic uplift is favoured. The passage of cold fronts plays a minor role in Australia compared to the United States in forcing thunderstorm development. While the incidence of storms in Australia is less than in the United States, thunderstorms over tropical Australia are some of the most intense in the world. They have been referred to as *stratospheric fountains* because their associated convection is strong enough to pierce the tropopause, and inject air from the troposphere into the normally isolated stratosphere. This process is highly effective during periods when Walker circulation is 'turned on'.

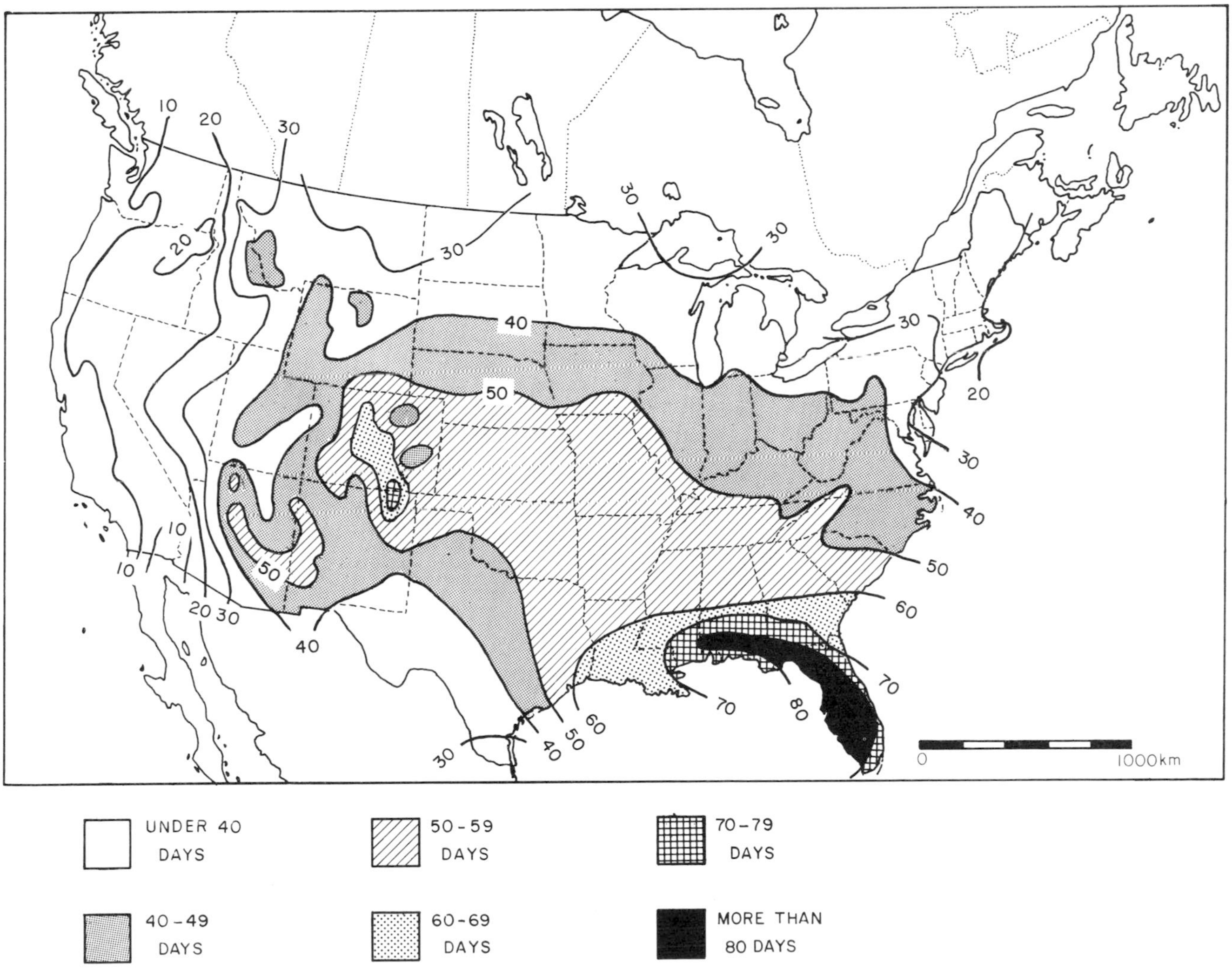

Fig. 7.1 Annual frequency of thunder days in the United States (from Eagleman, 1983)

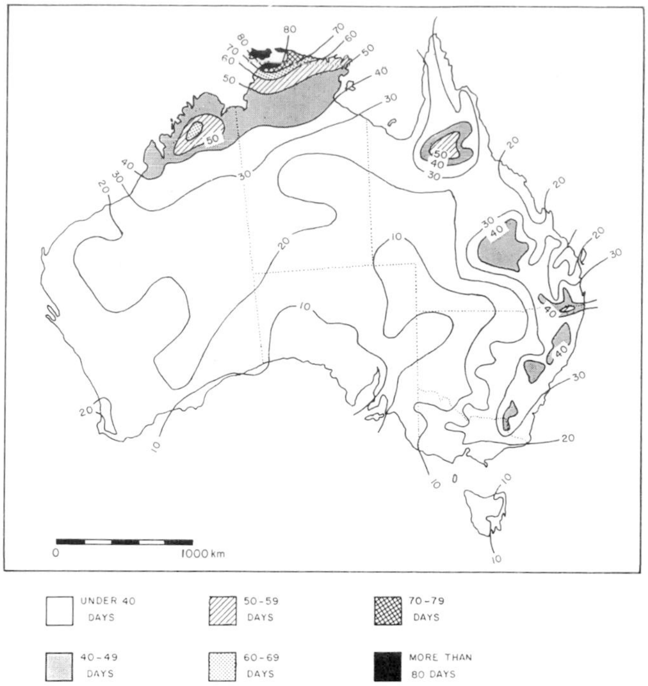

FIG. 7.2 Average annual thunder-days in Australia (based upon Oliver, 1986)

Thunderstorms globally reflect the leakage of electricity between a negative ground surface, and a more positively charged *electrosphere* at about 50 km elevation. There is, thus, a voltage gradient with altitude. Thunderstorm activity is increased if this gradient is steepened. This can be accomplished by increasing the amount of atmospheric aerosols such as smoke from bush fires, pollution or thermonuclear devices. The gradient can also be increased by solar activity through the interference of the solar wind with the Earth's magnetic field. During sunspot activity, the electrosphere becomes more positively charged. There is an association between the frequency of thunderstorms and sunspots, with thunderstorm frequency paralleling closely the 11-year sunspot cycle. Already as the number of sunspots begins to increase at a record rate towards its predicted 1993 peak, the frequency and intensity of thunderstorms worldwide has increased noticeably. In the summer of 1988 in the United States and Canada, storms were so violent that lightning charges blew out microwave ovens.

Because thunderstorms are so closely linked to tornado occurrence in the United States, the threat of thunderstorms is forecast daily by evaluating the degree of atmospheric instability. This is calculated at 93 stations using measured air temperature at the 500 hPa level, and comparing this to a theoretical value assuming that air has been forced up at the forecast temperature and humidity for that day. A difference in temperature of only 4°C has been associated with tornado development. The largest tornado occurrence in the United States (when over 148 tornadoes were recorded in 11 states on 3–4 April 1974) was associated with a predicted temperature difference of 6°C in the Mississippi valley. As well, the movement of thunderstorms is predictable. Most move in the direction of, and to the right of the mean wind. The effect is due mainly to the strength of cyclonic rotation in the thunderstorm. In the United States mid-west thunderstorm cells also tend to originate in the same place because of topographic effects. Thunderstorms, like tornadoes, tend to follow lower lying topography. Cities which develop an urban heat island also attract thunderstorms. The heat island tends to initiate updrafts over the city, which then draw in any thunderstorms developing in the area. This phenomenon has been observed, for example, both in Kansas City, United States, and London, England.

Thunderstorms by themselves are not a major hazard. No one has been killed directly by thunder. Thunderstorms do initiate other hazards such as tornadoes, which have already been discussed. They are also the main mechanism generating lightning and hail, and are a major climatic anomaly, giving rise to flash flooding. Lightning and hail as hazards will be described in the following sections, while flash flooding will be discussed at length in chapter 8. There is one aspect about thunder that is hazardous and often ignored. Thunder produces a pronounced shock wave in the atmosphere that can be transmitted to the ground. In areas of highly variable relief, this can trigger landslides. For example, along the Illawarra escarpment south of Sydney, Australia, rockfalls can be triggered along the coastal cliffs by thunder from thunderstorms passing over the adjacent ocean.

Lightning

Thunder is generated by lightning briefly heating the air to temperatures of $30\,000^\circ$K or five times the surface temperature of the sun. As a result, air in the conducting channel expands from a few millimeters to a few centimeters in a few millionths of a second. This expansion produces a shock wave and noise. Lightning is caused by a buildup of charged ions in the thundercloud (figure 7.3). The process is known as charge separation or

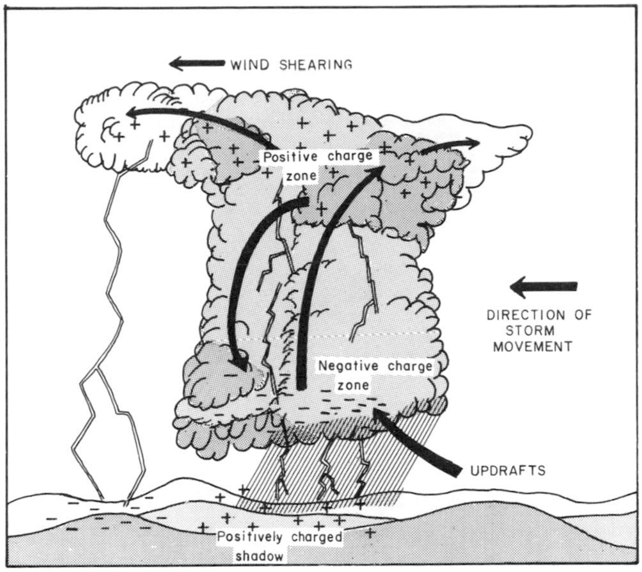

FIG. 7.3 Schematic representation of charge separation. The separation of charged water droplets by updrafts in a thundercloud gives rise to lightning.

polarization and is aided by updrafts and the presence of ice. There is no firm theory as to the cause of this charge separation; however, it results in positive ions migrating to the top of the cloud leaving negative ions at the bottom. Positive ions can also build up in a shadow zone at the ground under the thundercloud. The latter buildup, with its concomitant depletion of negative ions, has been postulated as the reason for an increase in suicides, murders and traffic accidents before the occurrence of thunderstorms. This distribution in ions produces an electrical gradient, both within the cloud, and between the base of the cloud and the ground. If the gradient reaches a critical threshold, it is reduced by sparking between the area of negative ions and the two areas of positive ion buildup. The amount of energy involved in a lightning discharge varies greatly, but is quite small compared to the volume of air included in the thunderstorm. For instance, the energy density of the charge is 100 million times smaller than that produced in the same volume of the air by convection or latent heat due to condensation. However, the total energy of the bolt, because it is concentrated in such a small channel, can reach 250 kilowatt hours. Under extreme wind-shear, the area of positive ions at the top of the atmosphere may move downwind faster than the central core of negative ions. If negative ions build up at the Earth's surface, a discharge can take place from the top of the atmosphere to the ground with ten times the energy of a normal lightning flash. Often the imbalance in ions adds an attractive force to moisture in the cloud. The opposing charges literally hold the cloud together. A lightning spark may weaken this hold, and lightning flashes are often accompanied by localized increases in precipitation. Initially, the sparking originates in the area of negative ions at the base of the cloud, and surges in jerky steps towards the top of the cloud or to the ground at the rate of 500 km s^{-1}. Once a leader channel no more than a few millimeters wide but up to 100 km long has been established, positive ions race along this channel at a speed of 100 000 km s^{-1}. On average, a channel, once established, lasts long enough to carry three distinct pulses of ions between the positive and negative ionized areas.

With modern technology, it is possible to track the occurrence of thunderstorms that produce lightning anywhere in the world. Lightning discharges generate high frequency radio noise that can be monitored by satellites, and located to a high degree of accuracy. There are on average 100–300 flashes per second globally, representing a continuous power flow of about 4000 million kilowatts. Because of the risk lightning poses in starting forest fires, a network of automatic direction-finding, wideband *VHF*, magnetic receivers has been established across North America to locate and measure the characteristics of each lightning flash. The greatest daily incidence of discharges to date occurred in June 1984 when 50 836 flashes were recorded, mainly in and around New York State.

Lightning is a localized, repetitive hazard. Because the lightning discharge interacts with the best conductors on the ground, tall objects such as radio and television towers are continually struck. When lightning strikes a building or house, then fire is likely as the lightning grounds itself. However, this effect, while noticeable, is not severe. In the United States, about $US25 million in damage a year is attributable to lightning. Insurance companies consider this a low risk. However, because most of the damage can be avoided by the use of properly grounded lightning rods on buildings, insurance companies in North America either will not insure, or will charge higher rates for unprotected buildings. Lightning, besides igniting forest fires, is also a severe hazard to trees and other vegetation. In North America tree mortality in some areas can reach 0.7–1.0 per cent per year because of lightning strikes, and up to 70 per cent of tree damage is directly attributable to lightning. Scorching is even more widespread. A single lightning bolt can affect an area 0.1–10 hectares in size, inducing physiological trauma in plants and triggering die-offs in crops and stands of trees. Lightning can also be beneficial to vegetation, in that oxygen and nitrogen are combined and then dissolved in rain to form a nitrogen fertilizer.

The greatest threat from lightning is the fact that the high voltage kills people. As shown by figure 7.4, which lists the annual number of deaths in the United States between 1940 and 1975 due to lightning, tornadoes, floods and tropical cyclones, more people die in the United States from lightning than from any other meteorological event. About 40 per cent more people die from lightning than from the next worse meteorological hazard, tornadoes. Note that, despite the fact that tornadoes and lightning have the same origin, there is a poor correspondence in the number of deaths between the two hazards each year. Figure 7.4 also indicates a decrease in the number of deaths by lightning over time, even though the American population continues to grow. This may indicate that public education on the risks of lightning are paying off. Despite the misconception that standing under a tree during a lightning storm is playing with death, statistics indicate that the number of people who die from lightning-induced charges inside houses or barns is four times the number of those who die outside standing under trees. This statistic may only reflect the fact that most people know about the risks of standing under trees during a storm, but not about other hazardous activities—something that is exemplified by experience in Australia. While most people avoid standing under trees, being in the open or climbing steel towers during thunderstorms, Telecom had to mount a major campaign in 1983 warning subscribers about the risk of using telephones during thunderstorms. In Australia or elsewhere in the world, few people would know that standing near electrical wiring or metal piping (including toilets) inside homes, or standing at the edges of forests or mountains during thunderstorms is as dangerous as any other activity generally perceived as risky during such storms. Most people would know that being in the water is dangerous during lightning; however, in Australia, one can still find people surfboard riding during thunderstorms. Also, few people realise that standing on or at the back of a beach during a thunderstorm is as risky, especially if the storm is over the ocean. In this case, the shoreline represents the preferred location for positive ion buildup and lightning discharge.

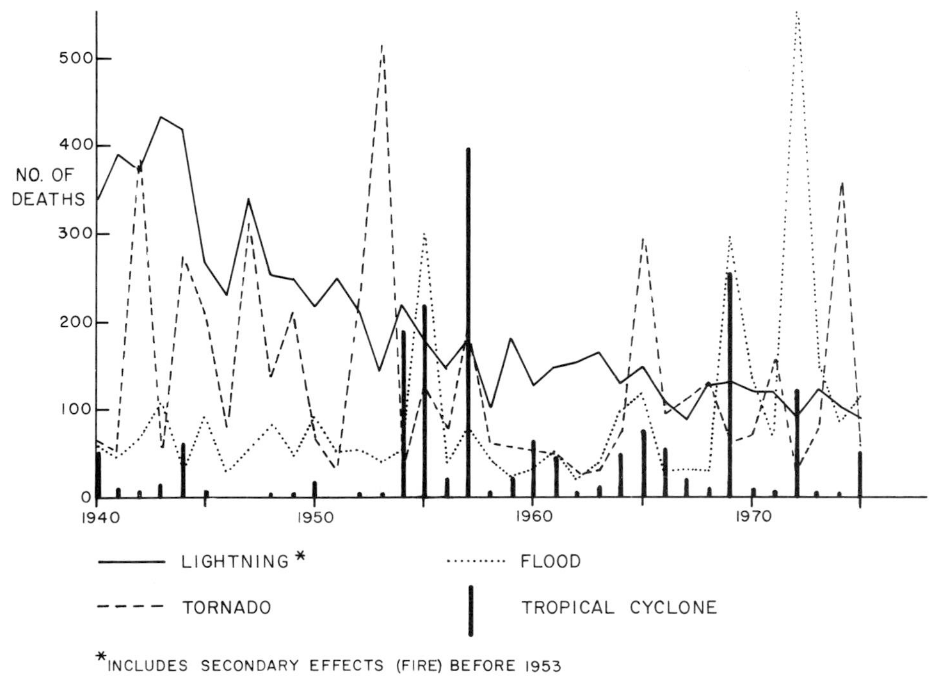

FIG. 7.4 Number of deaths due to lightning, tornadoes, floods and tropical cyclones in the United States, 1940–75 (data from Eagleman, 1983)

Hail

Hail is usually not a major cause of death in most parts of the world. In the United States there are records of only two people having been killed directly by hailstones. Bangladesh, however, appears to be unique in this regard because, on two occasions there in the last 20 years hailstorms have killed over 300 people. Bangladesh experiences the most frequent and intense hailstorms of any location in the world. Although not noted for its deathtoll, hail is one of the most destructive hazards to agricultural crops and animals, and even a light fall of hail can cause widespread damage to buildings and vehicles. About 2 per cent of the United States crop production is damaged by hail each year. On the Great Plains of the United States, losses in some years have amounted to 20 per cent of the value of the crop. In Australia in 1985, 20 per cent of the apple crop was wiped out by a single hailstorm around Orange, New South Wales. Following the drought of 1982–83, some Australian wheat farmers suffered the misfortune of seeing bumper wheat crops destroyed by hail in the eastern States. Property damage is less extensive than crop losses; however, hailstorms are the major cause of window breakage in buildings, and the major natural cause of automobile damage. The largest single insurance payout made by Munich Reinsurance Corporation, a major international insurance underwriter, occurred following the 12 July 1984 hailstorm that struck the city of Munich, West Germany. This company paid out \$US500 million dollars mainly on damage to automobiles. Insurance brokers in Australia call hailstorms the 'glaziers' picnic', because the large number of broken windows inevitably means there will be plenty of work for glaziers. In the Sydney region hailstorms are one of the major reasons for insurance claims. Over the past decade, three to four localized storms in the Sydney metropolitan area have led to claims amounting to \$A40–60 million each time. The most recent deluge, on 18 March 1990, dropped hail up to 9 cm in diameter, breaking windows, roofing tiles and fibro cladding. The damage bill reached \$A500 million.

Hail locations do not coincide with areas of maximum thunderstorm development, nor with areas of maximum tornadoes. Figure 7.5 illustrates the annual frequency of hailstorms in the United

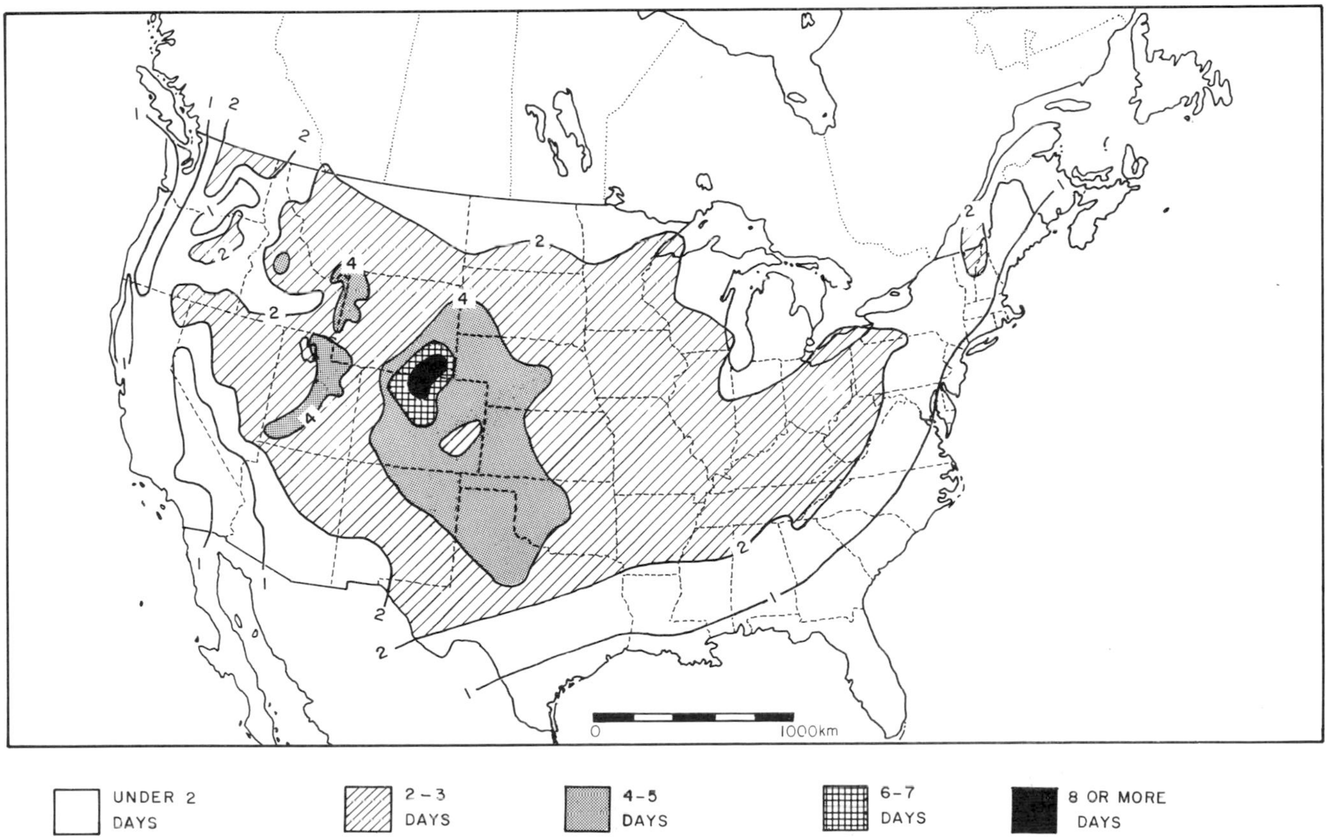

Fig. 7.5 Annual frequency of hailstorms in the United States (from Eagleman, 1983)

States. The middle of the Great Plains, underlying the zone of seasonal migration of the jet stream, is the most frequently affected region. Areas with the highest incidence of hailstorms on this map do not correspond well to those areas of maximum thunderstorm activity shown on figure 7.1. Most hailstorms in the United States occur in late spring and early summer as the jet stream moves north. A similar seasonal preference for hailstorms also occurs in Australia.

The formation of hail depends upon the strength of updrafts, which in turn depends upon the amount of surface heating. The more intense the heating at the Earth's surface, and the steeper the temperature gradient with elevation, then the more likely the occurrence of hail. The surface heating produces the updrafts, while the cooling aloft ensures hail formation. Almost all hailstorms form along a cold front shepherded aloft by the jet stream. The jet stream provides the mechanism for creating updrafts. Wind shear, which represents a large change in wind speed over a short altitude, also facilitates hail formation. The degree of uplift in a thunderstorm can be assessed by the height to which the storm grows; in colder climates, however, this height is less than in more temperate or subtropical latitudes.

The size of hail is a direct function of the severity and size of the thunderstorm. General physics indicates that a hailstone 2–3 cm in diameter requires updraft velocities in excess of 96 km hr^{-1} in order to keep the hail in suspension. An 8-centimeter hailstone would require wind speeds in excess of 200 km hr^{-1}, and the largest stones measured (>13 cm) require wind velocities greater than 375 km hr^{-1}. Multiple layered hailstones indicate that the stone has been lifted up into the thunderstorm repetitively. Ice accumulates less when the stone is being uplifted because it often passes through air cooler than the freezing point. The ice in this process can be recognized from its milky appearance. When the stone descends again, water freezes rapidly to the surface forming a clear layer.

Because of the tremendous damage hail can do to crops, there has been substantial research on hail suppression in the United States and the Soviet Union wheat- and corn-growing areas. Super-cooled water (water cooled below the freezing point but still liquid) is very conducive to hail formation, and attempts have been made to precipitate this water out of the atmosphere before large hailstones have a chance to form. This process is accomplished by seeding clouds containing super-cooled water with silver iodide nuclei, under the assumption that the more nuclei present in a cloud, the greater will be the competition for water and the smaller the resulting hailstones. In the United States seeding is carried out using airplanes, while in the Soviet Union and Italy, it is injected into clouds using rockets. In the Soviet Union seeding has reduced hail damage in protected areas by a factor of 3–5 at a cost of 2–3 per cent of the value of the crops saved. In South Africa crop-loss reductions on the order of 20 per cent have been achieved by cloud seeding. In the United States hail suppression programs were terminated at the end of the 1970s because of political controversy. The main reason was marginal evidence that cloud seeding experiments actually increased the amount of hail, especially if thunderstorms had reached a mature or supercell stage. However, data collected over a three-year period found that crop losses had been reduced by 48 per cent in western Texas and by 20 per cent in South Dakota. Despite the American experience, the Soviet Union considers hail suppression beyond the experimental stage and regularly uses it in areas prone to hailstorms. In Australia major programs for hail suppression have not progressed beyond the experimental stage. There is some question of cost effectiveness, given the lower crop yields over large areas. However, in the Griffith and Murrumbidgee irrigation areas, the cost of hail suppression would certainly be less than the losses experienced. To date, often the emphasis on cloud seeding in Australia is not for hail suppression, but for rain enhancement to relieve drought in New South Wales, and to prevent bushfires in Victoria.

Snowstorms, blizzards and freezing rain (Rooney, 1973; Whittow, 1980; Eagleman, 1983; Nalivkin, 1983; Muller and Oberlander, 1984)

Snowstorms

While many mid-latitude cyclonic depressions can give rise to exceptionally heavy rain and widespread flooding, conditions will always be worse if the precipitation falls in the form of snow. The fact that snow volume exceeds rainfall by a factor of 7–10 implies that even small amounts of precipitated water falling as snow can totally paralyze large sections of a continent. Even if the snow in individual storms is small, it can accumulate over many falls to present a serious flood hazard lasting one to two weeks during the spring melting season.

If snowstorms occur too frequently, then the effects of the previous storm may not be cleared away and urban transport systems can be slowly crippled. Whereas snow generally incapacitates transportation, if the precipitation falls as freezing rain, then severe and widespread damage can be caused to power transmission lines.

Most large snowstorms originate as mid-latitude depressions following the *meandering* path of the jet stream across the continents. The jet stream has a preferred location in winter across North America and Europe, so that the area affected by snowstorms tends to be persistent over periods of several weeks. There is also the risk that abnormal or aseasonal jet-stream paths can cause snowstorms to occur in areas totally unprepared for them. In North America the preferred path for winter snowstorms follows jet stream looping down over the United States mid-west towards Texas, and then northwards parallel to the Appalachians, Great Lakes-St Lawrence River valley, eastwards over Newfoundland. This jet-stream path is topographically controlled by the Appalachian mountains. Movement of the jet to the eastern side of the Appalachians can bring exceptionally heavy snowfalls to the east coast of the United States. In Europe the jet stream tends to loop down over England and northern Europe. Movement southwards is partially hampered by the Alps. Extended looping southwards can bring very cold conditions to northern Europe, and aseasonal snowfalls to the Mediterranean region of Europe.

Mid-latitude cyclonic depressions can affect very large areas of the continental United States in winter. Figure 7.6 shows the development of a low-pressure cell with the areas of intense snowfall marked. In this diagram, rainfall is falling in that part of the air mass which is warm. As the warm air is pushed up over the cold air, there is a small zone of freezing rain (to be discussed in detail later). As the air is forced higher along the warm front, temperatures drop below the freezing point and precipitation falls as snow. The area in front, and to the poleward side of the advancing warm front is the area of heaviest snowfall, and can cover a distance of 1000 km. Behind the warm air mass, cold air is rapidly pushed into the low. In this region there is rapid uplift of moist warm air. However, condensing moisture is rapidly turned to snow. Here, snow may fall during thunderstorms, but the amounts are generally less than that accumulating in the path of the warm front. In exceptionally intense cases, another cyclonic frontal system can develop to the equatorial side of the first, forming twin low-pressure cells. Figure 7.7 illustrates such a cyclonic situation that brought snow to the eastern half of the United States at the end of February 1984, paralyzing transport. The eastern low was associated with very warm air from the Gulf of Mexico that brought widespread rain to the southern part of the United States. The western low was dominated by uplift of moist air along the polar front. Heavy snow extended across 1500 km, from New York to Chicago, and from Canada to the southern states. In some cases, as the cyclonic depression moves down the St Lawrence valley and occludes, spiralling winds

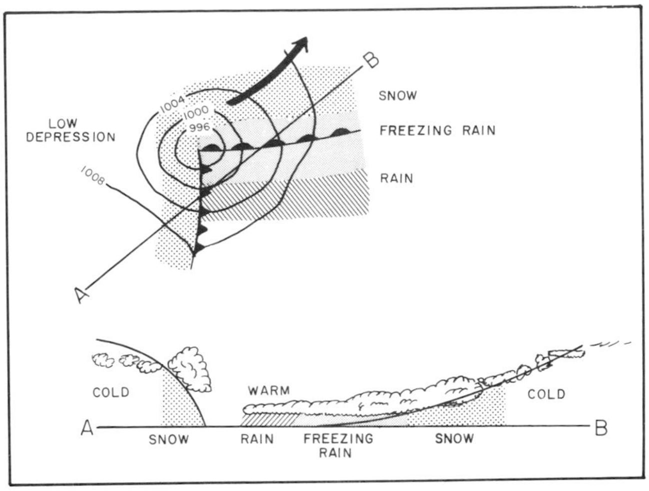

Fig. 7.6 Schematic representation of precipitation patterns for a mid-latitude cyclonic depression in winter in the northern hemisphere

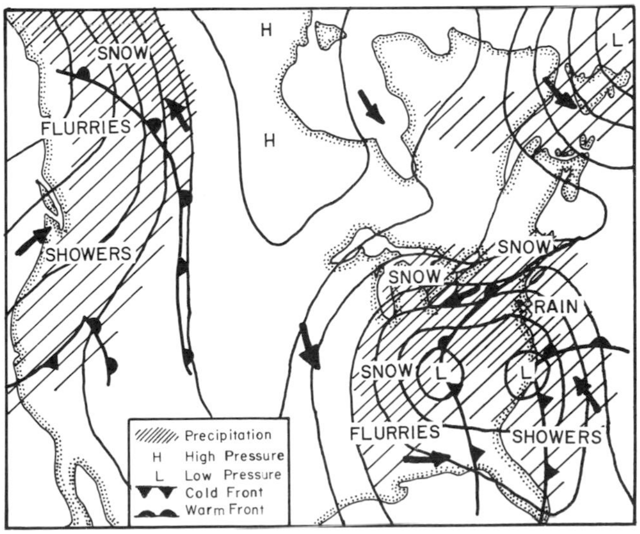

Fig. 7.7 Pressure patterns for the snowstorm that paralyzed the eastern United States, February 1984.

may turn the snow-bearing cloud back over parts of the continent already affected by the main storm, and prolong the fall of snow.

The winters of 1976–78 were particularly severe for heavy snowstorms in the eastern part of North America and western Europe. Snowfalls in excess of 1 m were typical. Similar winters had been experienced in the eastern United States in 1960, and 1914–15. In some cases, drifting snow associated with the strong winds that accompany these winter storms can exceed 10 meters or more, burying houses and telegraph poles. The United States mid-west was hit by one particularly severe storm in March 1966 that buried trains and crippled transportation for weeks.

In Australia very little snowfall is produced by mid-latitude depressions. Instead, snow often accompanies strong outbursts of Antarctic air in the winter months. These southwesterly winds blow over the southern ocean, and are orographically uplifted over the highlands of Tasmania, the Snowy Mountains on the mainland, and along the Great Divide to Queensland. The amounts of snow depend upon the temperature and strength of the cold outburst of air. Figure 7.8 illustrates a typical winter weather pattern that produced snow in eastern Australia. An intense high-pressure cell, with central pressure above 1028 hPa, pushed two cold fronts northeast across the southeastern part of Australia. Heavy falls of snow occurred from the 14–18 July in the Snowy Mountains and across the Central Tablelands. The unusual aspect about snowfalls in Australia is the fact that outbursts of cold Antarctic air are not always tied to the winter season. Patterns similar to those in figure 7.8 can occur quite late in the winter and even in summer. The highest amount of snow ever recorded in Canberra, the capital of Australia, was 15 cm, which fell in the middle of summer, on Christmas Day, 1966. For bushwalkers, not dressed for such weather conditions, aseasonal cold snaps and snow can pose a serious threat because of hypothermia (death caused by body temperature falling 5°C below normal). For graziers, such inclement weather can completely kill that year's spring lambing.

The effect of heavy snowstorms in northern hemisphere winters is substantial. As soon as snow begins to fall, transportation networks in the United States are disrupted. Local roads can be completely closed off within hours of a storm striking and, depending upon the severity of the storm, closed for several days (figure 7.9). Accident rates increase 200 per cent above average within a few hours of a storm. Motor vehicle insurance rates spatially reflect the incidence of storms likely to affect an area. During moderate falls, vehicles caught on roads can become stranded and impede snow removal. In extreme cases, stranded vehicles can become death traps. Motors left idling for

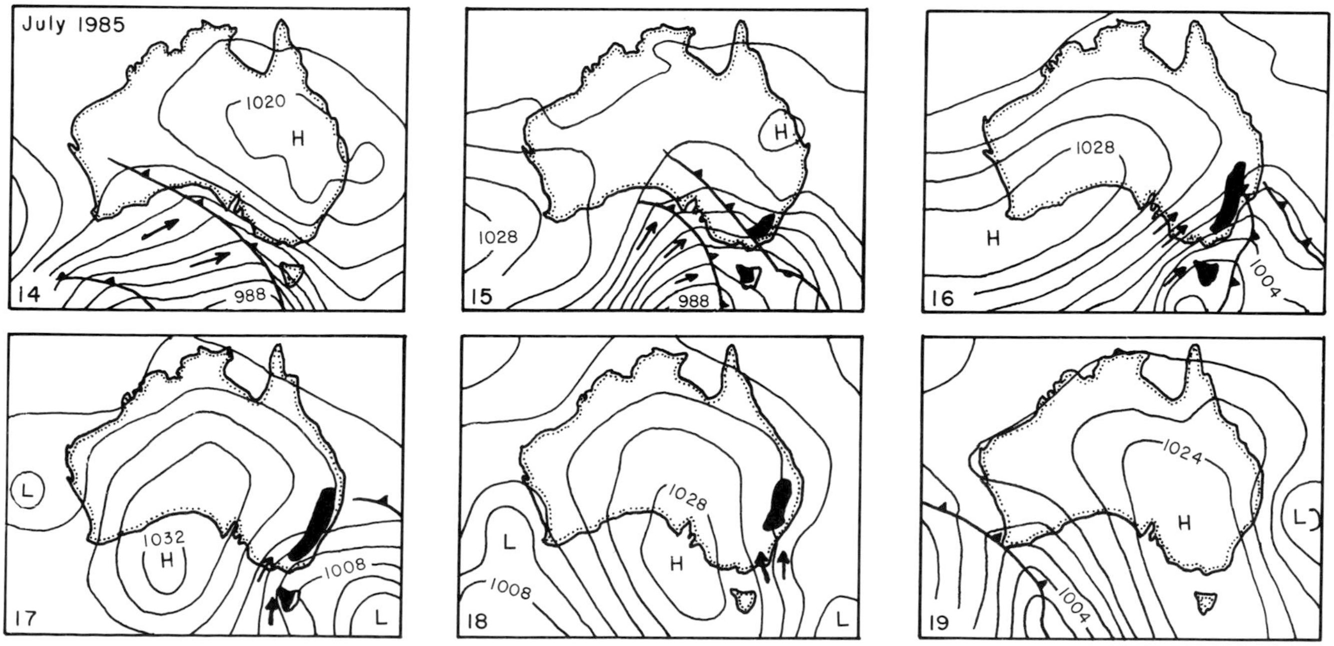

Fig. 7.8 Pressure patterns that brought snow to eastern Australia, 14–18 July 1985. Areas affected by snowfall are shaded.

FIG. 7.9 Effect of a heavy snowstorm on roadways in St John's, Newfoundland (photograph from Canadian Atmospheric Environment Service)

warmth can suffocate occupants with carbon monoxide, while accompanying cold temperatures can freeze people to death. In the severe winter of 1977–78 in Illinois, 24 people died in motor cars from these causes. Airports are also affected within hours, and can be kept closed for several days because of the time required to clear long, wide runways. Closure of major airports such as Chicago can disrupt airline schedules nationwide. Blocking of roads leads to closure of schools, industry and retail trade. The effect on retail trade, however, is often short term because shoppers usually postpone purchases until after the storm. In fact, the seasonal onset of snow is seen in many northern states as a blessing, because it triggers consumer spending for Christmas. Purchase of winter clothes, snow tires, and heaters can also increase. Motel business can boom during major storms; surprisingly, garage stations record heavier sales, too, because people unable to reach work can manage to travel to the local service station to complete delayed vehicle maintenance. For most industry, a major storm means lost production. Contingencies must be made by certain industries, such as steelworks, to begin shutting down blast furnaces before a storm affects the incoming shift of workers. Unless production is made up, industry can lose profits and contracts, workers can lose pay and governments income-tax revenue. Worker and student absenteeism can overload domestic power demand, as these people are at home during the day utilizing heating and appliances not normally used at that time. Severe, rapidly developing, or unpredicted snowstorms may even prevent the workers required for snow removal reaching equipment depots.

Many cities in North America annually budget millions of dollars to remove snow from city streets. Most cities have programs of sanding and salting roads and expressways to keep the snow from being compacted by traffic and turning into ice. The salt volumes, put on roads by cities surrounding Lake Ontario in winter, are so large that increasing lake salinity now threatens freshwater life. Such snow-removal programs necessitate a bureaucracy which can plan effective removal of snow at any time of day or night during the winter months. Inefficiency in this aspect has severe economic repercussions on industry and retail trading.

All of these disruptions depend upon the perception of the storm. Cities which experience large annual snowfalls generally are more prepared to deal with them as a hazard. Snowfalls under 10 cm are considered minor in the northeastern part of the United States, but will paralyze southern states. However, an awareness of snowstorms can be counter-productive. In the western United States, where snowfall on a personal level is considered as an everyday element of winter life, large snowfalls can cripple many major cities because they do not prepare for these higher magnitude events. In the eastern states large storms are expected, but continuous small falls are not. Here, snow clearing operations and responses are geared to dealing with these larger storms.

The variation in perception and response to the snow hazard is best illustrated around the Great Lakes. The Great Lakes remain unfrozen long into the winter because of their heat capacity. Cool air blowing over these bodies of relatively warm water can reach their saturation point quickly causing rapid accumulation of snow downwind under strong winds. As the prevailing winds are westerly in the southern lakes, and northerly around Lake Michigan and Lake Superior, snowbelts develop on the eastern or southern sides of the Lakes (figure 7.10). In these snowbelts, average annual snowfalls can exceed 4 m, double the regional average. Often this snow falls in a few events lasting several days each. Cities, such as Buffalo on the eastern side of Lake Erie, are prepared for the clearing of this type of snowfall, on top of which may be

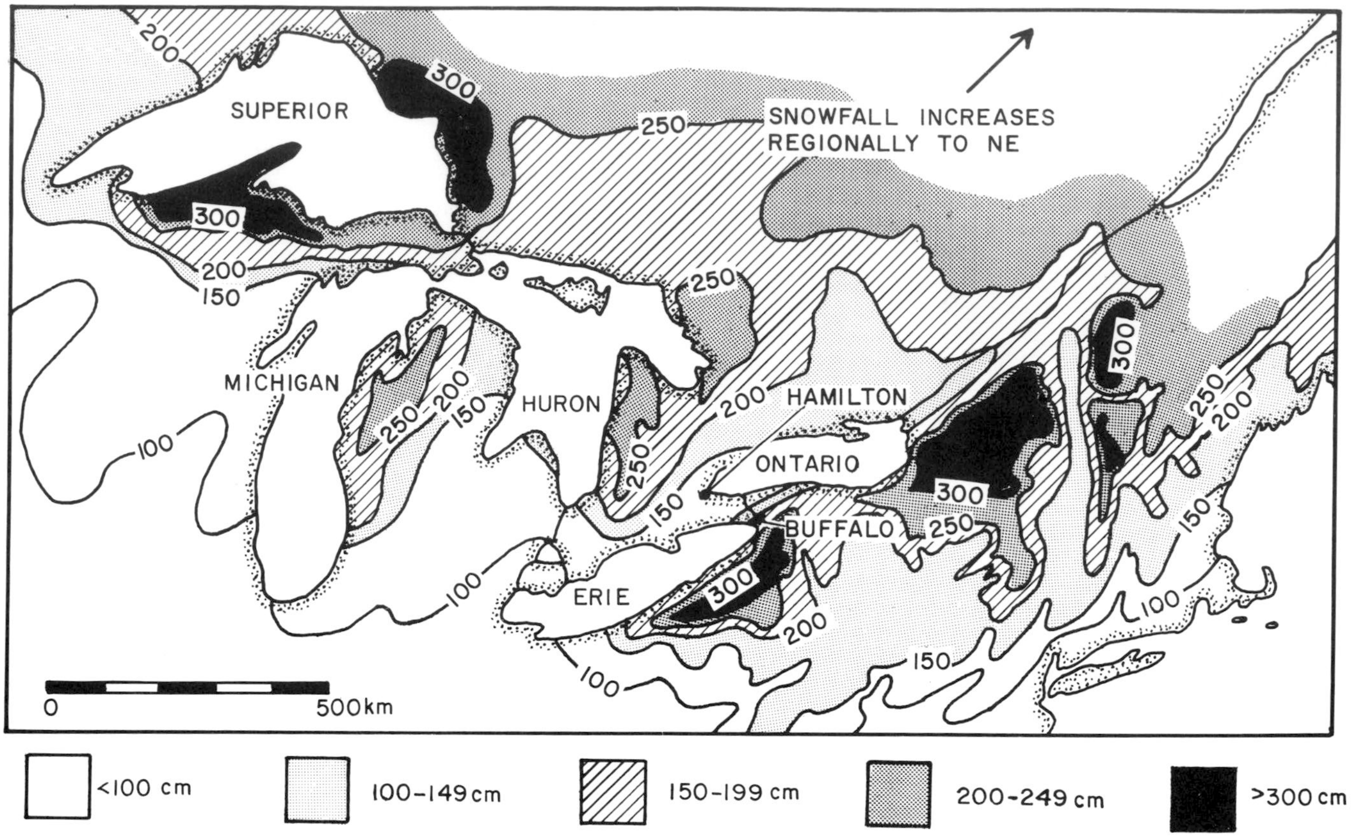

FIG. 7.10 Mean annual snowfall around the Great Lakes snowbelt (adapted from Muller and Oberlander, 1984)

superimposed major mid-latitude storms. Cities, such as Hamilton at the western end of Lake Ontario, only have to cope with the irregular occurrence of mid-latitude storms. Each city has built up a different response to snow as a hazard, and each city allocates different sums of money for snow removal operations. Buffalo removes its snow from roads and dumps it into Lake Erie, because if it did not, it soon would be overwhelmed by snow banked up along roadsides. Hamilton uses sanding and salting operations to melt smaller amounts of snow, and only resorts to snow removal for the larger storms. These two cities, within 60 km of each other, thus have totally different responses to snow as a hazard. However, average winter conditions may not appear. Too many mid-latitude cells, even of low intensity, tracking across the Great Lakes will diminish the time of westerly winds and increase the time of easterly winds. Hamilton then becomes part of a localized snowbelt at the west end of Lake Ontario, while Buffalo receives little snowfall. Buffalo may end the winter with millions of unspent dollars in its snow removal budget, while Hamilton has to request provincial government assistance, or increase its rates to cope with the changed conditions of one winter's snowfall.

The winter of 1976–77 was a dramatic exception for Buffalo. Abnormally cold temperatures set records throughout the northeast. Buffalo's well-planned snow removal budget was decimated by a succession of blizzards. On 28 January one of the worst blizzards to hit an American city swept off Lake Erie through the city. The storm raged for five days and occurred so suddenly that 17 000 people were trapped at work. Nine deaths occurred on expressways cutting through the city, when cars became stalled and people were unable to walk to nearby houses before freezing to death. Over 3.5 m of snow fell. Winds formed drifts 6–9 m high. For the first time in United States history, the federal government proclaimed a disaster area because of a blizzard. A military operation had to be mounted using 2000 troops to dig out the city. Not only was Buffalo cut off from the rest of the world, it was also completely immobilized within. Over 1 000 000 people were stranded in streets, offices, their own homes, or on isolated farms. It took two weeks to clear the streets. Snow had to be transported by train to other parts of the country to minimize the local spring flooding risk. The final clearance bill totalled more than $US200 million, more than five times the budgeted amount for snow removal. In contrast, the city of Hamilton was relatively unaffected by this blizzard, and ended up loaning snow removal equipment to Buffalo as part of an international disaster rescue operation.

The Buffalo blizzards of 1977 illustrate the additional hazard caused by abnormal snow amounts—flooding in the subsequent spring melting period. While snow accumulation can be easily measured, the rate of melting in spring is to some degree unpredictable. If rapid spring warming is accompanied by rainfall, then most of the winter's accumulation of snow will melt within a few days, flooding major rivers. The historic flooding of the Mississippi River in the spring of 1973 occurred as the result of such conditions, after a heavy snowfall season within its drainage basin. Snowmelt flooding is a major problem in the drainage basins of rivers flowing from mountains, and the interior of continents with significant winter snow accumulation. This aspect is particularly severe in the Po basin of northern Italy, along the lower Rhine, and downstream from the Sierra Nevada mountains of California. On a continental scale, the flooding of the Mississippi River and its tributaries is well known; however, for river systems draining north to the Arctic Ocean, snowmelt flooding can be a catastrophic hazard. Because the headwaters of these rivers lie in southern latitudes, melt runoff and ice thawing of the river channel occur upstream before occurring downstream at higher latitudes. The ice-dammed rivers, swollen by meltwater, easily flood adjacent floodplains and low-lying topography. In Canada flooding in spring of the Red River passing through Winnipeg, and the Mackenzie leading to the Arctic Ocean is an annual hazard. Nowhere is the problem more severe than on the Ob River system in the Soviet Union. All northward flowing rivers in the Soviet Union experience an annual spring flooding hazard, but on the Ob system the problem is exacerbated by the swampy nature of the low-lying countryside.

Blizzards

There is one special case of snowfall which does not depend upon the amount of snow falling but upon the strength of the winds blowing. The term 'blizzard' is given to any event with winds exceeding 60 km hr^{-1} and temperatures below −6° C. A severe blizzard occurs if winds exceed 75 km hr^{-1} with temperatures below −12° C. Strong mid-latitude depressions with central pressures below 960 hPa (less than some tropical cyclones), and accompanied by snow, will produce blizzard conditions. On the prairies of North America strong, dry northwesterly winds without snow are a common feature of winter outbreaks of cold Arctic air. These winds can persist for several days. Snow is picked up from the ground and blown at high velocity in a manner similar to that produced by dust storms described in chapter 3. Surprisingly, the deathtoll annually from blizzards in North America is the same as that caused by tornadoes. The 28 January 1977 blizzard at Buffalo, described above, resulted in over 100 deaths across the eastern United States, and had winds in excess of 134 km hr^{-1}. In March 1888 a spring blizzard with 110 km hr^{-1} gusts in New York, left 400 dead in the city and surrounding region.

The main hazard of blizzards is the strong wind which can drop the *wind-chill factor*, the index measuring equivalent still air temperature due to the combined effects of wind and temperature. Figure 7.11 summarizes the wind-chill factor. For a severe blizzard (wind speeds greater than 75 km hr^{-1} and temperatures less than −12° C), the equivalent temperature is −40° C. Exposed flesh will freeze at equivalent temperatures less than −14° C and frostbite is likely at wind-chill

Temperature °C	Wind Speed (km hr^{-1}) 10	20	30	40	50	60	70	80	90	100
	Equivalent wind chill temperature									
15	13	10	8	6	6	5	5	4	4	4
10	7	3	0	-2	-3	-4	-4	-5	-5	-5
5	1	-4	-7	-10	-11	-12	-13	-13	-13	-13
0	-4	-10	-14	-17	-18	-19	-20	-21	-21	-21
-5	-9	-16	-21	-24	-26	-27	-28	-28	-28	-28
-10	-15	-23	-28	-32	-34	-35	-37	-37	-37	-37
-15	-21	-30	-36	-40	-42	-44	-45	-46	-46	-46
-20	-26	-36	-43	-47	-49	-51	-52	-53	-53	-53
-25	-32	-43	-49	-54	-57	-59	-60	-61	-61	-61
-30	-37	-49	-57	-62	-65	-67	-68	-70	-70	-70
-35	-43	-56	-64	-70	-73	-76	-77	-78	-78	-78
-40	-49	-63	-70	-77	-80	-83	-85	-86	-86	-86
-45	-54	-69	-78	-84	-88	-90	-92	-93	-93	-93
-50	-60	-75	-85	-92	-96	-99	-101	-102	-102	-102

Blizzard Severe Blizzard

FIG. 7.11 Equivalent temperature or wind chill for given air temperatures and wind speeds temperatures of less than −35°C. At this latter temperature, lightly clothed skin or bare flesh will freeze in 60 seconds. If one's internal body temperature drops by more than 5°C, hypothermia and death will result. Given these constraints, anyone travelling on open roadways in a blizzard would be unable to live for more than 30 minutes in a vehicle if it stalled. If shelter was less than a couple of minutes away (and walking would be extremely difficult at wind speeds greater than 50 km hr^{-1}), death from exposure would be likely unless a person was clothed to withstand these temperature extremes.

Freezing rain

In some cases in winter or early spring, mid-latitude depressions will develop rain instead of snow along the uplifted warm front. In order to reach the ground, this rain must fall through the colder air mass. If this colder air is below freezing and the rain does not freeze before it reaches the ground, then it will immediately turn to ice upon contacting any object with a temperature below 0°C. If the warm front becomes stationary, then continual freezing rain will occur over a period of several days, and ice can accumulate to thicknesses of tens of millimeters around objects. The added weighted on telephone and power lines will snap the lines wherever they connect to a pole. Tree branches will be brought down, disrupting road transport and ensuring total collapse of the power transmission system (figure 7.12). On 4 March 1976 a front stalled in a line running from Texas to Maine. Warm Gulf air moved upward over this front, dropping rain through colder polar air, and producing one of the worst ice storms to affect the United States. The disruption to power lines took months to repair, and the damage bill ran into the hundreds of millions of dollars.

FIG. 7.12 Effect of freezing precipitation and wet snow on wires and tree branches in Ontario, Canada (photograph from Canadian Atmospheric Environment Service)

Concluding comments

Hazards associated with precipitation span a wide range of diverse topics. The common ground amongst them all involves the process of moisture condensation and the release of latent heat of evaporation. This release drives not only global air circulation but also the winds that characterize tropical cyclones and tornadoes. These latter two hazards were discussed in earlier chapters as separate phenomena because of their destructive potential. This chapter has attempted to summarize other smaller phenomena associated with this precipitation process and precipitation or flooding events not uniquely associated with large scale storms and tornadoes. It is recognized that many of these precipitation hazards are limited in scope. For instance, hailstorms do not operate in polar regions nor during winter in temperate climates. Freezing rain is only restricted to a few continental locations underlying the path of the jet stream in the northern hemisphere. However, some of these smaller hazards have enormous economic consequences or have considerable disruptive potential. For instance, while hailstorms are limited spatially, they are one of the most expensive hazards recurring over time in both Australia and the United States. Such smaller scale hazards can also be deadly: at present, in the United States lightning accounts for the most deaths of any natural hazard. The following chapter will discuss the larger consequences of flooding, involving the minor precipitation hazards discussed in this chapter together with those resulting from large-scale storms.

References

Eagleman, J. R. 1983. *Severe and unusual weather.* Van Nostrand Reinhold, NY.

Muller, R. A. and Oberlander, T. M. 1984. *Physical Geography Today* (3rd edn). Random House, NY.

Oliver, J. 1986. 'Natural hazards'. In Jeans, D. N. (ed.) *Australia: a geography*. Sydney University Press, Sydney, pp. 283–314.

Further reading

Geophysics Study Committee 1987. *The Earth's Electrical environment*. United States National Research Council, National Academy Press.

Linacre, E. and Hobbs, J. 1977. *The Australian Climatic Environment*. Wiley, Brisbane.

Nalivkin, D. V. 1983. *Hurricanes, storms and tornadoes*. Balkema, Rotterdam.

Rooney, J. F. 1973. 'The urban snow hazard in the United States: an appraisal of disruption'. In McBoyle, G. (ed.) *Climate in review*. Houghton Mifflin, Boston, pp. 294–307.

Whipple, A. B. C. 1982. *Storm*. Time-Life Books, Amsterdam.

Whittow, J. 1980. *Disasters: the anatomy of environmental hazards*. Penguin, Harmondsworth.

8

Flooding as a Hazard

Introduction

The earlier chapter on large-scale storms did not consider in detail the effects of flooding associated with tropical cyclones, extra-tropical storms or east-coast lows. Nor was it appropriate to discuss large-scale flooding in that chapter because some of the worst regional flooding in the northern hemisphere has occurred in spring in association with snow-melt. In addition, rainfall associated with thunderstorms is a major cause of flash flooding. Having covered these precipitation hazards in the previous chapter, it is now appropriate to discuss flooding phenomena in depth. This chapter examines flash flooding events first. Flash flooding refers to intense falls of rain in a relatively short period of time. Usually the spatial effect is localized. Modification of the overall landscape during a single event is minor in most vegetated landscapes; however, in arid, semi-arid, cultivated (where a high portion of the land is fallow) or urban areas, such events can be a major cause of erosion and damage. Steep drainage basins are also particularly prone to modification by flash floods because of the high stream-power values generated. Large-scale regional flooding represents the response of a major continental drainage basin to high magnitude, low frequency events. This regional flooding has been responsible historically for some of the largest deathtolls attributable to any hazard. When river systems alter their course as a response to floods, then disruption to transport, agriculture and urban land-use can occur over a large area. The chapter concludes with a discussion of the geological implications of flood diversion on major rivers in the world.

Flash floods

Magnitude and frequency of heavy rainfall

(Griffiths, 1976)

Short periods of heavy rainfall occur as a result of very unstable air with a high humidity. Such areas usually occur near warm oceans, near steep, high mountains in the path of moist winds, or in areas susceptible to thunderstorms. A selection of the heaviest rainfalls for given periods of time is compiled in figure 8.1. Generally, extreme rainfall within a few minutes or hours has occurred during thunderstorms. Areas with extreme rainfall over several hours, but less than a day, represent a transition from thunderstorm-derived rain to conditions of extreme atmospheric instability. The latter may involve several thunderstorm events. In the United States, the Balcones escarpment region of Texas has experienced some of the heaviest short-duration rainfalls in this category. Extreme rainfall over several days is usually associated with tropical cyclones, while extreme rainfall lasting several weeks to months usually occurs in areas subject to seasonal monsoonal rainfall, or conditions where orographic uplift persists. In the latter category, Cherrapunji in India, at the base of the Himalayas, dominates the data (9.3 m in July 1861 and 22.99 m for all of 1861). In this region monsoon winds sweep very unstable and moist air from the Bay of Bengal up over some of the highest mountains in the world. In Australia the highest rainfall in one year (11.3 m) occurred near Tully in Queensland, a location where orographic uplift of moist air from the Coral Sea is a common phenomenon in summer.

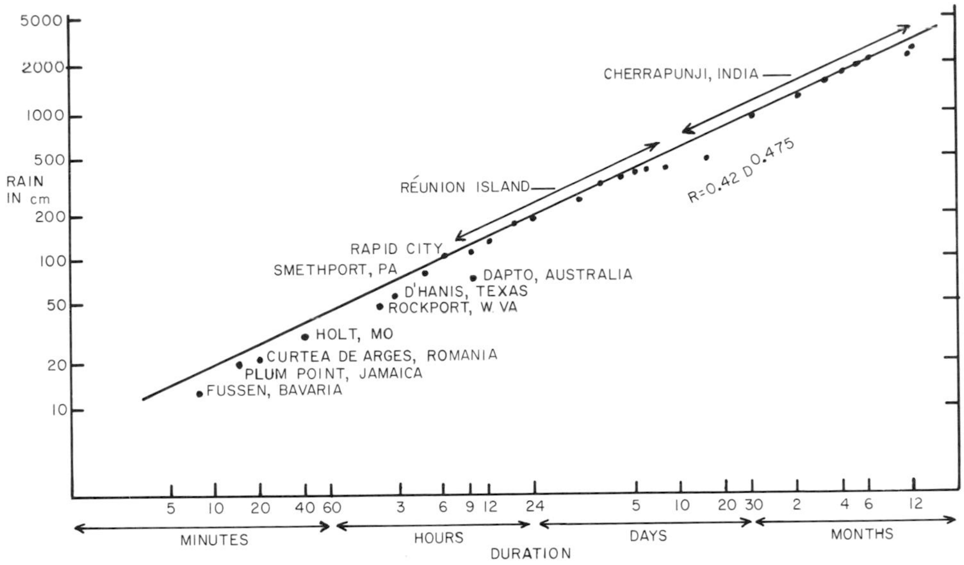

Fig. 8.1 Maximum rainfall amounts vs time (adapted from Griffiths, 1976)

The following equation (8.1) represents the line of best fit to the maximum rainfall data in figure 8.1.

$$R = 0.42\ D^{0.475} \quad (8.1)$$

where R = rainfall in meters
D = duration in hours

While this equation fits the data well for periods above one hour, the line overestimates the amount of rain that can fall in shorter periods. The latter may simply represent the scarcity of *pluviometers* able to record rainfall at this shorter timespan. The line probably represents a physical limit to the rate of condensation and droplet formation possible under the Earth's present temperature and pressure regime. The amounts of rainfall possible in less than one day are staggering. Within an hour, 0.42 m of rain are possible, and within a day, 1.91 m of rain can fall. The latter value represents more than the average yearly rainfall of any Australian capital city.

The large amounts of rainfall at Cherrapunji, India (or, for that matter, at Tully, Queensland), while exceptional, probably do not disturb the environment nearly as much as these intense but short falls. Over long time-periods, locations normally receiving heavy rainfall have evolved dense vegetation that can absorb the impact of falling rain, and have developed drainage patterns that can handle the expected runoff. Most extreme rainfalls over the short term can occur in places where vegetation is sparser, and where drainage systems may not be best adjusted to contain large volumes of runoff. Here, dislodgement of topsoil by raindrop impact can also be high. Sheetflow occurs within very short distances of drainage divides and overland channel flow will develop within several meters downhill. As a result, sediment erosion and transport is high and rapid. For these reasons, flash flooding in arid and semi-arid regions can become especially severe. In urban areas, where much of the ground is made impervious by roads or buildings and where drainage channels are fixed in location, flash flooding becomes more likely with much lower amounts of rainfall than indicated in the above graph. Flash flooding in semi-arid or urban catchments appears to be an increasing phenomena in the past two decades, both in the United States and Australia.

Flood power (Baker and Costa, 1987)

The amount of work that a flood can perform, and the destructive implications of a flash flood, are not necessarily produced by the high amounts of rainfall described above. Nor are high flood discharges a prerequisite for erosion. Rather, it is the amount of *shear stress*, and the stream *power*—or the amount of energy developed per unit time along the boundary of a channel—that are much more significant. It is because of stream power that floods in small drainage basins, as small as 10–50 km^2, can be more destructive than major floods on the Mississippi or Amazon rivers, with discharges 10–1000 times larger. This is particularly so where channels are narrow, deep and flow steeply through bedrock. Stream power per unit area of a channel is defined by the following equation:

$$\omega = \tau v \tag{8.2}$$

where ω = power per unit boundary area
v = velocity
τ = boundary shear stress
= $\gamma R S$
where γ = specific weight of the fluid ($9800\ \mathrm{N\,m^{-3}}$)
S = the energy slope of the flow
R = the hydraulic radius
= $A(2d + w)^{-1}$
where A = the cross-sectional area of the wetted channel
d = the mean water depth
w = the width of the channel

The parameters in equation 8.2 are not that difficult to measure, because many rivers and streams in the world have gauging stations that measure velocity and water depth during floods. Once the height of a flood has been determined, it is relatively simple to calculate the hydraulic radius. The slope of a channel is normally used in place of the energy slope; however, this may lead to variations from the true energy slope of 100 per cent during catastrophic flash floods. Problems may also arise in determining the specific weight of the fluid. The value quoted above is standard for clear water; during floods, however, waters can contain high concentrations of suspended material that can double the specific weight of floodwater.

Figure 8.2 plots stream power against drainage basin area for small flash floods, mainly in the United States, and for the largest floods measured in recent times. Some of these floods, such as the Teton River flood in Idaho on 5 June 1976, can be associated with the collapse of dams following heavy rains. The collapse of a dam can greatly increase the magnitude of stream power, because of its effect on hydraulic radius and the energy slope of the flow. In fact, dam collapses have led to the largest flash-flood deathtolls. For example, the Johnstown, Pennsylvania, flood of 31 May 1889 killed over 2200 people, and the Vaiont Dam failure in Italy on 9 October 1963 took 2000 lives. The failure of dams during heavy rains can be attributed to neglect, inadequate design for high-magnitude events, geological location (built across an undetected fault line) or mischance. The greatest stream power calculated for any flood in the United States was the 8 June 1964 Ousel Creek flood, in Montana. This flood obtained stream powers of $18\,600\ \mathrm{W\,m^{-2}}$. Most of the powerful floods in the United States have occurred on quite small drainage basins because they have steeper channels. In comparison to the Ousel Creek flood, the largest flood on the Mississippi River in 1973 had a stream power of only $12\ \mathrm{W\,m^{-2}}$.

For larger drainage basins, two facets are required to get high stream powers. Firstly, the channel must be constrained and incised in bedrock. This ensures that high velocities are obtained because water cannot spill out across a wide floodplain. Secondly, the upstream drainage basin must be subject either to high-magnitude rainfall events, or to high discharges. These two features severely limit the number of high stream-power events witnessed on large rivers. The Teton dam failure in Idaho in 1976 obtained a stream power in excess of $10\,000\ \mathrm{W\,m^{-2}}$, because it flowed through an incised river channel. Similar magnitudes have been obtained through the narrow Katherine River gorge in northern Australia. The Pecos River in Texas is also incised into a limestone escarpment subject to flash flooding, mainly during tropical cyclones such as Hurricane Alice, which penetrated inland in 1954. Finally, the Three Gorges section of the Chang Jiang (Yangtze) River, which drains an area of $1\,000\,000\ \mathrm{km^2}$, is the only large river to obtain stream power values as high as flash floods on streams with drainage basins less than $1000\ \mathrm{km^2}$. At this point, the Chang Jiang River concentrates snowmelt and rainfall from the Tibetan plateau in a narrow gorge, which, in the 1870 flood, flowed at a depth of 85 m with velocities of $11.8\ \mathrm{m\,s^{-1}}$.

Figure 8.2 does not plot the largest known floods. Stream power values for cataclysmic prehistoric floods associated with Lake Missoula at the margin of the Laurentian ice-sheet on the Great Plains, and with Lake Bonneville, were an order of magnitude larger. These resulted from the sudden bursting of *ice-dammed lakes* as meltwater built up in valleys temporarily blocked by glacier ice during the Wisconsin glaciation. The Missoula floods obtained velocities of $30\ \mathrm{m\,s^{-1}}$ and depths of 175 m, producing stream powers in excess of $100\,000\ \mathrm{W\,m^{-2}}$, the largest yet calculated on Earth. Similar types of floods may have emptied from the Laurentian ice-sheet into the Mississippi River, and from Lake Agassiz eastwards into the Great Lakes-St Lawrence drainage system. Equivalent large floods also existed on Mars at some time in its geological history; however, in order to reach the high stream powers recorded on Earth, the flows may have been as deep as 500 m to compensate for the lower gravity on Mars.

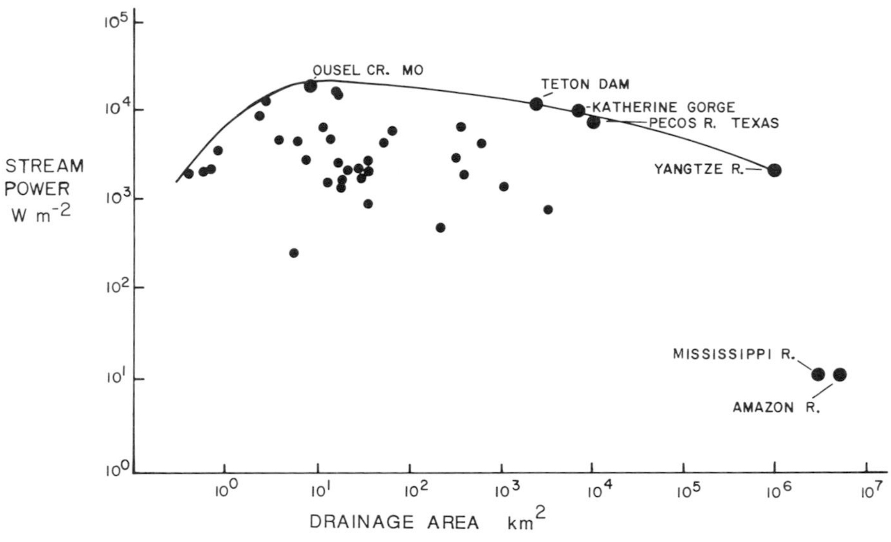

FIG. 8.2 Plot of stream power per unit boundary area vs drainage basin area (adapted from Baker and Costa, 1987). The curve delineates the upper limit in flood power.

The ability of these catastrophic floods to erode and transport sediment is enormous. Figure 8.3 plots mean water depth against the mean velocity for these flood events. Also superimposed on this diagram are the points where flow becomes supercritical and begins to jet, and where water will begin to vaporize or cavitate. *Supercritical flow* is highly erosive, and cannot be sustained in alluvial channels, because, the bed will be eroded so rapidly. It is also rare in bedrock channels because, once flood waters deepen a small amount, the flow will then revert to being *subcritical*. *Cavitation* is a process whereby water velocities over a rigid surface are so high, that the vapour pressure of water is exceeded, and bubbles begin to form at the contact. A return to lower velocities will cause these bubbles to collapse with a force in excess of 30 000 times normal atmospheric pressure. Next to impact cratering, cavitating flows are the most erosive process known on Earth and, when sustained, are capable of eroding bedrock. Fortunately, such flows are also rare. They can be witnessed on dam spillways during

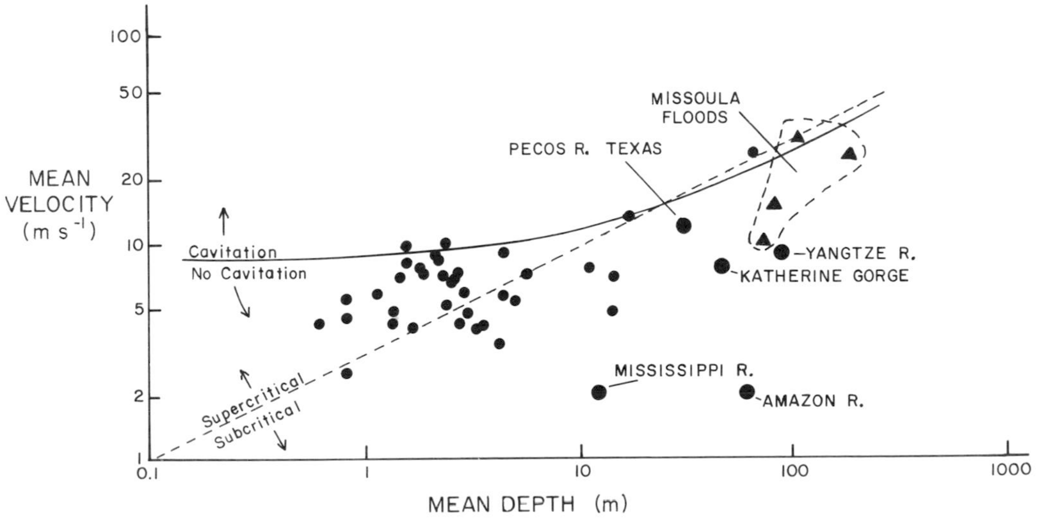

FIG. 8.3 Current velocity vs channel depth for various floods plotted together with boundaries for cavitation and supercritical flow (adapted from Baker and Costa, 1987)

large floods as the mass of white water that develops against the concrete surface towards the bottom of the spillway. Note that the phenomenon differs from the white water caused by air entrainment by *turbulence* under normal flow conditions at the base of waterfalls, or other drop structures. Cavitation is of major concern on spillways because it will shorten the lifespan of any dam, and may be a major process in the back-cutting of waterfalls into bedrock. Only a few isolated flash floods in channels have generated cavitation; however, some of the Missoula floods obtained cavitation levels over long distances.

Large flood events have enormous potential for modifying the landscape. The stream powers of cataclysmic floods, as well as many other common flash floods in the United States, are substantial enough to transport sand-sized material in wash load, gravel 10–30 cm in size in suspension, and boulders several meters in diameter as bedload. This material can also be moved in copious quantities. Even stream powers of 1000 $W m^{-2}$ are sufficient to move boulders 1.5 m in diameter. Many flash floods can reach this capacity. If cavitation levels are reached, bedrock channels can be quickly excavated. The Missoula floods were able to pluck basaltic columns of bedrock, each weighing several tonnes, from stream beds, and carve canyons through bedrock in the western United States in a matter of days. More importantly, as will be discussed below, present-day flood events are capable of completely eroding in a matter of days floodplains that may have taken centuries to form.

Flash flood events

United States (Bolt et al., 1975; Cornell, 1976; Maddox et al., 1980; Hirschboeck, 1987)

Figure 8.4 plots the location of significant flash flood events in the United States between 1935 and 1985. Not included on this map are larger flood events due to snowmelt or the effects of tropical cyclones. For instance, the Mississippi flooding of 1973, which will be discussed in detail later in this chapter, is not represented. Nor is the flooding of the Susquehanna River in 1972 following Hurricane Agnes. Noteworthy on this map is the clustering of flooding in central Texas, and in the central part of the United States from Nevada to Iowa. The degree of severity of a catastrophic flood is ultimately determined by basin physiography, local thunderstorm movement, past soil moisture conditions and degree of vegetation clearing. However, most floods originate in anomalous large-scale atmospheric circulations that can be grouped into four categories as follows: aseasonal occurrence or anomalous location of a definable weather pattern, a rare concurrence of several commonly experienced meteorological processes, a rare upper atmospheric pattern, or prolonged persistence in space and time of a general meteorological pattern.

The degree of abnormality of weather patterns is very much a function of three types of pressure and wind patterns that are often part of the general atmospheric circulation. These patterns are shown in figure 8.5. *Meridional flow* represents north-south air movement and is usually associated with Rossby wave formation linked to the meandering path of the polar jet stream, especially over continents in the northern hemisphere. As stated in chapter 5, the jet-stream path can be perturbed by El Niño-Southern Oscillation events, and can be associated with persistent weather patterns giving rise to droughts. Exaggerated meridional flow also appears to be one of the most frequent atmospheric patterns generating flash flooding in the United States. Frequently, it is associated with blocking of a low-pressure cell or, more often, a high-pressure cell gets stranded in situ with jet streams splitting and passing around the cell. Blocking in North America commonly occurs over the north Atlantic and Pacific. The latter occurrence is associated with warmer sea surface temperatures often linked to ENSO events. In the north Atlantic blocking increases extra-tropical storm activity and enhances persistence of patterns.

Within this general picture of anomalous circulation are smaller features which are responsible for localized flash floods. These features can be classified into four categories as follows: synoptic, frontal, mesohigh, and western events. Synoptic events occur with an intense mid-latitude cyclone and a semi-stationary front that are linked to an intense, low-pressure trough aloft. Rainfall is widespread, persistent and, in local cases, heavy as thunderstorms develop repeatedly in the same general area. Such storms occur in the seasonal transitions between winter and summer pressure patterns over North America. Thus, they are common from spring to early summer, and in autumn. Frontal events are generated by a stationary or very slow moving front with zonal air circulation. Upper air stability may exist, and rainfall is usually heavy on the cool side of any warm front, as warm air is lifted aloft. Such events basically occur in July–August in the United States. Mesohigh floods

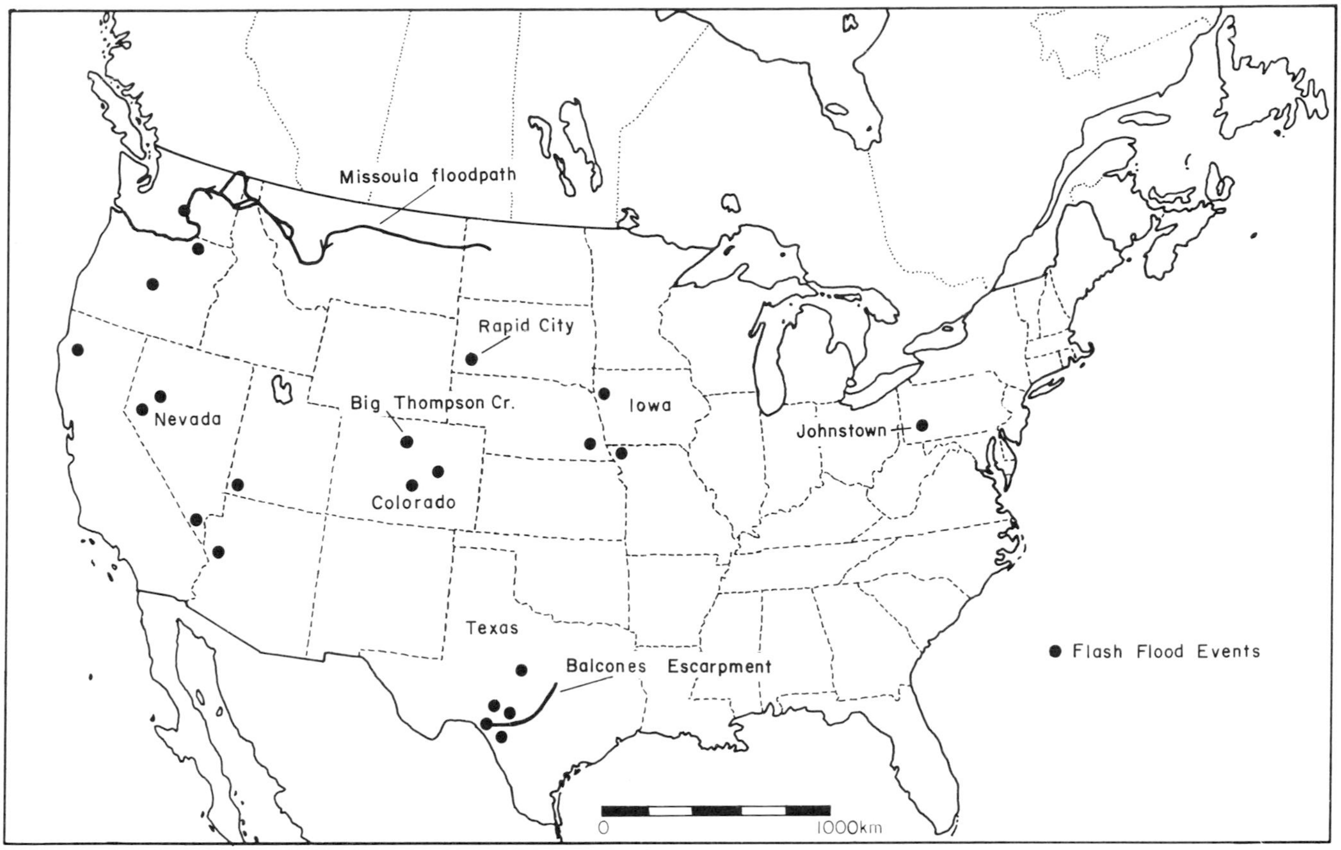

Fig. 8.4 Location map of catastrophic flash flood events in the United States, 1935–85 (based upon Hirschboeck, 1987)

are caused by instability and convection following the outbreak of cold air that may, or may not be, associated with a front. For instance, a stationary front may develop *supercell thunderstorms* that force out, in a bubble fashion, a high-pressure cell. These storms occur in the late afternoon or evening and are basically a summer feature.

Western-type events refer to a range of regional circulation patterns over the Rocky Mountains. These events tend to peak in late summer and are associated with either extremely meridional or zonal circulation. In the southwest of the United States, they tend to be associated with the summer monsoon, while elsewhere they have teleconnections

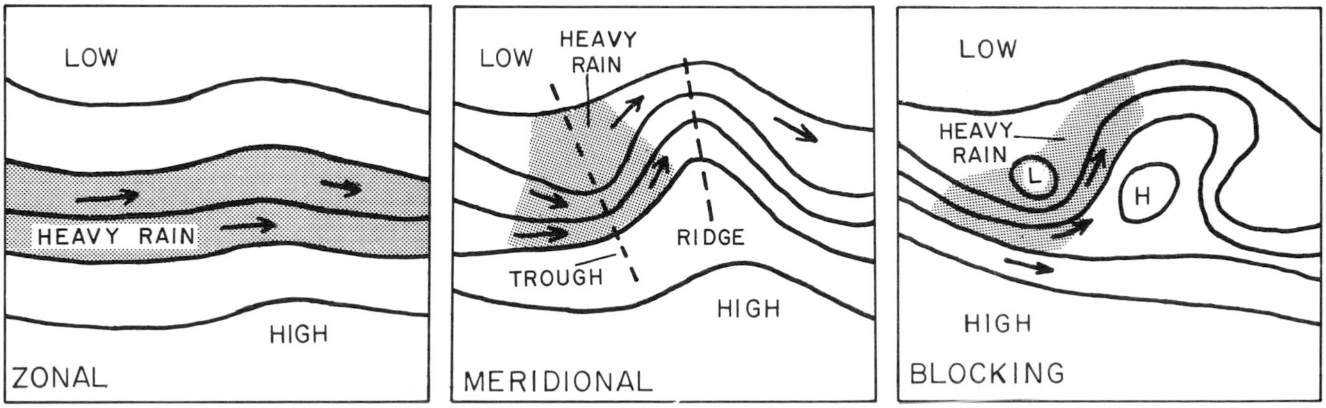

Fig 8.5 Abnormal patterns of air flow, within general air circulation, leading to flash flooding (based upon Hirshboeck, 1987).

with pressure patterns over the north Pacific. One of the better examples of the severity of this type of flooding was the occurrence of the Rapid City flood in the Black Hills of South Dakota, 9–10 June 1972. Air circulation was very meridional around a cut-off low. One meter of rain fell in the space of six hours, an amount that had a recurrence interval of 1:2000 years. Flash floods swept down all major streams including Rapid Creek, where an old dam—built 40 years earlier as a depression-relief project—collapsed, sending a wall of water through the downtown section of Rapid City. Over 230 people lost their lives and 2900 people were injured. Property damage, exceeding $US90 million, included over 750 homes and 1000–2000 cars. The possibility of future flash floods in this region has seen private property on urbanized floodplains bought up by the municipality, and converted to parkland.

The above classification does not totally characterize flash floods in Texas. Here, deep troughs and cold fronts can sweep across the state in winter, and interact with warm, humid air from the Gulf of Mexico. In addition, the area is also influenced by the intrusion of tropical cyclones and easterly waves embedded in trade winds. Strong orographic uplift of these westerly moving systems can occur along the Balcones escarpment. As a result, Texas has the reputation for being the most flood-prone region in the United States, and has recorded some of the highest rainfall intensities in the world (figure 8.1). For example, during the Pecos River flood of June 1954, Hurricane Alice moved up the Rio Grande valley and stalled over the Balcones escarpment. The hurricane reacted to an upper air disturbance, and turned into an intense, extra-tropical, cold-core cyclone. Over 1 m of rain was dumped in the lower drainage basin, resulting in floodwaters 20 m deep, with a recurrence interval of 1:2000 years. To date this is the largest recorded flood event in Texas.

In summary, flash flooding in United States has regional characteristics that are linked overall to general atmospheric anomalies. In the west flooding is controlled by the strength of blocking in the north Pacific Ocean, while on the Great Plains, thunderstorm convection appears to be important. In the Mississippi valley, the abnormal persistence of unusual trough systems is responsible for most flooding at least for larger drainage basins. In the eastern United States, blocking in the north Atlantic, in association with sea-surface temperature anomalies, directs both tropical and extra-tropical rain, producing cyclones inland.

Australia (Nanson and Hean, 1984; Bureau of Meteorology, 1985; Riley et al., 1986a, b)

The breaking of the long drought in eastern Australia following the 1982–83 ENSO event heralded the onset of extraordinary flash flooding in the Sydney-Wollongong region, as Walker circulation 'turned on' again over the following two years. The flooding was very localized and can be separated into two components: the Dapto flood of 17–18 February 1984 and the Sydney thunderstorms of 5–9 November 1984. In each case, slow moving or stationary convective cells developed in association with east-coast lows. Neither the localized convection or the east-coast lows were unusual. Initially, the events represented the rare concurrence of two commonly experienced meteorological processes. Unfortunately, the pattern has redeveloped continually in the 1980s with alarming frequency.

The Dapto flood event began in the late evening of 17 February 1984 as a cold front was pushed through the area by high pressure centered southeast of Tasmania. As the front passed through Wollongong, cold air raced rapidly into the Tasman Sea, orientating isobars perpendicular to the coastline. As a result, moist onshore air flowed into the area at the same time as a small east-coast low developed south of Newcastle. This low tracked down the coast and then stabilized over the 500 m high Illawarra escarpment, north of the town of Dapto, at 7 am on 18 February. This was followed by development of a complex, upper level trough over the Wollongong region for the rest of the morning. These lows produced extreme instability and caused marked convergence of moist northeast and southeast airflow into the Wollongong area. Orographic uplift along the 500–700 m high escarpment dumped copious amounts of rain over the Lake Illawarra drainage basin over 24 hours (figure 8.6). In a one-hour period on 18 February 123 mm fell west of Dapto. Over a nine-hour period, 640 mm and 800 mm were recorded, respectively, at the base and crest of the escarpment. The heavier rainfall had a 48-hour recurrence interval of 1:250 years.

The resulting flooding of Lake Illawarra, however, had a 1:10-year frequency of occurrence. Lake Illawarra was previously flooded as badly in March 1975. This case supports the United States evidence, showing that flash flooding with severe consequences can occur in parts of a drainage basin with areas less than 50 km^2. Figure 8.6 shows the highly localized nature of the event. Within 5 km

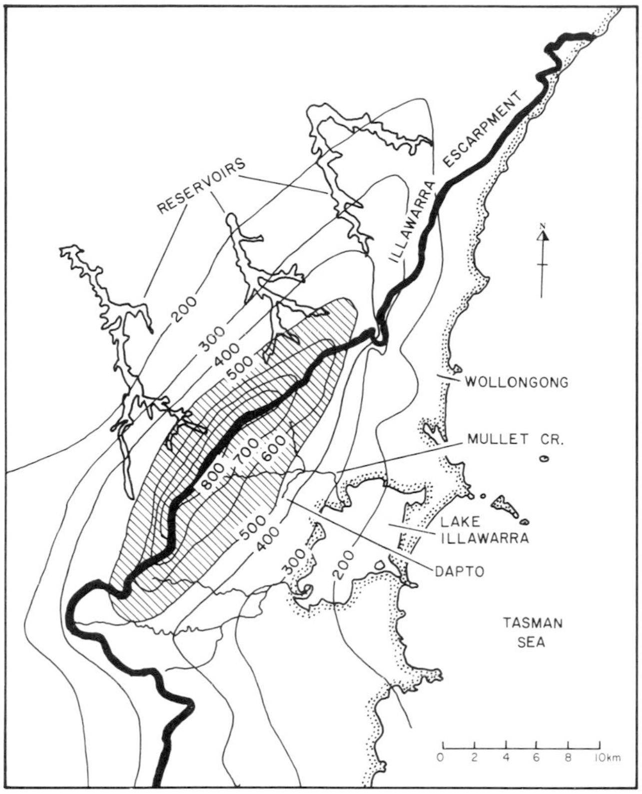

FIG. 8.6 48-hour isohyets for the Dapto flash flood, Wollongong, Australia, 18 February 1984 (based upon Nanson and Hean, 1984)

of the escarpment, rainfall amounts had dropped to less than 400 mm; within 10 km, amounts were less than 200 mm. Because any area shown on the map can be subject to this type of flooding, albeit rarely, it has been estimated that the Wollongong urban area could experience a Dapto-type event every 25 years. In fact, two subsequent flash flood events along the Illawarra escarpment, 10 km north of Wollongong, have occurred within four years of this flood. Emergency organizations should, thus, be prepared to handle high magnitude, localized rare floods in this region, with a much shorter expectancy rate than evidenced by individual events.

The heavy rainfall led immediately to flooding of creeks flowing into Lake Illawarra, peaking in the Dapto urban area between 1 am and 10 am on 18 February. Pleistocene floodplain terraces adjacent to these streams were flooded, and urbanized areas around Lake Illawarra inundated as waters drained through a long, constricting tidal inlet. The upper part of channels on steep slopes were extensively incised, both laterally and vertically, in response to high stream powers. In some cases, there was a fourfold increase in channel size (figure 8.7). Stream powers decreased rapidly downstream as gradients decreased and floodwaters spilt over the floodplain. Damage to the landscape was not restricted to channels and floodplains. Before the storm, the Illawarra escarpment was well vegetated and stable in this area. Landslides occurred along 25 per cent of the length of the escarpment in the heavy rainfall area. In addition, sediment-laden floodwaters deposited the equivalent of three to five years of normal sedimentation in Lake Illawarra.

The Sydney metropolitan flooding was not a single event, but rather a series of intense thunderstorms which struck various parts of the region on 5–9 November 1984. Accompanying the synoptic weather situation was a series of tornadoes and water spouts which caused over $A1 million damage at Norah Head to the north. The floods themselves cost the insurance industry $A40 million. One death resulted from lightning. This deathtoll is small, considering the fact that most observers would class the lightning activity on the night of 8 November as the worst in living memory. Over the five-day period, rainfall totalled 550 mm south-southeast of the Sydney central business district. Most of the eastern and northern suburbs, as well as the Royal National Park, south of Sydney, received totals in excess of 300 mm.

The floods originated from a low-pressure cell centered southeast of Brisbane on 5 November. This low moved slowly down the coast, rotating counterclockwise around a strong, blocking high-pressure cell (maximum pressure 1032 hPa) that drifted from Adelaide to New Zealand over the period (figure 8.8). The high-pressure cell directed

FIG. 8.7 New channel, Mullet Creek, Wollongong, Australia. This channel was incised through colluvium and into bedrock on the creek's upper slopes by the Dapto flash flood event, 18 February 1984.

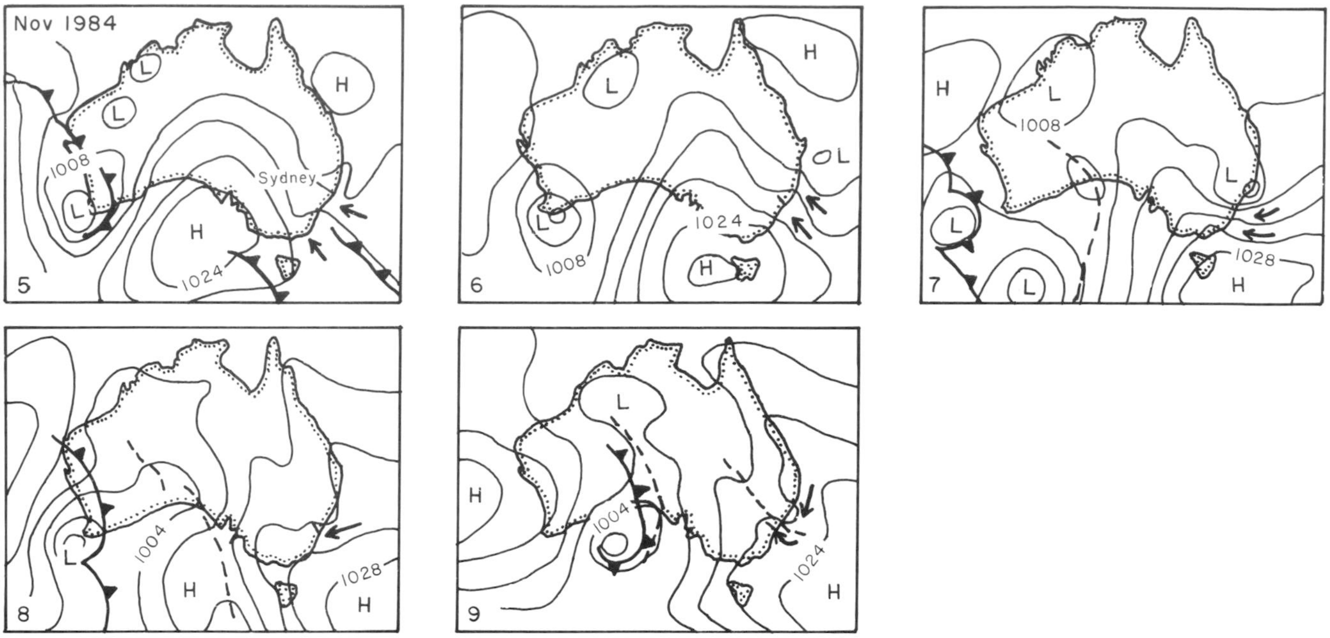

FIG. 8.8 Synoptic pattern for Sydney, Australia, flash floods: 5–9 November 1984

consistent, moist, zonal flow onto the coastline. Concomitant with the movement of the high, a low-pressure trough developed parallel to the coastline. This trough spawned low-pressure cells that converged meridional airflow over Sydney, which resulted in intense atmospheric instability. The flooding occurred as three separate events during this period.

The first event on Monday, 5 November began as thunderstorm activity at Cronulla (figure 8.9A) and slowly moved northward over the city center. This thunderstorm activity originated as a series of intense cells associated with a low-pressure trough parallel to the coast and a separate low-pressure cell 50–100 km south of Sydney. A second trough existed in the interior of the continent (west of the Dividing Range), and drifted over Sydney in the early evening. Both of these troughs were responsible for separate rainfall events. Over 220 mm of rainfall fell within 24 hours over Randwick. Rainfall intensities for timespans under nine hours exceeded the 1:100-year event. The most serious flooding was caused by a storm drain at Randwick race course that was unable to cope with the runoff. Adjacent streets in Kensingston bore the brunt of substantial flooding, which entered houses to depths of 0.5–1.0 m. Evening rush hour was thrown into chaos.

In the second event on 8 November, a low-pressure trough again was present, this time directly over Sydney. Airflow converged into this trough over the northern suburbs. Seventeen convective cells were responsible for rainfall during the morning with the first two producing the heaviest falls. Over 230 mm of rain fell at Turramurra, in the northern suburbs, with falls of 125 mm between 7:15 and 8:15 am (figure 8.9B). This latter intensity exceeded the 1:100-year event. The 20-minute rainfall intensity record at Ryde had a recurrence interval of between 50 and 500 years. Flooding during the second event began within 20 minutes of the commencement of heavy rainfall. Peak discharges occurred within five minutes of flooding. Many of the catchments which flooded were no more than 1 km long; however, flood depths ranging up to 1.5 m became common, both in channels, and in shallow depressions which formed part of the drainage network. Many residents were unaware that homes in these depressions occupied part of the flood stream network. Flooding occurred in the middle reaches of catchments as stormwater drains reached capacity and surcharged through access-hole covers. This *surcharging* capacity is deliberately built into Sydney's stormwater system, as it is prohibitively expensive to build an underground drainage network large enough to cope with rare events. Some of the flooding was due to storm inlet blockage by trees, loose debris or, in some cases, cars transported in floodwaters. It was even found that backyard paling fences could significantly block, divert or concentrate flows, hence increasing flood levels locally by as much as 40 cm.

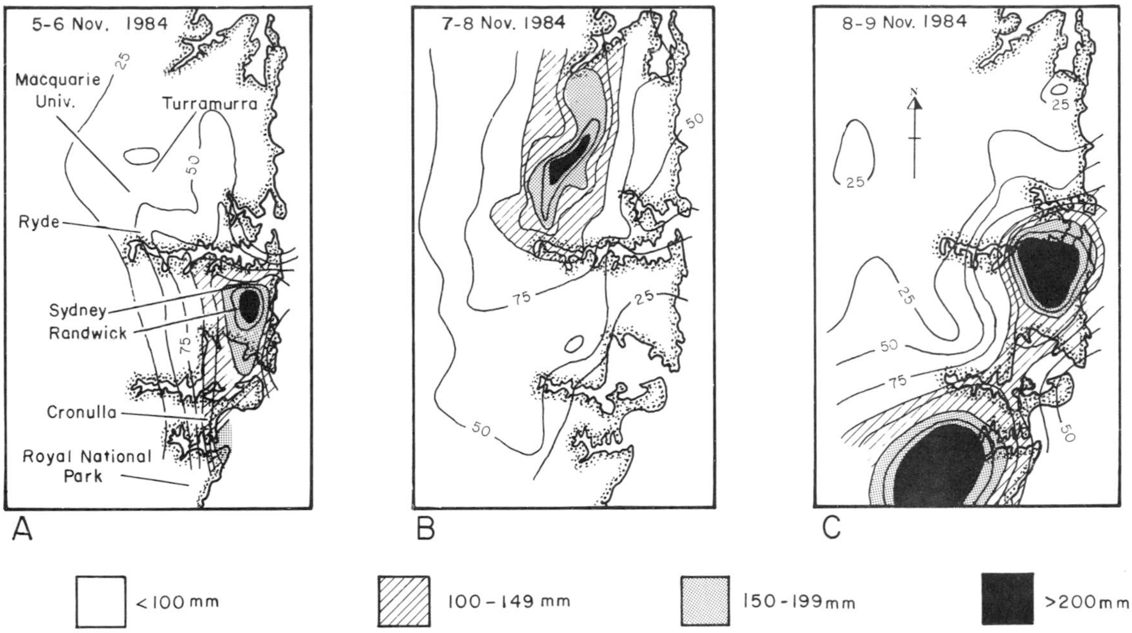

FIG. 8.9 Isohyets for three flash flood events in Sydney, Australia: 5–9 November 1984 (modified from Bureau of Meteorology, Australia, 1985)
A. 5–6 November.
B. 7–8 November.
C. 8–9 November.

The third event occurred on the night of 8 November. Synoptically, it was very similar to the second event, with a trough cutting through Sydney and converging air into the city region. Rainfalls of 249 mm were measured at Sydney's Botanical Gardens, and values near 300 mm in and around the Royal National Park to the south (figure 8.9C). Again, several cells were involved, and again peak intensities for timespans under nine hours exceeded the 1:100-year event. Many of the suburbs affected by this event had experienced their previous 1:100-year rainfall only four days previously. During this event, the road network supplanted natural drainage and took flood waters away from planned storm drains normally expected to carry runoff. Because roadways have reduced frictional coefficients, floodpeaks were almost instantaneous throughout the drainage system. Most of the damage during the third flash flood was caused by floodwaters reaching shopping centers at the base of steep slopes, sending walls of water and debris, in some instances including expensive cars, cascading through shops. In addition, the Tank Stream, which runs beneath the lower central business district of Sydney was reactivated, flooding the Sydney Stock Exchange and causing several millions dollars worth of damage. The State Library also was flooded with damage to rare exhibits.

There are three points about flash flooding in urban areas to be drawn from the Dapto and Sydney events. Firstly, the probability of occurrence of a high magnitude, localized rainfall event in a region can be an order of magnitude greater than the probability of the event itself. For example, the Dapto Flood was a 1:250-year event. The fact that any part of the Illawarra can experience such a localized storm, indicates that the probability of a similar flood occurrence in the Wollongong region is about 1:25 years. Sydney experienced at least five 1:100-year events in the space of five days. In some cases, emergency services had to respond to all five events. Since then, Sydney has received the heaviest 48-hour rainfall on record. On 5–6 August 1986 an east-coast low dropped over 430 mm of rain in three days causing rivers to flood, from Bathurst, in the Blue Mountains, to the Georges River, in south Sydney. At least six people died, from drowning or being electrocuted by fallen power lines. Damage exceeded $A100 million, which included over 3000 motor vehicles swamped by flood waters. In this case, rainfall with a recurrence frequency of 1:100 to 1:200 years was experienced over a large area, much of which had been affected by sporadic 1:100-year events two years previously.

Secondly, urban flooding in most cases peaked within half an hour of the onset of intense rain.

Much of the flash flooding was exacerbated by the structure of the urban drainage system. In the less densely built northwestern suburbs of Sydney, surcharging of the storm drains along existing natural drainage routes caused flooding to homes and shops built in these areas. In the more densely built up eastern suburbs, the road network often took the place of the natural drainage system, causing damage to areas that normally would not expect the type of flooding they received.

Finally, it is very difficult to comprehend what the increased frequency in flooding means. While some might see it as a signal of Greenhouse warming, it appears to represent a shift towards extreme rainfall events in this region. Not only did the climatic patterns giving rise to one catastrophic flood reappear days later in the same location, but the patterns also tended to repeat themselves at other locations along the coast, and recur several times over the next few years. Since 1984, Wollongong has experienced three very localized flash floods while the Sydney area has experienced at least ten similar events. Heavy rainfall in 1989 witnessed the spread of such events to Adelaide and semi-arid parts of Australia centered on Broken Hill. In the case of Wollongong, a coroner's inquiry into one recent event has shown that local authorities are almost unaware of this change in the rainfall regime. In New South Wales the State government has realized that urban flooding has become more prevalent, and has planned a $A25 surcharge on rate payers to cover clean-up expenses and recover relief costs. Even this is a misconception of the nature of the climatic shift, because the surcharge is being applied to the western suburbs of Sydney, which in fact have not experienced the greatest number of flash floods. Whatever the reason for flash flooding along this stretch of coastline, the events represent a significant increase in the occurrence of a natural hazard that has economical and personal ramifications for most of the 3 000 000 residents in the region. Neither the significance of present flooding, nor its consequences, have been fully appreciated.

High magnitude, regional floods

High magnitude, regional floods usually present a disaster of national or international importance. Usually whole, large-scale drainage basins, or many smaller rivers in the same region are flooded. Larger drainage basins of importance include the Mississippi, the Chang Jiang in China and the Murray-Darling in Australia. The implications of such flooding may have catastrophic effects on the landscape, accounting for major geomorphic realignment of the river channel, or even course diversion. On the Mississippi and the Chang Jiang, the possibility of large deathtolls or disruption to national economies is a distinct possibility. This section will look at the effects of two major flooding events of national importance to their respective countries, namely the Mississippi River flood of the 1973, and the eastern Australian floods of 1974.

Mississippi River floods of 1973 (Bolt et al., 1975; Noble, 1980)

The Mississippi River drainage basin is the third largest in the world, and the largest in North America. The basin covers 3 224 000 km^2 representing 41 per cent of the conterminous United States (figure 8.10). Its headwaters can be divided into two areas, one originating at the continental divide in the Rocky Mountains, and the other in the Great Lakes lowlands. The latter network has only been developed since the last glaciation, and includes an area which can accumulate large, winter snowfalls.

Flooding in the Mississippi River can be divided into two periods: before and after 1927. Before 1927, flood control was considered a local responsibility. The flood of 1927 brought in the realization that the Mississippi, which crosses state boundaries, could only be controlled at the national level. Since 1927, over 3000 km of *levees* and floodwalls, averaging 9 m in height in the lower part of the basin, have been built to control flooding. Four floodways were also built to divert excess floodwaters into large storage areas, or into the Gulf of Mexico. Some parts of the channel susceptible to erosion were also stabilized with concrete mats, which cover the complete channel boundary. Finally, reservoirs were constructed on many of the tributaries to delay passage of floodwaters into the Mississippi. The 1973 flood represents the severest test of these flood mitigation controls since they were constructed.

The 1973 flood was the greatest to hit the basin in the 200 years since records began. The event started in the summer of 1972, when storage reservoirs on tributaries began to fill and watersheds became saturated with persistent rainfall. Rainfall continued throughout the winter in all tributary basins. The precipitation patterns paralleled the winter of 1926 preceding the 1927 record flooding. Heavier than usual snowfalls in the mid-west and

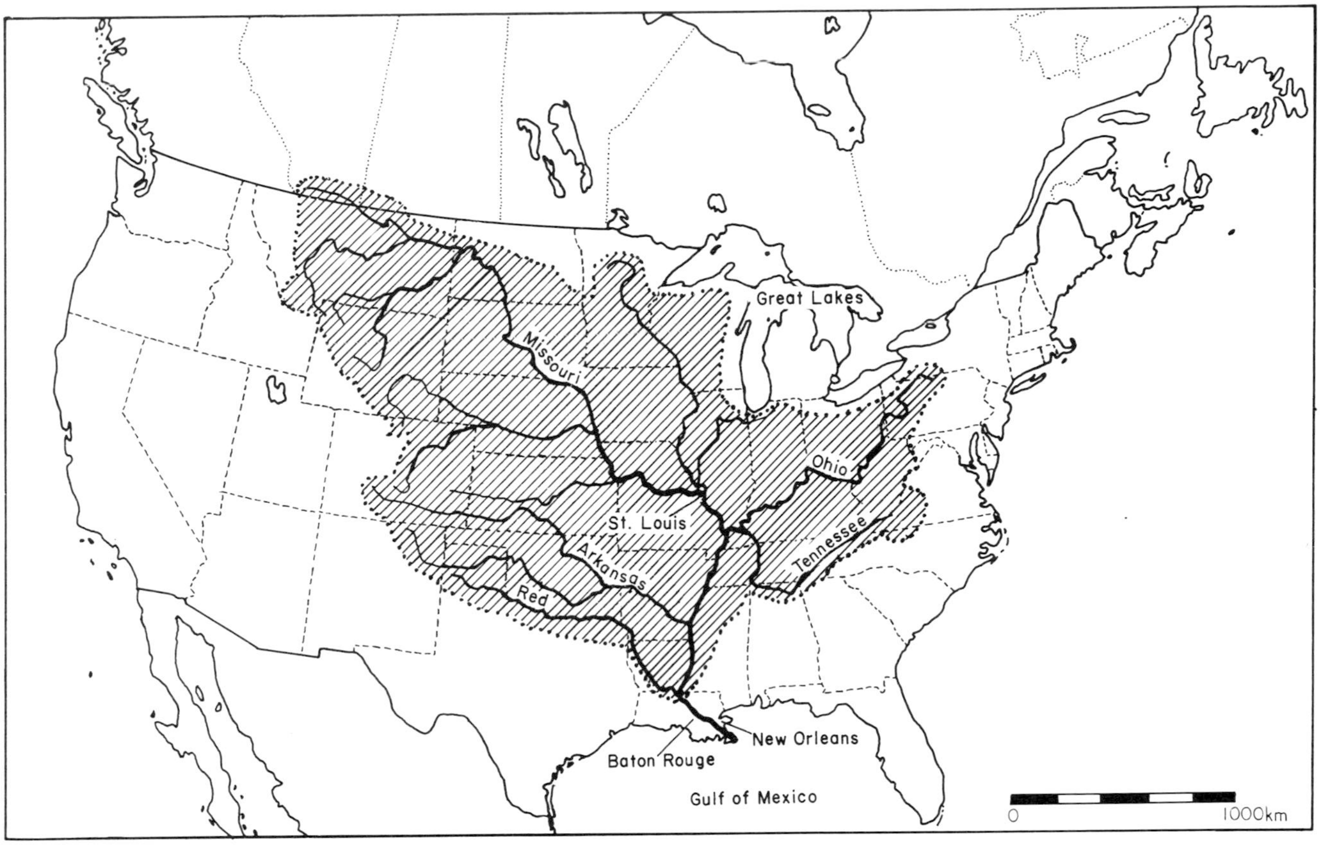

FIG. 8.10 Major tributaries of the Mississippi River drainage basin

the Great Lakes region also exacerbated the problem and, by 11 March 1973, flood stage was reached at St Louis before spring melting was complete. Within the next six weeks, the Mississippi River began to spill over levees along its middle reaches. The river maintained flood stage throughout April and May as heavy rainfall continued throughout the basin. This spring rainfall resulted from the repetitive passage of cyclonic lows through the central basin, and is a classic example of flooding in the United States due to the abnormal persistence of an expected air circulation pattern. Total rainfall for the six-month period exceeded by 45 per cent the norm in all parts of the basin. In one 30-hour period beginning 15 March, up to 220 mm of rain fell throughout the northern basin. The Mississippi River exceeded its average levels by 16 m from the confluence of the Ohio, downstream to Baton Rouge. Backwater flooding began along rivers entering the lower reaches of the river. There was some fear that at Baton Rouge the river might break its bank, and form a major and permanent diversion of the delta that would have stranded New Orleans as a major river port. On 28 April the river peaked at a 200-year historical high at St Louis, and residents were told in many places that levees would have to be raised higher in backwaters to prevent flooding. Unfortunately, in many towns, there was no sediment available to build higher levees, and 30 000 people were forced to evacuate. In mid-May, the river crested in the lower reaches but it was not until 20 June that the river returned to within its banks at Vicksburg, in the middle of the drainage basin.

Over 6.5 million hectares of land in seven states were inundated. Damage amounts exceeded \$US750 million, and 69 000 people were made homeless for periods up to three months. Without the flood mitigation works built since 1927, damage would have been as high as \$US10 000 million, and 12.5 million hectares of land would have been flooded. This flooding would have exceeded that resulting from the 1927 flood. The 3000 km of levees and floodwalls built since 1926 were subject to considerable river scour, wave wash and boiling. *Boiling* is a process whereby the head of water in the river channel forces pressurized water through, or under, permeable artificial levee banks. The process is innocuous enough as long as clear water seeps out; but if sediment becomes entrained in the flow, the seepage sorts this sediment, winnowing out the finer material and causing levee failure. The problem is particularly severe where levees overlay remnant, permeable point bars. Over 39 levee banks failed north of St Louis, with extensive flooding of farmland, and damage to buildings and communities relying on the protection of the levees. In some cases these communities had to be evacuated for periods of several months. Bank failures also occurred when river levels dropped suddenly. Water-saturated material turned liquid and flowed into the river, causing bank failure. Fortunately, many of the drops in water levels were temporary; however, points of failure present future locations of potential flood breakout. In the lower basin, flooding blocked tributaries, resulting in higher river levels and difficult-to-control flooding in at least three states.

The flooding of the Mississippi was a gradual process in which April and May flood levels had been predicted as early as January. The term of the flooding illustrates the time period over which the hazard had to be dealt with. Three flood crests passed by St Louis as various tributaries contributed their floods to the lower Mississippi. This is very similar to the flooding that occurs downstream of the meeting of the Murray-Darling rivers in Australia. Because 'great floods' on alluvial rivers move substantial amounts of sand, it is possible for the actual depth of the river to vary during the flood as bedload material passes by a given point. Maximum discharges of $50\,000\,m^3\,s^{-1}$ were measured on the Mississippi several times at peak stage. Because the river's cross-sectional area varied so much, this discharge produced a difference in flood-crest height of 0.6 m. Note that the actual stream power during the flood was quite low, and never exceeded $12\,W\,m^{-2}$ (figure 8.2). Predictable flood peaks were difficult to determine. Because of the low gradient of the system, backwater flooding on tributary rivers was particularly severe. Because of localized decreases in river depth and higher flood heights, over 1300 km of main levees had to be raised after the 1973 flood. The long-term nature of the flooding stretched the resources of disaster relief organizations and government organizations responsible for their control, and tested the fortitude of many farmers trying to sow crops on flooded land.

Great Australian floods of 1974 (Bolt et al., 1975; Shields, 1979; Short, 1979; Mckay, 1979; Whittow, 1980; Holthouse, 1986)

The major rivers of east Australia originate within the Great Dividing Range, which parallels the east coast (figure 8.11). Two systems flow westward. The Murray-Darling and their tributaries—the Warrego, Condamine, Macintyre, Macquarie, Lachlan and Murrumbidgee—form the largest river system in Australia, entering the southern ocean south of Adelaide. The other system consists of the Diamantina-Coopers Creek rivers, which flow into landlocked Lake Eyre. On the eastern side of the Dividing Range are a number of much shorter, but just as impressive, systems. These rivers consist of the Hunter, Macleay, Clarence, Brisbane, Burnett, Fitzroy and Burdekin, Sequences of tropical cyclones or monsoonal troughs have in the past severely flooded all of these rivers, but never at the same time. Aboriginal legends from the Dreamtime imply that many river valleys have been flooded over widths of 50 km or more. In the summer of 1973–74, flooding to record heights occurred simultaneously on all major rivers in eastern Australia, bringing true those Aboriginal legends.

The summer of 1973–74 followed a major El Niño-Southern Oscillation event, which peaked at Christmas in 1972. The 'turning on' of Walker circulation across the equatorial Pacific brought two of the heaviest years of rainfall in eastern Australia. Rainfall was torrential and continuous throughout most of January 1974, as the intertropical convergence settled over tropical Australia. On 25 January, Cyclone Wanda moved into the interior of Queensland and New South Wales, dumping in excess of 300 mm of rain in 24 hours over a wide area, and triggering massive flooding of all river systems. Rain in the catchment of the Brisbane River produced one of the worst urban floods in Australian history. These January floods represented the largest natural disaster to occur in Australia to that point in time. Flooding covered an area of $3\,800\,000\,km^2$, larger than the drainage basin of the Mississippi River. From Alice Springs to the Pacific Ocean, and from the Gulf of Carpentaria to the Murray River, military airlifts had to be arranged to supply isolated towns, cut off by flood waters, with emergency food and stock fodder. Around the Gulf of Carpentaria, the tributaries of the Flinders River amalgamated to form a river 150 km wide. In northern New South Wales, torrential rainfall continued week after week, raising

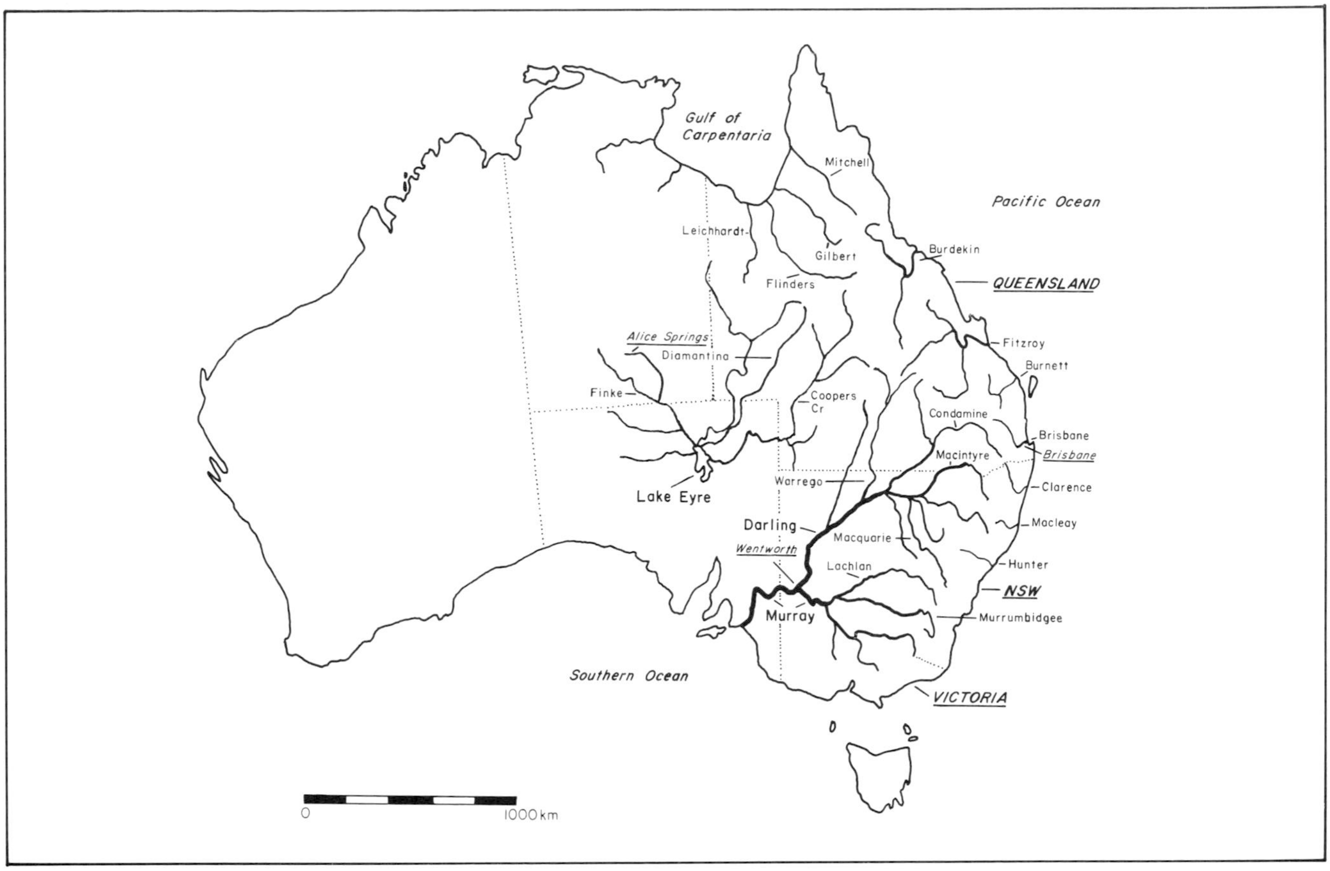

FIG. 8.11 Major rivers of eastern Australia

river levels in excess of 20 m. Irrigation and grazing lands were inundated for months. Most inland towns sustained severe damage as protecting levees were overtopped, despite frantic efforts by volunteers to raise their levels. Floodwaters slowly flowed down the Diamantina and Coopers Creek into Lake Eyre, filling it for only the fourth time this century. Stock losses to sheep alone in New South Wales totalled 500 000. The scale and magnitude of flooding was unprecedented. None of this flooding could have been prevented, because the construction of dams, weirs and levess necessary to prevent it, was not economically justified. In fact, on most rivers similar floods had occurred in the previous 20 years, but never simultaneously. While the Mississippi River in 1973 flooded some farmland for four to six months, at Wentworth in New South Wales floodwaters never left farmland for two years, as successive floods came down the Darling and Murray in 1974–75.

The most damaging flooding occurred in the city of Brisbane on the Pacific coast. The Brisbane flood was not extraordinary. Similar flooding had occurred four times previously in the city and was occurring repeatedly across eastern Australia at the time. However, the flood is an excellent example of a natural disaster striking an area lulled into a false sense of security by the rarity of such a disaster in recent times. The last major flood had occurred in 1893, and thus was outside the experience of most people. Since 1893, Brisbane had witnessed the commercial and residential development of many flood-prone areas. Not everyone was unaware of prior flooding events. Major insurance companies had mapped the flood limits of the 1893 flood, and had excluded flood insurance from policies issued to property owners living in those areas.

Synoptically, the flooding of the city was due to the persistent recurrence of cyclones tracking over eastern Queensland in the space of four weeks, culminating in Cyclone Wanda. However, many of the flood's effects were exacerbated by human factors. Somerset dam, built after the 1893 flood to contain similar events, was totally inadequate for this purpose; yet, many people believed that the dam protected them from flooding. The sprawling nature of growth in the Brisbane River basin had also led to large-scale clearing of forest in rugged terrain, an activity that enhanced runoff during the high intensity, rainfall periods of the 1974 flood. The character of the drainage basin was also altered. Urban development had seen some creeks filled in, while others had been lined with concrete. These latter channels were designed to contain only the 1:10-year flood. Large sections of the landscape had been sealed by roads, driveways and houses. These modifications resulted in calculated flood discharges being twice that of the 1:100-year flood for adjacent vegetated catchments. Some urban catchments experienced the 1:1000-year flood event. Additionally, there was very little delay between time of peak rainfall intensity and peak flooding. Flash flooding occurred in urban catchments with flood peaks cresting within one hour of peak rainfall intensities. Many areas also had inadequate rain-gauging stations, so that the exact amount of rain falling in some parts of the catchment was unknown, and the forecasting of flood levels impossible to determine.

The Brisbane flood also saw the collapse of organized disaster relief because of the unexpected occurrence and magnitude of the event. Apart from the scarcity of gauging stations in the catchment, communication breakdowns or evacuation of observers saw a further 50 per cent deterioration of the existing network. There was no central data-processing authority, so that local flooding in key areas went unnoticed. Flood warnings broadcast to the public were totally inadequate. Over 70 per cent of residents questioned afterwards about the flood had received no official warning. The media, in their efforts to report a major story from the field, sparked unnecessary rumours and clouded the true picture of flooding. The rapid series of flash floods, in isolated parts of the catchment 24 hours prior to the main flood, led to public confusion and committed the Disaster Relief Organization to what afterwards were evaluated as only minor disasters.

The damage from the flood was major. All bridges across the Brisbane River were damaged or destroyed and 35 people drowned. At its height, the river broke its bank and ran through the central business district of Brisbane. In Ipswich 1200 homes were destroyed; overall, 20 000 people were made homeless. The exact cost of property damage could not be assessed, because most properties were not covered by insurance; however, a figure in excess of $A30 million seems likely. Only 11 per cent of residents who experienced the main flood received assistance from emergency organizations in evacuation, and only 30 per cent acknowledged the help of any relief organization in cleaning up. Most relief came from friends and community groups. Over 40 per cent of victims received help from church contacts, and over 30 per cent of volunteers showing up in surveys stated that they belonged to no organized group. Generally, most people applied themselves to evacuation, alternative accommodation, clean-up and rehabilitation with little reliance on government or social organizations for help. This raised serious questions about the efficiency of such organizations in disaster relief in Australia, and led to significant changes in the response to disasters by both private relief organizations and federal government agencies. Eleven months later these revisions were tested to the upmost when Cyclone Tracy destroyed the city of Darwin.

Geological changes due to flooding of large rivers

A model for change (Wolman and Leopold, 1957; Nanson, 1986)

Changes in the course of large rivers pose a problem within the concepts of floodplain process and modification. It has generally been assumed, since the classic work by Wolman and Leopold, that floodplains accrete by lateral river meandering. Erosion of concave bank deposits is followed by emplacement on the opposite bank of point-bar deposits, overtopped with a thin veneer of *overbank deposits*. A river slowly reforms its floodplain by lateral migration. This view has been accepted as universal; however, it is riddled with apparent contradictions. For example, levee banks often escape flood events while pre-existing back-channels flood. Terraces are assumed to be relict from past climates or stranded by downcutting of the river, when in fact their age is often little different to the active floodplain. Some rivers lack coarse point-bar deposits, while others appear to have built up floodplains in situ over long periods of time without any migration. This latter process of *vertical accretion* is not rare. It is common of many rivers in the semi-arid part of the United States, and certainly dominates the high energy coastal streams of New South Wales, Australia. Even the Mississippi, which represents the classic example of a laterally accreting floodplain, does so in a narrow meander belt flanked by extensive, vertically accreted deposits.

While floodplains are assumed to be accretional deposits, little attention has been paid to the fact that they can also erode. In fact, erosion of the floodplains of New South Wales coastal rivers is significant; yet, it is *episodic*, non-contemporaneous between basins, and spatially non-coherent within the same drainage basin. Nanson (1986) has succinctly presented a model of catastrophic floodplain erosion based upon the New South Wales evidence. The model is shown in figure 8.12. For rivers that form floodplains by lateral accretion, the floodplain configuration is kept in equilibrium as the channel migrates laterally, consuming older deposits. If floods occur above bankfull, erosive energy is dispersed across the floodplain. These flood events tend to occur at regular frequency over most of the drainage basin. In floodplains where vertical accretion has dominated, the bank deposits for various reasons are not highly erosive but tend to contain flood flows. Floods only increase the height of the levee banks, which constrain flood events more to the channel, thus increasing stream power per unit area of the bed and banks. Very large floods direct most floodwaters on the floodplain through back channels, which have much steeper gradients than the main river channel. Eventually, the bankfull stream power (that is, the stream power of the river full to the top of its banks) will exceed the boundary resistance and the channel will severely erode.

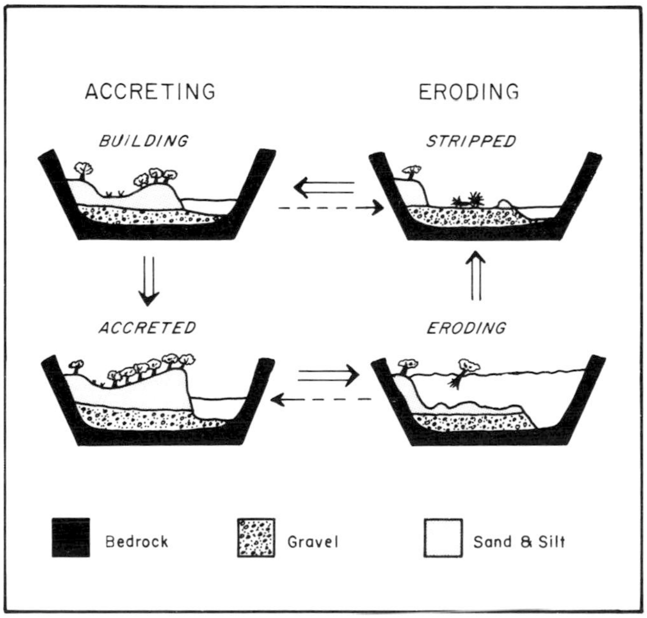

FIG. 8.12 Models for floodplain accretion and stripping (based upon Nanson, 1986)

Similarly, if large floods overwhelm the back channels on the floodplain along these steep gradients, then the complete floodplain of fine-grained sediment can be stripped away, leaving only a basal lag of coarse gravel. Cyclically, the floodplain will gradually reform by vertical accretion until flood events are again largely contained within the main channel, gradually increasing the channel's erosive energy.

Floodplain erosional events are not continuous along a river because of independent formation of floodplain segments. The periodicity of stripping varies as a result of local energy gradients and valley confinement. Steep, narrow valleys will experience stripping more often than flat, wide ones. These factors, together with the fact that flooding events are to a certain degree unpredictable, imply that floodplain stripping is a random process in time and space. Measurement of the bedload or suspension loads in rivers is no indication of the long-term volumes of sediment moving in a river system. Sediment volumes passing by any point in a river depend upon the severity, timing and location of floodplain stripping events upstream. While bankfull flooding in laterally accreting systems tends to occur every one to two years, bankfull flooding in vertically accreting rivers is more infrequent, at intervals of four to ten years in New South Wales coastal rivers. The frequency of bankfull flooding not only decreases with time but can vary along a river depending upon its stage of development in the cycle outlined in figure 8.12. Floodplain stripping does not necessarily result from infrequent events, but may occur at the end of a period of clustered flood events. For instance, on one stretch of the Manning River in New South Wales, over 200 000 m^3 of sediment was stripped by a 1:50-year event; however, this stripping was preceded in the previous decade by five preparatory events with recurrence intervals of 14, 4.9, 3.8, 4.2 and 9 years.

The model of vertically accreting floodplains, in fact, summarizes a spectrum of floodplain morphologies and behaviours, representing a balance between erosive energy distribution and sediment resistances. At the high energy end of the spectrum are narrow gorges devoid of alluvial deposits. As gorges widen, coarse-grained floodplains develop that are eroded and reformed only during extreme events. In wider and less steep rivers, with silty-to-gravelly sediment, floodplains may accrete vertically if channel migration is minimized by river incision, vegetation or resistant banks. Flood frequency is episodic in time and space, such that change is virtually unpredictable. If channel migration is possible, floodplains develop by lateral accretion, and bankfull events can be predicted, to a certain degree, by monitoring the amount and location of rainfall or snowmelt in the drainage basin. This is the case with all flooding in the Mississippi drainage basin. At the extremely low-energy end of the spectrum, there is insufficient stream power to allow even gradual channel migration, even if sediments are erodable. Streams remain locked in position for exceptionally long periods of time, and flooding events occur as overbank deposition, where water volumes cannot be contained within the channel.

Historical changes to the Mississippi River (Coleman, 1988)

From the above, a common, if widely unrecognized characteristic of even large rivers is the fact that a river can build its channel above the level of the floodplain, as it continues to overflow levees during flood events. Sediment is added to the height of levees each time the river floods. As long as the flood event is minor, and the sides of levees are not scoured or undermined by boiling, then the course of the river will remain between its natural levees. Human assistance can add to the process of stabilization by reinforcing levee walls, so that the river remains in its channel and does not spill across the adjacent floodplain. This is an artificial process because levees naturally never contain a river's floodwaters. This distortion of the natural levee-building process means that the levees must be regularly maintained. If the river is *aggrading* its bed, the levees must be built to even higher elevations. The river then becomes much higher than the adjacent floodplain and, when it breaks out of its channel, is more likely to cause a permanent shift in its course. The change in river position is no longer a process of gradual lateral meandering, but one of major breakouts at very infrequent intervals.

This process of levee building has been a natural one over the past 6000 years on the lower reaches of the Mississippi. The Mississippi has broken through its levees four times during this timespan, and formed a new course on the floodplain. These events can be seen in the upper part of figure 8.13. As the river nears the Gulf of Mexico, breaches and diversion of the river channel have occurred more frequently, forming seven distinct phases of sub-delta growth in the last 6000 years. The modern sub-delta south of New Orleans is a recent feature and has a small volume compared to some of the positions that have existed in the past. Breaches of the levees which give rise to different sub-deltas have tended to occur south of Baton Rouge. Each breach has represented a major diversion of the river's course at the upstream end of the overall delta. The first sub-delta to form, the Sale Cypremort splay, is almost unrecognizable today, because subsidence and wave attack have managed to erode and submerge much of its lobate feature. However, because this location now has the lowest relief and is closest to the existing channel, it is the most likely location for any future break-out of the Mississippi River. This break-out almost occurred during the 1973 flood. As sub-deltas become younger in age and less eroded, submergence and wave action have produced landward retreating barrier islands at their seaward ends. The more recent the sub-delta is, the larger and further from shore the islands are. These barrier islands are most noticeable on the Teche, St Bernard and Lafourche sub-deltas. More recently formed sub-deltas have tended to extend the St Bernard delta into the Gulf, and have tended to constrain the river to narrow levee-bordered channels. Barrier islands have not developed on these newer sub-deltas. The present bird's-foot sub-delta developed about AD 1500, and is bordered by levees which appear to fail at 20-year intervals. The longer this delta is maintained, the more probable the chance of break out of the Mississippi River around Baton Rouge. The artificial maintenance and building up of levees at this point ensures that any such break out will be extremely difficult to reverse. This will mean the stranding of New Orleans as a major United States port, and have dire economic consequences as the region tries to adjust to a new river channel.

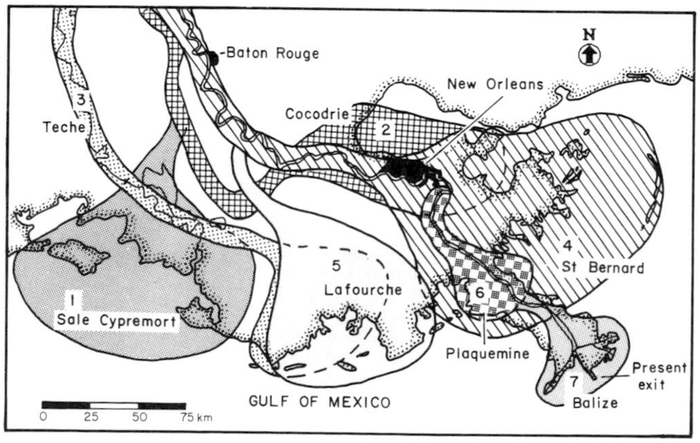

FIG. 8.13 Probable locations of major deltaic lobes of the Mississippi River during the Holocene (after Coleman, 1988). In the diagram, 1 is oldest, 7 is youngest.

Historical changes to the Hwang Ho River (Bolt et al., 1975; Czaya, 1981; Milne, 1986)

The Hwang Ho (Yellow) river of central China differs from the Mississippi in that attempts to stabilize the Hwang Ho have been going on for centuries. The Mississippi has only undergone channel stabilization this century. The Hwang Ho river drains 1 250 000 km^2 and flows over a distance of 4200 km. Throughout much of its upper course, it erodes yellow loess, and for 800 km of its lower length, the Hwang Ho flows without any tributaries on a raised bed of silt. Much of this material has entered the river because of poor land-use practices in the upper drainage basin in the last 2000 years. Today, the suspended load reaches 40 per cent by weight, making the Hwang Ho one of the muddiest rivers in the world. Where the river breaks onto the flatter coastal plain, vertical aggradation has built up a large alluvial fan-like delta from Peking (Beijing) in the north, to Nanking (Nanjing) in the south. The river thus flows on the crest of a cone-shaped delta and builds up its bed very rapidly during flood events. As early as 2356 BC the river channel was dredged to prevent silting. In 602 BC the first series of levees were built to contain its channel. Despite these efforts, in the last 2500 years the Hwang Ho has broken its banks ten times, often switching its exit to the Yellow Sea over a distance of 1100 km on each side of the Shantung peninsula (figure 8.14). Associated with these break outs has been massive flooding with the result that China historically has suffered the greatest loss of life in the world from flood hazards. The Hwang Ho is commonly referred to as the 'River of Sorrow'.

Between 2300 BC and 602 BC the Hwang Ho flowed through the extreme northern part of its delta in the vicinity of Peking. In 602 BC the river moved slightly south to the region of Tientsin, and stayed there until 361 BC, whereupon it underwent a catastrophic shift south of the Shantung peninsula. For the next 150 years, the Chinese expended considerable effort shifting it back north. In 132 BC the Hwang Ho switched to a course near the present entrance, where dike building tended to keep it stabilized in place, with minor channel switching, on the northern side of the Shantung peninsula until AD 1289. For this period, the northern part of the alluvial plain aggraded, a process that steepened the topographic gradient southwards. With the fall of the Sung dynasty, dike maintenance lapsed, and in AD 1289 the Hwang Ho again switched south of the peninsula, where it was allowed to remain for the next 600 years. Its

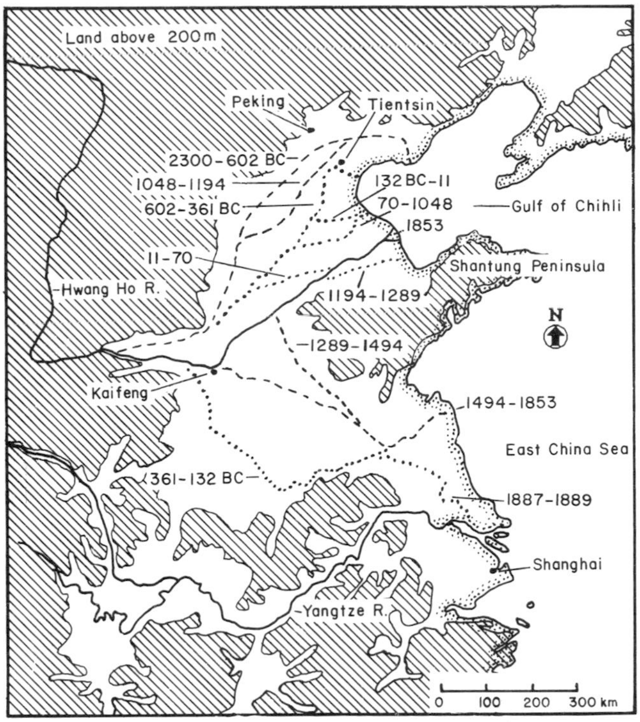

FIG. 8.14 Location of the Hwang-Ho River channel: 2300 BC to the present (after Czaya, 1981). Dates shown are AD unless otherwise indicated.

most destructive flooding occurred during this period in AD 1332, with 7 million people drowning and over 10 million people dying from subsequent famine and disease. By 1851, the Hwang Ho had aggraded its bed in the south to such an extent that it threatened to engulf the town of Kaifeng. Measured rates of aggradation exceeded 2 m yr^{-1}, and dykes more than 7 m high were built to contain its flow. Between 1851 and 1853 the Hwang Ho broke out northwards and took on its present course, which it has been trying to leave ever since. In 1887, a temporary breach saw the river move the furthest south it has ever been. It temporarily linked up with the Chang Jiang (Yangtze) River and, for a brief time, both rivers flowed out to sea through the same channel. The 1887 flood breached 22 m high embankments and flooded 22 000 km^2 of land to depths of 8 m. Over 1 000 000 people lost their lives. Not only was the deathtoll severe but the river also buried flooded land in meters of silt. Farmers relying upon the river in the northern part of the delta suddenly found themselves without irrigation water and faced starvation.

The natural tendency of the the Hwang Ho to shift course has been greatly exacerbated during wartime. In 1938 General Chiang Kai-Shek, in an attempt to prevent advancing Japanese armies from taking the city of Chengchow, ordered the levee dykes to be dynamited. The diversion failed to stop the Japanese advance, but the ensuing floods left 1 million unsuspecting Chinese dead, and possibly caused the death of 11 million people through ensuing famines and disease. It was not until 1947 that extensive engineering works forced the Hwang Ho permanently back into its pre-war channel. Today, the Hwang Ho is stabilized for much of its length, and is kept to the north of the Shantung peninsula. Because the river transports such large quantities of silt, its bed is continuing to aggrade, and it now sits 20 meters above its flood-plain, the latter protected by an inner and outer set of dikes spaced 10 km apart. The Chinese government has embarked on a program of dam and silt basin construction to minimize the flooding effects of the Hwang Ho; however, the river still is not controlled, and will probably breach its dikes in the future during a large flood.

References

Baker, V. R. and Costa, J. E. 1987. 'Flood power'. In Mayer, L. and Nash, D. (eds) *Catastrophic flooding*. Allen and Unwin, London, pp. 1–21.

Bureau of Meteorology, Australia. 1985. *A Report on the flash floods in the Sydney metropolitan area over the period 5 to 9 November* 1984. Special Report, Bureau of Meteorology, Melbourne.

Coleman, J. M. 1988. 'Dynamic changes and processes in the Mississippi River delta'. *Geological Society of American Bulletin* v. 100 pp. 999–1015.

Czaya, E. 1981. *Rivers of the world*. Edition, Leipzig.

Griffiths, J. F. 1976. *Climate and the environment: the atmospheric impact on man*. Paul Elek, London.

Hirschboeck, K. K. 1987. 'Catastrophic flooding and atmospheric circulation anomalies'. In Mayer, L. and Nash, D. (eds) *Catastrophic flooding*. Allen and Unwin, London, pp. 23–56.

Nanson, G. C. 1986. 'Episodes of vertical accretion and catastrophic stripping: a model of disequilibrium flood-plain development'. *Geological Society of America Bulletin* v. 97 pp. 1467–75.

Nanson, G. C. and Hean D. S. 1984. 'The West Dapto Flood of February 1984: rainfall characteristics and channel changes'. *Dept. Geography, University Wollongong, Occasional Paper No. 3*.

Wolman, M. G. and Leopold, L. B. 1957. 'River flood plains: some observations on their formation'. *U.S. Geological Survey Professional Paper* No. 282–C pp. 87–107.

Further reading

Bolt, B. A., Horn, W. L., MacDonald, G. A. and Scott, R. F. 1975. *Geological hazards*. Springer-Verlag, Berlin.

Cornell, J. 1976. *The great international disaster book*. Scribner's, NY.

Holthouse, H. 1986. *Cyclone: a century of cyclonic destruction*. Angus and Robertson, Sydney.

McKay, G. R. 1979. 'Brisbane floods: the paradox of urban drainage'. In Heathcote, R. L. and Thom, B. G. (eds) *Natural Hazards in Australia*. Australian Academy of Science, Canberra, pp. 460–70.

Maddox, R. A., Canova, F. and Hoxit, L. R. 1980. 'Meteorological characteristics of flash flood events over the western United States'. *Monthly Weather Review* v. 108 pp. 1866–77.

Milne, A. 1986. *Floodshock: the drowning of planet Earth*. Sutton, Gloucester.

Noble, C. C. 1980. 'The Mississippi river flood of 1973'. In Coates, D. R. (ed.) *Geomorphology and Engineering*. Allen and Unwin, London, pp. 79–98.

Riley, S. J., Luscombe, G. B. and Williams, A. J. (eds). 1986a. 'Proceedings Urban Flooding Conference: a conference on the storms of 8th November 1984'. *Geographical Society N.S.W. Conference Papers* No. 5.

——. 1986b. 'Urban stormwater design: lessons from the 8 November 1984 Sydney storm'. *Australian Geographer* v. 17 pp. 40–50.

Shields, A. J. 1979. 'The Brisbane floods of January 1974', In Heathcote and Thom (eds) *Natural Hazards in Australia*. pp. 439–47.

Short, P. 1979. '"Victims" and "helpers"'. In Heathcote and Thom (eds) *Natural Hazards in Australia*. pp. 448–59.

Whittow, J. 1980. *Disasters: the anatomy of environmental hazards*. Penguin, Harmondsworth.

9

Fires in Nature

Introduction

Of all natural hazards the most insidious is drought. But for some countries such as Australia and United States, droughts have not led to starvation, but to spectacular fires as tinder-dry forests ignite, grasslands burn and eucalypt bushland erupts in flame. Of all single natural hazard events in Australia, bushfires are the most feared. The litany of disasters over the past 150 years reads like the membership list of some satanic cult—'Black Thursday', 'Black Friday', 'Ash Wednesday'. For firefighters, State emergency personnel and victims, each name will invoke stories of an inferno unlike any other. North America has suffered just as badly in the past from fires, as has Europe and the Soviet Union. Many urban dwellers would consider that the forest fire threat near cities terminated with wide-scale deforestation. Fires are perceived to be a hazard only in virgin forest or, perhaps in southern California or southern France, a hazard where urban expansion has encroached upon vegetated mountains. In fact, any country where climate is influenced by Hadley cells or by periodic droughts is susceptible to natural fires. This is particularly true for countries experiencing a *Mediterranean climate*, especially those where highly flammable *eucalypts* have supplanted more fire-resistant vegetation. More importantly, large areas of the northern hemisphere have undergone reforestation this century, with the abandonment of economically marginal farmland. Here, forest fires may again pose the severe hazard they were under initial settlement more than a century ago.

This chapter will examine the hazard of drought-induced natural fires. Those types of

Fig. 9.1 William's Strutt's painting *Black Thursday* (LaTrobe Collection, © and with permission of the State Library of Victoria). This title refers to the fires of 6 February 1851, which covered one quarter of the colony of Victoria, Australia.

vegetation most susceptible to burning will be described first, followed by a discussion on the synoptic patterns favouring intense fires, and the main causes of fire ignition. Major bushfire disasters will then be described, with particular emphasis on the Australian environment, where some of the most intense *conflagrations* have been witnessed (figure 9.1).

Worldwide distribution of vegetation

Natural fires will occur in any environment where there is vegetation. Figure 9.2 maps the natural distribution of vegetation worldwide. This distribution may be vastly different from existing vegetation patterns because of human alteration; however, it is relevant when discussing historical fires. While tropical rainforests are not known for their fire potential, they can burn and have burnt. For example, the U Minh forest in Vietnam (see figure 9.3 for the location of major place names mentioned in this chapter) ignited naturally in 1968. That fire burnt for six weeks aided by incendiaries dropped on it by the United States air force. Large fires have also burnt through tropical vegetation in the Brazilian highlands and the Amazon basin, especially in association with land clearing. Arid and semi-arid regions may appear devoid of vegetation, but after wet spells can develop high biomass in the form of grasses and shrub foliage. The largest fires in Australia occur as grass fires in semi-arid regions. Much of the northern hemisphere is covered in coniferous forest called *taiga*. These forests are evergreen and are particularly susceptible to fire in dry conditions. While these forests typically cover colder regions of the world with a winter snow-cover, they can virtually burn (and have done so) at any time of the year. Once burnt, fire-scarred areas take decades to regenerate. Regrowth in these areas is particularly susceptible to future burning. In North America large burnouts such as the Tillamook in Oregon, burnt originally in 1933, spawned numerous fires for the next 30 years. Temperate, mainly deciduous forests characteristically lose their leaves in winter and are too humid to burn in summer. However,

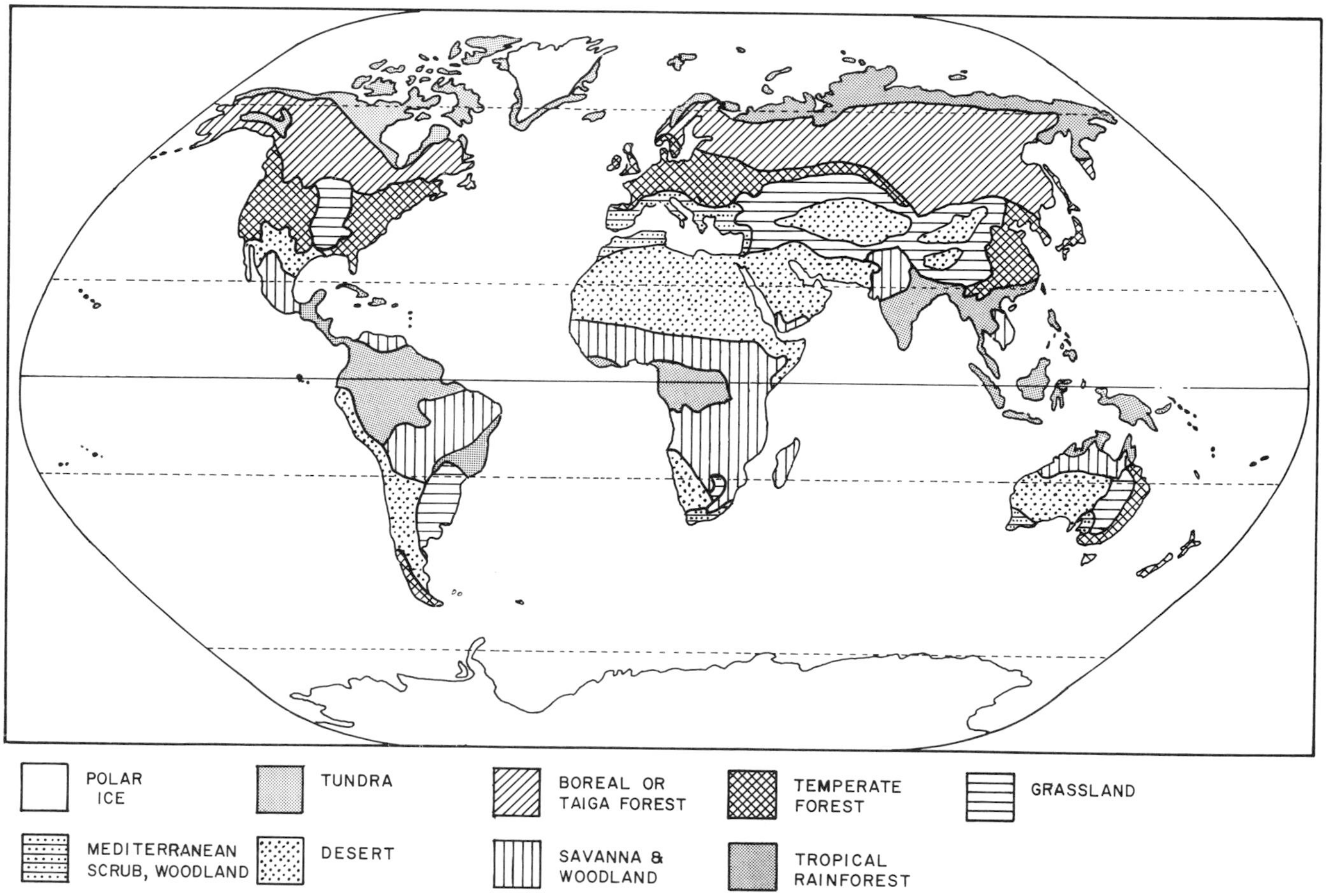

Fig. 9.2 Major vegetation zones of the world (adapted from White, Mottershead and Harrison, 1984)

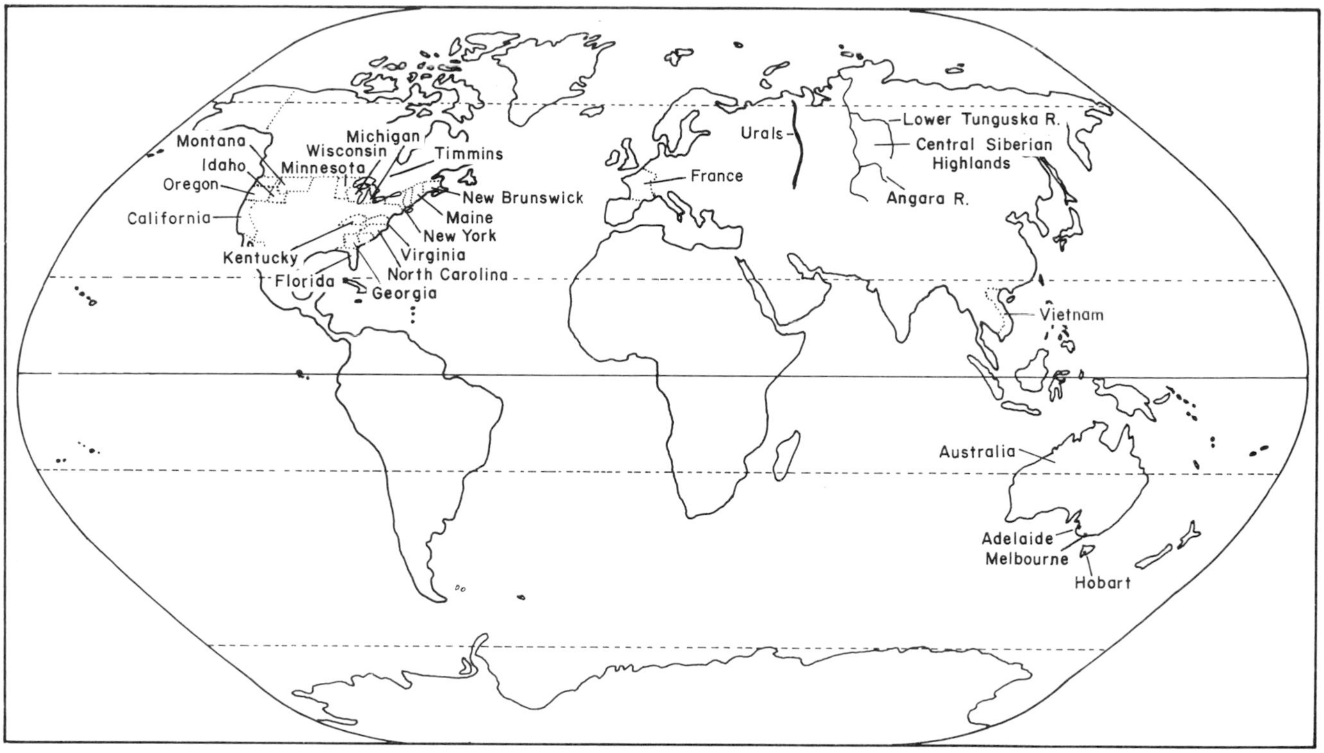

FIG. 9.3 Location map

in autumn when foliage dries out, even these trees can catch fire, especially during extreme drought conditions.

A different problem exists with eucalypts, whose leaves have a high oil content, encouraging burning. Such trees can regrow foliage rapidly through epicormic shoots on the main trunk, or through shoots on buried and protected roots. After only a few years, they can become sites of conflagrations as severe as the original fire. While eucalypts originated in Australia, they were exported successfully to other semi-arid countries, and now cover significant areas of southern California, northern Africa, the Middle East and India. Thus, one of the most fire-prone types of vegetation in the world now plagues these semi-arid environments with intense burning every couple of years. Other fire-prone vegetation dominated by scrub oak, chamise and manzanita exists in Mediterranean climates. Such vegetation communities are labelled *chaparral* in California, 'garrigue' or 'maquis' in France, 'phrygana' in the Balkans, 'matarral' in South America, 'macchin fynbosch' in South Africa and 'mallee' or 'mulga' scrub in Australia.

Under the right conditions, most of the world's vegetation can burn. Tundra and rainforest would appear to have the least chance of catching fire, while taiga and eucalypt forests have the best prospects of developing major fires. In addition, it is also possible to burn off the humus layer at the soil surface. This includes peat in bogs, which can dry out and burn after prolonged droughts. It should also be noted that just because an area has been cleared of natural vegetation, it does not mean it cannot experience fire. Most grain crops such as wheat, barley and rye, when drying out before harvest, are particularly vulnerable to fire. Some of the worst fires in Australia in terms of property loss have occurred in wheat country.

Conditions favouring intense bushfires (Vines, 1974; Luke and McArthur, 1978; Powell, 1983; Voice and Gauntlett, 1984; Yool et al., 1985)

The potential for a major fire hazard depends upon the characteristics of the fuel, the climate and fire behaviour. In Australia fuel characteristics are very important to the spread and control of fires. Fuel characteristics can be separated into two categories, grass and bush (mainly eucalypts). Grasslands produce a fine fuel both on and above the ground surface. Such material easily ignites and burns. In forests the fine fuel consists of leaves, twigs, bark and stems under 6 mm in diameter. Apart from eucalypt leaf litter with its high oil content, most of this forest fuel is more difficult to ignite. The arrangement of this fuel is also important. Compacted litter is more difficult to burn than litter with a lot of air mass. Where vegetation is spread out, fire is difficult to maintain. Forests with dense undergrowth growing up towards the forest crown, especially in pine plantations, will burn more readily than ones where the undergrowth has been culled. Again, the density of vegetation also determines the fuel quantity. The moisture content of living and dead material also dictates the possibility of ignition and fire spread. Dead grasses absorb moisture readily, and can have moisture contents which depend very much upon the diurnal humidity cycle. Humidity increases at night, so grasses have their highest moisture contents in the early morning. Monitoring of the proportion of cured grass and its moisture content can give a good index of the susceptibility of that fuel to ignition and burning. In eucalypt forest the type of tree species conditions fire behaviour. Stringy barks permit fire to reach tree tops, thus favouring spot fires, whereas smooth barks do not generate flying embers.

The climate before severe fire seasons also conditions the fire threat. If fuel has built up, then climatic conditioning is less important; however, it still takes dry weather to cure and dehydrate grass and bush litter. This desiccation process is exacerbated by prolonged drought in the summer season. Typically, El Niño-Southern Oscillation (ENSO) events occurring at Christmas coincide with Australia's summer season. During these events, below-normal rainfall can be expected in eastern Australia until March or April. Hence, in this type of drought year the fire season in southern bushland extends from October to April. ENSO events ensure that fuels dry out during the summer. Because such events can now be predicted months in advance, it is possible to warn of severe fire seasons before summer arrives.

The prediction of bush or forest fires may even extend beyond this timespan. As pointed out in chapter 5, the 11-year sunspot cycle appears in rainfall records for many countries, while the 18.6-year lunar cycle dominates the coincidence of worldwide drought or rainfall. Because natural conflagrations are most likely during major droughts, it is to be expected that forest fires evidence one of these astronomical cycles. Vines (1974) has performed detailed work on the cyclicity of bushfires in Australia and Canada, and believes that cyclicities of 6–7 years and 10–11 years appearing in these countries directly reflect sunspot cycles. The 10–11-year periodicity also appears in the fire records of the southern United States since 1930, although at sunspot minima. In the Lake states, major conflagrations between 1870 and 1920 coincide very well with peaks in the 11-year sunspot cycle. To date, little research has been carried out relating fire occurrence to the 18.6-year lunar cycle; however, there is no reason to believe that this signature will not be found in long-term fire records.

The typical daily weather pattern leading to extreme bushfires in southeastern Australia is presented in figure 9.4. A high-pressure system moves slowly southeast of Australia and ridges back over the continent. This directs strong, desiccating, northwest winds from the interior of the continent across the southeast corner of the continent. These winds are followed by a strong southerly change that may drop humidities, but maintain high wind speeds. Winds blowing out of the interior warm adiabatically as they drop in elevation from the highland areas of Victoria to the coast, producing what is commonly known as a *föhn* wind. This adiabatic warming can push temperatures to 40° C or above at the coast. The process also reduces relative humidities to levels below 20 per cent. Similar winds exist elsewhere especially in mountainous Mediterranean climates dominated by Hadley cells. Local variants of föhn winds are labelled mono in central California, Santa Ana in southern California, sirocco in Southern Europe, bora in the Balkans. In Australia strengthening of föhn winds is aided by convergence of isobars, and reinforced by a fast moving subtropical jet stream in the upper troposphere on the western side of the

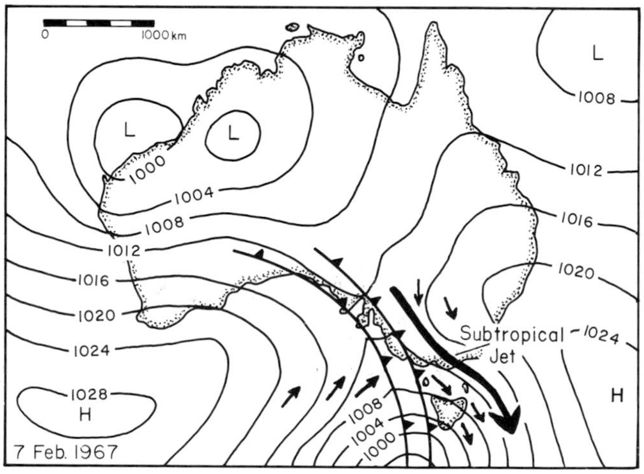

FIG. 9.4 Typical synoptic weather pattern conducive to bushfire conflagrations in southeast Australia

high-pressure cell. Because this sequence of winds moves easterly across the continent over several days, there is a gradual shift in the bushfire threat from South Australia to Victoria, and then to New South Wales. In Western Australia positioning of high-pressure cells with northeasterly winds out of the interior characterizes fire weather.

Fire behaviour and local topography determine the final severity of the fire hazard. Often fires accelerate as the day progresses, because the moisture content of fuels drops, and wind speed and atmospheric instability increase owing to diurnal heating. Additionally, during the day as the fire intensifies, convection is generated by the extra heat being supplied to the atmosphere. Radiation from the fire can dry out fuel in front of the fire, and *spot fires* can prepare forward areas for the passage of the main fire. Spotting in front of the fires is quite common in Australia because of the presence of stringy barks. Spotting ensures that convection occurs in front of the main fire, thus drawing the main fire-front towards the spot fires.

As wind speed increases, fires increase their rate of movement exponentially. A wind speed of 10 km hr^{-1} will move a fire in Australian bushland of average fuel content at the rate of 0.5 km hr^{-1}. At 20 km hr^{-1}, this rate increases to 0.75 km hr^{-1}; at 40 km hr^{-1}, the rate rises to 1.75 km hr^{-1}. If the fire actually reaches treetops and burns through the crowns of trees (figure 9.5), it can accelerate and travel at speeds of 20 km hr^{-1}. The rate at which a bushfire will move does not taper off with increased wind velocities; in grassland, however, fire movement decelerates with winds above 50 km hr^{-1} because of fragmentation of the fire head. This is not significant when one considers that grass fires can move about eight times faster than a bushfire under the same wind speed. If fires move upslope, their rate of advance will increase because rising heat from combustion dries out fuel. Fires will also move faster upslope because the wind velocity profile steepens without the influence of ground friction. In rugged terrain, spot fires are more likely because embers can be caught by winds blowing at higher speeds at the tops of hills or ridges.

Atmospheric stability also controls fire behaviour. If the atmosphere is unstable, then convective instability can be established by the fire's heating of the atmosphere. Once initiated, convective instability will continue even after the heat source is removed. Under conditions of extreme instability,

Fig. 9.5 Crown fires. *On the left*, in eucalypt forest, Australia (photograph by N. P. Cheney, National Bushfire Research Unit, CSIRO, Canberra). *On the right*, in boreal forest in northern Ontario, Canada (photograph by Jim Bryant, Ontario Department of the Environment). Note the similarities between the fires. Eucalypts are perceived as more flammable, but boreal forests produce three to four times more energy output per unit time because of their greater biomass.

fire whirlwinds are generated and can obtain speeds exceeding 250 km hr^{-1}. These whirlwinds consist of three types. The first is related to high combustion rates and generates a rotational updraft, which can move embers a couple of hundred meters beyond the fire front. The second type represents a thermally induced tornado. This type originates in the atmosphere on the downwind side of the convective column or the lee side of topographic relief. The winds are capable of lifting large logs and producing gaseous explosions in the atmosphere. These vortices are particularly dangerous because wind is sucked towards the fire tornado from all directions. The third type of whirlwind is the afterburn vortex (figure 9.6). It forms because of the heat emanating from a burnt area up to a day after the fire. These vortices are usually less than 20 m in diameter, but they can carry embers from burnt to unburnt areas.

FIG. 9.6 Whirlwind associated with a clearing after burning (photograph by A. G. Edward, National Bushfire Research Unit, CSIRO, Canberra)

Causes of bushfires (Luke and McArthur, 1978; Pyne, 1982; van Nao, 1982)

No matter how dry a forest or grassland area becomes, there can be no fire unless it is ignited. The exact cause of a fire depends upon how rigorously fire statistics are recorded in individual countries. For instance, in European countries, 45 per cent of all fires have unknown causes; however, this figure drops to 5 per cent in Canada. In Europe only 2 per cent of the known causes of bushfires can be attributed to natural causes such as lightning. Over half are the result of arson and 40 per cent, of human carelessness. The latter group includes out-of-control burning, smoking, children playing, railways, and camp fires. In Canada lightning is responsible for 32 per cent of all forest fires, whereas in the United States lightning accounts for less than 8 per cent of fires. The latter figure is misleading because lightning has played a major role in igniting fires in isolated areas of the United States. Between 1940 and 1975, over 200 000 fires were started by lightning in the western United States alone. This represents an average of 17.8 fires per day. Single storm events or pressure patterns can also produce a large number of lightning fires in a short space of time. Between 1960 and 1971, the western United States experienced six events that produced 500–800 lightning fires each. The largest such event, in June 1940, started 1488 fires.

Humans are the greatest cause of fires. Arson by mentally disturbed people accounts for 32 per cent of all fires in the United States and 7 per cent in Canada. In southern California, nearly a third of all fires are started by children, the majority of whom are boys playing with matches. Throughout the twentieth century in the United States, economic downturns have brought increased incendiarism, because firefighting affords temporary job opportunities for the unemployed. The major difference between arson and lightning as a cause of fire is the fact that lightning-induced fires tend to become conflagrations because they occur in isolated areas, while arson-induced fires tend to occur in accessible areas that are easy to monitor and control.

In Australia the causes of fire are more numerous because of the variety in vegetation, climate and land use. Additionally, burning-off as a fire prevention procedure is practised more in this country than any other part of the world; although, historically, prescribed burning was used in the United States southeast for clearing undergrowth and minimizing the outbreak of major fires. Between

1966 and 1971 in Western Australia, 36 per cent of all fires resulted from burning-off; only 8 per cent, from lightning. Vehicles, farm machinery and trains ignited 15 per cent of all fires, with carelessness accounting for 21 per cent. In the Northern Territory most fires are either caused by lightning or burning-off. Here, Aborigines traditionally light fires to clear the landscape so they can catch game. In the eastern States, because of a higher population, arson has become a major cause of fires in recent years. The problem is made more acute because many of the fires are lit by a single individual deliberately on high fire-risk days. Because the burning-off season in August and September has encroached upon early warm weather in spring, notably during drought years, the incidence of fires caused by burning-off has also increased in recent years. While burning-off has caused few major fires, it was the primary reason for the extent of the devastating Hobart bushfires in 1967. A significant number of fires are started by power lines touching each other or rubbing against tree branches during strong winds. Following the disastrous Ash Wednesday bushfires of 1983, the Victorian Electricity Commission was accused of allowing transmission line deterioration, siting and maintenance to reach a state favouring this cause of ignition. The Commission subsequently paid out an unspecified amount of money as compensation to landowners who were burnt out in areas where this was a suspected cause. It has even been suggested that power transmission should be cut off during severe bushfire weather conditions. However, this suggestion ignores the chaos that would result to communications during any fire-fighting operation.

Bushfire disasters

The world (Cornell, 1976; Luke and McArthur, 1978; van Nao, 1982; Seitz, 1986)

Bushfires as a hazard are usually evaluated in terms of loss of life, property damage (burning of buildings or complete settlements), and the loss in timber. While loss of life can still occur in northern hemisphere temperate forests today, most destruction is evaluated in terms of the loss of marketable timber. This statement needs qualification because timber has no value unless it can be accessed, and there is no need to access it unless it is close to centers of consumption. In North America it is mainly the forests that lie within 1000 km of dense urban development that are used to supply building timber, unless there is a need for timber with specific qualities—such as the West Coast Redwood. Beyond this distance forests are mainly used to supply pulp and paper in the form of newsprint. There are large areas in the Canadian and Alaskan boreal forest where no harvesting of trees takes place, because it is uneconomic. These areas generally are uninhabited and no attempt is made here to suppress forest fires. Fires are allowed to burn uncontrolled, and may take weeks or the complete summer season to burn out. In northern Canada it is not uncommon for such fires to cover thousands of square kilometers.

Historically, large sections of Europe in the Middle Ages suffered from forest fires. Damage and loss of life has gone virtually undocumented. Today, large tracts of forest still remain and forest fires pose a serious threat in the deciduous, mixed and chaparral forests of Europe. As recently as 1978, in Europe, over 43 000 fires consumed 440 000 hectares of forest and as much again in grasslands and crops. Losses exceeded $US650 million. In the same year in North America 153 000 fires swept through 1 000 000 hectares, resulting in a similar damage bill. These figures are not remarkable. No European country had more than 2 per cent of its total forest cover burnt. Individual fires on a much larger scale have occurred throughout the northern hemisphere. On 7 October 1825 over 1 200 000 hectares of North American forest burned in the Miramichi region of New Brunswick, together with 320 000 hectares in Piscataquis County, Maine. The Wisconsin and Michigan forest fires of 1871 consumed 1 700 000 hectares and killed 2200 people. Areas of similar magnitude were burnt in Wisconsin in 1894, and in Idaho and northwestern Montana in 1910. The Porcupine Fire of 1911 raced unchecked from Timmins, Canada, north through virgin forest for over 200 km, killing scores of settlers on isolated farms. During the 1980 Canadian fire season, virtually the whole mixed forest belt from the Rocky Mountains to Lake Superior was threatened. Over 4 800 000 hectares of forest were destroyed, with a potential loss equivalent to the timber supply required to supply 340 newsprint mills for one year.

All of these fires pale in significance when compared to the Great Siberian fires of July–August 1915. In that year, some 0.1–1.0 million km^2 of Siberia, from the Ural mountains to the Central Siberian Highlands burned (an area 2–20 times greater than the extent of the 1980 fires in Canada). An area the size of Germany covering 250 000 km^2 was completely devastated between the Angara and Lower Tunguska rivers. These gigantic fires were brought about by one of the worst droughts ever recorded in Siberia. Most of the taiga and larch forests were desiccated to a combustible state, producing crown fires. Over 500 000 km^2 of peat dried out and burnt to a depth of 2 m. The amount of smoke generated was $20-180 \times 10^{12}$ g, a figure that is equivalent to the estimated amount of smoke that would be produced in a limited-to-extreme nuclear war leading to a nuclear winter. Thick smoke was lifted more than 12 km into the sky. In the immediate area visibility fell to 4–20 m, and to less than 100 m as far as 1500 km away. As a result, the solar radiation flux was significantly attenuated at the ground, dropping average temperatures by 10°C. At the same time, longwave emission at night was suppressed such that the diurnal temperature range varied by less than 2°C over a large area. These below-average temperatures persisted for several weeks; however, harvests in the area were delayed by no more than two weeks. The smoke and temperature effects were totally regional. Little evidence of the fires was detected outside Siberia. Because of the isolation of the area and the low population densities, reported loss of life was minimal. The above documentation of large fires implies that conflagrations of this magnitude are frequent occurrences in the boreal forests of the northern hemisphere. If so, forest fires presently pose the most serious natural hazard in the two largest countries of the world, Canada and the Soviet Union.

United States (Pyne, 1982)

In North America at least 13 fires have, in the past, burned more than 400 000 hectares. In the twentieth century, almost all mega-fires have begun as controlled burns, or as wildfires that at some point could be considered controllable. And almost all severe fire seasons were preceded by drought. The intensity of any fire is not necessarily dependent upon the growth of biomass during wet seasons, but on the availability of dried litter or debris generated by disease or insect infestation, on windstorms, previous fires or landclearing. This point is well recognized. For example, the 1938 New England hurricane mentioned in chapter 2 left a trail of devastation in the form of uprooted trees and felled branches. Considerable effort was subsequently spent cleaning up this debris before it could contribute to major forest fires.

Much of the early fire history of North America was conditioned by Indians, who cleared or fired forests for agriculture, firewood, hunting, and defense. Thus, the original forests that met white settlers were park-like and open. While European settlement adopted many of the fire practices of the Indians, the demise of such practices made possible the reforestation of much of the northeast, and began the period of large-scale conflagrations. A similar process followed the decline of prescribed burning by, first, the turpentine operators and then the loggers in the pine forests of the southeast. The most dramatic effects were seen on the Prairies, where a grassland habitat was reclaimed by forests as land was converted to cultivation or managed rangeland. European civilization supplanted short cyclical fires with a longer cycle that became extremely hazardous.

The onset of forest fires in reforested regions was prodigious. In the northeast disastrous fires swept through the region in 1880, and it became a policy to restrict settlement in forest preserves. That policy led to a suppression of burn-off and a buildup of forest fuel that resulted in 400 000 hectares burning from New York to Maine. The fires were perceived as a threat to timber reserves and watershed flood-mitigation. They resulted in a concerted effort to set up fire lookouts in the northeast, and prompted the passage of the Weeks Act in 1911 to protect the watersheds of navigable streams, and to subsidize state fire-protection efforts. These efforts along with the formation of the Civilian Conservation Corps (CCC) in the 1930s saw the demise of large wildland fires; however, urbanization had expanded into forest areas. The nation's first urban forest fire occurred in October 1947 at Bar Harbour, Maine. Over 200 structures were burnt and 16 lives lost. This scene was repeated again over a wide area in 1963 when fires threatened suburbs in New York and Philadelphia, and destroyed over 600 structures on Long Island and in New Jersey. These fires led to incorporation of municipal fire units into the state protection system, and the signing of inter-state agreements for fire suppression assistance. The latter eventually spread nationally and internationally by the 1970s. Today urbanized brush fires also plague southern California, which has seen urban expansion since the 1950s encroach upon fire-prone scrubland, with little thought to building design and layout, and a jumble of organizations responsible at various levels for fire suppression. The costs of firefighting operations in southern California as a result have reached high levels. In 1979 the Hollywood Hills and Santa Monica mountain fires involved 7000 firefighters at a cost of a million dollars per day for a month.

In the southeast of the United States, reforestation and regeneration of undergrowth presented, by 1930, large reserves of biomass that fuelled major fires during droughts. In 1930–32, Kentucky, Virginia, Georgia and Florida suffered devastating fires. Further fires between 1941 and 1943 saw the reintroduction of prescribed burning. In 1952, 800 000 hectares of forest burned across Kentucky and West Virginia; in 1954–55, 200 000 hectares burned in the Okefenokee Swamp; and in 1955, 240 000 hectares burned in North Carolina. These occurrences led to the establishment, in North Carolina, of one of the most efficient fire suppression organizations in the country and, in Georgia, of the first fire research station. This station produced the nation's leading methods in prescribed burning.

There is one exception to the above picture, and this occurred in the virgin forests of the Lake states. Here, disastrous fires occurred as settlers first entered the region, together with logging companies and railways. The logging produced enormous quantities of fuel in the form of slash on the forest floor; the railways provided sources of ignition from smokestacks; and the settlers provided bodies for the resulting disaster. Between 1870 and 1930, one large fire after another swept through the region in an identical pattern. Most of the conflagrations occurred in autumn following a summer drought. The worse fires occurred in 1871, 1881, 1894, 1908, 1910, 1911 and 1918. The Peshtigo and Humboldt fires of 1871 took, respectively, 1500 and 750 lives in Wisconsin; the Michigan fires of 1881 killed several hundred; the Hinckley, Minnesota, fire in 1894 took 418 lives; and finally the Cloquet fire in Minnesota in 1918 killed 551 people. The fires were so common, that most residents took fire warnings nonchalantly, to their own detriment.

The great fires of 1871 were the first, and by far the worse, to sweep the Lake states. Ironically they occurred in October on the same days as the Great Chicago fire. In the weeks proceeding the fires, much of the countryside was continually ablaze. Almost all able-bodied men were employed in fire suppression, and many inhabitants had inaugurated procedures to protect buildings. On 8–9 October the fires broke into *firestorms* only witnessed on such a large scale again during the fire bombing of Tokyo, Hamburg and Dresden in the Second World War, and during the August 1910 fires in the northern Rockies, the Tillamook Fire in Oregon in 1933 and the Ash Wednesday bushfires in Australia in 1983. Flames lept 60 m into the air and were driven ahead at a steady rate of 10–16 km hr^{-1} by spot fires, intense radiation and updrafts being sucked into the maelstrom. The approach of the flames was heralded by dense smoke, a rain of firebrands and ash, and by combusting fireballs of exploding gases. Winds of 100–130 km hr^{-1} picked up trees, wagons and flaming corpses. The fire marched with an overpowering roar like continual thunder or artillery barrages. Hundreds were asphyxiated hiding in cellars, burned to death as they sheltered in former lake beds converted into flammable marshes by the drought, or trampled by livestock competing for the same refuges. The Lake fires were certainly the worst in documented history and show how vulnerable any forest can be under the right fuel and climatic conditions.

Australia (Luke and McArthur, 1978; Cheney, 1979; Powell, 1983; Voice and Gauntlett, 1984; Oliver et al., 1984; Butler, 1985; Webster, 1986)

Conditions

The occurrence of Australian bushfires is unique, because it represents the present occurrence of deadly fires only witnessed previously in the United States and Canada before the 1940s. Australian bushfires are also unique because of the highly flammable nature of Australian forest vegetation. On a world basis, Australian bushfires compare in size to the largest recorded in North America. While loss of life is now minimal in North America because of the nature of settlement and efficient evacuation procedures, in Australia it appears to be increasing because of expansion of urban populations into rugged bush country. Mention has been made of the free-burning nature of forest fires in northern Canada. While attempts are made in Australia to put out fires, many in inaccessible hill country are left to burn. These fires pose a threat directly if they change direction under strong winds, and indirectly because they provide a source for fire spotting beyond the limits of the main fire front. Australian bushfires are also impossible to control once they become wildfires, because of the flammable nature of the bush. Some of the most intense fires to occur have been the conflagrations in Hobart in 1967 and southeastern Australia in 1983.

The map in figure 9.7 shows that the spatial occurrence of fires in Australia is seasonally controlled. The tropical zone has a winter-spring season, because the summer is usually too wet and grasses will not cure and dry out before spring. The fire season progresses southward and across the continent along the coasts through the spring. The timing of movement virtually parallels the seasonal migration of the Hadley cell poleward. By late spring-summer, the fire season has peaked in a line stretching from Geraldton across to Canberra

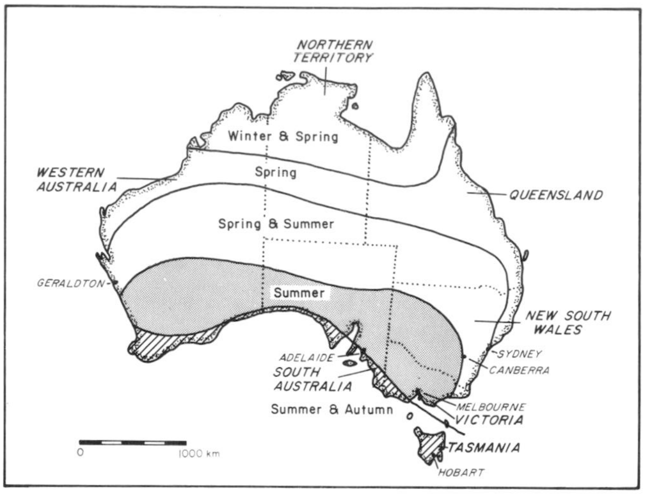

Fig. 9.7 Pattern of seasonal bushfires in Australia (after Luke and McArthur, 1978)

and Sydney. By mid-summer, the zone has moved to the southern part of the continent, concomitantly with the maximum poleward displacement of Hadley cells. Hot, desiccating winds from the interior blow towards the coast as these cells move eastward across the continent. Historically, this has been the time of severest bushfires and greatest loss of life in Australia. Finally, towards autumn, the southern extremities of the continent experience bushfires as vegetation dries out at the end of summer. Bushfires are not necessarily restricted to the seasons shown in figure 9.7. In the central spring and summer zone, severe bushfires have been known to occur in the autumn. For example, in 1986 the worst bushfires of the season broke out in early April following a dry summer, and record-breaking temperatures for the month.

Historic disasters

The worst fires in Australian history are well documented and have occurred in the fire-prone belt of southern South Australia and Victoria. On 14 February 1926 fires ignited under temperatures in excess of 38°C, with relative humidities of less than 15 per cent. Sixty lives were lost and an untold number of farms, houses and sawmills burnt out. In the 1931–32 fire season, 20 lives were lost. The 13 January 1939 fires occurred under record-breaking weather conditions. Temperatures over 46°C with humidities as low as 8 per cent sent fires racing through extremely dry bush. Over 70 lives were lost in numerous fires in the southeastern part of the country. The 1943–44 season was a repeat. Over 49 people died in grass and bushfires under oppressively hot and dry weather. Fires in the Dandenongs killed 14 people and burnt 450 houses in 1962. On 8 January 1969, 23 people died in bushfires as temperatures exceeded 40°C and winds topped 100 km hr^{-1}.

There are however three fires or fire seasons since the Second World War that stand out, the 1974–75 fire season, the Hobart fires of 1967, and the Ash Wednesday fires of 1983. The 1974–75 fire season caused the least loss of life, but was the most widespread of any fire season in Australia. It followed the record-breaking rainfalls of 1974, which covered most of Australia as a result of enhanced Walker circulation. Most of the desert and semi-arid regions of Australia were covered in a continuous blanket of grasses and dense shrub. The total area of rural Australia burnt represented 15.2 per cent of the land area of the continent or 117 million hectares. This is 24 times the area burnt out in the bad forest fire season of 1980 in Canada, and is equivalent to the maximum area affected by the Great Siberian fires of 1915. Over one quarter of this area was deliberately burnt to try and control the spread or occurrence of fires during that season. While stock losses overall were small, the burning of complete properties brought instant drought to many graziers trying to restock from the 1972 drought and recover from the 1974 floods. The greatest property loss was to wood fencing, which for many properties had to be completely replaced.

The Hobart, Tasmania, bushfires of 7 February 1967 were preceded by a season of above-average rainfall and heavy growth of grasses. These grasses cured and dried out in the summer of 1966–67, which was dry. Numerous controlled burns were lit to diminish the fire risk; however, 81 fires around Hobart were left burning on 7 February under the misconception that they would be extinguished by rains (figure 9.8). On that day, temperatures reached 39°C with humidity as low as 18 per cent. The fires were driven by northwest winds in excess of 100 km hr^{-1} generated by the movement of a high-pressure system out into the Tasman Sea. By noon, fires had raced through the outer districts of Hobart and were entering the suburbs. By the time the fires were controlled, 67 people were dead, 1400 homes and 1000 farm buildings destroyed, and 50 000 sheep killed. Insurance claims totalled $A15 million dollars.

The fires in Hobart are an example of a natural hazard exacerbated by humans. People in Tasmania had not experienced a major bushfire before, and were under the illusion that mainland fires could

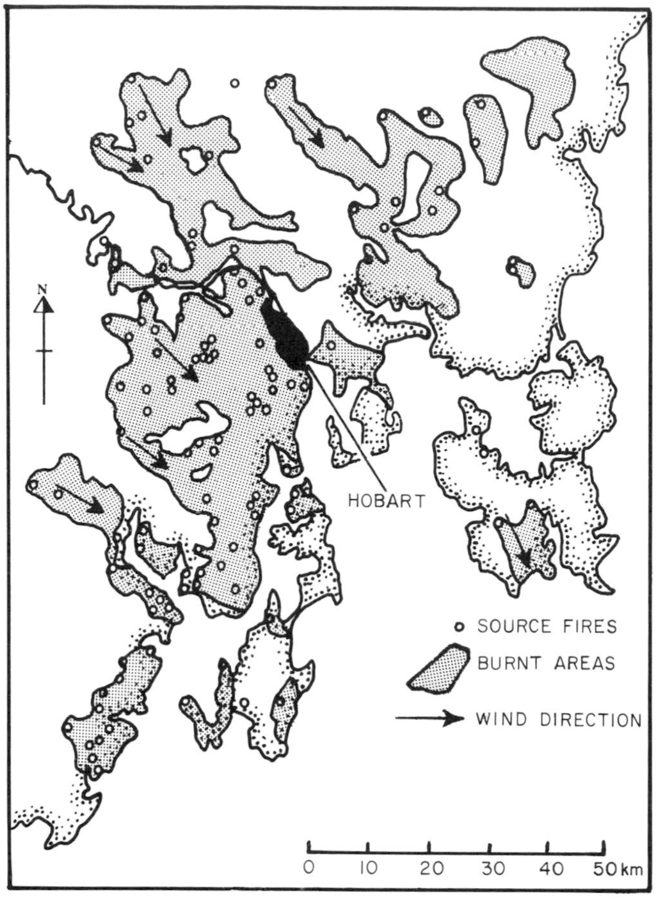

Fig. 9.8 Location of spot fires and extent of subsequent bushfires in Hobart, Tasmania, 7 February 1967 (adapted from Cheney, 1979)

not occur because the State was too wet and cool. Farmers around Hobart, in burning off grass, usually let the fires extinguish themselves under rain. Residents had built newer homes away from the city of Hobart in hilly bushland. The climatic conditions leading up to the fires were unusual in that rainfall was below average, and temperatures on the day of the fire were the hottest this century. Forewarnings of the disaster had been given in the previous two days when numerous fires had threatened homes and farm buildings in outlying areas. Many of these fires had not been completely extinguished because they were in inaccessible areas and, again, rainfall was being relied upon to put them out. While most spot fires were not responsible for the burning of Hobart, their occurrence did nothing to ease the pressure put on firefighters and communications to contain major outbreaks. While the extreme weather conditions were rare, the expansion of urban dwelling into bushland can only dictate that such disasters will occur again in Hobart.

Ash Wednesday fires of 16 February 1983

The summer of 1982–83 was the culmination of the worst drought in recorded history in Australia. The 1982 El Niño-Southern Oscillation event was the strongest in 100 years, and its drought producing consequences had exempted much of eastern Australia from appreciable rainfall throughout the summer. Bushland was desiccated, and grassland consisted of either paddocks denuded of vegetation, or grasses completely cured and dried out. The weather conditions of 16 February had been repeated often throughout the summer. Temperatures in excess of 40° C had been accompanied by strong north or northwesterly winds. These winds usually preceded strong cold fronts, which did little to lower temperatures and brought strong southwesterly winds. The week before the fires, one such cold front whipped up soil from barren farmland in western Victoria and swept it as a dust storm through Melbourne for only the second time this century (figure 3.7).

The weather conditions on the day of the fires were predictable and extreme. By the early morning, temperatures had already reached 35–40° C and winds were gusting over 60 km hr^{-1}. Weather forecasts accurately predicted one of the hottest and driest days of the summer. Figure 9.4 typifies the weather pattern for the day of the fires. It was very similar to the conditions that enhanced the Hobart fires. Hot, dry air flowed from the interior of the continent on strong winds, reinforced by the subtropical jet stream in the upper atmosphere. A strong, cool change was forecast to cross the southern continent in the afternoon and early evening, bringing some relief to the desiccating conditions. Total fire bans had been enforced and broadcast hourly over much of southeast Australia for weeks. Bushfire organizations and volunteer firefighters in South Australia, Victoria and New South Wales were put on alert.

There was little anyone could do to stop the fires. By noon, much of Adelaide was ringed by fires (figure 9.9) and at 3:30 pm the cold change swept through. The change dropped temperatures 10° C, but the winds increased and drove the fires to the northeast. In Victoria fires broke out around Mount Macedon to the north of Melbourne and in the Dandenong ranges to the east, with hot, dry northwesterly winds gusting up to 70 km hr^{-1}. Other fires developed in rural shrubland in western Victoria and in the coastal resort strip west of Geelong. The firefighters were alert to the conditions, and all but 8 of the 93 fires reported had

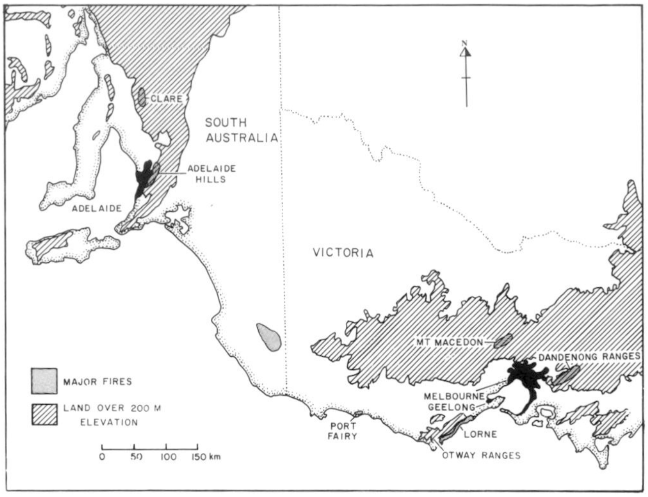

Fig. 9.9 Location of major Ash Wednesday bushfires in South Australia and Victoria, 16 February 1983

been brought speedily under control. The cool change swept through in the early evening and, instead of bringing the weather conditions which should have allowed these remaining fires to be brought under control, it turned them into raging infernos. Fire fronts were shifted 60–90 degrees and pushed forward by winds averaging 70 km hr^{-1} and gusting at times to 100–170 km hr^{-1}. The conflagrations which swept the Adelaide and Melbourne areas were unprecedented. Only three to four times per century could such a combination of conditions be expected anywhere in Australia. The strong winds brought by the cool change generated firestorms, which snapped the trunks of whole stands of trees at ground level. It is physically impossible under the most ideal fuel conditions to generate fire intensities in the Australian bush greater than 60 000–100 000 kW m^{-1} (boreal forests can have values up to 250 000 kW m^{-1} because of their greater biomass). Houses, farms and towns repetitively witnessed the maximum limit of these intensities as fires bore down on them at velocities close to 20 km hr^{-1}. Aluminum tire rims and sidings on houses melted into puddles. The intense heat generated thermal tornadoes. Trees in front of the fire-front exploded first into flames because of the intense radiation and then were consumed by the maelstrom. Such conditions had not been witnessed since the firestorms which swept through the cities of Hamburg and Tokyo as the result of fire bombing in the Second World War. People fleeing in cars were outrun and incinerated. Hundreds of people only survived by quick thinking and luck. At McMahons Creek, east of Melbourne, 85 people escaped the inferno by sheltering 30 m underground in the 1 m spacing between water pipes servicing a dam. At the town of Cockatoo, in the Dandenongs, over 150 townspeople survived in the local kindergarten, in the middle of a playground as their town burned down around them. And along the Otway coast people fled to the safety of the local beach and then into the ocean as the sand became too hot to bear or was windblasted around them.

By the end of the fires 76 people had died; 3500 were injured, many with burns that would cripple them for the rest of their lives; over 300 000 sheep and 18 000 cattle had been killed; over 500 000 hectares of urban, forest and pasture land had burnt in a single day; 300 000 km of fencing were destroyed; and 1700 homes and buildings were consumed (figure 9.10). In South Australia alone, 40 per cent of the commercial pine forest had been burnt. Insurance claims reached $A200 million dollars, with total property losses exceeding $A500 million. The late evening national television news on Channel 7 simply ended its program by silently listing the 20 or more towns that had ceased to exist that day.

The Ash Wednesday bushfires were a rare occurrence in Australia. No modern technology, no prescribed burning, no preparation by homeowners or municipal councils, no improvement in prediction, no number of firefighters or amount of equipment could have stopped those fires from occurring or negated their effects. The weather conditions leading up to Ash Wednesday were only matched by those preceding the 1939 bushfires; however, the fires exemplify the fact that there is a preferred weather pattern for extreme bushfires in southeastern Australia that should be researched and predicted. Warnings of this particular hazard and research into the design of buildings and the structuring of communities to prevent fire intrusion must be undertaken to minimize the threat. Issues pitting conservationists against fire authorities on prescribed burning practices, in attempts at preventing another Ash Wednesday, must be resolved. The Ash Wednesday bushfires of 1983 were the worst natural disaster in Australia up to that time. Research, planning and preparation for major bushfires must be seen in this country as the prime challenge for natural hazard mitigation. These activities coupled with an effective education program on bushfire survival techniques (staying inside a house or car until the fire passes over is the safest behaviour) offer hope of reducing the deathtoll in Australia from this hazard in the future.

FIG. 9.10 The 1983 Ash Wednesday bushfires devastated whole communities. This was Fairhaven, along the Otway coastline of Victoria, following the bushfires (photograph © and reproduced courtesy of the *Age*, Melbourne).

Concluding comments

Of all natural hazards occurring in Australia, bushfires now have the potential of being the major cause of property damage and loss of life. The Ash Wednesday fires of 1983 took more lives, and destroyed more property value in scattered semi-rural communities, than Cyclone Tracy did in moving through the center of Darwin. The dense settlement which is now occurring in rugged bushland on the outskirts of major Australian cities is increasing the probability of occurrence of another major bushfire disaster, such as Ash Wednesday. While the image of Ash Wednesday still remains fresh in people's minds, the lesson of that fire is being ignored. A drive through Mount Macedon and the Dandenong ranges outside Melbourne will show that houses, destroyed in those fires because they were not fire resistant, have been reconstructed exactly as they were before the fire. Even in New South Wales, which has never witnessed an Ash Wednesday-type bushfire, 10–20 homes per year have been destroyed by bushfires around Sydney in the last 20 years. In the northern suburbs of Sydney, homes with wooden decks and exteriors have been built up the inaccessible slopes of steep gullies. Throughout this book the 1:100-year event has been emphasized as being an event that has not occurred yet in most of Australia. Overseas in older established countries, structures designed to withstand that same event have been shown to be inadequate over periods longer than 100 years. If 1:100-year bushfire maps were constructed for settled areas in the southern part of Australia, it would become obvious that many newer suburbs, as well as some of the older inner suburbs, lie within bushland which has undergone recurrent burning. Australia has the best developed volunteer organization for fighting bushfires in the world. The ability of that organization to respond to the escalating bushfire threat in southeastern Australia will be severely tested by the year 2000.

Nor is this threatening situation confined to Australia. The same picture is paralleled in southern California, which has similar fire weather, vegetation, and urban expansion. What is not easily recognized is the fact that urban settlement has extended into forest areas in other parts of the continent. The 1947 Maine fires were the first major fires on that continent to witness the burning of urbanized bushland. Since then, urbanization has continued at a rapid pace, and the forests have reclaimed completely large tracts of abandoned farmland. Since the mid-1950s, this urbanization has taken on a different aspect as millions of urban dwellers have chosen to flee major cities during summer to cottages in nearby woodlands. The northeastern Appalachians, northern Michigan and Wisconsin have all witnessed this process. In Canada up to 50 per cent of the population of Toronto and Montreal escape the summer's heat each weekend by travelling into the forest on the Canadian shield. Historically, these forests have witnessed large conflagrations and, given the propensity of humans to play with fire, the regrowth of forests, and the tendency for warmer summers, will undoubtedly witness the occurrence of large fires again.

Finally, the United States is witnessing a reintroduction of prescribed and free burning in its national forests that is creating a major controversy. The fires in Yellowstone Park in the summer of 1988 attest to the dilemma. Fires were permitted to burn unchecked until half the Park was consumed, and then a major effort at suppression was mounted to save the rest of the Park. Here, fire is presently viewed as a natural occurrence that is inseparable from the cyclical development of a forest. Prescribed burning is being reintroduced in other forests to maintain low fuel levels and thus prevent conflagrations. America is, therefore, witnessing the natural burning of parks and reserves, and the preventative low-intensity burning of undergrowth in productive forests. Both processes represent a return to conditions experienced by the first settlers. However, the prescribed burning policy in the United States is not completely thought out. For example, in 1980, a prescribed burn to improve the habitat of the Kirtland warbler eventually rampaged across 18 000 hectares, killing one person. Oddly, just when America is reintroducing prescribed burning, it is being abandoned in Australia, where it has been under attack for years. Frequent prescribed fires have altered the species character of bushland without necessarily offering protection to urban dwellers. Prescribed burning in many Australian States is giving way to the concept of unchecked burning, except near suburbs. The debate in Australia or the United States, whether to burn or not, will in the end be resolved by defining which policy best mitigates against loss of life. An examination of the history and the effects of fires within each of these countries will certainly influence that decision.

References

Cheney, N. P. 1979. 'Bushfire disasters in Australia 1945–1975'. In Heathcote, R. L. and Thom, B. G. (eds) *Natural Hazards in Australia.* Australian Academy of Science, Canberra, pp. 72–93.

Luke, R. H. and McArthur, A. G. 1978. *Bushfires in Australia.* AGPS, Canberra.

Vines, R. G. 1974. 'Weather patterns and bushfire cycles in Southern Australia'. *C.S.I.R.O. Division Chemical Technology Technical Paper* No. 2.

White, I. D., Mottershead, D. N. and Harrison, J. J. 1984. *Environmental systems: an introductory text.* Allen and Unwin, London.

Further reading

Butler, J. E. 1985. *Natural Disasters* (rev. edn). Heinemann Educational, Richmond, Victoria, pp. 83–96.

Cornell, J. 1976. *The great international disaster book.* Scribner's, NY.

Oliver, J., Britton, N. R. and James, M. K. 1984. 'The Ash Wednesday bushfires in Victoria, 16 February 1983'. *James Cook University of North Queensland Centre for Disaster Studies Disaster Investigation Report* No. 7.

Powell, F. A. 1983. Bushfire weather. *Weatherwise* v. 36 No. 3 pp. 126.

Pyne, S. J. 1982. *Fire in America: a cultural history of wildland and rural fire.* Princeton University Press, Princeton.

Seitz, R. 1986. 'Siberian fire as "nuclear winter" guide'. *Science* v. 233 pp. 116–17.

van Nao, T. (ed.). 1982. *Forest fire prevention and control.* Martinus Nijhoff, The Hague.

Voice, M. E. and Gauntlett, F. J. 1984. 'The 1983 Ash Wednesday fires in Australia'. *Monthly Weather Review* v. 112 No. 3 pp. 584–90.

Webster, J. K. 1986. *The Complete Australian Bushfire Book.* Nelson, Melbourne.

Yool, S. R., Eckhardt, D. W., Estes, J. E. and Cosentino, M. J. 1985. 'Describing the bushfire hazard in Southern California'. *Annals Association of American Geographers* v. 75 No. 3 pp. 417–30.

II

Geological Hazards

10

Causes and Prediction of Earthquakes and Volcanoes

Introduction

Of all natural hazards, earthquakes and volcanoes release the most energy in the shortest possible time. In the past 40 years scientists have realized that the distribution of earthquakes and volcanoes is not random across the Earth's surface, but tends to follow crustal plate boundaries. In the past 20 years research has been dedicated to monitoring these regions of crustal activity with the intention of predicting major, and possibly destructive, events by several days or months. At the same time, planetary studies have led to speculation that the clustering of earthquake or volcanic events over time is not random, but tends to be cyclic. This knowledge could lead to prediction of these hazards decades in advance. Before examining these aspects, it is essential to define how earthquake intensity is measured, because specific earthquakes are always characterized by their magnitude in any discussion. This aspect will be examined first in this chapter, followed by a description of the distribution of earthquakes and volcanoes over the Earth's surface and some of the common causes of these natural disasters. The chapter concludes with a discussion on the long- and short-term methods for forecasting earthquake and volcano occurrence.

Scales for measuring earthquake intensity (Holmes, 1965; Bolt et al., 1975; Wood, 1986)

Seismic studies were first undertaken as early as AD 132 in China, where crude instruments were made to detect the occurrence and location of earthquakes. It was not until the end of the nineteenth century that these instruments became accurate enough to measure the passage of individual *seismic* waves as they travelled through, and along the surface of the Earth. The characteristics of these waves will be described in more detail in the following chapter. The magnitude or intensity of these waves are commonly measured using either the Richter, or Mercalli scales. Magnitude on the Richter scale measures the total amount of elastic energy released by each shock wave produced by an earthquake. The relationship between magnitude (M) and energy (E) released is given by the following equation:

$$\log_{10}E = a + bM \qquad (10.1)$$

The values for a and b have been modified several times, but have quantities of approximately 5.8 and 2.4 respectively. These values defined energy in ergs. Each unit increase in magnitude on this scale represents 2.4 orders of magnitude change in terms of energy. For example, an earthquake of magnitude 2 releases $10^{10.6}$ ergs of energy, while one of magnitude 3 releases 10^{13} ergs of energy. The difference between the two earthquakes (13 − 10.6) represents 2.4 orders of magnitude or a 240-fold increase in energy. This change in scale is not to be confused with the change in wave amplitude of the shock wave, which is commonly believed to represent the increase in energy. Each unit increase in magnitude on the Richter scale represents an order of magnitude or a tenfold increase in wave amplitude. Very few earthquakes have exceeded a magnitude of 8.9 on the Richter scale. Table 10.1 lists major earthquakes, mainly this century, in terms of the Richter scale. The largest earthquakes appear to occur along the western edges of the North and South American plates, or in China

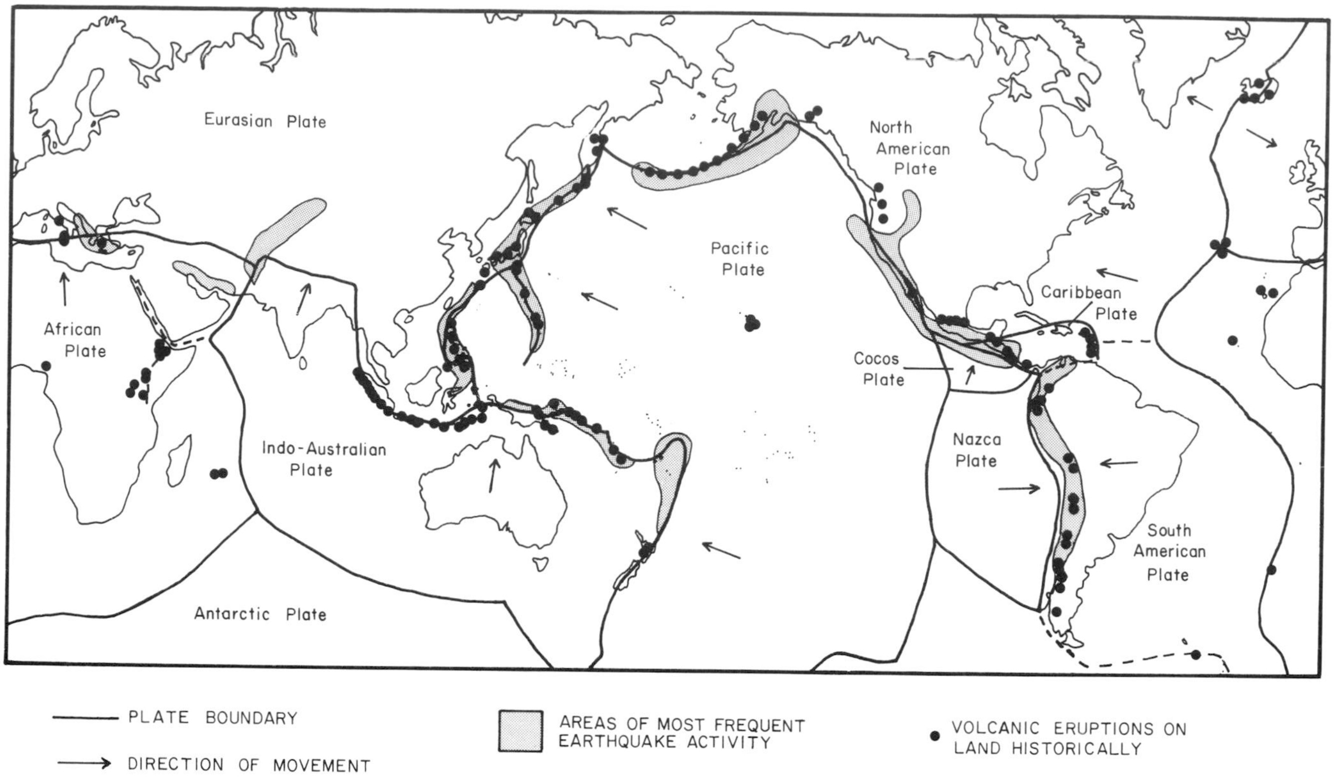

FIG. 10.1 Distribution of plate boundaries, intense earthquakes and historical land-based volcanic eruptions (based upon Bolt et al., 1975; Press and Siever, 1986)

(figure 10.1). Magnitudes based on the Richter scale can vary for the same event, because of differences in the pathways that seismic waves take travelling to recording stations, differences in seismograph equipment, and variations that still exist in the parameters used to define magnitude in equation 10.1. To overcome these limitations a new scale, called Moment, has recently been developed that measures the exact energy released by an earthquake; however, this scale relies upon surface expression of faulting associated with the earthquake. For many earthquakes, surface expression only occurs above 7 on the Richter scale. To date, the Richter scale still offers the best estimate of seismic energy for the full range of earthquakes.

TABLE 10.1 Some of the Largest Earthquakes, by magnitude on the Richter Scale

Year	*Location*	*Magnitude*
1755	Lisbon, Portugal	9.0
1906	Andes (Columbia)	8.6
1906	Valparaiso, Chile	8.4
1906	San Francisco, United States	8.25
1911	Tienshan, China	8.4
1920	Kansu, China	8.5
1923	Tokyo, Japan	8.2
1933	Japanese trench	8.5
1950	North Assam, India	8.6
1960	Chile	8.3–8.9
1964	Alaska	8.6
1977	Sumba, Indonesia	8.9

The Richter scale is not used as much by Europeans. Instead, the Mercalli scale or its modified form is used. This scale, summarized in table 10.2, can be related approximately to the Richter scale. The Mercalli scale qualitatively estimates the strength of an earthquake using descriptions of the type of damage that has occurred close to the origin or *epicenter* of the earthquake. Determining the magnitude of an earthquake on this scale does not depend upon seismographs, but instead uses actual field reports and pictures of damage to cultural features, mainly buildings. The scale has some quantitative basis in that the categories on the scale increase with increasing acceleration of the shock wave through the crust. The interval between each category is roughly proportional to $\log_2$. When the maximum acceleration exceeds $9800\,\mathrm{mm\,s^{-2}}$, which is the acceleration due to gravity, the shock wave of the earthquake hits the Earth's surface so hard that objects are tossed up into the air and trees can be physically hammered

TABLE 10.2 The Mercalli Scale of Earthquake Intensity

Scale	*Intensity*	*Description of Effect*	*Maximum Acceleration in mm sec^{-2}*	*Corresponding Richter Scale*
I	Instrumental	detected only on seismographs.	< 10	
II	Feeble	some people feel it.	< 25	
III	Slight	felt by people resting; like a large truck rumbling by.	< 50	<4.2
IV	Moderate	felt by people walking; loose objects rattle on shelves.	< 100	
V	Slightly Strong	sleepers awake; church bells ring.	< 250	<4.8
VI	Strong	trees sway; suspended objects swing; objects fall off shelves.	< 500	<5.4
VII	Very Strong	mild alarm; walls crack; plaster falls.	<1000	<6.1
VIII	Destructive	moving cars uncontrollable; chimneys fall and masonry fractures; poorly constructed buildings damaged.	<2500	
IX	Ruinous	some houses collapse; ground cracks; pipes break open.	<5000	<6.9
X	Disastrous	ground cracks profusely; many buildings destroyed; liquefaction and landslides widespread.	<7500	<7.3
XI	Very Disastrous	most buildings and bridges collapse; roads, railways, pipes and cables destroyed; general triggering of other hazards.	<9800	<8.1
XII	Catastrophic	total destruction; trees driven from ground; ground rises and falls in waves.	>9800	>8.1

from the ground. These accelerations can be used to construct maps of *seismic risks*, which will be described in more detail in the following chapter.

Distribution of earthquakes and volcanoes (Hodgson, 1964; Doyle et al., 1968; Denham, 1979; Bolt et al., 1975; Fielder and Wilson, 1975; Blong, 1984)

Figure 10.1 shows globally the distribution of areas of most frequent earthquake activity together with historically active volcanoes on land. Superimposed on these maps are the locations of active plate boundaries. The most striking feature is the correspondence between the occurrence of these two hazards and the boundaries of crustal plates. Active volcanoes occur in three locations: near convergent plate margins, at divergent plate margins and over *mantle* hot-spots. Briefly, convergent plate boundaries are either: *subduction zones*, where two sections of the Earth's crust are colliding, with one plate being consumed or subducting beneath the other; or mountain-building zones, where two or more plates are colliding with one overriding the other. The ocean trenches and the Pacific *island arcs* occur above subduction zones, while the Himalayas and European Alps occur within mountain-building zones. Although both these regions give rise to earthquakes, subduction zones are far more important in accounting for the majority of earthquakes and almost all explosive volcanism. While volcanoes at these locations produce only 10–13 per cent of the *magma* reaching the Earth's surface, they are responsible for 84 per cent of known eruptions and 88 per cent of eruptions with fatalities. The majority of these eruptions occur around the Pacific Ocean, along what is termed the Pacific 'ring of fire'. Plate boundaries through Indonesia, Italy and New Zealand have the longest record of activity. Between 1600 and 1982, 67 per cent of all known deaths resulting from volcanoes occurred in Indonesia, on the convergent boundary of the Australian and Eurasian plates. Mantle hot-spots are the next most important region of volcanic activity. The most notable hot-spot includes the Hawaiian Islands, which lie in the middle of the Pacific plate. Most divergent plate margin volcanoes occur along mid-ocean ridges where the crust is separating. The latter is the cause of volcanism in Iceland.

Volcanoes on plate boundaries or over hot spots are usually associated with earthquakes. About 75 per cent of earthquake energy is released in the upper 60 km of the crust along plate boundaries. These are termed *shallow-focus* events, compared to earthquakes that occur up to 700 km below the crust beneath subduction zones. Little if any seismic activity originates at depths greater than 700 km. There are however a significant number of earthquakes which are not associated with either volcanic activity or with plate boundaries. The historic incidence of earthquakes in China and North America, plotted in figures 10.2 and 10.3 respectively, illustrates this fact. The Chinese record goes back to 780 BC, and shows that some of the largest and most destructive earthquakes have occurred in a belt from Tibet to Korea, as much as 1000 km from the nearest plate boundary. The most destructive earthquake in terms of loss of life occurred on 23 January 1556 in Shensi province on the Hwang Ho River. These earthquakes result from crustal stresses due to continued uplift in the Himalayas, to sinking in the Shensi *geosyncline* or movement along the northwestern edge of the north China plain. The Tangshan earthquake of 28 July 1976, which took 250 000 lives, occurred in this latter region. In North America, where major earthquakes are generally perceived as occurring along the edge of the Pacific plate running the whole length of western North America, some of the largest earthquakes in fact have an *intraplate* origin. Examples include earthquakes at Timiskaming in 1935 in the middle of the Canadian shield, at New Madrid along the Mississippi River, between 16 December 1811 and 7 February 1812, and at Charleston, South Carolina, on 31 August 1886. Earthquakes in the region of the Canadian shield usually occur because of crustal readjustment to glacial deloading that began 10 000 years ago. The earthquake zone in the Mississippi valley occurs in a region of downwarping active since Cretaceous times. It is not an area of frequent, present-day activity; however, the New Madrid tremors were significant enough to alter the course of the Mississippi River, and to create new lakes such as Reelfoot Lake in Tennessee, and drain swamps. The South Carolina earthquake is not associated with any known surface faulting, and is to date unexplained.

Earthquakes can occur anywhere in the world, even on the Australian continent, which is generally perceived as being virtually *aseismic*. Until December 1989 no one had been killed by an earthquake in Australia and the property damage had never exceeded $A5 million dollars for any one event. The Newcastle earthquake of 28 December 1989 changed this dramatically. In ten seconds an earthquake of magnitude 5.5 on the Richter scale shook the center of Newcastle, New South Wales. It killed 12 people and left a damage bill of close to $A1000 million. Elsewhere in Australia earthquakes of moderately strong intensity have occurred at the

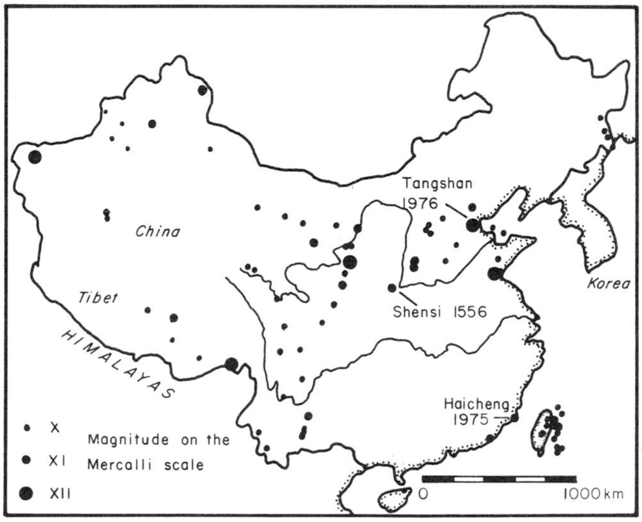

FIG. 10.2 Location of major earthquakes in China (adapted from Bolt et al., 1975)

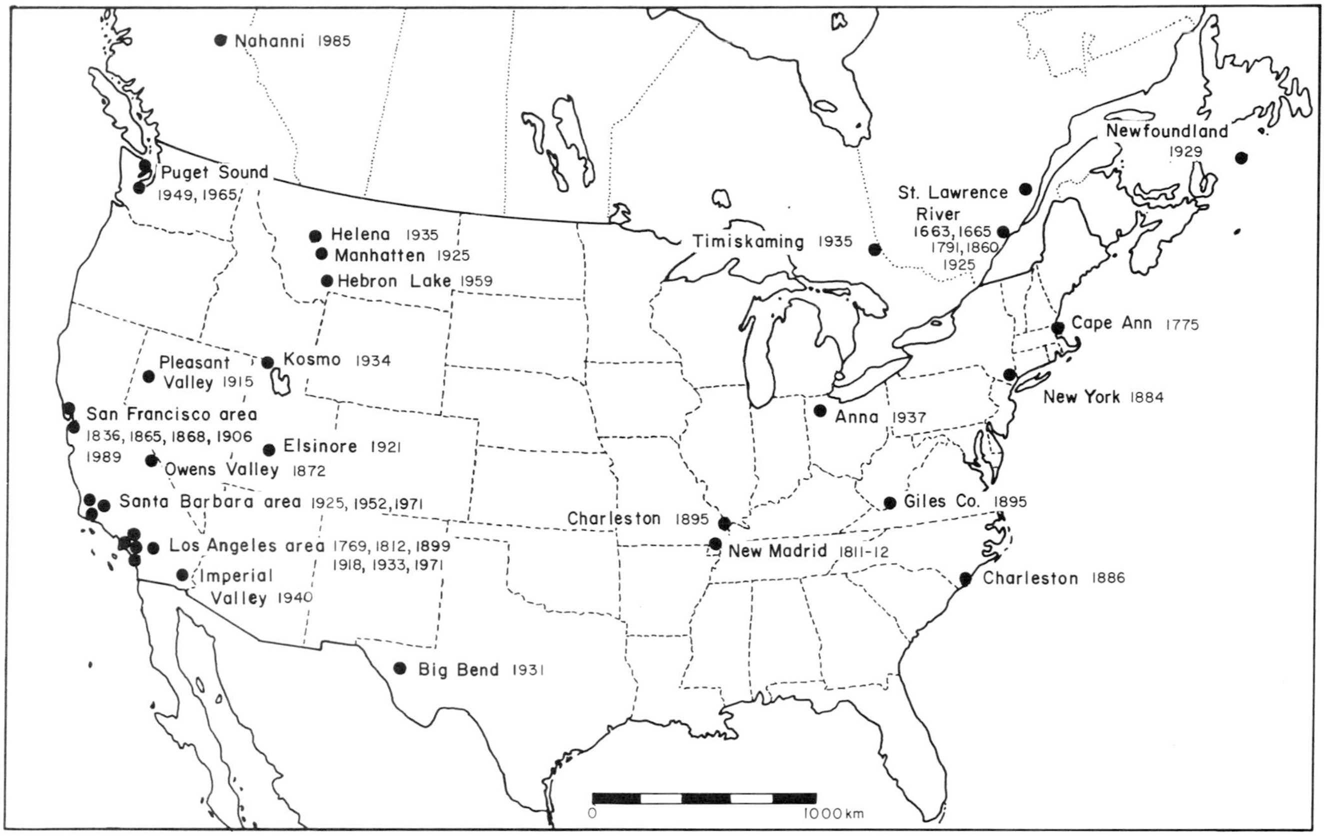

FIG. 10.3 Location of most major earthquakes in the conterminous United States region within recorded history (based upon Bolt et al., 1975; Oakeshott, 1983)

edge of the West Australian Pre-Cambrian shield, and along the Spencer Gulf-Lake Eyre *rift system* (figure 10.4). The Tennant Creek, Northern Territory, earthquakes of 22–23 January 1988 are the largest recorded in Australia, registering slightly more than 7.2 on the Richter scale. Previous to these, the Meckering earthquake of 1968 in Western Australia was the largest, registering 6.8. This was one of the few earthquakes in Australia to be associated with observed faulting. Considering the scarcity of earth tremors in this country, the picture of seismic areas is remarkably well defined. The locations of major registered earthquakes plotted on figure 10.4 cluster into three very broad regions. Little of the activity is associated with large, known *faults* such as the Darling fault in Western Australia. Almost all of the earthquakes occur within the upper crust. The seismic activity in the eastern highlands of Australia is similar to that of the Ozark-Appalachian-St Lawrence zone in eastern North America. Both border an aseismic Pre-Cambrian shield with rift zones giving way to *Palaeozoic* mountain chains with minor activity. A

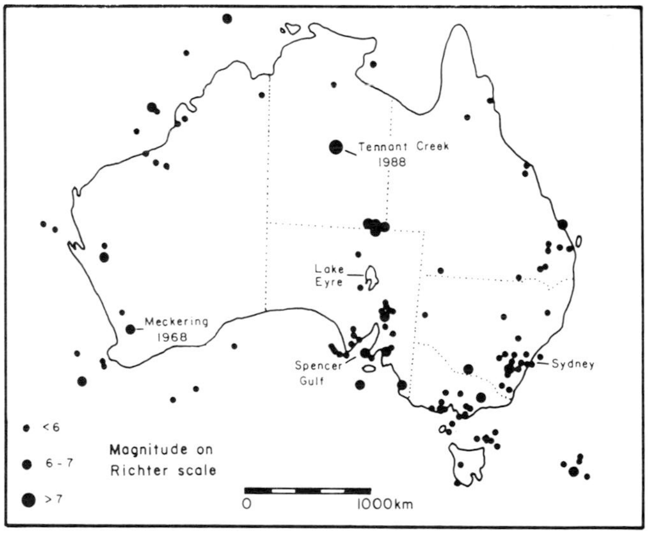

FIG. 10.4 Location of major earthquakes in Australia within recorded history (based on Doyle et al., 1968)

rift zone runs up through Spencer Gulf to Lake Eyre. Epicenters occur under the ranges rather than under the rifts. The whole area corresponds to the margin of the west Australian shield and was revived tectonically in the Cenozoic. Shallow earthquakes in Australia undoubtedly represent strain release of energy by fracturing. The source of stress probably originates from the thermal imbalance in the upper mantle and may be related to convective currents under Australia.

Figure 10.5 compares the return interval in years for earthquakes of various magnitudes between California, a seismically active region, and the supposedly aseismic Australian continent. The data for Australia do not fit a straight line because the most seismic part of this continent is so isolated, with some events going unrecorded. In any 100-year period, Australia can expect an earthquake of magnitude 6.5 or larger. This value has in fact been exceeded by the Meckering earthquake of 1968 and the Tennant Creek earthquake of 1988. In the last 75 years, there have been 18 earthquakes that have exceeded 6 on the Richter scale. The 1:5-year event has a magnitude of 5.8 on this scale. In contrast, California can expect an earthquake in excess of 8 on the Richter scale at least once in 100 years. Several earthquakes, including the San Francisco earthquake of 1906, have exceeded this value within historical time. In California, the 1:5-year event exceeds 6.4 on the Richter scale. The mean annual maximum earthquake in California is about 0.5 orders of magnitude larger than that in Australia. This difference is made that more substantial when it is realized that California is about one-twentieth the size of Australia.

This review on the distribution of earthquakes emphasizes three facts. Firstly, while earthquakes are most likely to occur along plate boundaries, the largest earthquakes in terms of deathtoll and destruction have not been associated with plate margins. Secondly, no continent or region can be considered aseismic. Even those that are perceived as being passive have a remarkably high incidence of earthquakes above 6 on the Richter scale. And finally, while a major damaging earthquake is expected along the San Andreas faultline before the end of the century, major earthquakes have occurred in the eastern part of North America which, if they occurred near populated centres today, would be just as destructive as forecast ones in California.

Causes of earthquakes and volcanoes (Holmes, 1965; Bolt et al., 1975; Fielder and Wilson, 1975; Whittow, 1980; Wood, 1986)

Plate boundaries

Most volcanism occurs along the plate boundaries in subduction zones, and along points of crustal spreading such as the mid-Atlantic ridge. The movement of the Earth's crustal plates can be explained either by convection currents within the mantle beneath the Earth's crust, or by expansion of the Earth. The exact reason is open to debate and beyond the scope of this book. If the theory of continental drift is to be believed, then the crust in the center of oceans spreads because of diverging convection cells in the mantle. These cells rise towards the Earth's crust and spread out at the surface, resulting in a ridge with a rift in the middle (figure 10.6). This rift is infilled with molten mantle material, which at weak points or fractures can form volcanoes. As the crust continues to spread apart, volcanoes migrate away from the center of

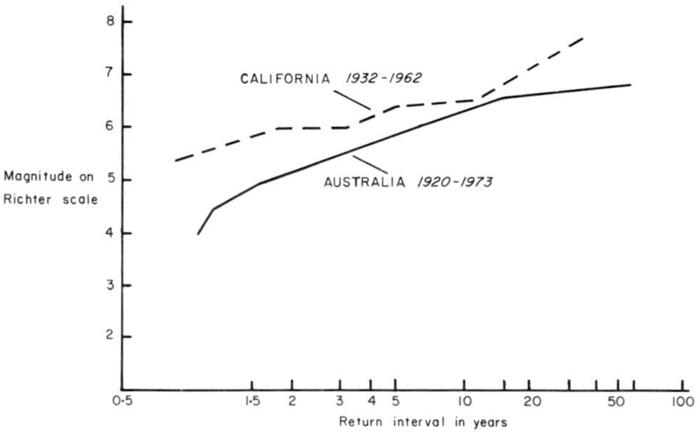

Fig. 10.5 Return interval of earthquakes of various magnitudes in California and Australia (after Denham, 1979)

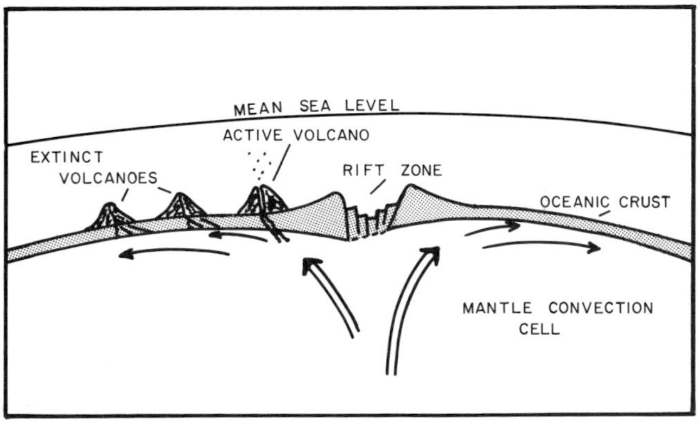

Fig. 10.6 Origin of volcanism associated with mid-ocean plate spreading

the ridging. If conduits joining the volcanoes to the magma pool can remain open, then the volcanoes will remain active. However, if the crust spreads faster than lava can be fed to the volcano, the conduits block, the volcano becomes extinct, only to be supplanted by newer ones closer to the rift.

Where plate boundaries are colliding, a different process occurs. Part of one plate may be driven beneath another (figure 10.7). This plate drags down with it crustal material and forms a *trench*. The crustal material is melted and incorporated into the mantle. A large zone of thrusting, termed the *Benioff zone*, forms to depths of 700 km with intense heat generation. This zone builds up heat because of magma extrusion, *friction* generated by earthquake activity or radioactive decay of crustal sediments. Beneath the overriding crust, less dense, heated magma (mainly *andesite* and rhyolite) rises from the Benioff zone into the upper layers of the mantle, before breaking sporadically through the crust to build volcanic cones. This magma contains gases derived from contact with groundwaters or from chemical reactions in dehydrated parts of the crust that are being subducted. The western Pacific Ocean ring of fire represents the consumption of the Pacific plate by the Eurasian and Indo-Australian plates. A chain of volcanoes, an island arc, forms parallel to the subduction zone. The Japanese, Indonesian and Aleutian Islands are classic examples of volcanic island arcs built in this manner.

The Benioff zone is also a major zone of earthquake activity because it represents thrust faulting of two plates over a wide area. Deep-focus earthquakes at 70–700 km depth occur mainly beneath the continental side of the islands surrounding the Pacific and Indian oceans. Over 75 per cent of all earthquakes occur in the plate rim of the Pacific Ocean. The Andes mountains can be included in this category as they represent an uplifted version of an island arc. As the depth of focus increases in the Benioff zone, the epicenters of deep earthquakes move further inland towards the centers of volcanic activity. At times of volcanic eruptions, earthquakes may occur because of volcanic explosions, shallow magma movements within the crust, or sympathetic tectonic earth movements. The first two factors result from release of pressures in the crust, or changes due to expansion or deflation of the magma volume. Sympathetic tectonism is rare.

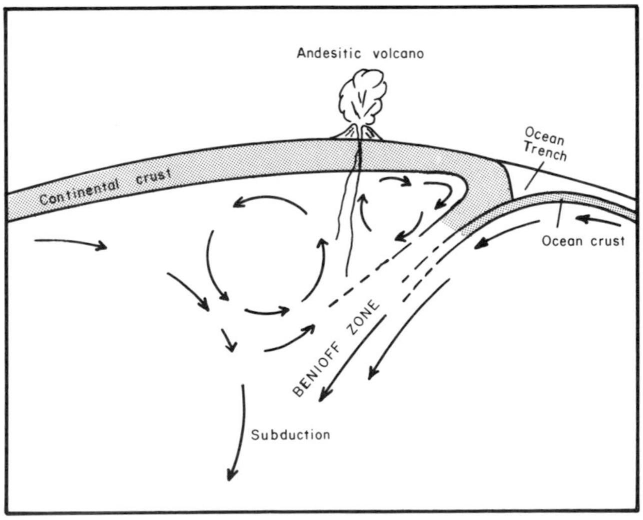

Fig. 10.7 Origin of volcanism over a subduction zone (after Holmes, 1965)

Plate collisions can also result in one continental crust overriding another, as is the case with the Indo-Australian plate overriding the Eurasian plate. Few if any volcanoes are generated in this situation because the crust of the Earth generally thickens at these locations, and fracturing down to the mantle is rare. Plates can also slide past each other as in the case of the Pacific and North American plates along the San Andreas faultline. Not only earthquakes but volcanoes can develop because fractures, called 'transcurrent faults', open up at right angles to the line of activity. Most of the volcanic activity in the Rocky Mountains and the Andes is of this nature.

Hot spots

A similar situation to volcano generation at separating plates can be produced if a hot spot in the mantle remains nearly stationary over time. Volcanoes will develop above the hot spot as crustal material is slowly melted from below. If a crustal plate drifts over the pseudo-stationary hot spot, then volcanoes drifting away from the spot become extinct, and newer volcanoes develop in their place. The Hawaiian Islands owe their origin to drifting of the Pacific plate westward over a hot spot (figure 10.8). Here, the active eastern island is about 3–4 million years younger than the furthermost western and dormant Kauai Island. In actual fact, the Hawaiian Island chain extends much further west as a series of *seamounts*, reappearing eventually as Midway Island. Rifting in the Hawaiian area ensures that fractures penetrate to the magma source below. The melting of crustal material above the hot spot, and subsequent fracturing and sinking of the weakened crust produces this rifting.

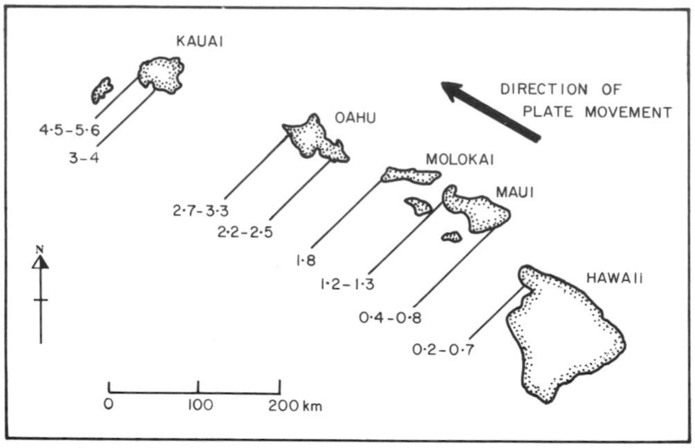

FIG. 10.8 Age and direction of movement of Hawaiian Islands (after Muller and Oberlander, 1984). The age for each of the islands, which are drifting over a hot spot, is shown in millions of years.

The main characteristic of the hot-spot formation of the Hawaiian Islands is the fact that the western islands are all older and extinct. There is a progressive decrease in age in the islands towards the main center of activity today. The indications are that the Pacific plate has drifted away from the East Pacific Rise northwest towards subduction zone trenches. Other island chains in the Pacific Ocean support this hypothesis. Figure 10.9 plots the location of main island strings in the Pacific

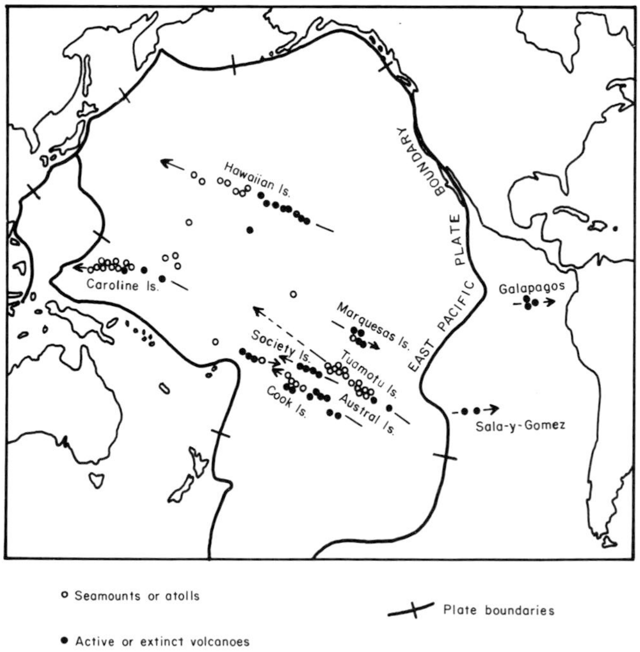

FIG. 10.9 Location of other hot spots and associated island strings on the Pacific plate (based on Holmes, 1965)

Ocean associated with drifting of the Pacific plate over different hot spots. These hot spots probably represent the location of upward flowing convection currents in the mantle that are responsible for the spreading of plates away from the East Pacific Rise. Note that not all volcanoes appear above the ocean surface: many are submerged. The Caroline, Tuamotu, Society and Austral volcanic island chains all parallel movement of the Hawaiian Islands, and thus indicate a northwest movement of the Pacific plate away from the East Pacific Rise. The Sala-y-Gomez Island and Galapagos Island chains on the east side of the rise indicate migration of plates eastwards.

Other faulting and dilatancy

The above factors account for the location of most volcanic activity and a significant number of earthquakes. However, not all earthquakes are restricted to plate boundaries or hot spots. Most damaging earthquakes simply represent the rapid release of strain energy stored within elastic rocks. Such earthquakes are termed *tectonic* earthquakes. The Earth's crust is being continually stretched and pulled in different directions as plates move relative to each other, and as forces act within the plates themselves. Even minor crustal movement sets up elastic strain within the surface of the Earth, which may be greater than the internal strength of the layers of bedrock. This strain builds up a reservoir of energy much like a coiled spring in a child's toy. Where this strain is excessive, ruptures or faults occur to relieve the pressures that are being built up. Fracturing may occur gradually, or as a sudden series of shocks that radiate outwards from the strain zone in an uneven fashion, dependent upon the spatial variation in rock strength. These variations form high-frequency waves, which travel through the Earth's crust at around $2-3\,km\,s^{-1}$, and are responsible for most of the damage.

The type of faulting or rupturing that occurs depends upon the characteristics of the fault. Figure 10.10 shows the typical range of faults which occur with earthquakes. The zone of earthquake influence is narrowest for the strike-slip fault, whereas normal or thrust faulting produces a wider zone of influence. However, the difference is really dependent upon the angle of the fault. For example, as the dip of the San Andreas fault is virtually vertical earthquake damage rapidly diminishes within 20 km of the faultline. This is why moderate earthquakes this century in southern California have not caused widespread damage. On the other hand,

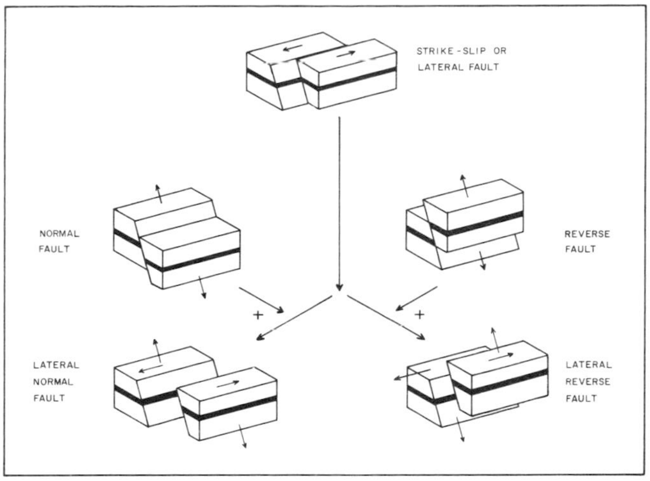

FIG. 10.10 Types of faults associated with earthquake activity

the high magnitude earthquakes in Chile in 1960, and Alaska in 1964 occurred on low-angled reverse faults (20° or less in the case of Alaska) associated with subduction zones, a feature that extended damage over a much larger area. Weathered material above bedrock can absorb considerable amounts of the fault displacement, such that faulting occurring in underlying bedrock may not be present at the Earth's surface. Elastic strain does not have to be built up in rocks, but can occur slowly over time as *tectonic creep*. Such creep can be monitored and used as a tool for earthquake prediction, a facet that will be discussed later.

Faulting and earthquakes can also operate via dilatancy in crustal rocks. At depths greater than 5 km, the pressure due to the weight of overlying rock is equal to the strength of unfractured rock. The shearing forces required to bring about sudden brittle failure and frictional slip can never be obtained, because the rock deforms plastically. However, the presence of water provides a mechanism for sudden rupture by reduction of the effective friction along any crack boundary. If crustal rocks in the upper crust strain without undergoing plastic deformation, they may crack locally and expand in volume. The process is known as *dilation*. Cracking may occur too quickly for immediate groundwater penetration, but water will eventually penetrate the cracks, providing lubrication for any remaining stresses to be released. This model has prospects for predicting earthquakes because it invokes measurable changes in ground levels, electrical conductivity and other physical factors preceding tremors.

Reservoirs or dams

The dilatancy theory can also explain the presence of earthquakes around reservoirs. Reservoir earthquakes occur in both seismic and non-seismic zones, and are unrelated to rock type or local faulting patterns. Interest in this phenomenon was instigated in 1935 with the building of the Hoover dam, which impounded Lake Mead on the Colorado River. The area was not noted for earthquake activity before the dam was constructed. Since that time, however, over 1000 earthquakes have been generated with enough strength to be felt by local inhabitants. Since 1935, this effect has been noted following the construction of 10–15 other reservoirs worldwide including the Kariba dam on the Zambezi River in Southern Africa and the Koyna dam near Bombay, India (see figure 10.14 for major place names). It should be noted that not all artificial reservoirs have generated earthquakes, even when they have been located in active seismic zones. However, enough moderate earthquakes registering 5–6.5 on the Richter scale have been produced at Koyna, India, to cause the loss of 177 lives. It appears that reservoir water penetrates the underlying bedrock, reducing effective frictional resistance along fractures, and permitting slippage significant enough to generate earthquakes. Mining, water and oil extraction, and waste-fluid disposal underground have also resulted in local seismic activity. For example, waste water from chemical warfare manufacturing was pumped underground at Denver, Colorado, between 1962–65 resulting in measurable tremors. However, the effects of these latter operations appear minor compared to the those produced so far by reservoir construction.

Prediction of earthquakes and volcanoes

Introduction

The prediction of earthquakes and volcanoes can be broken down into two timespans—long-term prediction beyond a period of several years, and short-term prediction from a couple of years to several days or hours before the event. Long-term prediction of volcano and earthquake activity involves the delineation of cycles in activity over decades or centuries. Most of the long-term prediction based on cycles is related to the gravitational ordering of planets and cycles in planetary alignment. Another plausible mechanism for earthquake prediction may originate with large-scale atmosphere-ocean interactions. This section will outline known cycles in volcanic and earthquake activity first. Next, the theories for long-term forcing will be discussed, followed by a separate review of short-term predictors for each hazard.

Clustering of volcanic and seismic events

Volcanoes (Gribbin, 1978; Lamb, 1972, 1982; Blong, 1984; Pandey and Negi, 1987)

Over geological time, periods of volcanism have not occurred as random events. Over the past 250 million years, volcanic activity appears to be enhanced every 33 million years with a minor peak occurring every 16.5 million years. The greatest peak in activity occurred 60–65 million years ago corresponding to the formation of the Deccan plateau in India, and the Cretaceous mass extinction, which wiped out the dinosaurs. Periods of volcanic activity correlate with other mass extinctions, major reversals in the Earth's magnetic field, meteor or comet impact-cratering and other terrestrial processes. The cause of these catastrophic events is related to the passage of the solar system through the *galactic disc*. Basically our galaxy, the Milky Way, is a flattened spiral of stars that, from the side, looks like a plate. The solar system rotates relative to the center of the galaxy, passing from one side of the disc to the other. A complete circuit takes 33 million years, such that the solar system passes through the axis of the galaxy every 16.5 million years. At present, the solar system is presently passing through the galactic disk and volcanism has been increasing over the last 2 million years. This period has also witnessed some of the most extensive glaciation that the Earth has experienced.

Passage of the solar system through the galactic disc also perturbs the orbits of comets in the *Oort belt* at the outer edge of the solar system. This may increase the frequency of comet impact on the Earth, leading to increased volcanism. However, the importance of comet impact in triggering catastrophic plant and animal extinctions through increased volcanism may be over-rated. For instance, the worldwide increase in iridium concentrations in thin sedimentary layers during the Cretaceous mass extinction, attributed to a catastrophic comet impact, can be linked to Deccan plateau volcanism. Volcanism, by itself, certainly represents a major change in the *isothermal*, circulatory and convective behaviour of the mantle. As a result, major plate movements take place, leading to faster crustal spreading rates, to broadening of oceanic ridges and to sea-level rises. More importantly, increased volcanic dust during major periods of volcanic activity screens out enough solar radiation, reducing *photosynthesis* and disturbing the food chain sufficiently to cause the disappearance of many land- and ocean-based biological families. Astronomical periodicities at the galactic level are sufficient to account for most periods of volcanism on the Earth concomitantly with increased comet impacts, *magnetic reversals*, climatic change and ultimately mass extinctions. Volcanism must be one of the most far reaching and profound processes affecting the Earth geologically.

The historical record of volcanic activity, based mainly on the amount of dust that volcanoes can, and have thrown into the atmosphere, illustrates the impact that volcanic eruptions have upon our climate. The amount of dust can be evaluated accurately by examining the attenuation of solar radiation through the atmosphere. These measurements are very accurate and have been compiled by various people worldwide since the seventeenth century. It should be pointed out that, while many explosive volcanoes can inject dust and aerosols into the upper troposphere and even the stratosphere, where dust can be suspended for several years, a number of explosive volcanoes have not injected dust to great heights. The Mount St Helens explosion in the United States in 1980 was triggered by a landslide on its northern flank. As a result, most of its ejected material blew out, ineffectively, sideways. There are also a number of volcanoes which emit large quantities of dust that do not go high into the atmosphere. Thus, the total dust content of the atmosphere is not necessarily a function of volcanic activity. Some readers may have realized that not all dust in the atmosphere has a volcanic origin. Today, methods of sampling dust in the atmosphere permit the dust to be separated into terrestrial or human-made components. It is thus possible to measure very accurately the percentage of dust in the atmosphere having a volcanic origin.

Figures 10.11 and 10.12 illustrate the pattern of volcanic dust migration following two very different types of volcanic eruptions: Krakatoa, which erupted on 27 August 1883; and Mount St Helens, which erupted on 18 May 1980. Both these eruptions will be described in more detail in chapter 12. Dust from Krakatoa encircled the globe within two weeks of the eruption, mainly within the zone of tropical easterlies. Within three months, dust had spread into the northern hemisphere, covering most of the United States and Europe. In France

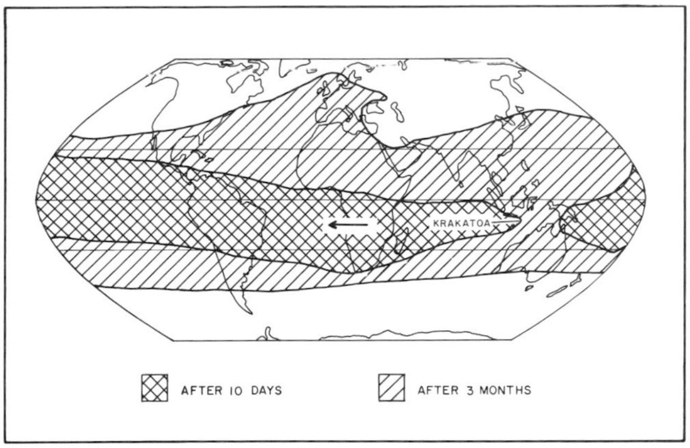

Fig. 10.11 Distribution of stratospheric dust following the eruption of Krakatoa, 27 August 1883 (based upon Gribbin, 1978)

solar radiation dropped 20 per cent below normal, and three years afterwards, was still 10 per cent below normal at the ground surface. Spectacular, prolonged sunsets were observed everywhere, caused by high altitude dust scattering incoming solar radiation as it entered the Earth's atmosphere at low angles. The first occurrence of such sunsets sparked false alarms to fire departments in the eastern United States as residents reported what appeared to be distant, but horrific fires.

The Mount St Helens eruption was different. Most of the dust was blasted out the side and settled to the ground within 700 km of the site. Ash was observed in eastern North America and over the Atlantic. Within 17 days, ash had encircled the globe at a height of 9–12 km, at the top of the troposphere. Some ash moved into the stratosphere, but this amount was minor. Temperatures were estimated to have decreased by 0.5°C for a few weeks directly downwind because of the reduced incoming solar radiation, but the worldwide effect on climate was virtually irrelevant.

Professor Lamb in the United Kingdom has compiled indices of volcanic activity going back to the seventeenth century. His index is referenced to the amount of dust produced by the Krakatoa explosion (base value of 1000) and is known as the *Dust Veil Index*. As a historical note, the Santorini eruption in the Aegean in 1628 BC produced an index value of 3000–10 000, while Mount Vesuvius in AD 79 generated a value of 1000–2000. Figure 10.13 summarizes this index at the decadal level

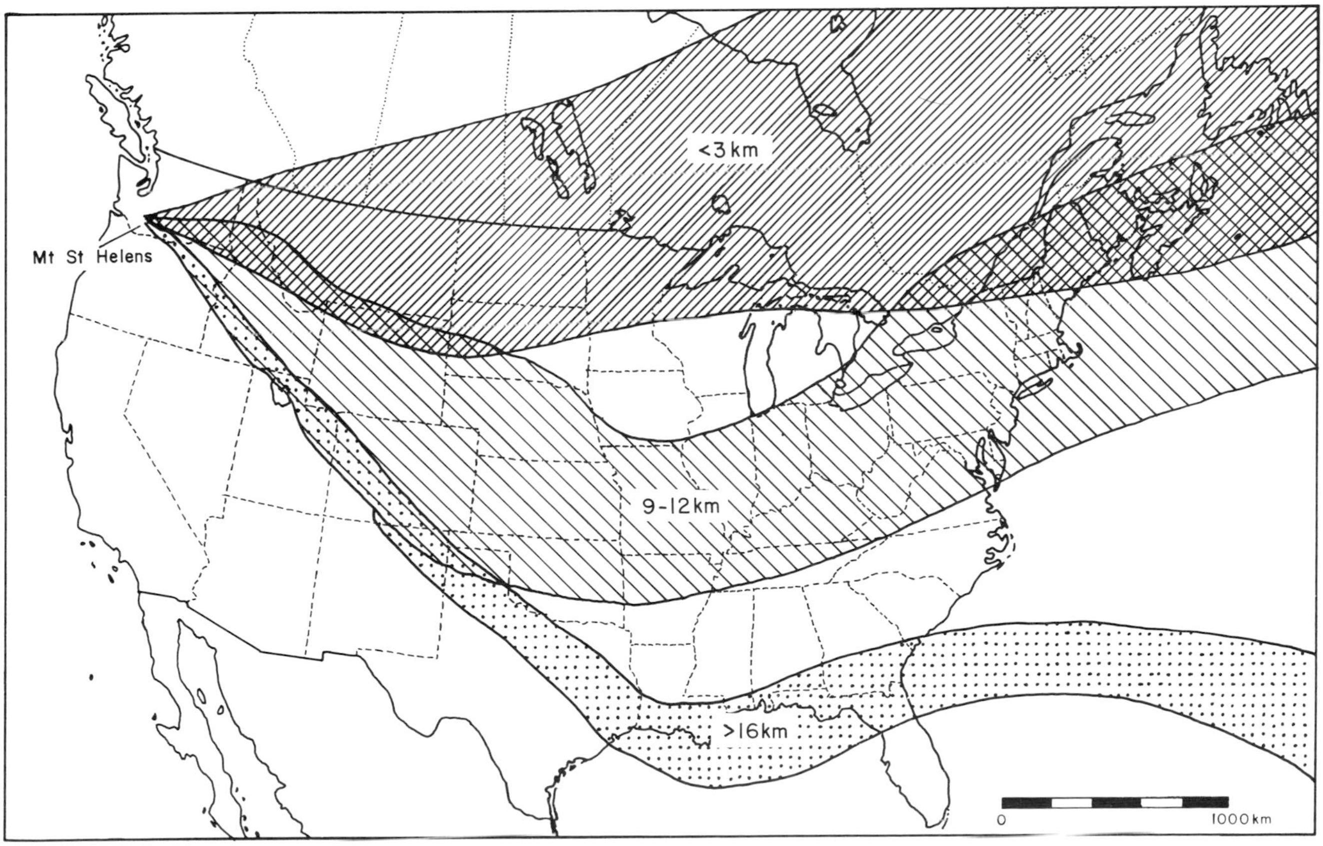

Fig. 10.12 Distribution of tropospheric dust at three elevations, following the eruption of Mount St Helens (from Hays, 1981 after *National Geographic*, January 1981)

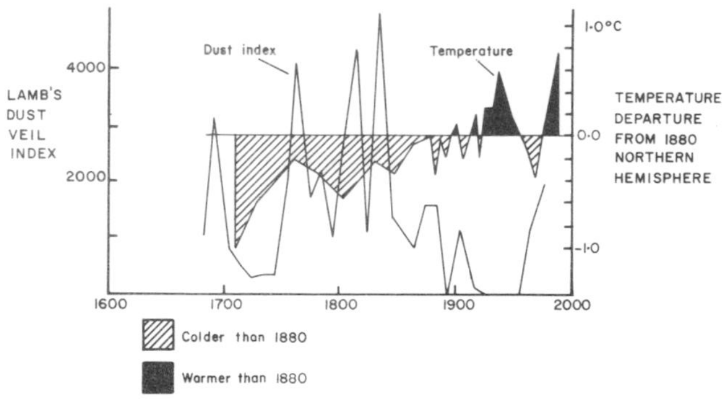

FIG. 10.13 Lamb's Dust Veil Index of volcanic activity vs global temperature change: 1700–1985 (based on data from Gribbin, 1978)

between 1700 and 1985. Also superimposed on the diagram, for meteorological interest, is the average temperature of the northern hemisphere referenced to 1880, a point that is significant in global climate because it represents the termination of the Little Ice Age, and the beginning of a general climatic warming that is still continuing today. There have been some truly cataclysmic eruptions over the past 300 years (major place names not found on regional maps are located on figure 10.14). The Tambora eruption on the island of Sumbawa in Indonesia during its main eruption in 1815, and the Cosequina eruption in Nicaragua in 1835, both released four times as much dust as Krakatoa. These two eruptions account for the two highest peaks in the record; however, the decades 1760 and 1770 also were notable for volcanic activity worldwide. More obvious is the decline in volcanic activity towards the twentieth century. Krakatoa was just a small hiccup in this decline. The last major volcanic eruptions were in 1902 in the Caribbean Sea area, the largest of which included Mount Pelée in Martinique, Soufrière on St Vincent Island and Santa Maria in Guatemala. After 1912, there was not one significant dust-producing volcanic eruption that affected the northern hemisphere climate until Mount Agung in Bali in 1961. Mount Hekla, Iceland, in 1947 produced 100 000 $m^3 s^{-1}$ of ash, which reached as far as Finland; however, the eruption was short-lived and did not inject significant debris into the stratosphere. Mount Agung represented a mild reawakening of activity worldwide, which did not become intense until the Mount St Helens eruption. Besides Mount St Helens, there have been major eruptions since 1980 of Galunggung, western Java, Indonesia, in April 1982; El Chichon, Mexico, in March 1982; Nevado del Ruiz, Columbia, in November 1985; and Alaska

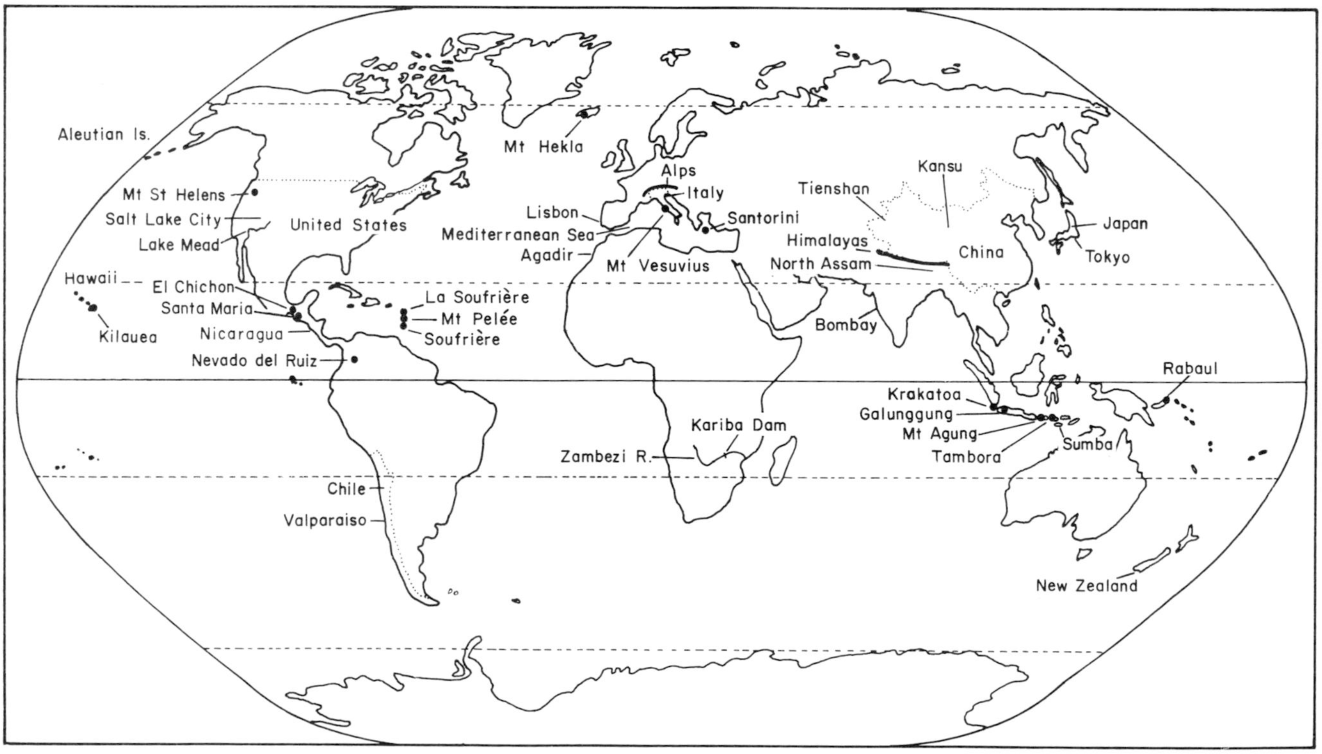

FIG. 10.14 Location map

in 1986. As a result of these eruptions the Dust Veil Index climbed above 2000 for the 1980s.

The correspondence between the Dust Veil Index and global temperature is impressive. Periods of increased volcanism have also been ones of decreased temperatures. The correlation coefficient between the two times series is 0.65, which is highly significant (0.01 level). More importantly, some of the climatic ramifications have had a dramatic impact on agriculture and the course of history. For instance, volcanism in the early 1780s was related to cold winters and drought throughout western Europe, particularly France, and to famine in northern Japan. In France the proportion of the poorer classes' earnings required to purchase daily bread needs reached 88 per cent. While these adverse climatic events did not cause the French Revolution in 1789, they certainly exacerbated conditions that finally triggered general revolt. In 1816 the dust from the Tambora eruption in the previous year produced temperature drops of 1°C over the northern hemisphere, resulting in 'the year without a summer'. Frost occurred in every month of the year in eastern North America, crops failed in Wales leading to famine, and the Southeast Asian monsoon was very intense. This volcano, and a series of others, contributed to climatic conditions that triggered the first global epidemics of typhus and cholera between 1816 and 1819.

Since 1950 volcanic activity has not paralleled temperature changes. Even the increased volcanic activity that began with Mount St Helens has not resulted in decreased temperatures globally. In fact, the 1980s have witnessed the northern hemisphere reaching its warmest summer temperatures in 1000 years. The effect of volcanic dust in the atmosphere appears at present to be overriden by the exponential increase in CO_2 and other gases building up in the atmosphere as a result of industrialization and agricultural practices. On the other hand, there have been significant changes in the Southern Oscillation, favouring warmer land temperatures globally. While the correlation in figure 10.13 appears so exact, it is restricted mainly to northern hemisphere climate. Whereas volcanism subsided worldwide in the period 1910–1960, the greatest peak of ash-producing volcanism in the southern hemisphere occurred in the period 1925–45. The northern hemisphere warmed, especially in the Arctic, during this period; but southern hemisphere temperatures dipped slightly because of this latter phase of volcanic activity.

Blong (1984) has summarized some interesting data on the frequency of volcanic eruptions over the past 10 000 years. Over the last 500 years, individual volcanoes have erupted at the median rate of once every 220 years. The number of eruptions per century, when plotted on log probability paper, forms a straight line relationship similar to the frequency of occurrence of most natural hazard events (figure 10.15). About 20 per cent of volcanoes erupt less than once every 100 years, and 2 per cent less than once every 10 000 years. The El Chichon eruption in Mexico in March 1982 exemplifies this latter frequency. El Chichon had no historic record of eruption and appeared extinct. The 25 most violent volcanic eruptions occurred with a median frequency of 865 years. Of the more than 5500 eruptions by 1340 volcanoes since the last glaciation, only 40 per cent are known to have erupted in the historic past. Thus, most of the world's extinct volcanoes could become active in the future. On average, one extinct volcano erupts every five years. These statistics reinforce the view that, despite an apparent relationship between increased Greenhouse gases and global warming, the aperiodic eruption of a few volcanoes could easily and quickly direct temperatures towards a colder regime.

Earthquakes (Whittow, 1980)

Figure 10.16 plots the magnitude of earthquakes based on the Richter scale in Japan since 684, Europe since 1902, and the United States since 1857. The results before 1900 must be treated with caution since a well-established, seismic recording

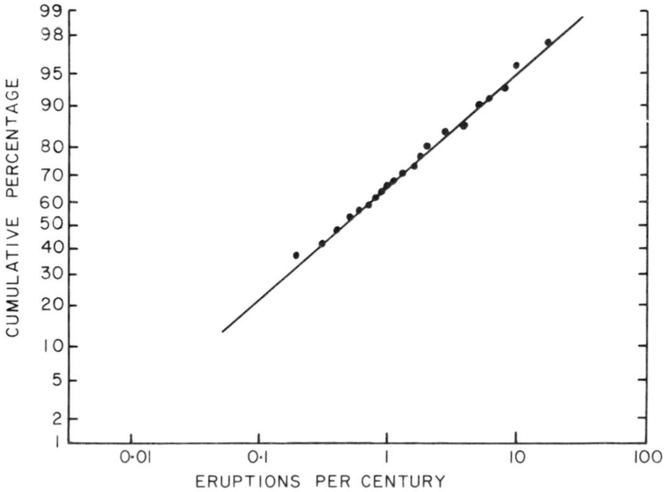

Fig. 10.15 Probability diagram of eruptions of individual volcanoes per century (Blong, ©1984, with permission Harcourt Brace Jovanovich Group, Australia)

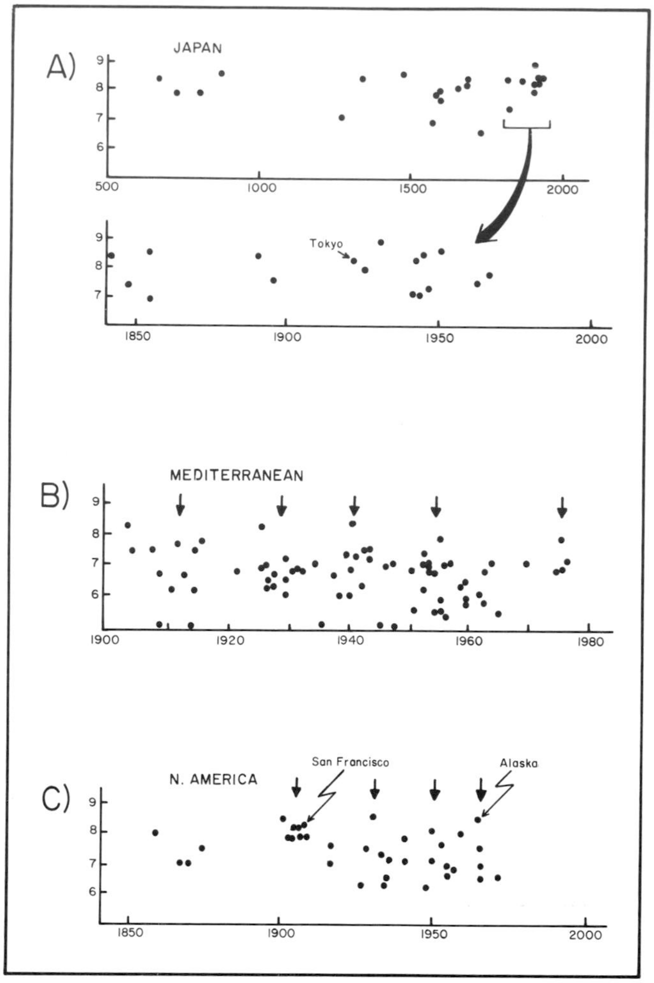

FIG. 10.16 Plot of magnitude on the Richter scale of major earthquakes: A. in Japan, B. in the Mediterranean, and C. in the United States (all data from Whittow, 1980). Arrows indicate clustering.

network did not exist in any of these areas. The most obvious aspect about the data is the fact that earthquakes appear to occur in clusters, and that the spacing of these clusters in some cases is regular. Japan, which is seismically very active, has experienced a series of high-magnitude earthquakes centered around the eighth, sixteenth and twentieth centuries. In the nineteenth century earthquakes appear to be clustering every 50 years. The most seismic period in the record occurred after the Second World War but, by far the most destructive earthquake in Japan in recent times was the 1923 earthquake that destroyed Tokyo.

In the Mediterranean earthquakes occur more regularly (figure 10.16B). Since 1920, there have been four clusters of earthquakes, centered around 1930, 1941, 1953–1956, 1977. It might be tempting to see this regularity as approximating an 11-year cycle, which could be linked to sunspots; however, the pattern is broken in the late 1960s and early 1970s by a period of relative inactivity. The clustering of earthquakes in the Mediterranean region cannot be ignored. If reasons for the pattern could be determined, then it might be possible to forecast up to a decade in advance increased seismic activity in this region.

In North America clustering of earthquakes has also occurred along the western continental margin. The most intense occurrence of seismic events occurred just after 1900 and culminated in the San Francisco earthquake of 1906. Most of the early earthquakes in this cluster took place in Alaska. Other periods of clustering occurred around 1930, 1951, and 1964; however, the clustering is not nearly as pronounced as in the Mediterranean region. Because the North American record presented in figure 10.16C concentrates mainly on large events, inclusion of smaller events could delineate a pattern similar to that experienced in the Mediterranean.

If earthquake occurrence is dominated by some force external to the Earth, then one would expect clustering to be taking place at the same time worldwide: the data, however, do not support this. While the lull in activity in the late 1960s and early 1970s appears to be worldwide, as does the activity centered around 1930, there are many times when each of the regions shown in figure 10.16 had heightened earthquake activity in isolation from the rest of the world. The data, while supporting clustering of events, simply do not substantiate worldwide occurrence at the same time. Furthermore, if earthquake and volcanic activity are linked, then one could expect to find evidence of increased volcanism at the same time as major earthquakes for timespans of decades or centuries. This certainly has not been the case during the twentieth century. Volcanic activity has been minor during the middle part of the twentieth century; however, this period has witnessed some very high magnitude, destructive earthquakes.

Long-term prediction of seismic activity

Length of Day, the Southern Oscillation and Solar activity (Gerety et al., 1977; Gribbin, 1983; Salstein and Rosen, 1984)

There is compelling evidence that as solar activity varies so does the rate of spin of the Earth and seismic activity. In 1971 Dr Challinor of the

University of Toronto noticed that changes in the length of day were accompanied by bursts of earthquake activity. The *length of day* (LOD) refers to the time it takes the Earth to complete one revolution about its axis. Over the last 10 000 years, the LOD has lengthened by 0.16 seconds because of the tidal drag induced by the moon's force of gravity. Seasonally, LOD varies by 0.0025 seconds because of changes in wind speed induced by seasonal variation in air-mass intensity and location around the globe. Changes in global wind patterns affect the rate of spin of the Earth through the transferal of *angular momentum* from the atmosphere to the Earth. Like a spinning skater being able to slow their motion by varying the distance of the arms from the body, the thickness of the atmosphere and direction of air motion can slow down or speed up the Earth. Any process which changes the overall angular momentum of the atmosphere changes the rotation of the Earth. The angular momentum of the atmosphere is equal to the sum of its mass, times wind strength (taking into account the easterly or westerly direction), times the distance from the Earth's rotational axis for all parcels of air in the atmosphere. This is represented mathematically by the following equation:

$$M = \sum_{i=1}^{n} (m \ u \ r) \qquad (10.2)$$

where M = the angular momentum of the atmosphere
m = the mass of the atmosphere
u = a particle of air's wind speed relative to the Earth's rotation
r = distance from the Earth's rotational axis
n = the number of individual partitionings of air flow in the atmosphere

Figure 10.17 illustrates the effect. Excess westerly winds cause the Earth to slow down, while excess easterlies cause it to speed up. Since westerlies predominate in the northern hemisphere winter, angular momentum in the atmosphere increases, causing the Earth to rotate slower in January.

The Southern Oscillation also has an effect on LOD (figure 10.18). The 1982–83 El Niño-Southern Oscillation (ENSO) event caused the Earth to slow down 8 per cent (0.2 milliseconds) more than the previously recorded decrease for January. The main reason was that easterlies were decreased in the southern hemisphere during this event, and the westerly polar jet stream in the

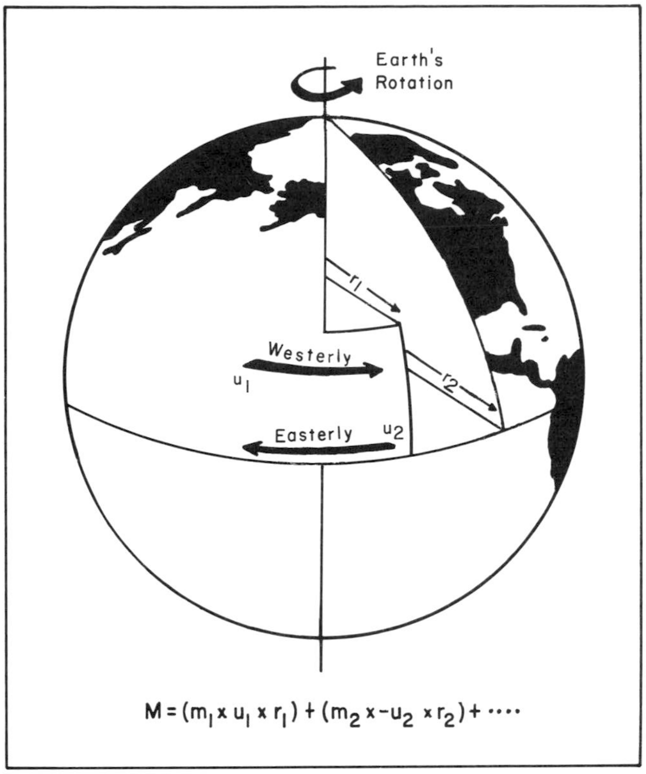

Fig. 10.17 Schematic representation of equation 10.2 (based upon Salstein and Rosen, 1984). This equation relates angular momentum of the atmosphere to mass, wind speed and distance from the Earth's axis of rotation.

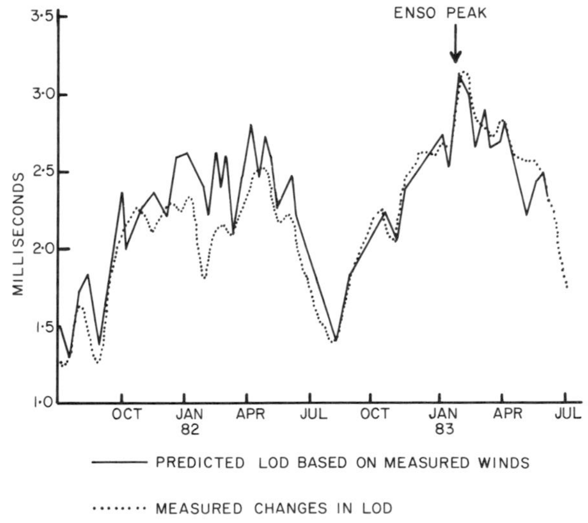

Fig. 10.18 How the 1982–83 ENSO event slowed the Earth's rotation and, hence, length of day, by decreasing the strength of easterlies (adapted from Salstein and Rosen, 1984).

northern hemisphere was greatly increased in strength. The fact that the jet stream shifted towards the equator also increased the distance of this wind from the Earth's rotational axis, and exacerbated the effect. ENSO events also control the rotation of ocean gyres in the north and south Pacific ocean. Ocean gyres also carry angular momentum. As they rotate faster their angular momentum increases. The *thermocline*—the line dividing warm surface water from colder water below—increases in depth in the centre of the gyre or ocean and sea level as a result is raised. The two gyres, in the north and south Pacific respectively, fluctuate in speed out-of-phase at around four-year intervals, such that as one gyre speeds up, the other decelerates. The change in gyre rotation is caused by the massive redistribution of warm water east-west across the ocean, associated with the Southern Oscillation. The speed of the gyres is out of phase between hemispheres because of the redistribution of warm water between hemispheres during an ENSO event. The warm water fluctuations are also accompanied by changes in sea-level which reinforce changes in gyre rotation. There are, thus, two mechanisms—one in the ocean and one in the atmosphere—for changing the LOD. The magnitude of changes in LOD does not necesssarily correlate with the occurrence of earthquakes, but the rate of change in LOD does. Both rapid accelerations and decelerations are correlated with earthquakes. Because earthquakes occur when the buildup of strain between two rock masses can no longer be contained, it is possible that sudden deceleration or acceleration of the Earth's rotation can trigger earthquakes.

Change in the rotational speed of ocean gyres has an interesting side-effect. Because the ocean contains salt, movement in the gyre can generate weak electric currents, which interact with the Earth's magnetic field through the *dynamo effect*. The interaction is reciprocal: changes in ocean-gyre rotation can affect the intensity of the Earth's magnetic field, while changes in the Earth's magnetic field can affect the rotation of ocean gyres. It is appropriate here to re-examine the clustering of earthquake events in the Mediterranean region shown in figure 10.16B. Figure 10.19 replots this time series and compares it to an index of *geomagnetic activity* that appears to lag slightly in time maxima in the 11-year sunspot cycle. Mediterranean earthquake clusters tend to correspond to peaks in this geomagnetic activity, especially for the period centered on 1930 and 1941. While the above associations are speculative, they may provide an intriguing method for forecasting the times of most likely earthquake activity in some parts of the world.

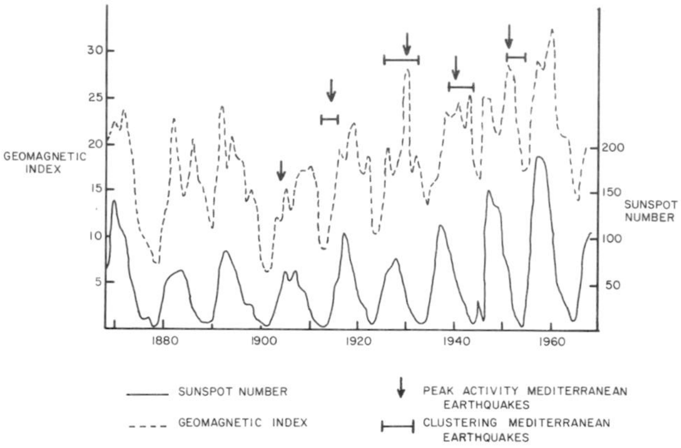

FIG. 10.19 Geomagnetic activity, sunspot frequency and the clustering of Mediterranean earthquakes: 1870–1970 (geomagnetic index from Gerety et al., 1977)

Earthquake cycles (Wood, 1986)

In 1975 research into earthquake movements in swamp deposits in the Los Angeles area indicated that there had been eight major earthquakes since AD 565, spaced at intervals between 55 and 275 years. The average return period was 160 years, with the last major earthquake occurring in 1857. Based upon this periodicity, the next major earthquake above eight on the Richter scale is imminent in the Los Angeles area within the next 20–30 years. This forecast is not illogical given the fact that most earthquakes occur along plate boundaries, which are slowly moving past each other at rates averaging 10 cm yr^{-1}. If the frictional drag between two plates moving relative to each other is not released continually, then forces will build up over time, culminating in a single major jump of the plates along a faultline. Because the plates are moving at a continual rate, the historic record of fault displacements should provide the return period or frequency of earthquakes in a region. For instance, if the average displacement along a fault is 5 m for each earthquake, then the fault should be active every 50 years, given the above rate of plate movement, at 10 cm yr^{-1}. Around the Pacific Ocean, the return period for earthquakes is between

75 and 300 years. More significantly, once a plate has moved along a faultline in an area, the probability of a similar sized event occurring in the next few decades after the earthquake is extremely small. In the area of the 1960 Chilean earthquake, the return period of large earthquakes for the past 300 years averages 80 years. The next major earthquake here is not expected until 2040 ± 20 years.

Away from plate boundaries, the overall rate of movement along faults is an order of magnitude slower than around the Pacific rim. Here, the repeat time of an earthquake becomes much larger. This has significance for isolated areas in the middle of plates that have experienced isolated earthquakes in historical times. For instance, the New Madrid earthquake in the Mississippi valley in 1811–12, and the Charleston, South Carolina, earthquake in 1886 appear to be anomalous (figure 10.3). The concept of earthquake repeat-cycles would indicate that these regions are highly unlikely to experience a similar intensity earthquake for hundreds, if not thousands of years. There are also active faultlines where little or no major earthquake activity has been recorded historically. For instance, the Wasatch fault, which passes through Salt Lake City, appears to be more active than others in the region. Yet, historically it has never experienced a major earthquake. For these types of faults, major earthquakes may occur at 500–1000-year frequencies.

The concept of earthquake cycles depends upon crustal movement occurring at constant rates over geological time, and the buildup of frictional drag forces along the faultline. Around the Pacific region, plates are moving so consistently that anomalous regions that are not moving can be easily detected. For instance, in the Alaskan region, the Pacific and North American plate movements have generated continual earthquake activity over the past 150 years, as stresses build up to crucial limits and are periodically released at various points along the plate margin. However, at some locations, the stresses may not be released easily; these points appear in the historical record as abnormally aseismic, while adjacent regions are seismically active. These locations, called *seismic gaps*, are prime sites for future earthquake activity. The Alaskan earthquake of 1964 filled in one of these gaps, and a major gap now exists in the Los Angeles area. Seismic gaps will be discussed in more detail in the following section.

Short-term prediction of seismic and volcanic activity

Earthquakes (Scheidegger, 1975; Rikitake, 1976; Ward, 1976; Whittow, 1980; Simpson and Richards, 1981; Coates, 1985; Kisslinger and Rikitake, 1985)

Of all natural hazards, earthquakes generate the largest and longest range of associated phenomena that can be used to foreshadow the timing of the impending catastrophe. Research into earthquake prediction has been carried out since the late nineteenth century. In 1891 the Japanese government set up an earthquake investigation committee, which collected historic records of earthquakes in Japan as far back as AD 416. In 1925 the Earthquake Research Institute was established at Tokyo University to give earthquake research a scientific input. Following Japan's economic recovery in the 1950s and 1960s, earthquake prediction was undertaken seriously in the late 1960s. The Chinese were later in establishing earthquake research facilities. Historical records going back 3000 years were compiled only in the 1950s. Between 1960 and 1965 studies on earthquake *precursors* were undertaken. On 5 February 1975 the city of Haicheng (figure 10.2) became the first major inhabited area to be evacuated before the occurrence of a significantly large earthquake. In the months leading up to the earthquake, ground tilting was noticed together with unusual animal behaviour and water spurting from the ground. The city was evacuated 12 hours before the earthquake struck. The earthquake measured 7.2 on the Richter scale and, despite the destruction of 90 per cent of the city, only a few lives were lost. In the Soviet Union research into earthquake prediction was well under way by the late 1960s. In 1965 forecasts of earthquake activity within the next decade were made for the Kuril Islands. These forecasts came to fruition eight years later. It is only since 1971 that the United States has undertaken serious research into earthquake forecasting, with the Geological Survey establishing the Earthquake Hazards Program. Most of this activity has centered around the forecasting of earthquakes in the region of the San Andreas faultline in California.

Precursors for earthquakes group into the following five categories: land deformation, seismic activity, geomagnetic and *geo-electric activity*, ground-water, and natural phenomena. Land deformation studies are based on the concept that the strain energy building up in the Earth's crust will evidence itself in minor lateral or vertical distortions at the Earth's surface. Much of the Earth's surface has been surveyed with benchmarks or triangulation stations. In some countries, these stations have been accurately surveyed to form an extensive triangulation network. By resurveying stations within this network every decade, it is possible to detect gross movement, ranging from centimeters to meters, in the Earth's crust. It is also possible to relate these stations to tide gauges, in order to look for land movements relative to sea-level in coastal areas. On a smaller scale, it is now possible with infra-red and laser survey equipment to mount survey stations on opposite sides of active fault zones, and to measure earth movements as small as 1 mm over distances of several kilometers. This has been carried out along key sections of the San Andreas faultline. Plots evidencing rapid changes in earth movement along the fault may signal future earthquake activity. Crustal deformations can also be measured using tiltmeters and strain gauges. Increases in strain signal earthquake activity within months, while rapid increases in tilting of the Earth's crust (called *tilt steps*) give warnings of imminent activity within hours.

Seismic activity also increases in an area before major shocks. It has already been noted that earthquake activity along the western North American plate boundary increased in the years leading up to the San Francisco earthquake of 1906. A predominance of magnitude 5 earthquakes in a region usually indicates the imminent occurrence of a much larger earthquake several months in advance. These smaller earthquakes often ring the future epicenter, forming what is known as a *Mogi doughnut*, named after the discoverer of the effect. The lack of earthquake activity in an earthquake zone may also indicate future earthquake activity. These regions, seismic gaps, are most significant where earthquake activity has not occurred in the previous 30 years, along fault zones where activity has taken place in adjacent areas. Major gaps as of 1988 are shown in figure 10.20. The Los Angeles area of California is located at a seismic gap, as is much of the Caribbean Sea. On a smaller time scale, *foreshocks* or *microseisms* also indicate an impending earthquake. However, the nature of the warning is paradoxical and depends upon the area. Some earthquake zones give foreshocks days-to-hours before the main shock, while others evidence a decrease in microseismic activity before the shock.

Changes in seismic wave velocities through rock can also be used to foreshadow an earthquake. Strain changes the velocity at which a shock will travel through the ground. By measuring

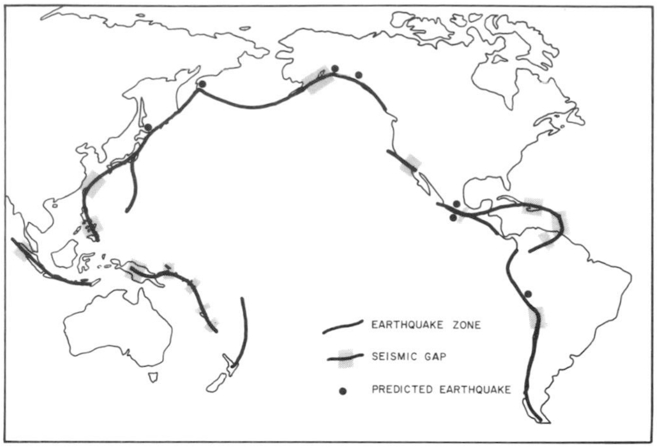

FIG. 10.20 Location of major seismic gaps in the Pacific ring of fire. (Coates, 1985)

the velocities of human-induced shocks across known earthquake zones, it is possible to forecast major earthquakes years in advance. Decreases in strain of up to 20 per cent have been measured for compression-type waves before earthquakes.

There are also geomagnetic and geo-electric precursors. It has already been pointed out that the Earth's geomagnetic activity varies with the sunspot cycle, a relationship that may be linked to earthquake activity. Geomagnetic activity also varies spatially and temporally over the surface of the Earth. Anomalous changes in geomagnetic activity on the order of 4–20 gammas have been measured up to ten years before earthquake activity in local regions. Electrical currents, termed *telluric currents* (or Earth currents), are also continually flowing through the Earth's crust. Changes on the order of 2 mV have been observed hours before small earthquakes. Unfortunately, readings can be affected by natural and human-made interference, such as secular variations in the Earth's geomagnetic and electrical activity caused by solar flares or thunderstorms, and stray noise generated by electricity lines. The Earth's ability to transmit currents, or its *electric resistivity*, can also foreshadow earthquake activity. Changes in resistivity are monitored by sending electrical pulses through the ground and measuring their strength some distance away. Decreases in resistivity of 10–15 per cent have been detected several months before major earthquakes, while stepped increases have been detected several hours in advance, even for earthquakes some distance from the monitoring site.

Groundwater fluctuations in wells can foreshadow earthquakes with magnitudes greater than five on the Richter scale, by half a day up to ten days before the event. This technique is used extensively now as a forecasting tool in China. The longer the existence of anomalous water levels and the wider the anomaly, the larger the impending earthquake. Changes in groundwater may be related to tectonic strain induced by crustal movements, or to the dilatancy phenomena. An associated feature with changes in groundwater level is the change in the water's radon content. Radon, especially in mineral waters, will increase exponentially years before an earthquake, and then decrease rapidly to previous levels after the event. The increase in radon may reflect increased water movement through cracks opened by dilatancy, permitting more of this isotope to be incorporated into groundwaters.

There are also natural indicators of impending earthquakes. Catfish in Japan become very active before earthquakes and even jump out of the water. In addition, local fish catches around Japan increase significantly just before earthquake activity. Just as rats leave a sinking ship, the fish appear to leave an area about to experience an earthquake. Ground animals will leave tunnels, stabled animals will become restless, and dogs will start barking hours before an earthquake. Snakes, weasels and worms deserted the ancient city of Helice, Greece, days before it was completely destroyed by an earthquake in 373 BC. Before the Lisbon earthquake of 1755, worms crawled from the ground and covered the surface. The evacuation of the city of Haicheng, China, in February 1975, just before it was destroyed by an earthquake, followed observations of strange animal behaviour: geese flew into trees, pigs became aggressive, chickens refused to enter coops, and rats appeared to have drunken behaviour. In the San Francisco zoo, animals have been found to group together into species about half an hour before an earthquake. Animal behaviour is now being monitored daily in zoos and marinelands around San Francisco to incorporate animal behaviour into an earthquake prediction system.

The exact reasons why animals forewarn earthquakes is not clear. A response to longwave electromagnetic radiation has been given as one reason for animal panic. Animals may also be supersensitive to vibrations and *ultrasound* generated by small earthquakes preceding the main event. However, reptiles and birds which exhibit unusual behaviour, do not hear ultrasound. In addition, ultrasound from deep earthquakes is absorbed by rock, leaving only very low frequency sound to reach the surface. In this case humans as well as animals should hear the sounds. Another reason for odd animal behaviour may be an animal's extreme sense of smell to methane, which is known to leak from the ground during tremors. However, many burrowing animals, which continually tolerate natural methane as they dig, also panic before earthquakes. Finally, tremors may be preceded by a barrage of *electrostatic* particles emanating from the ground. Animals with fur or feathers are very sensitive to electrostatic charges, and the sudden increase in quantity of such particles may simply irritate them to the point of panic or flight. Whatever the cause, scientists in China and other parts of the world now view the monitoring of animal behaviour as the best means of forecasting the occurrence of major earthquakes.

Unusual weather changes may also be a harbinger of impending disaster. Mists close to the ground, as well as glowing skies, have been reported before earthquakes. In some cases, the light appears to emanate from the ground as a flash of flame. While this luminosity has often been dismissed, there is a new theory that suggests that the Earth is *degassing* over geological time because of large methane deposits trapped beneath the Earth's crust. Fracturing of the crust that occurs before, or at the time of a large earthquake, releases this gas which then ignites. Alternatively, laboratory studies have shown that rock crushed under pressure gives off luminescence.

The prediction of earthquakes is still not an exact science. China, because it faces such large deathtolls from earthquakes, has developed the most advanced techniques for prediction. Five major earthquakes struck China in the early-to-mid 1970s. All but one were predicted early enough to permit orderly evacuation of people from buildings with minimal death or injury. The exception, the Tangshan earthquake of 28 July 1976, took the lives of 250 000 people—the highest number of lives lost in an earthquake in two centuries. While the early warning signs of that earthquake were observed, they were too faint to arouse concern. Foolproof prediction of large earthquakes is still not possible.

Volcanoes (Scheidegger, 1975; Decker, 1976; Tazieff and Sabroux 1983; Smith, 1985; Coates, 1985)

While the first outburst of volcanic activity can now be predicted, it is at present almost impossible to predict the direction or intensity of activity that follows. To date, only a handful of eruptions have been forecast, the earliest being the renewed activity of Kilauea, Hawaii, in November 1959. Most of these predictions so far have been for those fluid magma eruptions which usually are not a threat to life. Volcanoes that consist of *viscous* magmas, or that become explosive, still cannot be predicted. One of the most recent examples illustrating this fact was the eruption of Nevado del Ruiz, Columbia, in November 1985. While the renewed activity of this volcano was noted, the eruption was not predicted, and 20 000 people lost their lives in the ensuing heated mud flows. There have also been some notable and costly false alarms for these latter types of volcanoes. On 12 April 1976 the residents or Guadeloupe in the West Indies were told to evacuate because of an imminent eruption of La Soufrière volcano. (Note that this is not the same volcano which erupted before Mount Pelée in 1902. That was Soufrière volcano, on St Vincent Island.) Over 75 000 people heeded the warning and were evacuated. They waited for 15 weeks until the volcano finally produced a small, totally harmless eruption on 8 July. It cost 500 million dollars to maintain the evacuation and brought economic ruin to many. In 1985 the residents of Rabaul, New Britain, in Papua New Guinea, were warned about the explosive eruption within a matter of months of a volcano near their city. To date, however, nothing has happened. Such predictions, while soundly based, only tend to weaken the believability of subsequent warnings.

The techniques for predicting volcanic eruptions or activity are just as varied as, but more technical than those for predicting earthquakes. Precursors for volcanoes group into the following five categories: land deformation, seismic activity, geomagnetic and geo-electric effects, temperature and gases. Ground deformations around volcanoes are due to the subterranean movements of molten magma. These movements can be vertical, lateral or oblique. They evidence themselves at the surface by tilting of the Earth's surface, which can be measured using tiltmeters. The movements can be fast or slow, positive or negative. Differences in movement can occur over short distances. Large, sudden tilts generally herald a violent eruption while slight tilts of increasing frequency foreshadow the movement of magma closer to the Earth's surface. Tiltmeters on the northern side of Mount St Helens indicated dramatic inflation at the rate of $0.5-1.5\,m\ day^{-1}$ preceding the eventual explosion at that location. Tiltmeters have also been used in Hawaii to forecast correctly eruptions of Kilauea.

Studies on seismology have dominated volcanology since the early 1900s. As magmas flow through subterranean channels, they apply stress to rocks, which can fracture and set off seismic activity. This applies mostly to volcanoes characterized by fluid magma. Earthquakes set off by volcanoes differ from tectonic ones in that they occur at depths of less than 10 km, and are low in magnitude. Volcanic tremors can be divided into two groups. The first group consists of prolonged, continuous volcanic vibrations induced by the flow of fluid magmas. Some explosive volcanoes, notably Mount St Helens, are also beset with this type of tremor. The second group consists of spasmodic or regular vibrations, the frequency of which indicates the origin and nature of the magma. All forecast eruptions of Kilauea, Hawaii, have been based upon tilting and earthquake precursors. However, not all seismic activity associated with volcanism can be used with certainty to predict subsequent activity. While most eruptions are usually, but not always, proceeded by swarms of earthquakes, the presence of tremors can also represent collapse of rock into emptying magma channels, a process which may indicate cessation of activity. It was seismic patterns that were used in 1986 at La Soufrière, Guadeloupe, as the main indicator of an impending volcanic eruption that eventually proved minor. In a few cases, the subsidence of earthquake activity may also signal an eruption.

Active volcanoes are dominated by temporally and spatially variable geomagnetic fields. Volcanoes contain a high content of ferromagnetic minerals which can set up changes in the local magnetic field. Magnetization however is reduced by increasing temperature, vanishing completely above 600°C. The magnetic field of a volcano is thus reliant upon the proximity and temperature of molten magma near the surface. As hot magma between 200 and 600°C approaches the surface, the geomagnetic field should therefore decrease. Magnetization can also be enhanced by increasing pressure and stress exerted by flowing magma as it approaches the surface. This process is termed *piezomagnetism*, and at present is being researched extensively as a new technique for prediction.

Geo-electrical measurements involving the resistivity of the sub-surface layers of a volcano and the change in the telluric currents can give an image of the behaviour of magma at depth. Resistivity depends upon the nature of the rock, its water content, salinity and temperature. Telluric currents depend upon geological structure, *lithology*, moisture content and temperature. At present, these methods are used to define the structure of the volcano, including the presence of natural conduits, which may become the preferred pathways for continued magma movement.

The analysis of gaseous constituents exhaled from a volcano constitutes one of the best techniques for understanding and forecasting eruptive activity. Unfortunately, while the technique is so informative, it is restricted by the need to chemically analyze the gas constituents when they are immediately vented from the volcano. Many gas studies are still in their infancy, mainly because techniques for sampling and analyzing gases from venting volcanoes are still being developed. The most common gases vented by a volcano are H_2O, CO_2, SO_2, H_2, CO, CH_4, COS, CS_2, HCl, H_2S, S_2, HF, N_2 and the rare gases helium, argon, xeon, neon and krypton. The chemical nature of these gases depends upon the maturity of the magma melt, because not all gas components have the same solubility at a given pressure. The type of gas also depends upon the amount of crustal material relative to original magma that is incorporated into the melt. If the magma is chemically stable, the partial pressure ratios between various gases can give an indication of the pressure and temperature of the magma en route to the surface. For example, the partial pressure ratio of $CO:CO_2$, HF:HCl and $H_2O:CO_2$ can all be used as qualitative geothermometers. The first two ratios increase, and the last one decreases, as temperature increases. The ratios of $CO:CO_2$, $H_2:H_2O$, and $H_2S:SO_2$ are sensitive to changes in the conditions that control the thermodynamic equilibrium of the gas phase, and can be used to distinguish between *hydrothermal* and magma flows. The ratios $SO_2:CO_2$ and S:Cl, plus the absolute amount of HCl, increase immediately prior to eruptions, while the ratio $He:CO_2$ decreases. These changes result from the different solubilities of individual gases, and from the ascent of fresh magma to the surface. Isotope studies of various gases also can be used to delineate the different sources of material reaching the magma pool from below.

Concluding comments

This chapter has shown that earthquakes and volcanoes are no longer poorly understood phenomena that invoke fear, superstition and pagan attempts at control because of a lack of knowledge about their origin and behaviour. While earthquakes and volcanoes can still kill a large number of people, and can occur with extreme suddenness, our knowledge is reaching the stage where a significant number of these events can be predicted in space and time to permit measures to be taken to minimize the loss of life. Since the widespread acceptance of continental drift theory, and the accurate measurement of thousands of earthquake epicenters, the most likely locations of future earthquake activity have been mapped for the globe. While cosmic/planetary links with earthquake and volcanic activity would be considered by most to be highly speculative, sound scientific observations and instrumental measurements are defining the best precursors for forecasting imminent tremors or eruptions. The success of such techniques in China in the 1970s is exemplary; the failure of authorities in California to install sensitive strain gauges along the San Andreas faultline, which could have forecast the 1989 San Francisco earthquake, is alarming. Complacency or disbelief will always dog our attempts to predict both earthquakes and volcanic eruptions. Both traits can be justified because there is still a large random component to the occurrence of these hazards. The Tangshan earthquake of 1976 and the eruptions of Mount St Helens in 1980 and El Chichon in 1982 are only a few of the examples to have occurred in recent years supporting this statement.

The next two chapters will describe in more detail the exact nature and associated phenomena of earthquakes and volcanoes. As well, some of the more significant events in recorded history will be described to reinforce the view that these two hazards still are amongst the most destructive natural events to afflict mankind.

References

Blong, R. J. 1984. *Volcanic hazards: a sourcebook on the effects of eruptions*. Academic Press, Sydney.

Bolt, B. A., Horn, W. L., MacDonald, G. A. and Scott, R. F. 1975. *Geological hazards*. Springer-Verlag, Berlin.

Coates, D. R. 1985. *Geology and Society*. Chapman and Hall, NY.

Denham, D. 1979. 'Earthquake hazard in Australia'. In Heathcote, R. L. and Thom, B. G. (eds) *Natural Hazards in Australia*. Australian Academy of Science, Canberra, pp. 94–118.

Doyle, H. A., Everingham, I. B. and Sutton, D. J. 1968. 'Seismicity of the Australian continent'. *Journal Geological Society Australia* v. 15 No. 2 pp. 295–312.

Gerety, E. J., Wallace, J. M. and Zerefos, C. S. 1977. 'Sunspots, Geomagnetic Indices and the weather: a cross-spectral analysis between sunspots, geomagnetic activity and global weather data'. *Journal Atmospheric Science* v. 34 pp. 673–78.

Gribbin, J. 1978. *The Climatic Threat*. Fontana, Glasgow.

——. 1983. *Future Weather: the causes and effects of climatic change*. Penguin, Harmondsworth.

Hays, W. W. 1981. 'Facing geologic and hydrologic hazards: Earth-science considerations'. *United States Geological Survey Professional Paper* 1240–B pp. 86–109.

Holmes, A. 1965. *Principles of Physical Geology*. Nelson, London.

Muller, R. A. and Oberlander, T. M. 1984. *Physical Geography Today*. (3rd edn). Random House, NY.

Oakeshott, G. G. 1983. 'San Andreas Fault: Geologic and Earthquake History'. In Tank, R. W. (ed.) *Environmental Geology*. Oxford University Press, Oxford, pp. 100–107.

Press, F. and Siever, R. 1986. *Earth* (4th edn.) Freeman, New York.

Salstein, D. A. and Rosen, R. D. 1984. 'El Niño and the Earth's Rotation'. *Oceanus* v. 27 No. 2 pp. 52–57.

Walker, G. P. L. 1974. 'Volcanic hazards and the prediction of volcanic eruptions'. *Geological Society of London Miscellaneous Publication* v. 3 pp. 23–41.

Whittow, J. 1980. *Disasters: the anatomy of environmental hazards*. Pelican, Harmondsworth.

Further reading

Decker, R. W. 1976. 'State of the art in volcano forecasting'. In *Geophysical Predictions*. National Academy of Science, Washington, pp. 47–57.

Fielder, G. and Wilson, L. (eds). 1975. *Volcanoes of the Earth, Moon and Mars*. Elek, London.

Hodgson, J. H. 1964. *Earthquakes and Earth Structure*. Prentice-Hall, Englewood Cliffs, New Jersey.

Kisslinger, C. and Rikitake, T. (eds). 1985. *Practical approaches to earthquake prediction and warning*. Reidel, Berlin.

Lamb, H. H. 1972. *Climate: Present, past and future*. v. 1 & 2, Methuen, London.

——. 1982. *Climate, history and the modern world*. Methuen, London.

Pandey, O. P. and Negi, J. G. 1987. 'Global volcanism, biological mass extinctions and the galactic vertical motion of the solar system'. *Geophysical Journal* v. 89 No. 3 pp. 857–68.

Rikitake, T. 1976. *Earthquake Prediction*. Elsevier, Amsterdam.

Scheidegger, A. E. 1975. *Physical aspects of natural catastrophes*. Elsevier, Amsterdam.

Simpson, D. W. and Richards, P. G. (eds). 1981. *Earthquake Prediction*. American Geophysical Union, Washington.

Smith, J. V. 1985. 'Protection of the human race against natural hazards (asteroids, comets, volcanoes, earthquakes)'. *Geology* v. 13 pp. 675–78.

Tazieff, H. and Sabroux, J. C. (eds). 1983. *Forecasting volcanic events*. Lange and Springer, Berlin.

Ward, P. L. 1976. 'Earthquake prediction'. In *Geophysical Predictions*. National Academy of Science, Washington, pp. 37–46.

Wood, R. M. 1986. *Earthquakes and Volcanoes*. Mitchell Beazley, London.

11

EARTHQUAKES AND TSUNAMI AS HAZARDS

Types of shock waves (Hodgson, 1964; Holmes, 1965; Whittow, 1980)

Earthquakes are basically shock waves that are transmitted from an epicenter, which can lie anywhere from the surface to 700 km beneath the Earth's crust. Earthquakes generate a number of types of waves, which are illustrated in figure 11.1. A primary or P-wave is a compressional wave which spreads out from the center of the earthquake; it consists of alternating compression and dilation, similar to waves produced by sound travelling through air. These waves can pass through gases, liquids and solids, and undergo refraction effects at boundaries between fluids and solids. P-waves can thus travel through the center of the earth; however, at the core-mantle boundary they are refracted producing two shadow zones, each 3000 km wide, without any detectable P-waves on the opposite side of the globe.

The second type of wave is a shear or S-wave, which behaves very much like the propagation of a wave down a skipping rope that has been shaken up and down. These waves travel 0.6 times slower than primary waves. While the velocity of a primary wave through Earth depends upon rock density and compressibility, the rate of travel of a shear wave depends upon rock density and rigidity. Shear waves will travel through the liquid mantle, but not through the Earth's rigid core. Thus, there is a shadow zone on the opposite side of the Earth that doesn't record S-waves. This zone overlaps completely the two shadow zones produced by refraction of P-waves at the core-mantle boundary. The spatial distribution and time separation between the arrival of P- and S-waves at a seismograph station can be used to determine the location and intensity of an earthquake. Three stations receiving both P- and

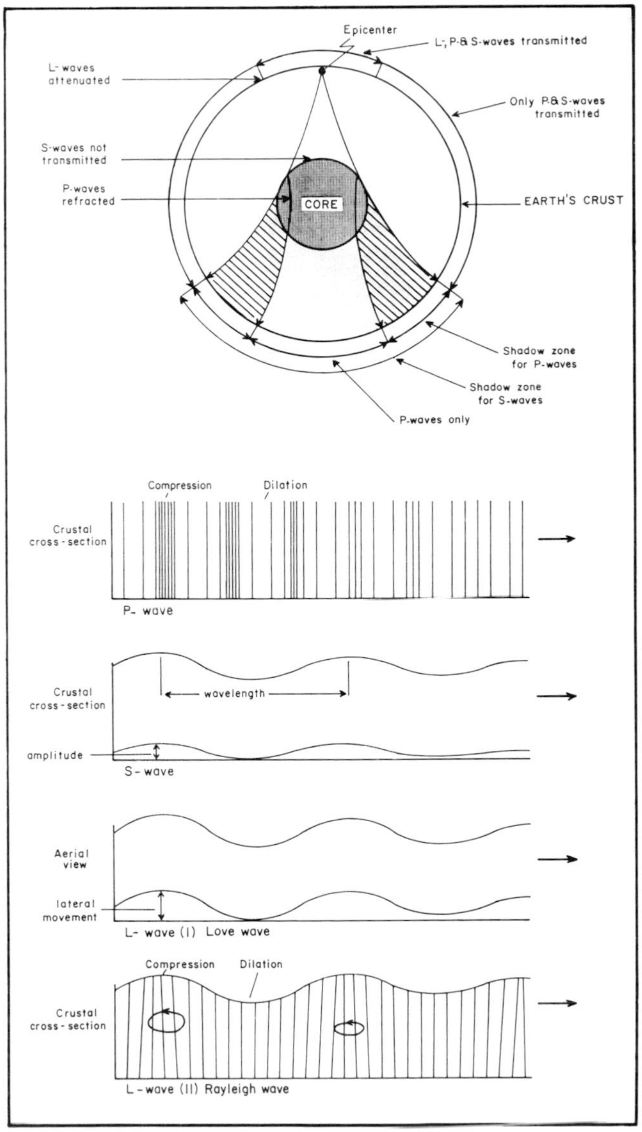

FIG. 11.1 Description of types of seismic wave and their travel behaviour through the Earth (adapted from Whittow, 1980)

S-waves are necessary to determine the exact position of any epicenter.

Earthquakes also generate several different surface, long or L-waves, trapped between the surface of the Earth and the crustal layers lower down. These waves are not transmitted through the mantle or core, but spread slowly outwards from the epicenter along the surface of the globe. Their energy is dissipated progressively from the focal point of the earthquake. One type of L-wave, the Rayleigh wave, behaves similarly to an ocean wave, while another type, the Love wave, literally slithers back and forth through the crust. This latter type of wave is responsible for much of the damage witnessed during earthquakes, and produces the shaking that makes it impossible to stand up. Long waves are the slowest wave of all seismic waves, taking about 20 minutes to travel a distance of 5000 km along the Earth's crust. All three types of waves, because they travel differently through dissimilar rock densities and states of matter, have been used to delineate the structure of the Earth's core, mantle and crust.

Seismic risk maps (Denham 1979; Keller, 1972; Stevens, 1988)

Peak accelerations of long waves have been defined for hundreds of earthquakes covering most of North America and Australia (see figure 11.2 for the location of all major place names in this chapter not shown on individual site maps). As the gravitational constant is a crucial parameter defining the limit at which total destruction becomes apparent, peak horizontal or vertical accelerations are usually expressed as a percentage of the gravitational constant ($9.8 \, m \, s^{-2}$). Hence, the most destructive earthquake on the Mercalli scale would have a peak acceleration in excess of 100 per cent. The rate of decay of these accelerations from an epicenter for any earthquake can also be determined. Thus, it is possible to map, over large areas of a country, the probability of occurrence of any peak acceleration. It is standard procedure to map the peak acceleration value that will be exceeded a certain percentage of time over a specific period—for example 10 per cent of the time in a 50-year period. Because mean horizontal

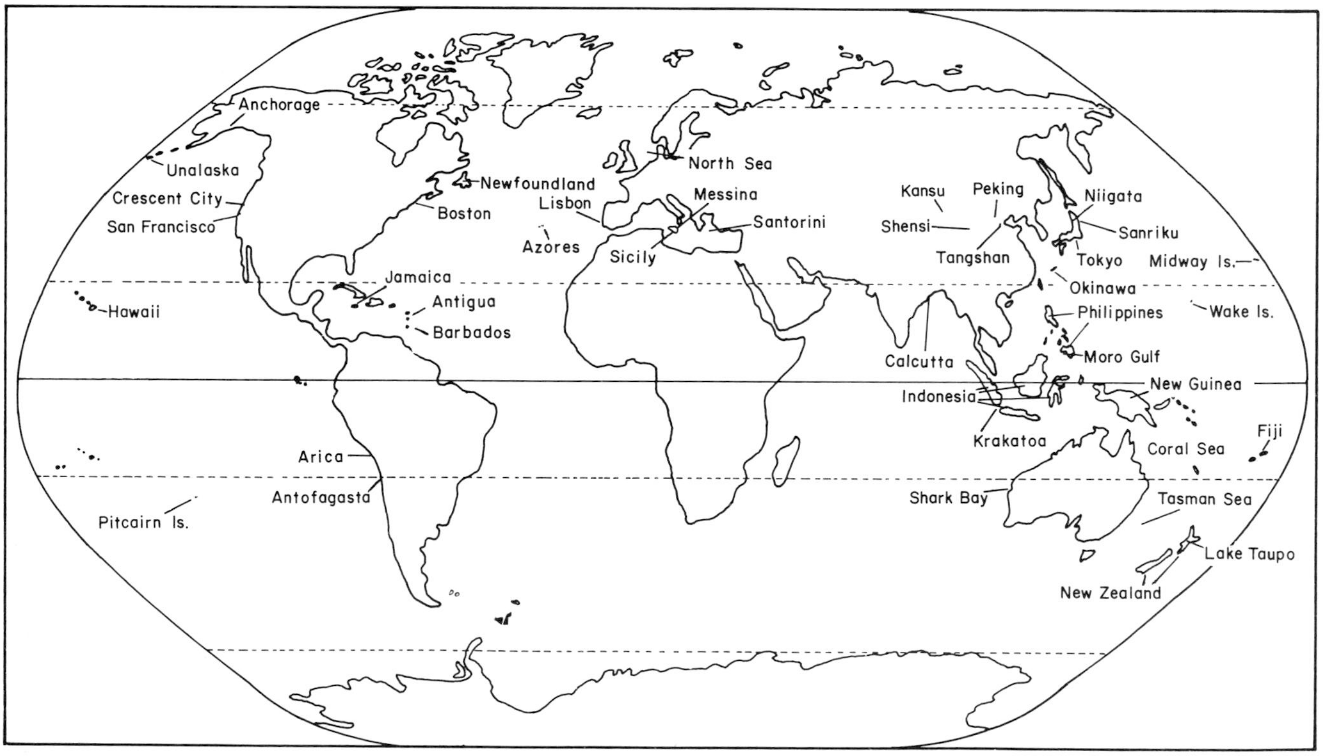

FIG. 11.2 Location map

accelerations are twice the value of peak vertical accelerations, horizontal accelerations are usually plotted on seismic risk maps. Engineers can use such maps to assess the lateral forces acting on objects, such as rigid buildings, to determine the risk of their collapse during earthquakes.

Seismic risk maps for North America and Australia are contoured in figures 11.3 and 11.4 respectively. Similar types of maps have been constructed for Southeast Asia, China, the Philippines and Indonesia. The North American map plots the peak horizontal acceleration of seismic waves that could be expected to be exceeded 10 per cent of the time within 50 years. In North America, the southern Californian region around the San Andreas fault zone has the greatest risk of destructive shock waves of any region. However, this area is closely followed by a zone along the eastern foothills of the Rockies in the United States, along the Pacific Islands off the west coast of Canada, and the lower St Lawrence estuary. These latter regions have a 20 per cent probability in 100 years of a destructive earthquake corresponding to an intensity of eight or greater on the Mercalli scale, and 6.5 or greater on the Richter scale. While Californian cities are generally perceived as the only North American cities at risk from destructive earthquakes, large urban centers such as Salt Lake City, Vancouver, Montreal and Quebec City are all seismic. Additional areas at high risk also exist in relatively unpopulated areas such as the Canadian high Arctic (beyond the bounds of the map), the edge of the continental shelf south of Newfoundland (where the only tsunami recorded in eastern North America killed 27 people, on 18 November 1929), and along the southern Alaskan coastline (not plotted on figure 11.3). This seismic risk map permits buildings to be designed to withstand earthquake

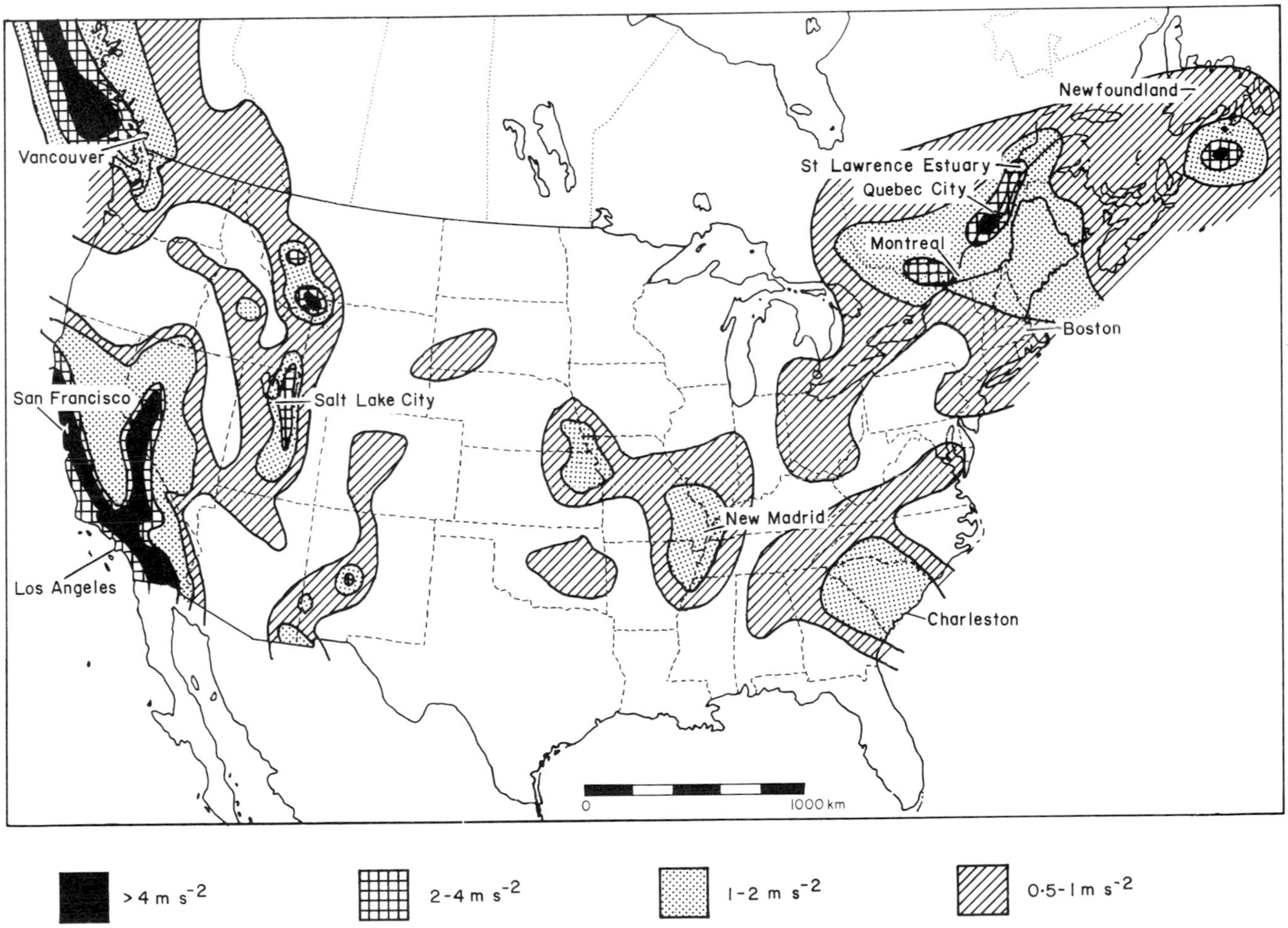

FIG. 11.3 Seismic risk map for the conterminous United States and adjacent region. Contour lines define peak horizontal accelerations with a probability of exceedence of 10 per cent in 50 years (based upon Keller, 1982; Stevens, 1988).

damage in most parts of North America. However, the maps are incomplete and not foolproof. For instance, the New Madrid, Missouri, earthquakes of 1811–12 and the Charleston, South Carolina, earthquake of 1886 both caused destruction beyond what could be deduced from the probabilities shown in figure 11.3. In addition, two of the largest earthquakes to occur recently in Canada, took place on the Nahanni River on 5 October and 23 December 1985. Both earthquakes registered over 6.5 on the Richter scale and plot in an area on figure 11.3 having only a marginal risk of earthquake activity.

The seismic risk map for Australia (figure 11.4) shows that most of this continent is aseismic. There are still a few areas marked as having no risk where minor earthquake activity has been reported and suspected. These include a large area east of Kalgoorlie in Western Australia, in Tasmania and the northern foldbelt in north New South Wales. The Simpson desert and Avon River area of Western Australia are the two most seismic areas in Australia, where destructive earthquakes are possible. Fortunately, both areas are situated in sparsely populated, or unpopulated areas so that earthquakes pose little threat of loss of life. The belt running from Spencer Gulf to the Simpson desert, the western section of Western Australia, the southern Sydney basin, and the southern tips of Victoria are all areas where moderate earthquake activity greater than five on the Richter scale has occurred. In these areas earthquakes can be expected within a hundred-year-period that would crack walls in buildings and cause minor structural damage. Two large cities—Adelaide (800 000 people) and the Sydney metropolitan region (>3 000 000 people)—lie within these regions. Here, some attention should be paid to designing buildings to withstand repetitive shaking over long periods of time.

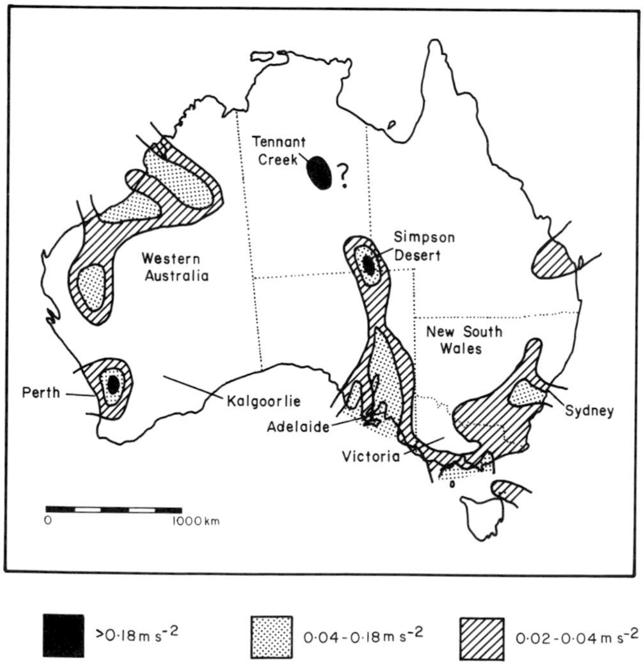

FIG. 11.4 Seismic risk map for Australia. Contour lines define peak horizontal accelerations with a probability of exceedence of 2 per cent in 50 years (based upon Denham, 1979).

Earthquake disasters

General (Holmes, 1965; Cornell 1976; Whittow, 1980; Coates, 1985)

On average, each year earthquakes kill 10 000 people and cause $US400 million property damage. In the period 1964–1978, over 445 000 people lost their lives from earthquakes or earthquake-related hazards. Table 11.1 lists the worst recorded earthquakes in terms of loss of life since AD 800. As with tropical cyclones, there have been some notable events. In 1290 in China two earthquakes—in the Gulf of Chihli, and at Peking—took 100 000 lives. The Shensi event in China, on 23 January 1556 took 830 000 lives. That earthquake struck loess country, and caused the collapse of thousands of human-made caves dug into cliffs and used as homes. Another earthquake at Kansu, China, in 1920 killed 200 000 people. The Calcutta earthquake of 11 October 1737 wiped out half the city, killing 300 000 people. In 1923 the Tokyo earthquake, which is discussed below in more detail,

TABLE 11.1 Major Historical Earthquakes, by deathtoll or damage

Location	*Date (AD)*	*Deathtoll*
Corinth, Greece	856	45 000
Cilicia, Asia Minor	1268	60 000
Chihli, China	1290	100 000
Shensi, China	1556	830 000
Caucasia, Shemaka	1667	80 000
Calcutta, India	1737	300 000
San Francisco, United States	1906	700
Messina, Italy	1908	75 000
Kansu, China	1920	200 000
Tokyo, Japan	1923	143 000
Northern Peru	1960	70 000
Tangshan, China	1976	250 000
Armenia, Soviet Union	1988	25 000

killed 143 000 inhabitants. Italy has been struck many times by earthquakes, with ones near Naples and Catania, Sicily, in 1693, and at Messina on 29 December 1908 taking between 60 000 and 150 000 lives each. The San Francisco earthquake on 18 April 1906 was the worst American earthquake, killing 700 people, with much of the city being destroyed by ensuing fires. Most recently, there was the 1976 Tangshan earthquake, with its deathtoll of approximately 250 000 people making it one of the worst earthquakes on record.

While many earthquakes stand out historically, very little research has been performed on the causes of some of the largest such as the Calcutta earthquake. In modern times four to five events stand out because of their magnitude and the nature of the disaster. The 1906 San Francisco earthquake occurred in the vulnerable San Andreas fault area of California and wiped out a major United States city in a country which believed itself to be immune from such disasters. The Tokyo event of 1923 totally destroyed the largest city in Japan, and is known for its severity. The Chilean earthquake of 1960 marked a revival in large earthquake activity after a decade of small events, and produced one of the largest widespread tsunami events in the Pacific Ocean. The Alaskan earthquake of 1964 took few lives, but literally shook the globe like a bell. The 1976 Tangshan earthquake, notable for one of the most massive deathtolls in the twentieth century, struck in an area not renowned for seismic activity, at a time when the Chinese believed that they could predict major earthquakes. The event is also notable for the cloak of secrecy surrounding its aftermath. Not only did China refuse international aid, but it also released very few details of the event until five years afterwards. Two of these events, the 1923 Tokyo earthquake and the 1964 Alaskan earthquake, are discussed in more detail below.

Tokyo earthquake of 1 September 1923

(Hodgson, 1964; Holmes, 1965)

The Tokyo earthquake was not centered in the city, but in Sagami Bay 50 km to the southwest (figure 11.5). The crustal movements twisted the mainland in a clockwise direction, with a maximum horizontal displacement of 4.5 m and a vertical drop of 2 m. Until the Alaskan earthquake, these were some of the greatest crustal shifts recorded. Within the bay itself, changes in elevation were much larger owing to both faulting and compaction. Measurements after the earthquake indicated that parts of the bay had deepened by 100–200 m with a maximum displacement of 400 m. Such distortions

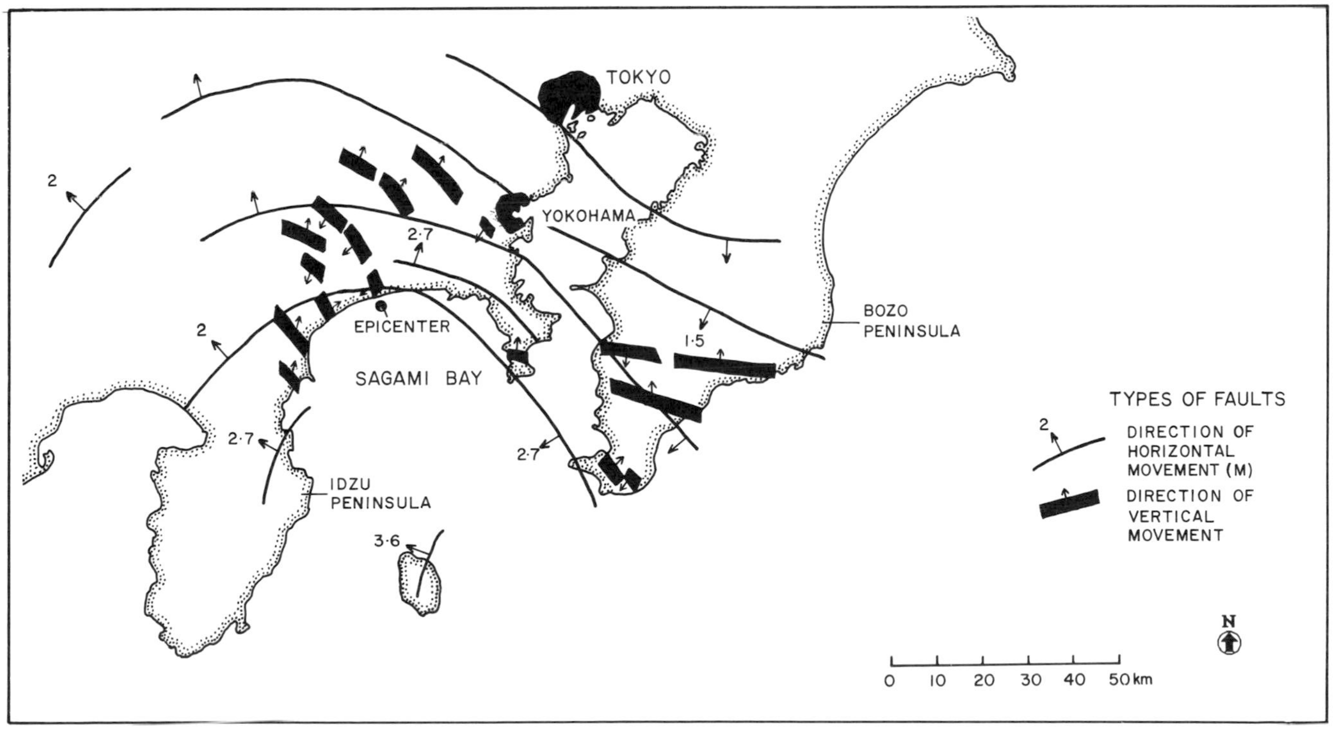

Fig. 11.5 Distribution of faulting for the Tokyo earthquake of 1 September 1923 (from Holmes, 1965)

are not unheard of, but difficult to explain, unless the sea bed near ocean trenches responds differently to tectonic movement, compared to the crust beneath continents. The Tokyo earthquake immediately collapsed over a half million buildings and threw up a tsunami 11 m in height around the sides of Sagami Bay. However, these events were not responsible for the eventually large deathtoll, because many of the collapsed buildings consisted of lightweight materials. The deaths resulted from the fires that immediately broke out in the cities of Tokyo and Yokohama, and raged for three days destroying over 50 per cent of both cities. The fires rate as one of the major urban conflagrations in history—on the same scale as fires that destroyed London in 1666, Chicago in 1871, Moscow during the Napoleonic Wars in 1812, San Francisco during the 1906 earthquake, and Dresden and Hamburg during the Second World War. The outbreaks of fire were minor to begin with, but they occurred throughout both cities in large numbers, mainly because the earthquake occurred at lunchtime when many open cooking fires were being used. Within half an hour, over 200 small fires were burning in Yokohama and 136 were raging in Tokyo. The difficulty of dealing with the fires was compounded by the rubble of collapsed buildings in the streets and cracked watermains. Firefighters could neither get to the fires nor obtain a reliable water source to put them out. Once the fires raged, the cluttered streets hampered evacuation. In Yokohama 40 000 people fled to an open area around the Military Clothing Depot, which was subsequently engulfed in flames. All these people died from the searing heat or suffocation caused by the withdrawal of oxygen.

The spread of fire was aided initially by strong tropical cyclone winds. In the chapter on tropical storms, it was pointed out that the earthquake was probably triggered by the arrival of a tropical cyclone the previous day. The calculated change in load on the Earth's crust as a result of the pressure drop, and the increased weight of water from the storm surge, was 10 million tonnes km^{-2}. The fires were exacerbated by the development of strong winds generated by the cyclone in the lee of surrounding mountains. These föhn winds can lack precipitation and be desiccating. All descriptions of the fires allude to the strong winds which swept them out of control. In chapter 2, it was pointed out that the seemingly dichotomous occurrence of torrential rain and uncontrolled fire is not uncommon.

Alaskan earthquake of 27 March 1964

(Hansen, 1965; Bolt et al., 1975; Whittow, 1980; Hays, 1981)

The Alaskan earthquake struck on Good Friday, when schools and businesses were closed. It occurred in a seismically active zone running westward along the coast of Alaska, parallel to the Aleutian trench, south of the Aleutian Island arc (figure 11.6). The earthquake was only one in a series of earthquakes in the area beginning 10 million years ago in the late *Pliocene*. In 1912 and 1934 there had been earthquakes of magnitude 7.2 on the Richter scale adjacent to the 1964 epicenter. Other large earthquakes had occurred in neighbouring areas in 1937, 1943, 1947 and 1958. These earthquakes appear to be generated by the southward movement of Alaska over the Pacific plate at a shallow angle of 20 degrees.

The earthquake was one of the strongest to be

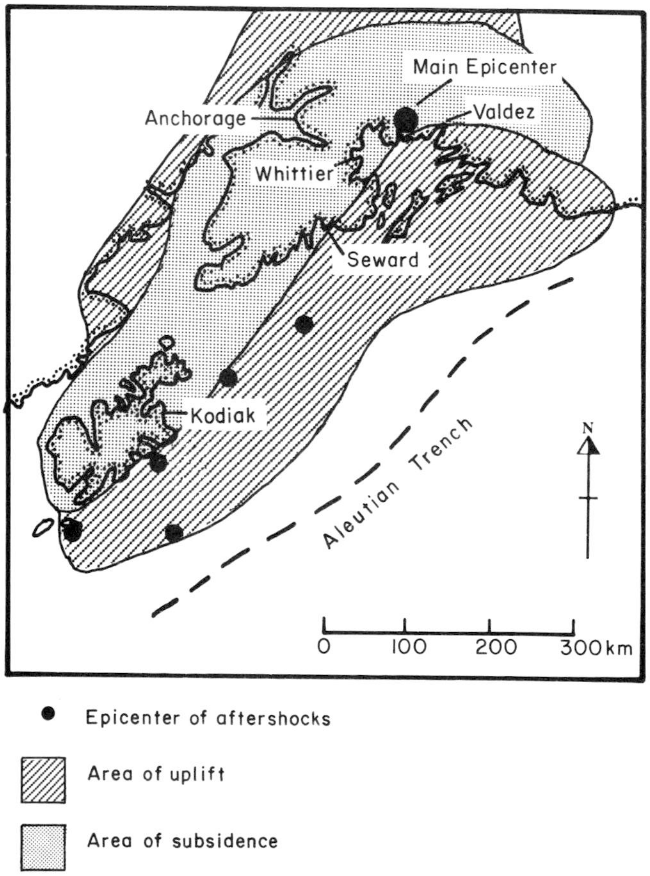

FIG. 11.6 Extent of Alaskan earthquake of 27 March 1964 (based upon Whittow, 1980). This shows both zones of uplift and subsidence.

Longwave vibrations lasted 4–7 minutes. The shock waves were so high in amplitude that seismographs ran off scale all over the world, thus preventing accurate measurement of the earthquake intensity (probably reaching 8.5 on the Richter scale). The earthquake rang the earth like a bell, and set up seiching in the Great Lakes of America 5000 km away. Water levels in wells in South Africa on the other side of the globe fluctuated. The ground motion was so severe that the tops of trees were snapped off. Earth displacements covered a distance of 800 km along the Danali fault system parallel to the Alaskan coastline. Over 1200 aftershocks were recorded along this faultline. Tsunami, exceeding 10 m in height along the Alaskan coast, swept across the Pacific Ocean at intervals of one hour. The west coast of North America as far south as San Diego was damaged. In Crescent City, California, over 30 city blocks were flooded, resulting in $US7 million damage.

Maximum uplift and down-throw amounted to over 15 m and 3 m respectively, the greatest deformation from an earthquake yet measured. In some locations, individual fault scarps measured 6 m in relief. Coastal cliffs along much of south Alaska, and along many relief features inland, were affected badly by landslides. Land subsidence occurred in many places underlain by clays, and in some areas liquefaction was noted (figure 11.7). Liquefaction is a process whereby firm, but water-saturated material, can be shaken and turned into a liquid, which then flows downslope under the effect of gravity. It will be discussed in more detail later in this chapter. About 60 per cent of the $US500 million damage (remembering that this earthquake, apart from the cities of Anchorage, Valdez and Kodiak, did not occur in a densely settled area) was caused by ground failures. Five landslides caused $US50 million damage in Anchorage alone. Most of the ground failure in this city occurred in the Bootlegger Cove clay, a glacial estuarine-marine deposit with low shear strength, high water-content and high sensitivity to vibration. In the ports of Seward, Whittier and Valdez the docks and warehouses sunk into the sea because of flow failures in marine sediments. Within an hour of the earthquake each of these towns was being overwhelmed by tsunami, 7–10 m in height, which continued coming over the next nine hours.

FIG. 11.7 Head scarp of one of numerous landslides that developed in Anchorage, Alaska, during the 27 March 1964 earthquake. Sand and silt lenses in the Bootlegger Cove Formation liquefied causing loss of soil strength (photograph courtesy of the United States Geological Survey, source Hays, 1981).

The Californian earthquake hazard (Keller, 1982; Oakeshott, 1983; Wood, 1986)

Of all seismic areas in the world, the San Andreas faultline running through southern California poses one of the most potentially hazardous zones for earthquakes. Its many subsidiary branches pass by, or through a region inhabited by 15 000 000 people (figure 11.8). The most devastating earthquake in this region, in terms of property loss, was the San Francisco earthquake of 18 April 1906 that killed 700 people. This quake was estimated at 8.25 on the Richter scale, and had a maximum vertical and lateral displacement of 2.5 and 6 m respectively. Fires which broke out following the earthquake were responsible for most of the damage. This earthquake was not unique. There have been earthquakes of similar size in recent geological times with the earthquake at Fort Tejon on 9 January 1857 being the largest yet recorded in southern California. The Owen valley tremor on 26 March 1872 is considered the third greatest earthquake to strike California in recent times. It produced surface faulting extending up to 150 km, with a maximum vertical displacement of 4 m. Other significant earthquakes occurred in 1769 southeast of Los Angeles, on 10 June 1836 in the East Bay area of San Francisco, in June 1838 at Santa Clara, on 8 October 1865 in the Santa Cruz mountains, on 21 October 1868 south of San Francisco, and on 24 April 1890 at Chittenden Pass.

The San Andreas faultline has been active since the *Jurassic*, 135 million years ago. Since then, the crust east of the fault has moved southward 550 km. During the Pleistocene (last 2 million years) horizontal displacement has amounted to 16 km.

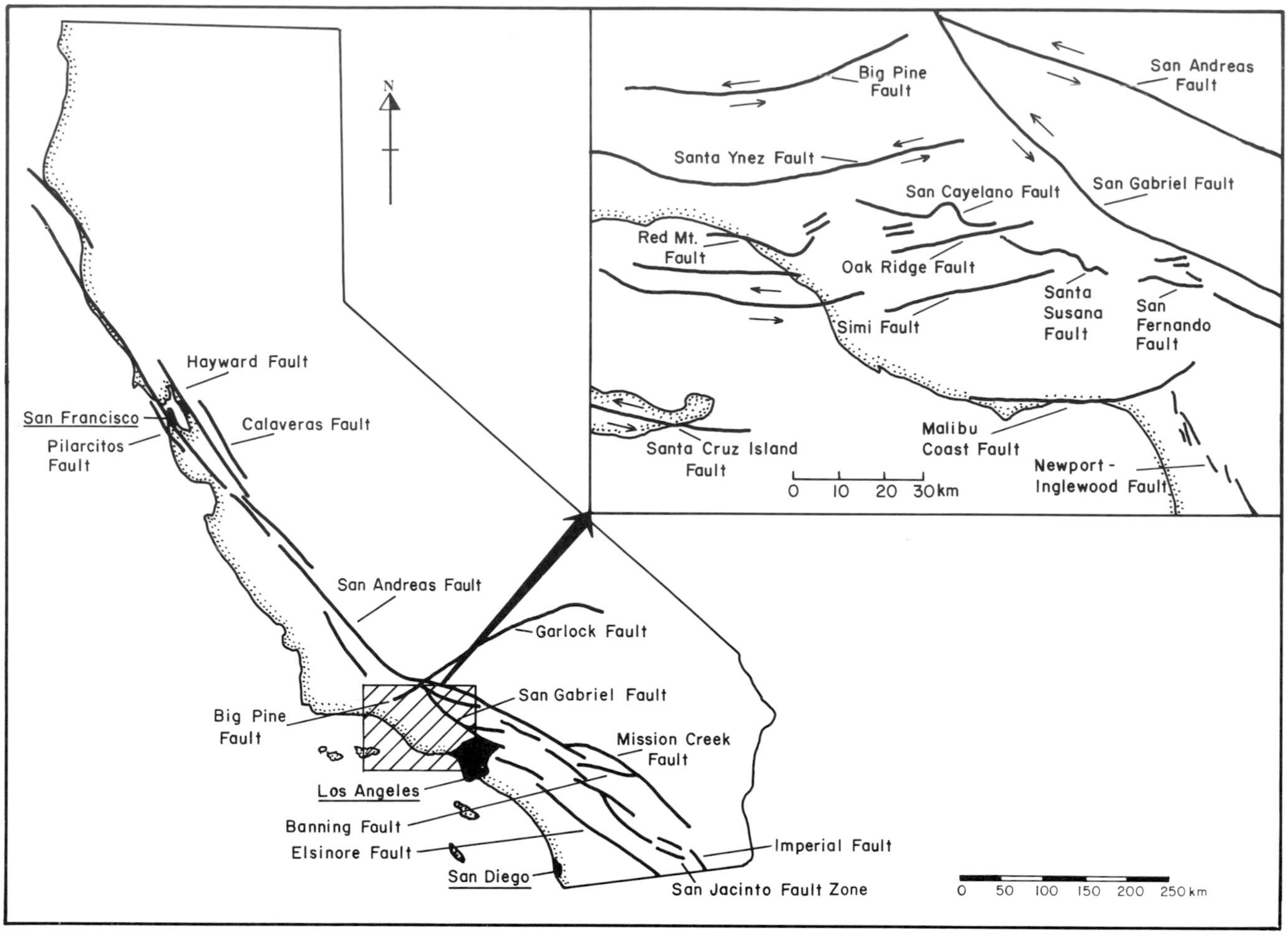

FIG. 11.8 Historically active fault systems of southern California (adapted from Oakeshott, 1983)

Dating of charcoal deposited over the past few thousand years over various faultlines indicates that major earthquakes occur at intervals of 50–300 years. The Pacific plate is moving slowly northwest relative to the North American plate at the rate of $5\,\text{cm}\,\text{yr}^{-1}$. Most of the movement is lateral, but in a few places there is a vertical component on the North American plate side. In most places this movement is continual; however, in some places the plates become locked together and the movement is released in one sudden jerk. The San Francisco earthquake gave rise to the greatest lateral displacement (6 m) yet measured along the San Andreas fault zone. Even where slow creep takes place, earthquakes can occur.

The strongest earthquake since 1911 occurred on 6 August 1979 at Hollister, 100 km south of San Francisco, where continuous movement is characteristic along the Calaveras fault zone. In southern California around Los Angeles, there has been no major movement since the 1857 earthquake. Based upon geological evidence, this area together with the San Francisco area is overdue for a major earthquake greater than eight on the Richter scale. Both cities exist in seismic gap areas, surrounded by zones which have experienced recent tremors in the past 30 years. In the next ten years the probability of an earthquake occurring in these gaps is 50 per cent. At Palmdale, 50 km northwest of Los Angeles, there already is evidence of the type of crustal activity which precedes a major earthquake. In the 15 years up until 1975, the crust rose here by 0.4 m and has since subsided 17 cm. This magnitude of movement was also recorded before the Alaskan earthquake of 1964 and the Chilean earthquake of 1960. Tests on rock stress in this region have indicated that the rocks adjacent to the faultline are fractured, water-saturated and at the limit of their natural fracturing point. It would appear that major populated segments of California lie in one of the most unstable zones in the world for earthquakes.

The consequences of such an earthquake would be catastrophic. Many of the past earthquakes in southern California, which produced little damage, occurred in unpopulated areas which now are densely settled. The San Fernando earthquake of 1971, along the Hayward fault, caused the collapse of public buildings, including two hospitals built to earthquake building code specifications. At the same time, sections of freeway overpasses built to the same standards also failed to withstand the shock waves. Subsequent to that earthquake, many public buildings and roadways have been reinforced or rebuilt. Building codes have been tightened to account for the type of damage produced by that relatively small earthquake. Two dams in the area almost failed. One, the Lower Van Norman earth-filled dam, came within inches of collapsing and flooding 80 000 people in the lower San Fernando valley. There are 226 dams in the San Francisco region alone with over 500 000 people living downstream, and many of these dams, together with others in southern California, have subsequently been strengthened. Some of these remedial measures have not worked. The 17 October 1989 earthquake in San Francisco, measuring only 6.9 on the Richter scale, collapsed a large section of the upper deck of the Nimitz Freeway in Oakland, even though the supports holding up this expressway were reinforced in the 1970s. Much of the housing construction in the San Francisco area since 1906 has occurred on unconsolidated landfill that could liquefy or amplify shock waves up to ten times, causing total destruction of buildings. This additional hazard was also brought home in the October 1989 earthquake, which caused the complete collapse of many houses in the Marina district for this reason.

It is not even known for certain whether all of the branches of the San Andreas fault have been mapped. For instance, the Diablo Canyon nuclear plant north of Los Angeles was built to withstand an earthquake of 8.5 on the Richter scale originating from a distant faultline. However, during construction a major and active faultline was discovered offshore within 5 km of the plant. While the plant's design was altered, there is still doubt whether or not it could withstand a major earthquake along the nearby faultline. The next major earthquake in southern California above seven on the Richter scale is inevitable, and will bring untold destruction. It has the potential to destroy large sections of major cities and to start numerous fires, which could turn into urban conflagrations similar to the 1906 earthquake. Numerous dams could fail causing major flooding and loss of life in downstream valleys, and the process of ground liquefaction, especially in the San Francisco area, could flatten suburbs in most of that city.

Liquefaction or thixotrophy

(Hansen, 1965; Bolt et al., 1975; Scheidegger, 1975; Hays, 1981; Lomnitz, 1988)

Liquefaction is a process whereby relatively firm clay-free sands and silts can become liquefied, and either flow as a fluid, or cause objects which they have been supporting to sink. The mechanics of liquefaction will be discussed in more detail in chapter 13. For now, liquefaction will be discussed qualitatively as an earthquake-induced hazard. Basically, the weight of a soil is supported by the grains which touch each other. Spaces or voids in the soil fill with water below the water-table. As long as the soil weight is borne by grain-to-grain contacts, the soil behaves as a rigid solid. However, if the pressure on water in the voids increases to the point that it equals or exceeds the weight of the soil, then individual grains may no longer be in contact with each other. At this point, the soil particles are suspended in a dense slurry that behaves as a fluid. It is this process, whereby a rigid soil becomes a liquid, that is liquefaction or *thixotrophy*.

Liquefaction can be induced by earthquake-generated shear or compressional waves. Shear waves distort the granular structure of soil causing some void spaces to collapse. These collapses suddenly transfer the ground-bearing load of the sediment from grain-to-grain contacts to pore water. Compressional waves increase the pore water pressure with each passage of a shock wave. If pore water pressure does not have time to return to normal before the arrival of the next shock wave, then the pore water increases incrementally with the passage of each wave (figure 11.9). The process can be illustrated quite easily by going to the wettest part of the beach foreshore at low tide. On flat slopes, one can usually run along the sand easily. It is even possible to drive vehicles along this zone on some fine sand beaches. However, if you tap the sand in this zone slowly with your foot, it will suddenly become soft and mushy, and spread outwards. If the tapping is light and continual, the bearing strength of the sand will decrease and your foot will sink into the sand.

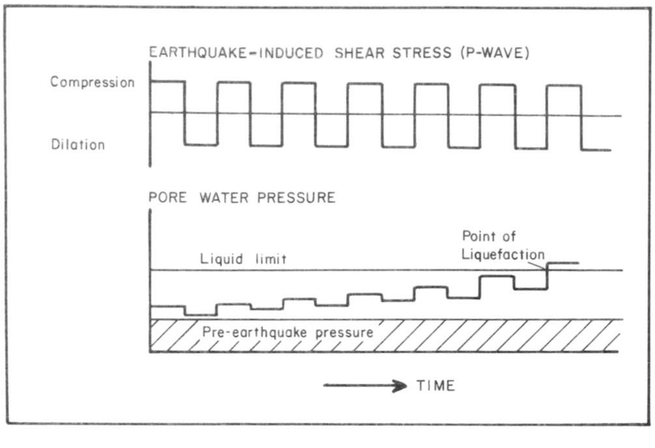

FIG. 11.9 Accumulative effect of compressional P-waves upon pore water pressure, leading to liquefaction (after Bolt et al., 1975)

Pore water pressure can only return to normal if there is free movement of water through the soil. Smaller particles inhibit water movement because capillary water tension between particles is greater. This *cohesion* is so large with very fine silt and clays that the pore water pressure cannot reach the point of liquefaction. Medium-to-fine grained sands satisfy both criteria for liquefaction to occur. They are fine enough to inhibit rapid internal water movement, yet coarse enough that *capillary cohesion* is no longer relevant. Since these sand sizes are common in river or marine sediments recently deposited in the last 10 000 years, liquefaction is an almost universal feature of earthquakes. The younger and looser the sediment deposit and the higher the water-table, the more susceptible it is to liquefaction.

The buildup of pore water pressure can cause water to vent from fissures in the ground resulting in surface sand boils or mud fountains. If pore water pressure approaches the weight of the overlying soil, then the soil at depth starts to behave like a fluid. The same process of liquefaction can also be induced by rising groundwaters through silt or sand material. Again, grain-to-grain contacts are minimized and the sand behaves like a fluid. Objects of any density will easily sink into this quicksand because the sub-surface has virtually no *bearing strength*. Thus, liquefaction causes three types of failure: lateral spreads, flow failures and loss of bearing strength. Lateral spreads involve the lateral movement of large blocks of soil as the result of liquefaction in sub-surface layers due to ground-shaking during earthquakes. It occurs on slope angles ranging from 0.3–3.0 degrees, with horizontal movements of 3–5 m. The longer the duration of ground-shaking, the greater the lateral movement and the more the surface layers will tend to fracture, forming fissures and scarps. In the Alaskan earthquake, floodplain sediments underwent this process, and virtually every bridge in the affected zone was destroyed by lateral spreading. Flow failures involve fluidized soils moving downslope, either as a slurry, or as surface blocks overtop a slurry at depth. These flows take place on slopes greater than 3 degrees. Some of these flows can move tens of kilometers at speeds in excess of 15 km hr^{-1}. More commonly, these flows can occur in marine sediment underwater. Loss of bearing strength simply represents the transference of the ground load from grain-to-grain contacts to the pore water. Any object which has been sitting on this material and using it for support is then susceptible to collapse or sinking. At present, the exact conditions responsible for earthquake-induced liquefaction have not been determined. Evidence indicates that the acceleration and number of shock waves control the process; however, prior history of the sediment cannot be ignored. Only mapping of sub-surface material in earthquake-prone areas will delineate the area of risk from this process during earthquakes.

As described above for the Alaskan earthquake, the effects of earthquake-induced liquefaction can be spectacular. For example, the Japanese earthquake at Niigata of 16 June 1964 registered 7.5 in magnitude on the Richter scale. The earthquake was located 60 km north of the city, and at Niigata, produced maximum ground accelerations of 0.16% g, values that cannot be considered excessive. The city itself had expanded since the Second World War onto reclaimed land on the Shinano River floodplain. The earthquake shock-waves did not destroy buildings but it liquefied unconsolidated, floodplain sediments, suddenly reducing their bearing strength. Large apartment blocks toppled over or sunk undamaged into the ground at all angles (figure 11.10). Afterwards, many were jacked back upright, underpinned with supports, and reused.

The San Francisco earthquake of 1906 produced some evidence of liquefaction around the city. The breaking of watermains, which hampered fire suppression, was generated by liquefaction-induced lateral spreads, mainly near the harbour. Much of the development of the city since has occurred on top of estuarine muds, which contain a high water content. In addition, much of the habourside construction in southern California occurs on landfill, which can easily undergo liquefaction if water-

Fig. 11.10 Liquefaction following the 16 June 1964 Niigata, Japan, earthquake (photograph courtesy of the United States Geological Survey, source in Catalogue of Disasters #B64F16-003). These worker's apartments suffered little damage from the earthquake, but toppled over because of liquefaction of unevenly distributed sand and silt lenses on the Shinano floodplain. Horizontal accelerations were as low as 0.16% g. The buildings subsequently were jacked up and reoccupied.

saturated. The San Fernando earthquake in 1971 produced liquefaction in many soils, and caused landslides on unstable slopes. Similar liquefaction problems were experienced during the October 1989 San Francisco earthquake. When a larger earthquake inevitably strikes this city, liquefaction will be the major cause of property damage.

The intensity of seismic waves does not have to be great to induce liquefaction—as illustrated in figure 11.10, where the bearing capacity of soils underlying the buildings failed but the buildings remained virtually undamaged. However, the intensity of compressional shock waves triggering liquefaction may be considerably lower than the values experienced at Niigata. For instance, the Meckering earthquake of 1968 in Western Australia induced liquefaction of sands underlying many parts of the expressway system in Perth, 100 km away from the epicenter. At Perth, the earthquake only registered 3 on the Mercalli scale. Not only did the edges of roadways sink 4–6 cm within hours of this seismic event, but the problem also persisted for weeks afterwards, eventually resulting in widespread damage to road verges.

The liquefaction process usually involves water movement in clay-free sand and silt deposits. In the loess deposits of China it can involve air as the suspending medium. The Kansu earthquake of 16 December 1920 broke down the *shearing* resistance between soil particles in loess hills. The low permeability of the material prevented air from escaping from the loess, which subsequently liquefied; this resulted in large landslides, which swept as flow failures over numerous towns and villages, killing over 200 000 people. The 1556 earthquake in Shensi Province may have generated similar loess failure, giving rise to a deathtoll in excess of 800 000.

Water-logged muds can also amplify the long period components of shock waves. Two recent earthquakes—at Newcastle, Australia, in 1989 and in Mexico in 1985—provide clear evidence of this process at work. At Newcastle, the seismic wave from a weak tremor, registering only 5.5 on the Richter scale, amplified in river clays beneath the city. As a result, what was really only a small tremor still managed to damage over 10 000 buildings, creating a damage bill of $A1000 million. With the Mexican earthquake, damage was even more extensive.

Mexico City was founded in AD 1325 by the Aztecs upon the deep, weathered ash deposits of a former lake bed, Lake Texcoco, which was then drained by the Spaniards. These deposits have weathered to form basically *montmorillonitic* clays, which have the ability of absorbing water into their internal structure, thus increasing their water content by 350–400 per cent. From a *geophysical* point of view, these clays can be effectively modelled as water, even though they have a solid appearance, upon which multi-storied buildings have been constructed. Today, Mexico City has grown to a population in excess of 12 million. Two earthquakes struck Mexico 400 km southwest of the city, offshore from Rio Balsas on 19–20 September 1985. The first earthquake measured 8.1 on the Richter scale, while the second aftershock, 36 hours later, measured 7.5. These shock waves were not exceptionally large. In Mexico City, because of the distance from the Mexican subduction zone where these earthquakes occurred, peak accelerations corresponded to a calculated return period of 50 years. Because the area surrounding the epicenter was sparsely populated, damage was minor. Jalisco, Michoacán and Guerrero, the three coastal states nearest the epicenter, suffered extensive damage, with loss of life around 10 000. Damage decreased sharply away from the epicenter, closer to Mexico City. However, when the shock waves traversed the lake beds underlying Mexico City, they were amplified sixfold by resonance, such that the underlying sediments shook like a bowl of jelly under complex seiching. Surface gravity waves, akin to waves on the surface of an ocean, criss-crossed the basin for up to three minutes. These waves had wavelengths of about 50 m length, the same as the base dimensions of many buildings, and caused the destruction of 10 per cent of the buildings in the city center, covering an area of 30 km^2. Of the 400 multi-storey buildings wrecked, all but 2 per cent were between 6 and 18 stories high. Many had become pliable with age and had developed a fundamental resonance period of two seconds, the same as that obtained for peak accelerations of surface seismic waves. The buildings folded like a stack of cards. Close to 20 000 people were killed in Mexico City, and the damage bill amounted to $US4.1 billion. The Mexican earthquake illustrates a fundamental lack in our knowledge regarding the nature of ground motion in sedimentary basins containing a high water content, or alternating dry and wet layers.

Tsunami

Description (Wiegel, 1964; Hodgson, 1964; Bolt et al., 1975; Murty, 1977; Shepard, 1977; Lida and Iwasaki, 1981; Myles, 1985; Lockridge, 1988)

Tsunami (both the singular and plural forms of the word are the same) are water wave phenomena generated by the shock waves associated with seismic activity, explosive volcanism or submarine landslides. These shock waves can be transmitted through oceans, lakes or reservoirs. The term 'tsunami' comes from Japan, where it means 'harbour wave', because such a wave often develops as resonant phenomenon in embayments after offshore earthquakes. Tsunami are analogous directly to wind waves, and behave like them. They have a wave length, a period, a deep-water height and can undergo shoaling, refraction and diffraction. Most tsunami originate from submarine seismic disturbances. The displacement of the Earth's crust by several meters during underwater earthquakes may cover tens of thousands of square kilometers, and impart tremendous potential energy to the overlying water. Tsunami are rare events, in that not all submarine earthquakes generate them. Between 1861 and 1948, only 124 tsunami were recorded from 15 000 earthquakes. Along the west coast of South America, 1098 offshore earthquakes have generated only 19 tsunami. This low frequency of occurrence may simply reflect the fact that most tsunami are small in amplitude and go unnoticed, or the fact that most earthquake-induced tsunami require a shallow focus seismic event greater than 6.5 on the Richter scale.

Tsunami can also have a volcanic origin. Of 92 documentable cases of tsunami generated by volcanoes, 16.5 per cent resulted from tectonic earthquakes associated with the eruption, 20 per cent from *pyroclastic* (ash) flows or surges hitting the ocean, and 14 per cent from submarine eruptions. Only 7 per cent resulted from the collapse of the volcano and subsequent *caldera* formation. Landslides or avalanches of cold rock accounted for 5 per cent; avalanches of hot material, 4.5 per cent; *lahars* (mud flows), 3 per cent; atmospheric shock waves, 3 per cent; and lava avalanching into the sea, 1 per cent. About 25 per cent had no discernible origin, but probably were produced by submerged volcanic eruptions. The eruptions of Krakatoa in 1883 and Santorini in 1628 BC were responsible for the largest and most significant volcano-induced tsunami.

A partial geographical distribution of tsunami is tabulated in table 11.2, while the source regions for tsunami in the Pacific for the period 1900–1983 are plotted in figure 11.11. The worst regions for tsunami occur along the stretch of islands from Indonesia, through to Japan and the coast of the Soviet Union. Most tsunami in this region originate locally. Of 104 damaging tsunami in the past century, only 9 have caused damage beyond their source regions. In the Pacific these originate between Japan and the Kamchatka peninsula inclusive, along the Aleutian Islands and south Alaskan coast, and along the coast of South America. For example, the Chilean earthquake in 1960 affected most of the coastline of the Pacific Ocean. Here, earthquakes greater in magnitude than 8.2, affect the entire Pacific Ocean and have a probability of occurrence of once every 25 years. The high number of tsunami in Oceania includes those at Hawaii, which receives about 85 per cent of all tsunami affecting the Pacific. In fact, Japan and Hawaii are the two most tsunami-prone regions in the world, each accounting for 19 per cent of all tsunami

TABLE 11.2 Percentage Distribution of Tsunami in the World's Oceans and Seas

Location	%
Atlantic East Coast	1.6
Mediterranean	10.1
Bay of Bengal	0.8
East Indies	20.3
Oceania	25.4
Japan-Soviet Union	18.6
Pacific East Coast	8.9
Caribbean	13.8
Atlantic West Coast	0.4

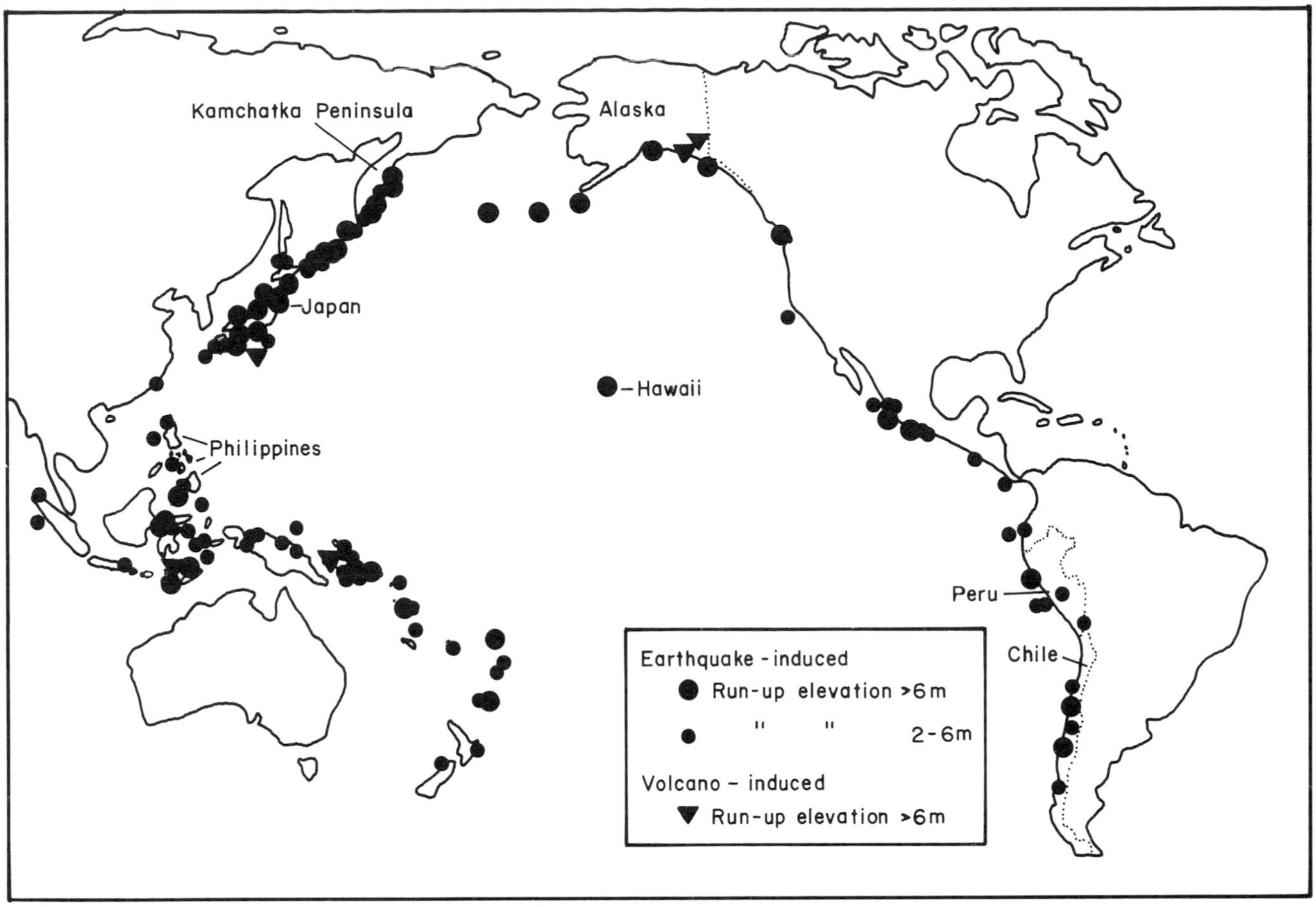

FIG. 11.11 Source areas for tsunami in the Pacific region generating run-up elevations greater than 2.5 m: 1900–83 (adapted from National Geophysical Data Center and World Data Center A for Solid Earth Geophysics, 1984)

measured. For this reason, Hawaii monitors closely all tsunami source regions in the Pacific.

Peru and Chile are most affected by local earthquakes. This coastline is also the only source area to affect the southwest Pacific region. Japan is mostly affected by locally generated tsunami, and directs its attention to predicting their occurrence locally. Since AD 684, in the Japan region 73 tsunami have resulted in over 100 000 deaths. Indonesia and the Philippines are also affected by locally originating tsunami. In the Philippines recorded tsunami have killed 50 000 people, mainly in two single events in 1863 and 1976. Indonesia has experienced a similar deathtoll in recorded times; however, half this total has been caused by tsunami associated with local volcanism. The Atlantic coastline is virtually devoid of tsunami; however, the Lisbon earthquake of 1755 produced a 3–4 m wave that was felt on all sides of the Atlantic. The continental slope off Newfoundland, Canada, is seismically active and has produced tsunami that have swept onto that coastline. One recorded tsunami from this site reached Boston, with a height of 0.4 m, on 18 November 1929. One of the longest records of tsunami occurrence exists in the eastern Mediterranean. Between 479 BC and AD 1981, 7 per cent of the 249 known earthquakes produced damaging or disastrous tsunami. Here, about 30 per cent of all earthquakes produce a measurable seismic wave.

Earthquake-generated tsunami are associated with events registering more than 6.5 on the Richter scale. Two-thirds of damaging tsunami in the Pacific region have been associated with earthquakes of magnitude 7.5 or greater. Most tsunami-generating earthquakes are shallow and occur at depths in the Earth's crust of 0–40 km. Studies have shown that the peak energy period of a tsunami is a function of the magnitude of the earthquake according to the following equation:

$$T_t = 0.57M - 2.85 \qquad (11.1)$$

where T_t = the period of the tsunami and
M = the magnitude of the earthquake on the Richter scale

Tsunami generated close to shore are not as large as those generated in deeper water. Tsunami have wave periods in the open ocean of seven minutes, with heights of 0.4 m. Waves with this period will travel at speeds of 600–800 km hr^{-1} in the deepest part of the ocean, and 150–300 km hr^{-1} across the continental shelf. In some cases in the Pacific Ocean, initial wave periods of 60 minutes have been recorded. Wave periods of this length in the open ocean will often degrade to less than 2.5 minutes in shallow water. Tsunami resonating in harbours typically have periods of 8–27 minutes. If the wave period is some harmonic of the natural frequency of the harbour or bay, then the amplitude of the tsunami wave can greatly increase over time. This resonance may cause the period of the wave to shift to higher wave periods (lower frequencies). Figure 11.12 plots typical tide-gauge records of tsunami at various locations in the Pacific Ocean. The concept that a tsunami consists of only one or

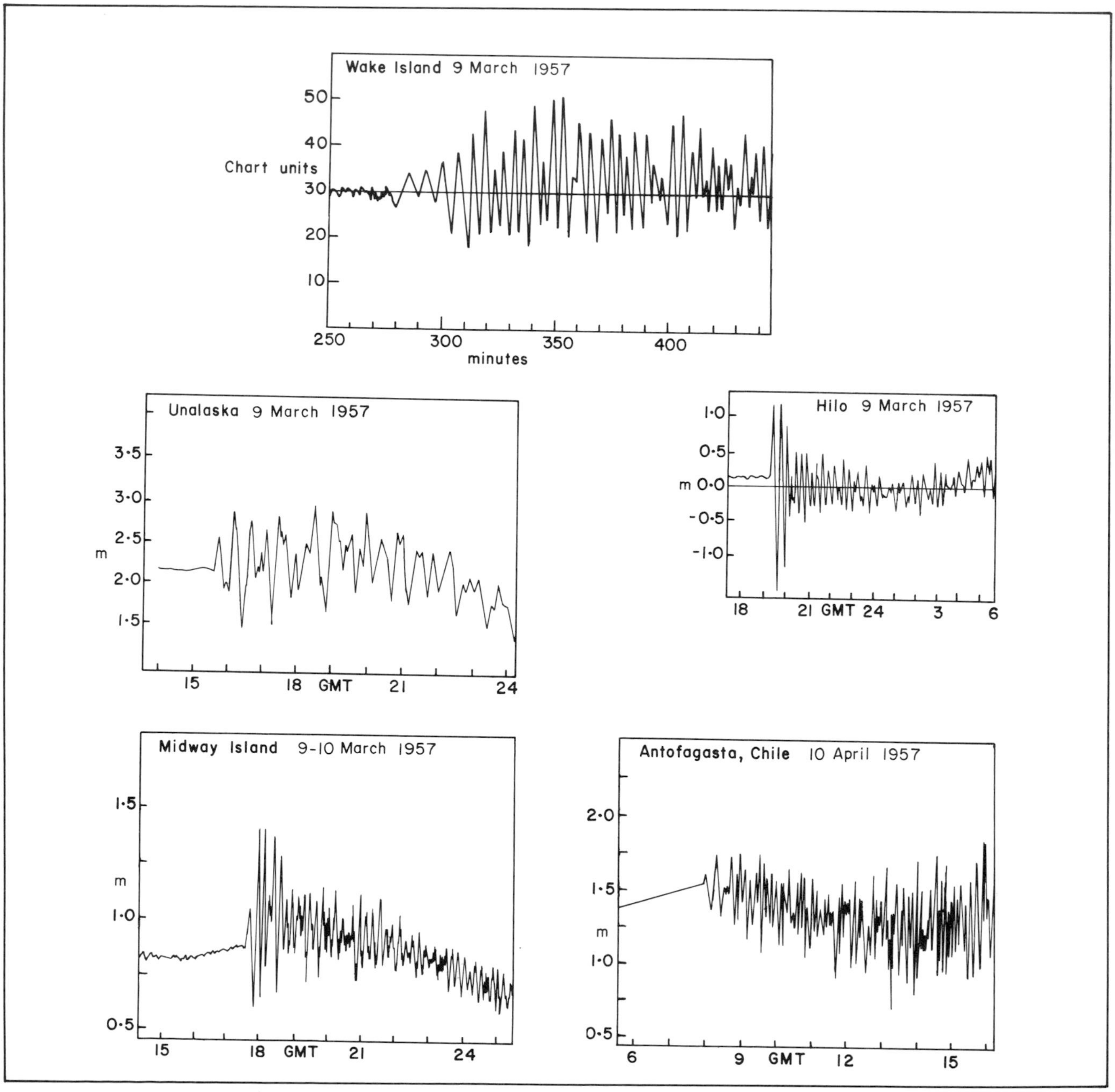

FIG. 11.12 Plots of tsunami wave trains at various tidal gauges in the Pacific region (after Salsman, 1959; Van Dorn, 1959; Wiegel, 1964)

two waves is not borne out by the records. Some wave periods range between 8 and 30 minutes and persist for over six hours. Wave characteristics are highly variable. In some cases, the waves consist of an initial peak which then tapers off in height exponentially over four to six hours. In other cases, the tsunami wave train consists of a maximum peak well back in the wave sequence.

The wave length of an average tsunami of eight minutes is about 360 km (refer to equation 4.3). As explained in chapter 4, with this wave length it is possible for the tsunami to feel the ocean bottom at any depth, and to undergo refraction effects. Refraction diagrams of tsunami give accurate prediction of arrival times across the Pacific, but not of run-up heights near shore. It has been shown that the spread of tsunami close to shore and around islands follows classic diffraction theory, similar to the spreading out of a wave into a harbour after it passes through a narrow entrance. Figure 11.13 illustrates the movement of the Chilean tsunami wave of 23 May 1960, which wreaked havoc throughout the Pacific region. It clearly shows that the wave radiated symmetrically away from the epicenter, and was unaffected by bottom topography until islands or seamounts were approached. The island chains in the west Pacific clearly broke up the wave front; however, much of Japan received the wave unaffected by refraction.

It is a very noticeable phenomenon that ships anchored several kilometers out to sea hardly notice the arrival of a tsunami wave; however, once the wave approaches shore, the wave very rapidly shoals and reaches its extreme height. This run-up height controls the extent of damage. Because most shorelines are relatively steep compared to the wavelength of the tsunami, the wave will not break or form a bore but will surge up over the foreshore

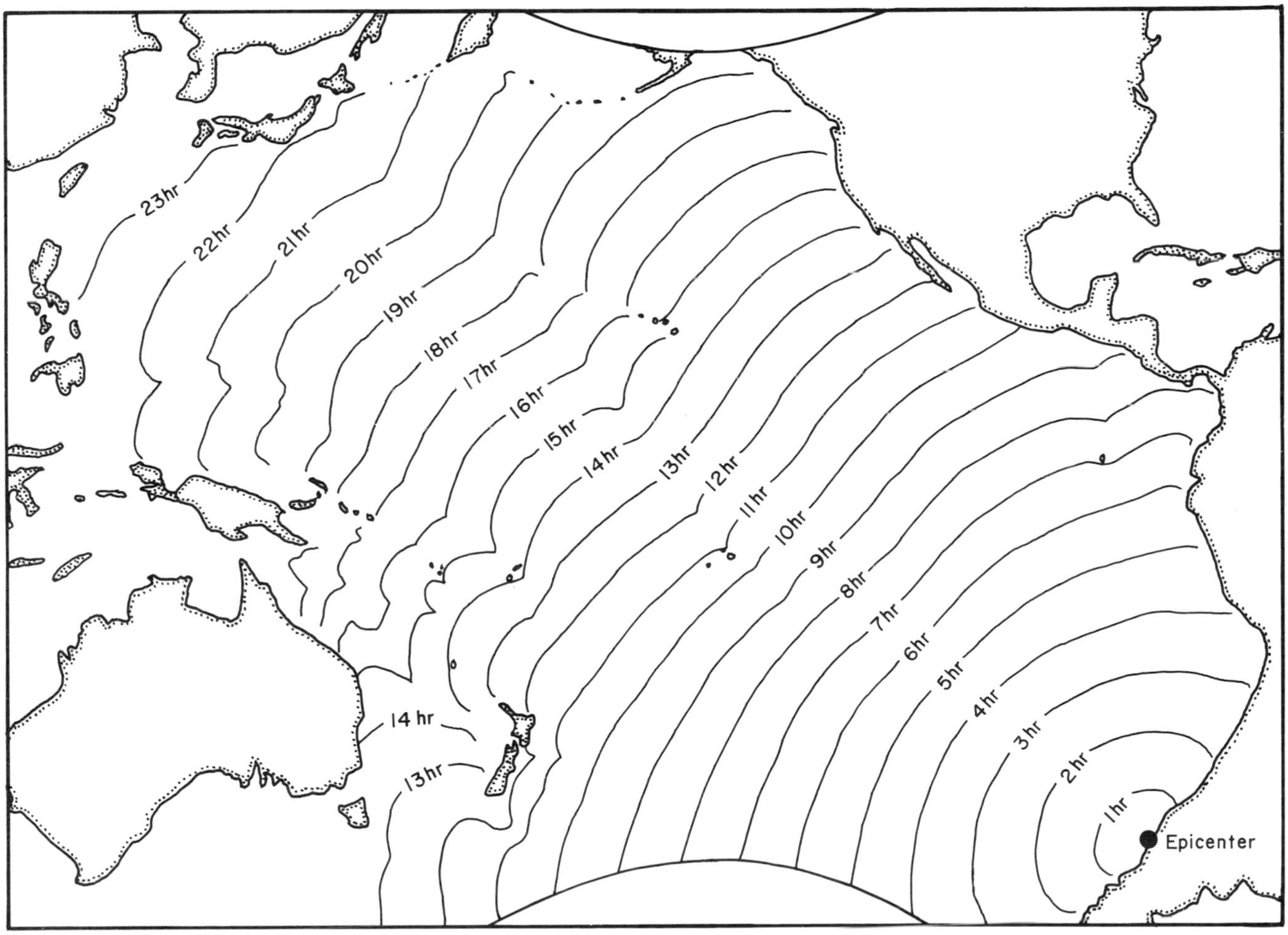

Fig. 11.13 Tsunami wave refraction pattern in the Pacific Ocean following the 23 May 1960 Chilean earthquake (after Wadati et al., 1963)

(figure 11.14). For example, the earthquake offshore from Gisborne, New Zealand, on 26 March 1947, generated a tsunami run-up 10 m high along a 13 km stretch of coast, while the eruption of Krakatoa in 1883 generated a wave that reached elevations up to 40 m high along the surrounding coastline. By far the largest run-up height recorded is that produced by an earthquake-triggered icefall into Lituya Bay, Alaska on 9 July 1958. The resulting wall of water surged 490 m up the shoreline on the opposite side of the bay. Figure 11.15 dramatically illustrates the potential extent of flooding inland such tsunami are able to generate. The American gunship *Wateree* and the Peruvian ship *America* were both standing offshore during the Arica, Chile, tsunami of 8 August 1868. They were both swept 5 km up the coast, and 3 km inland over the top of sand dunes by two successive tsunami waves. The ships came to rest 60 meters from a vertical cliff at the foot of the coastal range, against which the tsunami had surged to a height 14 m above sea level. Unfortunately, 25 000 deaths also resulted from this event along the coasts of Chile and Peru.

Run-up heights also depend upon the configuration of the shore, diffraction, characteristics of the wave and resonance. Within some embayments, it takes several waves to build up peak tsunami wave-heights. If the first wave in the tsunami wave train is the highest, then it can be dampened out. Figure 11.16 maps the heights of tsunami run-up around Hawaii for the 1946 Aleutian earthquake. The northern coastline facing the tsunami advance received the highest run up. However, there was also a tendency for waves to wrap around the islands and strike hardest at supposedly protected sides, especially on the islands of Kauai and Hawaii. Because of refraction effects, almost every promontory also experienced high run-ups that often exceeded 5 m. Coastlines where offshore depths dropped off steeply were hardest hit, because the tsunami waves could approach shore with minimal energy dissipation. For these reasons, run-up heights were spatially very variable. In some places, for example on the north shore of Molokai, heights exceeded 10 m, while several kilometers away heights did not reach 2.5 m.

In Japan, run-up heights have been measured as high as 32 m. The 1896 earthquake offshore from the Sanriku coast sent a wave 30 m high crashing into the towns of Yoshihama and Kamaishi. Because of the high frequency of occurrence of tsunami around Japan, extensive research has been carried

Fig. 11.14 Sequential photographs of the 9 March 1957 tsunami overriding the backshore at Laie Point, Oahu, Hawaii (photographs courtesy of the United States Geological Survey, source in Catalogue of Disasters #B57C09-002). Fifty-four people died in this tsunami, which was generated by an earthquake in the Aleutian Islands.

Fig. 11.15 The American warship *Wateree* and Peruvian warship *America* stranded 3 km inland after the tsunami of 8 August 1868 at Arica, Chile (photographs courtesy of the United States Geological Survey, source in Catalogue of Disasters #A68H08-002)

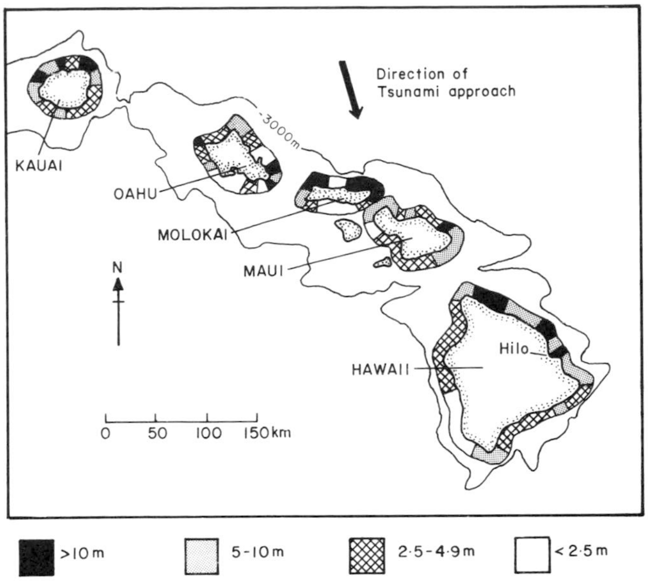

FIG. 11.16 Run-up heights of the 1 April 1946 tsunami around the Hawaiian Islands, following an Alaskan earthquake (after Shephard, 1977)

out there into predicting their characteristics and magnitude. A scale has been developed rating the magnitude of Japanese tsunami in relationship to earthquake magnitude on the Richter scale. Table 11.3 summarizes these relationships. Japanese tsunami have between 1 and 10 per cent the total energy of the source earthquake. Only earthquakes of magnitude seven or greater are responsible for significant tsunami waves with run-up heights in excess of 1 m. However, as earthquake magnitude rises above eight, then the run-up height and destructive energy of the wave dramatically increases. An earthquake of magnitude eight can produce a tsunami wave 4–6 m in height. The intensity needs only to increase to 8.75 to generate the 30 m wave height that crashed into Sanriku coast.

TABLE 11.3 Earthquake Magnitude versus Tsunami Run-up Height, in Japan

Earthquake Magnitude Richter Scale	*Tsunami Magnitude*	*Maximum run-up in meters*
6	−2	<0.3
6.5	−1	0.5–0.75
7	0	1.0–1.5
7.5	1	2.0–3.0
8	2	4.0–6.0
8.25	3	8.0–12.0
8.5	4	16.0–24.0
8.75	5	>32.0

Disaster descriptions (Bolt et al., 1975; Cornell 1976; Coates, 1985; Myles, 1985)

Table 11.4 outlines some of the major tsunami that have occurred in recorded history. While earthquakes are responsible for most destructive tsunami, the Santorini, Tamboro and Krakatoan volcanic eruptions, in total, killed 100 000 people. Tsunami have always appeared as impressive events, but the earthquake or volcanic event that caused them usually accounts for a larger deathtoll. The largest tsunami wave recorded occurred in 1737 on the Kamchatka peninsula when a 64 m high wave washed across the southern tip of the peninsula. Probably the most devastating event was the Santorini eruption of 1628 BC, generating a tsunami which must have destroyed all coastal towns in the eastern Mediterranean. The Santorini crater is five times larger in volume than that of Krakatoa, and twice as deep. On adjacent islands there is evidence of pumice stranded at elevations up to 250 m above sea-level. The initial tsunami waves may have been 60 m in height as they spread out from Santorini. The Lisbon earthquake of 1 November 1755, which is possibly the largest earthquake known (possibly nine on the Richter scale), resulted in the lower town being submerged under a tsunami 15 m in height and sent a wall of water across the Atlantic ocean that raised tide levels 3–4 m above normal in Barbados and Antigua, West Indies. Tsunami also ran up and down the west coast of Europe and along the Atlantic coast of Morocco. The Spanish port of Cadiz as well as Madeira in the Azores were also hit by waves 15 m high, while a 3–4 m high wave sunk ships along the English Channel. Water level oscillations were also noticed in the North Sea.

The Caribbean has also witnessed its share of devastating tsunami. Noted events occurred in 1842, 1907, 1918 and 1946. However, the worst tsunami recorded destroyed Port Royal, Jamaica in June 1692. The earthquake that triggered the tsunami collapsed the city, sending much of it sliding into the sea because of liquefaction. It produced a tsunami within the bay that flung ships standing in the harbour inland over two-storey buildings. While the deathtoll was only 2000, because of the small population of the town, the earthquake was a mirror image of the one that struck Lisbon.

TABLE 11.4 Major Historical Tsunami Disasters

Date	*Location of Tsunami*	*Source*	*Deathtoll*
1628 BC	Santorini Is, Aegean Sea	volcanic	[a]
497 BC	Potidaea, Greece	earthquake	[b]
1293	Sanriku Coast, Japan	earthquake	30 000
1703	Sanriku Coast, Japan	earthquake	~100 000
1737	Kamchatka, Soviet Union	earthquake	[c]
1755	Lisbon, Portugal	earthquake	~10 000
1815	Tamboro, Indonesia	volcanic	12 000
1868	Chile and Peru	earthquake (Chile)	25 000
1883	Krakatoa, Indonesia	volcanic	36 000
1896	Sanriku Coast, Japan	earthquake	27 000
1933	Sanriku Coast, Japan	earthquake	3 000
1946	Hilo, Hawaii	earthquake (Aleutian Islands)	173
1960	Pacific	earthquake (Chile)	200
1964	Pacific	earthquake (Alaska)	122
1976	Philippines	earthquake	3 000[d]
1978	Acajutla, Ecuador	earthquake	100

[a] Deathtoll not known, but the tsunami wave is reputed to have destroyed Minoan civilization.
[b] Deathtoll unknown.
[c] Deathtoll unknown; the tsunami wave reached a height of 64 m.
[d] Estimate.

The Sanriku coast of Japan has the misfortune of being the heaviest populated, most tsunami-prone, coast in the world. About once per century, this coastline has been swept by killer tsunami and within the span of 40 years between 1896 and 1933 it was hit twice. On 15 June 1896 a small earthquake on the ocean floor, 120 km southeast of the city of Kamaishi, sent a 30 m wall of water crashing into the coastline, killing 27 000 people. The same tsunami event was measured 10.5 hours later in San Francisco on the other side of the Pacific Ocean. In 1933 disaster struck again when a similar positioned earthquake sent ashore a wave that killed 3000 inhabitants.

The Krakatoa eruption of 27 August 1883, which will be discussed in more detail in the next chapter, sent out a tsunami wave measured around the world. The wave rounded the Cape of Good Hope in South Africa 6000 km away, and was recorded along the English Channel 37 hours later. On the other side of the Pacific Ocean, water levels were affected along the west coast of Panama and in San Francisco Bay. The waves which rounded the world were probably associated with the earthquakes that accompanied the eruption, or caused by the collapse of the caldera. However, a substantial atmospheric shock wave, which induced tsunami waves away from the site, must have been produced because the tsunami was detected in bodies of water which were not connected to each other. For instance, it is difficult to see how the tsunami wave could have got through the island archipelagos of the west Pacific to register at San Francisco. In addition, a 0.5 m oscillation was measured in Lake Taupo, in the center of New Zealand, coincidentally with the passage of the atmospheric shock wave.

The most active area seismically producing tsunami is situated along the eastern edge of the Nazca crustal plate, along the coastline of Chile and Peru. This region has been inundated by destructive tsunami at roughly 30-year intervals in recorded history in 1562, 1570, 1575, 1604, 1657, 1730, 1751, 1819, 1835, 1868, 1877, 1906, 1922 and, most recently, 1960. The tsunami events of 21–22 May 1960 were produced in Chile by over four dozen earthquakes with magnitudes of 7.8–8.9 on the Richter scale. A series of tsunami waves, spread over a period of 18 hours, took over 2500 lives across the Pacific, and produced property damage in such diverse places as Hawaii, Pitcairn Island, New Guinea, New Zealand, Japan, Okinawa, and the Philippines (see figure 11.13). The tsunami resulted from sudden land displacement of 2–4 m along the faultline during the earthquakes. Within 15 minutes of the faulting, three large waves devastated coastal towns in Chile, where over 1700 people lost their lives. The arrival of the first wave was predicted at Hilo, Hawaii, five hours in advance to an accuracy within one minute. Unfortunately 61 people, mainly sightseers, were killed as the wave hammered the

Hawaiian Islands. When the wave reached Japan 24 hours later, it still had enough energy to obtain a height of 3.5–6 m along the eastern coastline. Five thousand homes were washed away, hundreds of ships sunk and 190 people lost their lives. In Hawaii, this event illustrated the folly of some humans in the face of disaster. Despite plenty of warning, only 33 per cent of the residents of the affected area in Hilo evacuated. Over 50 per cent only evacuated after the first wave arrived, and 15 per cent stayed behind even after large waves had beached. Many of those killed were spectators who went back to the see the action of a tsunami hitting the coast. Another more recent large tsunami in the Pacific occurred on 26 May 1983 offshore from Japan. Over 100 people lost their lives and property damage amounted to $US700 million.

Prediction in the Pacific region (Wiegel, 1964; Bolt et al., 1975; Hays, 1981; Blong and Johnson, 1986)

Except for local events, the short-term prediction of tsunami is best developed in the Pacific Ocean region. Here, the warning of the arrival of a tsunami can be guaranteed 1–24 hours in advance, depending upon the location of sites relative to an earthquake epicenter. Because each tsunami has its own individuality, it is very difficult to compile long-term statistics on tsunami magnitude and frequency at any one location. Attempts to relate magnitude to frequency using probability plots do not show the classic straight line relationships that other natural events evidence. Figure 11.17 plots the probability of exceedence of maximum run-up of tsunami at Hilo, Hawaii. While larger events plot as a straight line, smaller events that are barely detectable are too common. The 1960 Chilean tsunami had a frequency of occurrence of 1:200 years. However, the data cannot be extrapolated to larger events, because there may be a physical limit to the magnitude of earthquakes and, thus, to the size of tsunami generated in the Pacific region. What is more relevant is the fact that refraction and diffraction effects control the eventual run-up height at any location. The probability plot shown in figure 11.17 is specific to Hilo and cannot be applied elsewhere. It also includes a limited number of tsunami originating from very specific locations around the Pacific. It is quite feasible that earthquakes occurring in areas different from those plotted on the diagram could change the nature of the diagram significantly.

Because the Pacific Ocean is the area most susceptible to earthquakes, the Seismic Sea Wave Warning System was established by the United States government following the Aleutian tsunami of 1946 which so badly affected Hawaii. This warning system developed into the International Tsunami Warning System, with its operational center situated at the Pacific Tsunami Warning Center in Hawaii, and staffed by the United States National Weather Service. Presently the system is responsible for detecting tsunami using 53 tide gauges scattered around, and throughout the Pacific. The warning system relies on Pacific Ocean earthquakes being detected at crucial seismic stations around the world that are on 24-hour alert. These stations are situated so that they do not lie in the shadow zones of any P- or S-wave originating from the Pacific region. Once a suspect earthquake has been detected, the system then relies upon detection of tsunami at any of the 53 tide gauges scattered around, and throughout the Pacific region. As soon as a tsunami train is detected, a tsunami warning is communicated to member nations. The tsunami's path is then monitored to obtain information on wave periods and heights. These data are then used to defined travel paths using refraction-diffraction diagrams, calculated beforehand for any possible tsunami originating in any part of the Pacific region. Warnings were initially issued for the United States and Hawaiian areas; but, following the 1960 Chilean earthquake, the scheme was extended to all countries bordering the Pacific Ocean. Japan, up until 1960, had its own warning network, believing at the time that all tsunami affecting Japan originated locally. The 1960 Chilean tsunami indicated that any submarine earthquake in the Pacific region could affect any Pacific coastline. As of 1986, 22 countries, including New

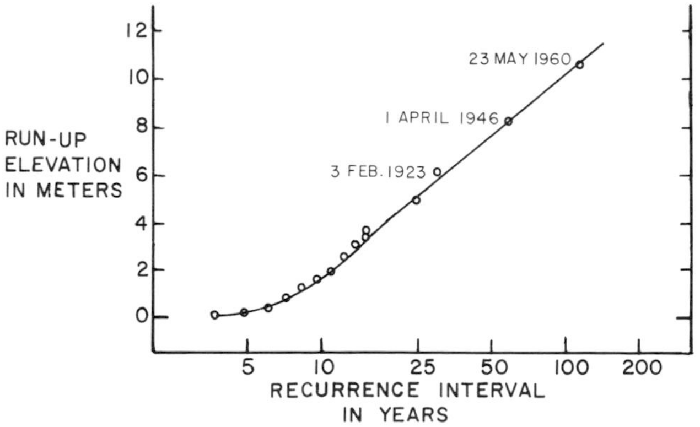

Fig. 11.17 Probability of run-up exceedence of tsunami at Hilo, Hawaii (Robert L. Wiegel, *Oceanographical Engineering*, ©1964, p. 108. Reprinted by permission of Prentice Hall, Inc., Englewood Cliffs, New Jersey)

Zealand and Australia, co-operate in the International Tsunami Warning System.

The system was significantly tested following the Alaskan earthquake of 1964. Within 46 minutes of the earthquake being detected by seismic stations, a Pacific tsunami warning was issued. In fact, the system was so efficient that Kodiak, Alaska, was forewarned of an impending tsunami within an hour of the earthquake. Following the arrival of the wave in California, warning sirens in Hawaii were sounded to alert the general population to evacuate threatened areas. Because it takes four hours for a tsunami to travel between Alaska and Hawaii, and about six hours between Alaska and Japan, the Pacific Warning System gave residents in these countries plenty of time to evacuate tsunami-prone coastal areas. In Hilo, Hawaii, these areas had been mapped following the 1946 tsunami, and converted to parkland to minimize future loss of life and property damage. Because tsunami are rare events, these warnings have now created an additional threat. The Alaskan earthquake warnings in California gave spectators enough time to flock to the coast to view the arrival of the tsunami. Over 10 000 people crowded San Francisco beaches to await the arrival of the wave. Ignorance of tsunami behaviour has also led to deaths even though warnings were given. In Hawaii, the deathtoll following the Alaskan earthquake was exacerbated by people returning to flooded areas before the arrival of the biggest wave. In California, some residents returned to clean up damage after the passage of the first two waves in the 1964 Alaskan tsunami. The third and fourth waves were the largest, and caught many of the returnees by surprise.

The Pacific Warning system is being updated to ensure that false alarms are not given, and that all tsunami are picked up. Satellite communications now speed up data collation and warning broadcasts. Other methods of detection are being investigated, including the monitoring of atmospheric pressure waves, as the initial disturbance giving rise to a tsunami also can generate a long-period pressure wave transmitted long distances through the atmosphere. It may even be possible to position sensitive pressure detectors on the ocean bottom adjacent to remote regions such as Alaska and Chile, where the highest magnitude earthquake-induced tsunami originate. Shorter period ocean swell will not be detected at these depths, but the longer period tsunami waves will. Such a scheme, called the THRUST program, has been established offshore from Valparaiso, Chile, and linked to an onshore computer to provide advanced warning of locally generated tsunami along this coastline. Today, there is little chance that areas surrounding the Pacific ocean would suffer loss of life from distant earthquakes. As of 1986, about three warnings a year are issued for the Pacific region. In the majority of cases, no tsunami ever eventuates; however, the fact remains that these warnings must always be taken seriously, even by countries in the Pacific region not affected by past tsunami events.

The Pacific Warning System is not flawless. The risk still exists in Japan, and on other island archipelagos along the western rim of the Pacific, that local earthquakes could generate tsunami too close to shore to permit advanced warning. The 7.8 magnitude (Richter scale) Philippines earthquake of 17 August 1976, in the Moro Gulf on the southwest part of the island of Mindanao, generated a local tsunami 3–4.5 m high that reportedly killed between 400 and 8000 people. The event was virtually unpredictable because the earthquake occurred within 20 km of a populated coastline. The event is unusual in that an accurate deathtoll could not be determined for the event, even though it occurred recently in a populated country. The Honolulu-based International Tsunami Information Center put the deathtoll at 8000, with 90 000 homeless; however, the actual loss of life was probably 400–500 people, with 12 000 homeless. While this tsunami is generally believed to be the last major one in recent years, three obscure tsunami killed 539–700 people on the island of Lembata, Indonesia, in 1979. The tsunami are obscure because the events responsible for their generation are unknown. Not only were these latter events poorly documented, they were totally unpredicted. They illustrate the point that once one enters the island groups at the western side of the Pacific, tsunami still represent a potentially unexpected hazard.

The threat in Australia (Oliver, 1988)

While Australia appears to lie exposed to tsunami sweeping across the Pacific, historically no part of this continent's coastline has been affected significantly by large tsunami. The closest sources for earthquake-generated tsunami lie along the Tonga-New Hebrides trenches, and in the Indonesian islands. An earthquake with a magnitude greater than 8.3 on the Richter scale can be generated in the southwest Pacific every 125 years. The highest tsunami to be recorded at Sydney since 1870, occurred on 10 May 1877 and had a height of 1.07 m.

The Chilean earthquake of 1868 produced a tsunami height of 1 m, while the 1960 Chilean earthquakes generated a tsunami height of 0.85 m. These heights are close to the maximum storm surge produced by mid-latitude coastal storms along this coast.

The path of this latter tsunami wave across the Pacific is shown in figure 11.13. This diagram illustrates the reasons why the east coast of Australia is rarely affected by large tsunami generated along the shorelines of the open Pacific. There is a line of islands and submerged ridges running from New Zealand north to Tonga, across to Fiji and New Caledonia, and north again to Vanuatu (New Hebrides) and Solomon islands, guarding the Tasman and Coral seas against the intrusion of tsunami waves. As any tsunami wave passes through these island gaps, it is dissipated and diffracted to the point that it rapidly loses energy coming into the Tasman Sea. The Chilean tsunami of 1960 actually entered the Tasman Sea from the south. Any long period wave entering the Tasman from this direction will tend to spread outwards as it moves north, because of the configuration of the Tasman Sea. Similarly, any wave coming from the east must first cross a rise running from New Zealand to New Caledonia before entering deeper water in the Tasman. Such a process causes long-period wave energy to be dampened. The presence of the Great Barrier Reef off the north Queensland coast protects that coastline from any tsunami wave entering the Coral Sea. This reef, plus others strung out eastwards across the Coral Sea, shield coastline southwards from tsunami which may originate from islands to the north.

On the west coast of Australia a different picture exists. It is possible that tsunami originating from Indonesia may affect the northwest coastline; however, the wider continental shelf tends to dissipate long-period waves crossing it. South of Shark Bay, the coastline deflects away from the travel paths of any tsunami spreading outwards into the Indian Ocean from Indonesia. Much of the coastline is also protected by an aeolianite reef, which parallels the shore several kilometers seaward. The eruption of Krakatoa in 1883 disturbed water levels worldwide but only produced a tsunami wave of 1.5–2.0 m north from Geraldton. There were no major accounts of damage. Apart from Indonesia there are no other source areas for tsunami in the Indian Ocean.

The above discussion does not discount the past existence of large tsunami along the Australian coastline. The proximity of the northwest coastline to the volcanically, and seismically active Indonesian archipelago makes large tsunami in the 10 m range a distinct possibility. Additionally, the east coast lies exposed to tsunami generated by earthquakes on seamounts in the Tasman Sea, and along the Alpine fault running down the west coast of the South Island of New Zealand. This latter fault has been significantly active during the Holocene, but has produced few seismic waves historically. Indeed, sporadic evidence does exist along the south coast of New South Wales to indicate that large magnitude waves have swept across the coastline. At Wollongong, a raised Holocene beach 2 m above present high tide is topped by isolated boulders which appear to have been swept shoreward along an adjacent platform, and thrown up over the beach. Large storm waves could have transported the boulders, but then such waves would probably have destroyed the underlying beach deposit. Further south at Haycock Point, a platform, also 2 m above present high tide, contains an assortment of angular boulders that have been wedged, and tightly stacked in a crevice at the back of the platform. Except for this material the platform is virtually devoid of debris. The boulders could not have fallen off the backing cliff and rebounded elastically into the crevice, because the drop from the cliff is too low. Nor is it likely that storm waves positioned the material, because the blocks are not rounded to the slightest degree. They lack the packing characteristics usually imparted by waves to boulder material as evidenced by boulder beaches along this coast. The above geomorphic evidence suggests that the historical high of 1.07 m for a tsunami wave at Sydney is much lower than the geologically recorded elevation for such waves along this coast.

Concluding comments

The seismic risk for many locations in the world can now be assessed to the point that many buildings and structures can be designed to withstand damage from tremors. However, two aspects have prevented this from happening in reality. The first is obvious in that the seismic risk in many locations, even in developed countries or ones where earthquakes have been recorded for centuries, cannot be assessed completely. For example, the Australian seismic risk map (figure 11.4) is very much a product of that country's distribution of cities. Most seismic recorders are located near those cities, and hence either do not adequately measure isolated earthquakes, or when they do, are not read because the earthquake has not been felt and thus reported by many people. Additionally, the seismic risk map for North America (figure 11.3) is still deficient because major earthquakes occur so infrequently in the eastern part of the continent that, even after centuries of settlement, the full extent of the hazard is not known.

This chapter has not highlighted recent earthquakes but concentrated upon two large events, the Alaskan earthquake of 1964 and the Tokyo earthquake of 1923, to show the power that can be released in a tremor and the destruction that can result. The Alaskan earthquake was one of the largest events this century. Even though it occurred in a relatively isolated part of the world, its magnitude and subsequent tsunami attracted world attention. The Tokyo earthquake is noteworthy because of the appalling deathtoll and the manner in which these deaths happened. The descriptions of these events could easily have been set within southern California, a region where earthquakes with the same magnitude as the Alaskan one can easily occur, and a region where the effects upon dense human settlement might be as dramatic. In discussing the California earthquake hazard, it was hoped to highlight the extent of the hazard and some of the inadequacies in people's attempts to minimize future damage and loss of life. This section was written before the San Francisco tremor of October 1989; but it took very little modification to work this event into the text. The San Francisco earthquake showed that the lessons of the San Fernando earthquake of 1971 had not been learnt fully. The southern Californian situation is not unique. The section could easily have been written for New Zealand, Japan or China. There are extensive regions prone to large earthquakes where people must live and adjust to this threat.

Finally, the discussion on tsunami was written to illustrate the fact that remote earthquakes can affect almost any coastal region in the world. Many of the areas where tsunami currently are recognized as a hazard were described. The discussion on the Australian tsunami hazard was included to illustrate just how extensive the threat from this hazard may be. Unfortunately, few scientists, politicians or planners in eastern Australia are aware of the threat or would give it credence, although some State emergency assistance groups now include it in their training programs. The Australian situation, again, is not unique. The lack of recognition of tsunami as a hazard could have been described just as easily for the east coast of the United States or even western Europe.

References

Bolt, B. A., Horn, W. L., MacDonald, G. A. and Scott, R. F. 1975. *Geological hazards*. Springer-Verlag, Berlin.

Denham, D. 1979. 'Earthquake hazard in Australia'. In Heathcote, R. L. and Thom, B. G. (eds) *Natural Hazards in Australia*. Australian Academy of Science, Canberra, pp. 94–118.

Hays, W. W. 1981. 'Facing geologic and hydrologic hazards: Earth-science considerations'. *United States Geological Survey Professional Paper* No. 1240–B pp. 54–85.

Holmes, A. 1965. *Principles of Physical Geology*. Nelson, London.

Keller, E. A. 1982. *Environmental Geology* (3rd edn). Merrill, Columbus, Ohio, pp. 133–67.

National Geophysical Data Center and World Data Center A for Solid Earth Geophysics. 1984. *Tsunamis in the Pacific Basin 1900–1983*. United States National Oceanic and Atmospheric Administration, Boulder, Colorado, map 1:17 000 000.

Oakeshott, G. G. 1983. 'San Andreas Fault: Geologic and Earthquake History'. In Tank, R. W. (ed.) *Environmental Geology*. Oxford University Press, Oxford, pp. 100–107.

Salsman, G. G. 1959. 'The tsunami of March 9, 1957, as recorded at tide stations'. *U.S. Coast and Geodetic Survey Technical Bulletin* No. 6.

Shepard, F. P. 1977. *Geological oceanography*. University of Queensland Press, St Lucia.

Stevens, A. E. 1988. 'Earthquake hazard and risk in Canada'. In El-Sabh, M. I. and Murty, T. S. (eds) *Natural and man-made hazards*. Reidel, Dordrecht, pp. 43–61.

Van Dorn, W. G. 1959. *Impulsively generated waves*. Scripps Institute Oceanography, Report No. II, Contract Number 233(35), (unpublished manuscript), cited in Wiegel (1964).

Wadati, K., Hirono, T. and Hisamoto, S. 1963. 'On the tsunami warning service in Japan'. *Proceedings Tsunami Meetings Associated with the Tenth Pacific Science Congress*, IUGG, Monograph No. 24, pp. 138–45.

Whittow, J. 1980. *Disasters: the anatomy of environmental hazards*. Pelican, Harmondsworth.

Wiegel, R. L. 1964. *Oceanographical Engineering*. Prentice-Hall, Englewood Cliffs, NJ, pp. 95–108.

Further reading

Blong, R. J. and Johnson, R. W. 1986. 'Geological hazards in the southwest Pacific and southeast Asian region; identification, assessment, and impact'. *Bureau Mineral Resources Journal of Australian Geology and Geophysics* v. 10 pp. 1–15.

Coates, D. R. 1985. *Geology and Society*. Chapman and Hall, NY, pp. 129–152.

Cornell, J. 1976. *The great international disaster book*. Scribner's, NY.

Hansen, W. R. 1965. 'Effects of the Earthquake of March 27, 1964 at Anchorage, Alaska'. *United States Geological Survey Professional Paper* No. 542–A.

Hodgson, J. H. 1964. *Earthquakes and Earth Structure*. Prentice-Hall, Englewood Cliffs, New Jersey.

Lida, K. and Iwasaki, T. 1981. *Tsunamis: their science and engineering*. Reidel, Dordrecht.

Lockridge, P. A. 1988. 'Historical tsunamis in the Pacific basin.' In El-Sabh , M. I. and Murty, T. S. (eds) *Natural and man-made hazards*. Reidel, Dordrecht, pp. 171–81.

Lomnitz, C. 1988. 'The 1985 Mexico earthquake'. In El-Sabh and Murty (eds) *Natural and man-made hazards*. pp. 63–79.

Murty, T. S. (ed.). 1977. 'Seismic sea waves: tsunamis'. *Bulletin Fisheries Research Board of Canada* No. 198.

Myles, D. 1985. *The Great Waves*. McGraw-Hill, NY.

Oliver, J. 1986. 'Natural hazards'. In Jeans, D. N. (ed.) *Australia: a geography*. Sydney University Press. pp. 283–314.

Scheidegger, A. E. 1975. *Physical Aspects of Natural Catastrophes*. Elsevier, Amsterdam.

Wood, R. M. 1986. *Earthquakes and Volcanoes*. Mitchell Beazley, London.

12

Volcanoes as a Hazard

Introduction (Bolt et al., 1975; Tazieff and Sabroux, 1983)

Of all natural hazards, volcanoes are the most complex. Whereas a tropical cyclone has a predictable structure, or a drought can generate a predictable sequence of events in rural communities, such predictions can not be made in respect of volcanoes. There are a multitude of volcanic forms, and almost each event appears unique in the way that it behaves, and the physical and human consequences it produces. This chapter will examine the different types of volcanoes and the secondary phenomena associated with their occurrence. It will conclude with a detailed description of some of the more spectacular volcanic disasters that have occurred in recorded history.

Volcanoes are conduits in the Earth's crust through which gas-enriched, molten silicate rock—magma—reaches the surface from beneath the crust. The origin of magma is still debated, but it is generally believed from seismic evidence that the mantle is partially liquefied 75–300 km below the Earth's surface. There are two types of magma. The first type consists of silica-poor material from the mantle, and forms *basaltic* volcanoes. The second type consists of silica-rich material originating from either the melting of the crust in subduction zones or the partial differentiation of liquefied mantle material. This second type forms 'andesitic' volcanoes, the largest group of which ring the western Pacific Ocean, where the Pacific crustal plate is subducting beneath the Eurasian plate (figures 10.1 and 10.7). Basaltic lavas have temperatures, on erupting, of between 1050 and 1200°C; while andesitic ones are 100–150°C cooler. For each 50°C decrease in temperature, fluidity decreases tenfold. For this reason, and because of their higher silica content, andesitic lavas are 200–2000 times more viscous than basaltic lavas. Andesitic lavas are also gas-rich, with 50–90 per cent of the gas consisting of water assimilated from dissolved crustal material. This high water content and low viscosity makes andesitic volcanoes, located over supposed subduction zones, much more explosive than basaltic volcanoes sitting over hot spots in the center of crustal plates.

Magma is not necessarily extruded at the Earth's surface by pressure originating in the mantle below. Instead, lava penetrates through tens of kilometers of weakened, fractured crust because of gas pressure. All magmas, be they andesitic or basaltic, contain various amounts of dissolved gas. These magmas can become over-saturated in gas due to crystallization of the magma, its cooling down, decreases in *hydrostatic pressure*, or enrichment of volatile molecules from outside the magma. The gas in over-saturated lavas begins to vesiculate or form bubbles, a process that immediately makes the magma lighter or more buoyant. If the magma is forced upwards quickly by pressure from below, then the hydrostatic pressure acting on the magma to dissolve these bubbles can decrease rapidly, causing a chain reaction of new bubble formation. If the magma is gas rich, as andesitic lavas are, and the chain reaction is rapid enough, an explosive eruption can occur. Once explosive magma reaches the atmosphere, it is fragmented into small particles by the gas and by the force of the eruption. These pieces are called *pyroclastic ejecta* or tephra and are classified according to size as follows: *bombs* (liquid, boulder-sized blobs), *blocks* (angular, solid boulders), *scoriae* (cinder material of any size), *lapilli* (fragments 2–60 mm in diameter), sands and ash (<0.1 mm). Depending upon the size and distance hurled, much of this material falls back to the ground around the vent to build up a cone.

The study of the gas phase of magma is crucial in determining whether or not the volcano will become explosive. If the magma is rising slowly, a small quantity of gas will continuously escape through vents to the surface through *fumaroles*. This movement tends to be slow, because the viscosity of the magma inhibits bubble release. Bubbles, once formed, tend to stick to liquid-solid interfaces no matter how buoyant they become. These gases also have sufficient time to dissolve in ground water. The monitoring of groundwater or escaping gas can give an accurate indication of the maturity of a magma, and its closeness to the surface. Some of these *geochemical* techniques for prediction purposes were discussed in chapter 10.

In some volcanic eruptions, only the upper magma is initially affected by this degassing process. As this magma is forced out, the magma immediately beneath degasses explosively. In this way, there may not be one big explosion, but a series of continual blasts, each lasting a few seconds-to-minutes, over several days. Degassed magma may also pour out of the volcano as lava which, depending upon viscosity, output and the slope of the ground, can flow from a few meters to several hundreds of kilometers from the vent. All lava soon solidifies to form hard rock, although extremely thick lava beds may take years to cool down completely to surrounding environmental temperatures. There are three types of basaltic lava that can be ejected from a volcano: pahoehoe, aa, and block lava. Pahoehoe is basaltic lava which solidifies at the top, but which is still fed from below by pipe vesicles. Such lava has a smooth, wrinkled surface, is less than 15 m thick and flows at rates of a few meters per minute. Aa forms as a thin river of lava, less than 10 m thick, with a spiky, cinder-like topping. Block lava consists solely of andesite or rhyolite up to 300 m thick, and has a surface coating consisting of smooth-sided fragments. It has the slowest travelling speed of all lava flows, moving only a few meters per day. Both aa and block lava behave in all respects like a river and can be bordered by levee banks.

The total *energy* of a volcanic event can be broken down into four categories as follows:

- energy released by volcanic tremors and earthquakes,
- energy necessary to fracture the overburden,
- energy expended in ejecting material,
- energy expended in producing atmospheric shock waves and, occasionally, tsunami.

Volcanic tremor and earthquake energy can be assessed in similar ways to any other earthquake event. These methods of measuring energy have been discussed under earthquake magnitude in chapter 10. Earthquakes of magnitude seven or larger on the Richter scale have been produced by major explosive eruptions, and contain more than 4×10^{15} joules of energy. The energy required to break up overburden is typically greater than 1×10^{11} joules. Energy spent in ejecting material consists of kinetic energy, potential energy and thermal energy. These terms can be calculated for most volcanic eruptions, and indicate that volcanic explosions release as much energy as large nuclear devices—typically 1×10^{15} joules. The energy used in generating the atmospheric shock wave can be expressed as follows:

$$E_s = 1.25 \times 10^{13} \sin \phi \, \Sigma \, 0.5 \, At^2 \quad (12.1)$$

where E_s = energy to generate atmospheric shock wave
$\sin \phi$ = distance from source of the explosion in degrees on the Earth
A = amplitude of shock wave in millibars
t = wave period in seconds

A similar type of equation as 12.1 can be used to measure the energy contained in any tsunami wave generated by an eruption. These shock waves require energy expenditure greater than 1×10^{15} joules. In total, volcanic eruptions contain energy in excess of 1×10^{16} joules. Some of the larger eruptions, such as Krakatoa in 1883, or Tambora in 1815, contained energy in excess of 1×10^{18} joules (see figure 12.1 for all major place names mentioned in this chapter and not located on individual site maps). By calculating the amount of energy partitioned amongst these various processes, it is possible to classify the type of volcano and its magnitude. This is of little help in predicting volcanic events, but it can give information on the type of eruption that has occurred. For instance, eruptive volcanoes expend considerable energy in easily measured atmospheric shock waves and tsunami. Knowledge of these values can be used to evaluate the overall energy of the volcano. In addition, the continual monitoring of energy expenditure, and the way it is partitioned can be used to predict whether a volcano is dying down, or increasing in activity.

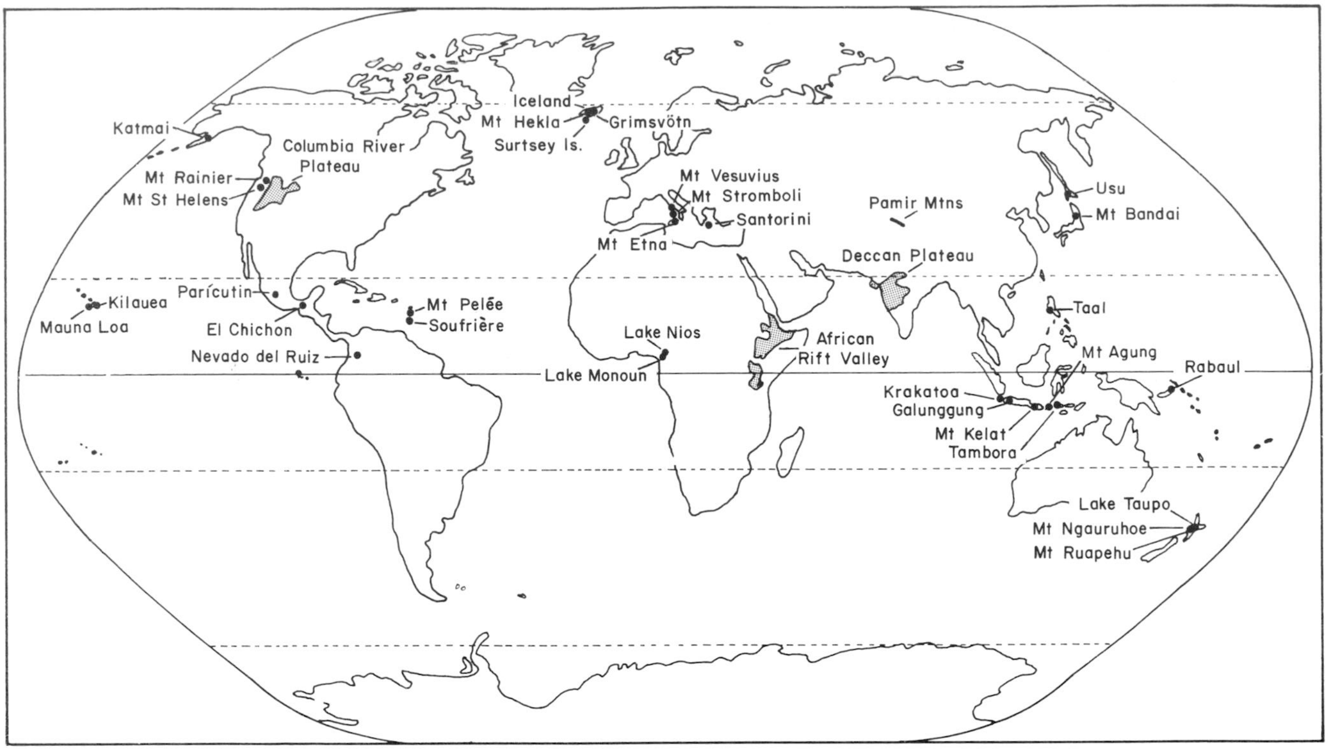

Fig. 12.1 Location map

Types of volcanic eruptions

(Fielder and Wilson, 1975; Blong, 1984; Wood, 1986)

While most classification schemes refer to a specific historical event characterizing an eruption sequence, it should be realized that each volcanic eruption is unique, and may over time take on the characteristics of more than one type. There are two types of non-explosive volcanoes. Volcanoes producing *flood lavas* are the least hazardous of all eruptions, because they occur very slowly, permitting plenty of time for evacuation. Flood lavas are basaltic in composition. At present, such eruptions are rare, only occurring in Iceland, where they happen with sufficient warning to permit safe evacuation. In the geological past, some volcanoes have had enormous magma chambers that emptied and flooded landscapes over vast distances in a short period of time. Some flood lavas have laid down deposits 500 m thick over distances of 300 km. The largest such deposits on Earth (flood lavas also occur on other planets and their moons) make up the Deccan plateau in India, the island of Iceland, the east African rift valley and the Columbia River plateau in the United States. 'Hawaiian'-type eruptions are very similar to flood lavas. However, at times they can produce tephra and faster moving, thin lava flows. Hawaiian eruptions tend to be aperiodic, building up successive deposits over time. These can form large cones, which, in the case of the Hawaiian Islands, can be tens of kilometers high and hundreds of kilometers wide at their base.

Explosive eruptions can be more complex. The simplest form begins with the tossing out of moderate amounts of molten debris. 'Strombolian' volcanoes, named after Mount Stromboli in Italy, throw up fluid lava material of all sizes. Most eruptions of this type toss out bombs a few hundreds of meters into the air, rhythmically, every few seconds in events that can go on for years. They can also produce moderate lava flows. Strombolian volcanoes are characterized by a very symmetrical cinder cone that grows in elevation around the vent. Strombolian volcanoes also represent an intermediate stage between basaltic and acidic magma extrusions. 'Vulcanian'-type eruptions are more acidic, and give off large blocks of very viscous, hot magma with few if any flows. This very viscous material may be tossed up to heights of 10 km (40 km in extreme cases). Activity is explosive for periods ranging up to several months. These volcanoes build up tephra and block cones. Usually the lava freezes in vents between eruptions, so that the strength of the plug ultimately determines the amount of pressure that can build up before the next eruption. 'Surtseyan' eruptions are violently explosive, because hot, fluid magma comes in contact with sea water. These types of eruptions are termed 'phreatomagmatic' and make up 8.8 per cent of all volcanic events. Because of the interaction of magma and water, large clouds of very fine dust are produced that can drift several kilometers upwards into the atmosphere. Coarser particles tend to form rings around the vent, rather than cones or large ash sheets. The rubbing together of fine dust particles builds up static electricity, leading to numerous and spectacular lightning displays around the rising column of ejecta.

'Plinian' events are more explosive, with wide dispersal of tephra. Over 1 km^3 of magma may be driven straight up, at velocities of 600–700 $m\,s^{-1}$, to heights of 25 km by a continuous gas jet and thermal expansion over the course of several hours. As the magma chamber slowly empties, the character of the magma changes, and the eruption becomes starved for material. Usually, the volcano then collapses internally under its own weight to form a caldera. The Mount Vesuvius eruption of AD 79 was of this type, and will be described later in more detail. While such events have been dramatic on Earth, on other planets with lower gravity and thinner atmospheres, Plinian eruptions have been truly cataclysmic.

If the eruptive ash column from an explosive volcano collapses under the effect of gravity, it can produce a debris avalanche of ash and hot gas that can race downslope at speeds reaching 60 $km\,hr^{-1}$. This type of eruption is termed a 'Peléean' eruption after the explosion of Mount Pelée in 1902. The debris clouds are called pyroclastic flows or *nuées ardentes*, and the resulting debris deposits are termed *ignimbrites*. The large deathtoll of Vesuvius in AD 79 can be attributed to an associated pyroclastic flow event. Pyroclastic flows will be examined later in this chapter when secondary volcanic hazards are examined in detail. The most destructive and largest Peléean eruptions are termed the 'Katmaian' type, after the Katmai eruption at the eastern end of the Alaska peninsula on 6 June 1912. This eruption ejected ten times the material of the 1902 Mount Pelée event. As the magma chamber emptied, a 2500-metre high mountain collapsed into a caldera 5 km across and 1 km deep. The eruption was heard 1200 km away, with dust being deposited over a 1500 km distance. The pyroclastic flow from this eruption flooded into a valley 20 km long, leaving an estimated 11 km^3 of hot, fused ash spread to a depth of 250 m.

Peléean eruptions should not be used to refer to all volcanic events that generate hot ash flows. Many destructive pyroclastic flows have not originated as ash columns collapsing under the effect of gravity, but have been preceded by a ground or basal surge caused by the lateral eruption of the volcano. This type of explosion is termed a 'Bandaian' eruption. The 1980 eruption of Mount St Helens in the United States was of this type, which is more properly known as a *base surge*. Because the force of the eruption shoots out particles at velocities in excess of 150 $km\,hr^{-1}$, the blasts have been likened to cyclones of ash. Objects in the path of such hot and violent blasts are destroyed by sandblasting. Base surges often precede a large and slower moving pyroclastic flow, and represent the most destructive of all types of volcanic eruptions.

Volcanic hazards (Bolt et al., 1975; Whittow, 1980; Hays, 1981; Tazieff and Sabroux, 1983; Blong, 1984; Wood, 1986; Symons et al., 1988)

There are many hazardous phenomenon produced directly, or as secondary effects by volcanic eruptions. The association of earthquakes and tsunami with volcanoes has already been covered in chapter 11, and will be discussed in detail for particular events at the end of this chapter. Excluding these two phenomena it is possible to group volcanic hazards into six categories as follows:

- lava flows,
- ballistics and tephra clouds,
- pyroclastic flows and base surges,
- gases and acid rain,
- lahars (mud flows), and
- *glacier bursts* (Jökulhlaups).

Lava flows

Since the sixteenth century, 20 km^2 of land per year has been buried by lava flows. There are about 60 lava-flow events per century which result in severe social and economic disruption. The main reason for this disruption is the fact that lava flows in volcanic areas are very fertile, and have attracted intense agricultural usage and dense settlement. Re-eruption of volcanoes in these areas has, thus, led to the destruction of this agricultural land and to loss of life. Much of this destruction can be avoided simply by mapping and avoiding those areas most frequently inundated by lava flows. Techniques also exist to modify flow behaviour, especially for pahoehoe and aa flows.

Based upon the more than 1000 flows which have occurred historically, low viscosity flows have a median length of 4.1 km and an average thickness of 10 m, while high viscosity flows have a median length of 1.3 km and an average thickness of 100 m. There is a 1 per cent probability that high and low viscosity lava flows will exceed respectively 11 km and 45 km in length. The fastest moving, hottest and most mobile flows are the low viscosity Hawaiian- and Icelandic-type eruptions. These flows would appear to be dangerous; however, they are thin and often consist of pahoehoe-type lava, which cools very rapidly. Pahoehoe lavas may develop large diameter lava tubes extending several kilometers under the flow. These can provide conduits for subsequent lava venting once the original flow has cooled at the surface. Such flows can travel great distances, but are the easiest to divert. Aa flows are thicker and more viscous. They tend to channelize to depths up to 30 m. Both aa and block lava flows are more difficult to divert.

The characteristics of flows can vary depending upon temperature, viscosity, yield strength and expulsion rate. Because of the high temperatures of flows (880–1130° C), all carbon materials such as cloth, paper and wood are easily ignited upon contact with lava. Once temperatures drop below 800° C, a skin forms on the magma which effectively inhibits further heat loss. Lavas have been known to have high internal temperatures 5–6 years after they have stopped flowing. As temperature decreases, viscosity increases, and it is this factor which stops most flows. While the velocity of a flow depends upon gravity or the slope of the terrain, the rate at which lava is expelled also controls the speed of movement. Most expulsion rates range between 15–45 $m^3 s^{-1}$, but values above 1000 $m^3 s^{-1}$ have been measured. The distance reached by lava is a direct function of this expulsion rate, such that the distance of travel in kilometers is about three times the expulsion rate measured in $m^3 s^{-1}$. The actual velocity of flows is in a range of 9–100 km hr^{-1}, with average values around 30 km hr^{-1}.

While most documented flows rarely exceed 4 km^3 in volume, some have caused impressive damage. The Laki fissure eruption of 1783–84 in Iceland covered 550 km^2, had a length of 88 km and a total volume of 12.3 km^3. This eruption caused melting of glaciers and flooding of valuable farmland. The sulphurous vapours from the lava hung over the countryside for weeks, destroying livestock and crops. The 'haze famine' that resulted killed 20 per cent of Iceland's population, and wiped out 50 per cent of its cattle and 75 per cent of its sheep and horses. Extensive flows sent down the slopes of Mount Vesuvius engulfed the town of Bosco Trecase at its base in 1906, and in 1944 the town of San Sebastiano. Lava from Mount Etna in 1669 reached the wall of the feudal city of Catania 25 km away, eventually overtopping it and moving slowly through the city. In 1928 the town of Mascali was also destroyed by lava from Mount Etna. Parícutin in Mexico, in 1946, buried 2400 hectares of forest and agricultural land before completely covering the town of San Juan Parangaricutiro, 5 km away from the main vent. Eruptions from Mauna Loa and Kilauea in Hawaii have consistently flooded agricultural land, destroyed small villages surrounding the cones, and threatened the major city of Hilo on several occasions.

The easiest way to stop lava movement is to enhance the processes that slow it down. Flow depends upon yield strength, which can be increased by lowering the temperature of the melt, increasing the rate of gas escape from the flow, stirring, and seeding the lava with foreign nuclei. In some cases, increasing the *yield strength* of the flow may only turn a fluid flow into a more viscous one, making it thicker and more difficult to stop. Because of the insulating properties of solid lava, cooling by water is not always effective. It has been observed that lava flows can continue to move even when submerged under the sea. One notable exception appears to be the attempts in 1973 to stop a block lava flow which threatened the town of Vestmannaeyjar in Iceland. Water was intensively pumped for months onto, and just behind the moving lava front, thus increasing the viscosity of the magma and slowing its rate of progress. Agitation of the flow can be difficult to achieve except by bombing. This was tried on one tube feeding a pahoehoe flow which threatened Hilo, Hawaii, in 1935. The flow was disrupted enough that it turned into a more viscous aa flow, and the tube became blocked. The bombing might not have been the reason for the flow termination, because it was already in the final stages of expulsion. However, there are still contingency plans for bombing lava flows if they threaten Hilo in the near future. The sides of a flow can also be breached, thus diverting lava to a safer path. However, what one person considers safe, another may consider a threat. This problem beset the unfortunate inhabitants of Catania in the 1669 eruption of Mount Etna. The citizens of the walled city of Catania made the first recorded attempt to alter a lava flow. They clad themselves in wetted animal skins, and dug into one of the lava levees controlling the route of the flow towards their city. Unfortunately, the diversion sent the flow towards the town of Paterno. Five hundred enraged residents of this latter city raced up the slopes to attack the citizens of Catania, and terminated their efforts after winning a pitched battle. The lava flow clogged the breach, proceeded to Catania, overflowed its 20 m high walls, and inundated large sections of the town. A similar attempt in Hawaii in 1942, rather than diverting the flow, only sent it on a parallel path.

Barrier construction has also been used to divert aa flows, or give more time for effective evacuation. Most lava flows will pond behind the flimsiest obstacle, and then simply overtop it, as eventually occurred in Catania. Barriers should not be used to stop a flow, but simply to divert it. Because lava flows can become very viscous as they cool down (10^6 times more viscous than water), they may only be flowing over very low grades and exerting very little force. They can, thus, be diverted successfully by putting guiding barriers diagonally in their path, as long as the diversion route is steep. However, the rapid movement of some flows rules out time-consuming barrier construction. Even construction of barriers long distances in front of the advancing lava may be impractical, because the flow could easily stop naturally before reaching the barrier. Barriers must also consist of denser material than the lava, be attached to the ground, or be three times wider at their base than the flow thickness itself. Less dense material will simply be buoyed up by the lava and incorporated into the flow. If barriers only divert a flow, there is always the problem of finding a safe diversion path that does not affect someone else's property or life. Only additional study in the mechanics and behaviour of lava flows will permit barrier construction to be used effectively.

Ballistics and tephra clouds

The eruption of Vesuvius in 1944 illustrates another major hazard of volcanic eruptions. At that time the Allied war effort in the area was severely hampered by the bombing of airfields, not by the Germans, but by liquid lava blobs tossed out by the volcano. Strombolian-, Vulcanian-, Surtseyan- and Plinian-type eruptions all shoot out ash to heights greater than 30 km. The larger particles consist of boulder-sized blobs of fluid magma and remnant blocks of the volcano walls. Measured velocities are in the range of 75–200 m s^{-1}, and maximum distances are obtained with trajectory angles of 45°. Measured distances for boulder material rarely exceed 5 km, although material weighing 8–30 tonnes has been projected over distances of 1 km or more. This type of debris can be voluminous and hot, and can fall over a small area. It can also be extremely destructive. The density of projectiles varies considerably so that the kinetic energy of impact is wide ranging. Larger bombs can behave exactly as human-made projectiles in their explosive effect when hitting the ground. Panoramic photographs of vegetated landscapes subject to sustained projectile bombardment look exactly like war bombing scenes.

The production of tephra or ash can also be just as destructive (figure 12.2). The eruption of Mount Hekla, Iceland, in 1947 ejected 100 000 m^3 s^{-1} of material, dropping ash as far away as Finland two days later. Tephra can rise at velocities of 8–30 m s^{-1}, with drift rates downwind of 20–100 km hr^{-1}. While much of this dust falls out locally, fine material <0.01 mm in size can be thrown up to heights in excess of 27 000 m well into the stratosphere. Here, residence times may last two or more years. The eruption of Tambora in Indonesia on 5–10 April 1815 is the largest recorded tephra eruption. It blasted 151 km^3 of material into the atmosphere as fine ash, which was responsible for a cooling of the Earth's surface

FIG. 12.2 Tephra ejecta from Mount St Helens, 22 July 1980 (photograph courtesy of the United States Geological Survey, source Hays, 1981). Pyroclastic flows generated the lower ash clouds (*to the right*). Note the spreading of the ash cloud at the tropopause and the fallout of coarse particles downwind.

temperature by 0.5–1.0°C. In comparison, the largest recorded landslide, at Usoy in the Pamir Mountains of what is now the Soviet Union, in 1911, contained only 2.5 km^3 of material. As already mentioned in chapter 10, volcanic dust exerts a dramatic control on the Earth's temperature, with many of the changes in temperature over the last three centuries correlating well with fluctuations in volcanic dust production.

The area covered by tephra deposits can be considerable because most tephra falls from atmospheric suspension within a short distance of the volcano. Fine ash settling out from Krakatoa in 1883 covered an area of 800 000 km^2. This volcanic material can be extremely fertile once incorporated into the soil; as a fine dust, however, it can have severe environmental consequences. Only a small proportion of volcano-related deaths can be attributed directly to ash fallout; nevertheless, ash fallout 70–80 km from Krakatoa was still hot enough to burn holes in clothing and vegetation. Tephra also can contain toxic fluoride compounds, which are very toxic to animals attempting to eat contaminated fodder. The 'haze famine' of 1784 in Iceland, and the death by famine of 80 000 people following the eruption of Tambora in 1815, can both be attributed to the destruction of vegetation by tephra. The glassy tephra from Mount St Helens in 1980 had a severe impact on road conditions and on motor vehicles. Tephra, from the 1982 eruption of Galunggung volcano in western Java, almost brought down two 747 jet passenger planes bound for Australia in separate incidents. The initial volcanic eruption had not been noticed or reported to aviation authorities. The first plane to succumb was a British Airways 747 flying between Singapore and Australia, when it entered the ash cloud at 11 000 m. All four engines cut out, one after the other, sending the plane into a 16-minute silent descent, through 7500 m, before the startled pilot could restart three of the engines. Upon landing, it was found that the cockpit windscreen had been pitted by its impact with the ash. Two weeks later, a Singapore Airlines 747 flying the same route unwaryingly strayed into another ash cloud and dropped 2500 m before the pilot regained control. Only after this second incident was a warning issued for planes to avoid the area.

Pyroclastic flows and base surges

Mention has already been made of the collapsing ash columns—termed 'nuées ardentes' or 'pyroclastic flows' (figure 12.3)—and the basal surges produced by the lateral explosion of volcanoes. Base surges are akin to the lateral blast waves accompanying a nuclear explosion, and can pick up considerable dust in addition to carrying the ash originating from the volcano itself. The Bandai eruption of 1888 saw a 300-metre high volcano pulverized and converted to a debris avalanche in this manner. The blast wave can reach speeds in excess of 150 km hr^{-1}, and the dust can sandblast objects. The Taal eruption in the Philippines in 1965 tore 15 cm of wood from the sides of trees within 1 km of the eruption. The basal surge may

FIG. 12.3 Beginning of a pyroclastic flow or nuée ardente on Mount Ngauruhoe, New Zealand, in January 1974 (photograph courtesy of the United States Geological Survey, source in Catalogue of Disasters #0401-09j). Tephra has been blasted vertically only to collapse under its own weight. The pyroclastic flow consists of ash suspended in hot carbon dioxide, and can be seen as the rapidly moving cloud closely hugging the ground (*to the left*).

be followed by a pyroclastic flow. The Mount St Helens eruption followed this sequence. The classic description of a pyroclastic flow is 'an avalanche of an exceedingly dense mass of hot, highly gas-charged and constantly gas-emitting fragmental lava, much of it finely divided, extraordinarily mobile, and practically frictionless, because each particle is separated from its neighbors by a cushion of compressed gas' (Perrett 1935). Not only can the blast from the flow be destructive, but the flow can be extremely hot. Pyroclastic flows are also called 'glowing avalanches' because of the presence of heated debris within the flow. The Mount St Helens flow produced temperatures around 350°C near the source and 50–200°C near the margins of the flow. These temperatures were sustained for less than two minutes. Temperatures in the 1902 Mount Pelée flow were as high as 1075°C, as evidenced by the partial melting of gold coins and bottles. Temperatures can vary as much as velocities do. For example, the Mount Vesuvius flow of AD 79 was hot enough to carbonize wood in places, while food in other locations was left uncooked. It is difficult to say whether this variation represents differences between several pyroclastic flows, the effect of fires ignited by the flow, or possible lightning effects accompanying the flow. The deposits from pyroclastic flows can remain hot for considerable periods of time. Temperatures in excess of 400°C were found within one deposit from the Augustine, Alaska, eruption of 1975, several days after the event.

Pyroclastic flows can also travel at high speeds because they behave like a density flow under the effects of gravity. Velocity measurements of small ash flows range between 10 and 30 $m\,s^{-1}$, while larger flows can obtain speeds of 200 $m\,s^{-1}$. At these velocities, the flows can override high topographic barriers. Turbulence can be generated over topographical obstacles and as gases change temperature. The destructive force of the flow can thus vary over short distances. The Mount Vesuvius pyroclastic flow of AD 79 knocked over walls and statues in some places, while short distances away furniture inside buildings was left undisturbed. The Mount Pelée flow of 1902 refracted around topographic obstacles, and appeared to increase in velocity seawards. Both basal surges and pyroclastic flows are exceedingly deadly because of their speed, and the fact that they are mixed with large amounts of hot, toxic gases. More mention will be made later of their deadly impact when detailed descriptions of the Vesuvius and Pelée eruptions are presented.

Historically, occurrences of pyroclastic flows have covered minor areas. The Katmai eruption in Alaska in 1912, one of the biggest flows measured, only covered an area of 126 km^2, while the Mount St Helens pyroclastic flows extended over 600 km^2. The Rabaul eruption, 1400 years ago, produced a flow 2 m deep over 1200 km^2. Eruptions in the Taupo volcanic zone of New Zealand are the largest to have occurred on Earth. There is geological evidence that Lake Taupo has discharged pyroclastic material at the rate of 1 000 000 $m^3\,min^{-1}$. The AD 186 eruption of Taupo produced 60–100 km^3 of tephra, an amount five times greater than that produced by Krakatoa in 1883. The tephra covered all of the North Island of New Zealand with at least 10 cm of ash, while intense pyroclastic flows overran 1000-metre high ridges around the crater and flooded an area of 15000 km^2.

From ignimbrites—deposits resulting from pyroclastic flows—there is some evidence that the flows have climbed ridges 500–1000 m high up to 50 km from the source of eruptions. At these high velocities, the flows can also be erosive gouging out channels in less resistant sediments. Ignimbrites produced by Peléean volcanoes are distinct from ash fallout deposits resulting from Plinian eruptions. Because particles settle from suspension in air, tephra deposits from the latter tend to be well sorted and spread evenly over the landscape. Ignimbrites on the other hand consist of particles of all sizes that have been transported with air as the interstitial medium. They are made up of volatile and gas-rich magmas of a rhyolitic or andesitic nature, are chaotically sorted, and tend to pool in depressions because their flow is controlled by gravity. The heat in a pyroclastic flow can weld ignimbrite material together, whereas Plinian deposits remain unconsolidated. Some ignimbrites tend to have an inverted size grading in which finer material is overlain by coarser blocks. For this to happen, the cloud must be very dense, travel at high velocities and be thin. Such a process is called *shear sorting*, and is analogous to large lumps in a sugar bag coming to the surface as the bag is shaken. Alternatively, these well-sorted deposits may represent evidence of base surges, which usually produce thin, well-sorted beds less than 10 m thick. Ground surges, thus, appear to precede a large and slower moving pyroclastic flow, resulting in chaotically sorted ignimbrites being deposited on top of well-sorted ground surge deposits. Because well-sorted ignimbrites are plentiful, basal surge events may be more common than previously thought.

Gases and acid rains

Many pyroclastic flows were believed at first to consist solely of hot gases. While this has been found not to be true, gases cannot be ruled out as a hazard. Gases in eruptions can be transported as acid aerosols, as compounds adsorbed on tephra particles and as salt particles. These gases do not come solely from explosive volcanoes. Passively degassing volcanoes release annually into the atmosphere 0.05–4.7 million tonnes of fluorine and 0.3–10 million tonnes of chlorine as, respectively, hydrofluoric and hydrochloric acid. The richest magmas may contain 0.5–1.0 per cent chlorine. The more explosive the eruption, the greater the transport of gas on adsorbed particles. This is very common for tephra clouds, which can scavenge gases and acids from the erupting plume. Hydrochloric, sulphuric, carbonic and hydrofluoric acids can be produced in the presence of water. These acids are easily leached out of the atmosphere by light rains. The hydrofluoric acid content in the Hekla, Iceland, eruption of 1970 reached 1700 ppm. Above 600° C, calcium fluorsilicate forms on the glassy surfaces of tephra particles. These particles, being fine, stick to, and subsequently burn vegetation. This aspect poses a serious problem to farmers following most Icelandic eruptions, which are notoriously high in fluorides.

Volcanoes giving off sulphurous fumes over a long time have done more damage to crops in adjacent areas by this means than by their own eruptions. The Katmai eruption of 1912 dispersed an acid rain that damaged clothes hung out to dry in Vancouver 2000 km to the south. Children in Anchorage were issued with gas masks as a precaution against the fumes. Carbon monoxide is especially dangerous, being toxic to mammals in low quantities, and it inhibits leaf respiration on plants. Even CO_2 can prove harmful. Volcanoes can produce large quantities of this gas (20–95 per cent of the gas discharge), which, being denser than air, can pool in depressions and suffocate livestock. On 21 August 1986 these gases plus sulphur dioxide and cyanide were released from the bottom sediments of Lake Nios, a dormant volcanic crater in Cameroon, Africa, by an earthquake or small eruption. Clouds of deadly gases hugged the ground and flowed under gravity down topographic lows killing all animal life in their path. Over 1500 people died within a matter of minutes, while 10 000 survivors suffered skin burns. A gas discharge in neighbouring Lake Monoun, Cameroon, killed 37 people in 1984, while a similar disaster on the Dieng plateau in Java in 1978 killed 180 people.

Of all gases produced by volcanoes, hydrochloric acid (HCl) is the most interesting given present day concern about depletion of the Antarctic *ozone hole* by free chlorine during the southern hemisphere spring. The speculated increasing growth of the hole has been attributed to chlorine derived from chlorofluorocarbons (CFCs). Approximately 2.28 million tonnes of chlorine per year were manufactured as CFCs and other halogens before the Montreal Protocol of 1987 limited production. In contrast, up to 10 million tonnes of chlorine as HCl can be released by passively degassing volcanoes in years with no great volcanic eruptions. For instance, Mount Erebus in the Antarctic in 1983 released 370 000 tonnes of chlorine, equivalent to the entire world's production of chlorine from CFCs that year. Other more recent volcanoes to have released significant amounts of chlorine as HCl are Agung, Bali (1.5 million tonnes), in 1963; Soufrière, Guadeloupe (1 million tonnes), in 1979; and Augustine, Alaska (525 000 million tonnes), in 1976. This passive release of hydrochloric acid has the same probability of reaching the stratosphere as heavier molecules of CFCs. Single explosive eruptions are even more voluminous in the amount of chlorine they can inject directly into the stratosphere. For instance, Tambora in 1815 and Krakatoa in 1883 ejected a minimum of 210 and 3.6 million tonnes of chlorine as HCl respectively. While much of this gas can be precipitated out of the atmosphere through condensation, scavenged by ash or dissolved in water-rich eruption plumes, there is evidence from recent eruptions that a significant fraction of chlorine as HCl enters the stratosphere as a gas. The eruption of El Chichon, Mexico, in 1982 released only 40 000 tonnes of HCl into the stratosphere; however this resulted in an estimated 40 per cent increase in stratospheric HCl between 20°–40° N latitude. Volcanoes are certainly significant contributors to the global chlorine flux and must be considered together with CFCs and halogens as a possible mechanism of stratospheric ozone depletion.

Lahars

One of the more unusual, though still hazardous, phenomena produced by volcanoes consists of lahars, or mud flows, that can occur at the time of the eruption (primary lahars) or several years afterwards (secondary lahars). About 5.6 per cent of volcanic eruptions in the last 10 000 years have produced mud flows at some time. Primary lahars can be generated by pyroclastic flows or by crater lake eruptions. In the former case, a pyroclastic flow can easily entrain water from streams and rivers as it moves down topographic lows. In the process, the gas-rich flow is slowly converted to a fast moving, heated mud-flow as more water is entrained in the mix. The Toutle River lahars (see figure 12.9) from Mount St Helens in 1980 had this origin. Volcanoes with crater lakes can produce mud flows at the time of any eruption, if the crater lake is ruptured. The size of the mud flow is then related directly to the volume of water in the lake. The 1919 eruption of Mount Kelat on Java expelled water from a crater lake, which covered 200 km^2 of farmland and killed over 5000 people. Attempts have been made to control the Kelat situation by drilling tunnels through the walls of the crater to lower the level of the lake. The first attempt was in 1929, when Dutch engineers reduced the lake volume from 21 to 1 million cubic meters. A subsequent eruption in 1951 blocked the tunnels and, even after repairs, an eruption in 1966 generated lahars that took hundreds of lives. Indonesian engineers have since replaced the tunnels and drained the lake completely to negate the threat of future lahars.

Secondary lahars are caused by freshly fallen, uncompacted tephra in depths in excess of 20 cm following rainfall. Such water-soaked material is very unstable, and can move downslope as a mud flow which entrains all loose debris in its path. If acids have not been leached out of the tephra, then lahars can become acidic enough to cause serious burns. Such flows have covered enormous areas. One on Mount Rainier in Washington State 5000 years ago had a volume of 1.9×10^9 m^3 and flowed 60 km from the summit. Other lahars preserved in the geological record in the United States have extended over 5000–31 000 km^2.

Because lahars are restricted to topographic lows, they can be predicted in advance. The buildup of tephra on a volcano's slopes can be easily measured, so that steps can be taken to evacuate people in threatened areas. The Galunggung eruption mentioned previously led to the evacuation of 35 000 people because of a lahar threat. A special Indonesian conference was held in September of 1982 to plan for the expected lahars that would occur in the following monsoon season. Fortunately, the monsoon season occurred in a drought year coinciding with the 1982–83 El Niño-Southern Oscillation event which, as mentioned in chapter 5, was one of the most intense on record. It is even possible to determine the rainfall required to initiate a secondary lahar. After the Usu eruption in 1977 in Japan, it was calculated that tephra deposits thicker than 0.5 m on slopes longer than 300 m, and steeper than 17–18 degrees, would require only 20 mm of rain at a rate of 10 mm hr^{-1} to initiate lahars. The 86 lahars that occurred afterwards closely followed these predictions.

Japan has also tried to inhibit lahar movement. In the most susceptible areas, numerous concrete barriers have been constructed down valleys to brake or deflect lahars from reaching settlements downslope. In New Zealand, on the Whakapapa ski-field situated on the northwest slopes of Mount Ruapehu, seismometers have been installed to detect small earth movements indicating the passage of a lahar on slopes where three other flows have been detected this century. Above a threshold signal of 3.8 on the Richter scale, alarms automatically go off on the ski-fields giving skiers 5–10 minutes to ski out of valleys. Downstream electronic triggering devices set off alarms in Whakapapa village, and a pre-recorded message warns inhabitants that possible lahar movement has been detected. Other devices set off warning signals that prevent trains on the main Wellington-Auckland railway line from entering the area, and initiate procedures to prevent mud flows from entering the Tongariro power scheme. While lahars are very common and can reach velocities in excess of 20 km hr^{-1}, they can be prevented in some cases. Unfortunately, lahars can also be initiated by erosion of old tephra deposits years, or decades after the original eruption has taken place. For instance, Mount Rainier was affected by a lahar in 1947 even though no volcanic eruption has been recorded historically.

Glacier bursts

Of particular danger during a volcanic eruption is the melting of glaciers or snowfields by hot lava or gases. The water can be subsequently heated and mixed with mud debris to form a steaming hot lahar. These events are termed 'glacier bursts', or 'Jökulhlaups' (an Icelandic word with the same meaning). Glacier bursts are restricted mainly to Iceland and the Andes; however, they can occur elsewhere. For example, the Mount St Helens lahars were probably exacerbated by melting of snow and ice by hot gases. In November 1985 over 20 000 people were killed by a glacier burst mud flow that swept off the Nevado del Ruiz volcano in Columbia. This was the worst such event this century. In some cases, in the matter of a few hours, the volume of meltwater discharging down streams may exceed the discharge of the greatest rivers in the world. In Iceland two icecaps, the Myrdalsjökull and the Vatnajökull, lie overtop active volcanoes. The Vatnajökull glacier caps the 40 km^2 Grimsvötn caldera, where hot gases are continually escaping. The thickness of these glaciers prevents heat escape, while their weight ponds water from flowing out of the caldera. Meltwater builds up slowly beneath the glacier until water depths are sufficiently deep to float the glacier. At this point, about once every five years, the meltwater escapes and flows beneath the glacier, coming out at its snout. In 1918 the discharge of a Jökulhlaups from the Myrdalsjökull icecap exceeded three times the discharge of the Amazon River, the largest flowing river in the world. The Katla volcano in Iceland has had glacier bursts from the Vatnajökull glacier exceeding 92 000 m^3 s^{-1}, with a total volume in excess of 6 km^3. The latter floods over time have formed an outwash plain 1000 km^2 in area. Events in Iceland are unusual, but predictable to a certain extent. By monitoring the height of the glacier, it is possible to forecast an impending event.

Water discharge can also accompany volcanic eruptions in areas without snow or ice. The 1902 eruptions of Mount Pelée and Soufrière in the West Indies were both accompanied by massive flooding of dry rivers. The 1947 eruption of Mount Hekla, Iceland, led to a discharge of 3 000 000 m^3 of water, which could not be attributed solely to melting of snow or ice. These latter events suggest that water may be expelled from the water-table by escaping gases during some volcanic eruptions. Nor do these discharges have to accompany the eruption. In 1945 the Ruapehu volcano in New Zealand erupted, emptying its crater lake. This lake slowly refilled inside a newly formed crater that became dammed by glacial ice. In 1953 an ice cave formed that began lowering the crater; however, on Christmas Eve of that year the crater wall suddenly collapsed near the cave causing the lake to drop 6 m in two hours at the rate of 900 m^3 s^{-1}. The floodwater swept down the Whangaehu River picking up mud and boulders, and forming a lahar. Two hours later at 10:00 pm, the debris-laden water swept away a section of the Tangiwai Rail bridge, three minutes before the night train from Wellington to Auckland arrived; the train plunged through the gap, killing 151 people.

Volcanic disasters

Volcanoes are visually one of the most spectacular natural hazards to occur, and probably one of the most devastating in terms of complete loss of human life. Even the Sahelian droughts, or earthquakes that completely flatten cities, leave more survivors than dead. Volcanoes have wiped out entire populations. There were only two survivors out of a population of 30 000 in the town of St Pierre, Martinique, following the Mount Pelée eruption of 1902. Four volcanic events stand out in the historic record. Three of these—Mount Vesuvius, Krakatoa and Mount Pelée—are significant because of either the enormity of the eruption, or the resulting deathtoll. The other, Mount St Helens, stands out because it represents a major eruption beginning a period of volcanic activity at the beginning of the 1980s.

Vesuvius (25 August, AD 79) (Bolt et al., 1975; Whittow, 1980; Sigurdsson et al., 1985; Blong, 1984)

Mount Vesuvius lies on the southeast corner of the bay of Naples on the west coast of Italy (figure 12.4). The Roman towns of Pompeii and Herculaneum were situated at the base of the southwestern slopes, 5 km from the mountaintop. Both

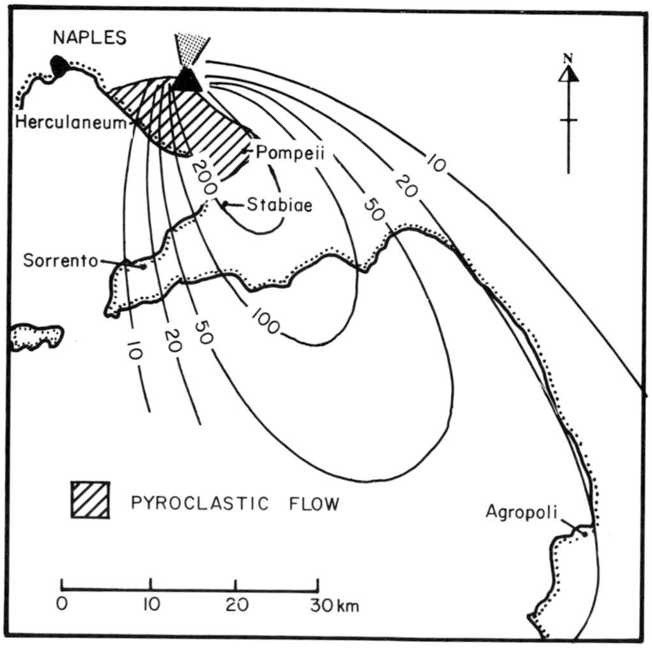

Fig. 12.4 Location map of Mount Vesuvius (Blong, ©1984, with permission Harcourt Brace Jovanovich Group, Australia). Isopleths show depth of ash in centimeters.

towns were prosperous regional centers serving as summer retreats for the wealthy of Rome. Mount Vesuvius, prior to its eruption, consisted of a single, flat-topped cone with a small crater. It was recognized by the Romans as a volcano, but there is no Roman or Etruscan record of it ever being active. In AD 63 a major earthquake struck the region with its epicenter at Pompeii. Both the towns of Pompeii and Herculaneum were severely damaged, and by the time of the eruption in AD 79, only minor rebuilding had been completed in Pompeii.

The description of the eruption comes from the eye-witness account of Pliny the Younger, who recounts the death of his father during the eruption. The writings, afterwards, of Cassius provide descriptions of the destruction of the two cities. Pliny the Elder was the commander of the Roman fleet at Misenum, 30 km to the west at the entrance to the Bay of Naples. On 24 August Vesuvius began to erupt, accompanied by violent earthquakes. Pliny sailed off to the scene to rescue inhabitants at the base of the mountain. He could not get near the shore because of a sudden retreat of the shoreline, and his boats were covered in a continuous ash fall. He sailed that night to Stabiae to stay with a friend. Unfortunately, Stabiae was downwind from Pompeii, and during the night the ash fall became so deep that people realized they would have to flee for their own safety. Pliny the Elder reached the coast during the daylight hours of 25 August, but the ash fall was so heavy, that it still appeared to be night. As the magma chamber within the volcano emptied, the top of the mountain collapsed inward, producing a caldera 3 km in diameter. About 50 per cent of the cone was involved. In the early hours of 25 August a series of six surges and pyroclastic flows swept the area. The first surge overwhelmed Herculaneum. All remaining inhabitants met a grisly death from asphyxiation as they sheltered on the beach. The fourth surge reached Pompeii at about 7 am. The last two surges were the largest, and reached the towns of Stabiae and Misenum. At Stabiae Pliny the Elder died on the beach, apparently from a heart attack. It took three days before it was light enough to recover his body. The pyroclastic ash falls buried Stabiae and Pompeii to depths of two and three meters respectively. Most of the residents of Pompeii appear to have fled, but 10 per cent were literally caught dead in their tracks by the pyroclastic flows that climaxed the eruption. Herculaneum was also affected by the tephra fall-out, but in the latter stages of the eruption, was engulfed by a mud flow 20 m thick. This lahar appears to have been caused by groundwater expulsion as the caldera collapsed.

The eruption marks one of the first volcanic disasters ever documented. The sequence of events at Vesuvius has been labelled a Plinian-type eruption after Pliny's descriptions. Many of the buildings in Pompeii and Herculaneum protruded above the ash. The buildings were ransacked of valuables, dismantled for building materials and subsequently buried by continued eruptions. The ash reverted to rich agricultural land, and Pompeii was forgotten. It was not until 1699 that Pompeii was rediscovered and excavated. Vesuvius has continued to erupt aperiodically until the present day. Subsequent major eruptions occurred in 203, 472, 512, 685, 787 and five times between 968 and 1037. It then remain dormant for 600 years until 1630, when it produced extensive lava flows and killed 700 people. The last major eruption in 1906 produced tephra up to 7 m thick on the northeast side of the mountain. About 300 people were killed, mainly by collapsing roofs.

Mount Pelée (8 May 1902) (Holmes, 1965; Bolt et al., 1975; Whittow, 1980; Blong, 1984)

The year 1902 was not a good year for the residents of the West Indies or the surrounding Caribbean (figure 12.5). The Pacific coast of Guatemala was struck by a strong earthquake on 18 January and again on 17 April. On 20 April Mount Pelée began to erupt on the island of Martinique and on 7 May Soufrière, on the nearby island of St Vincent, erupted killing 2000 people. Izalco in El Salvador erupted on 10 May, Masaya in Nicaragua erupted in July, and Santa Maria in Guatemala exploded on 24 October. The latter eruptions caused mainly property damage. The worst eruption in terms of loss of life was the explosion of Mount Pelée on 8 May. It ranks as one of the most catastrophic natural eruptions witnessed.

The eruption of Soufrière on St Vincent, 160 km south of Martinique, was a prelude to the Mount Pelée disaster. Soufrière became active in 1901, and in April of 1902, earthquake activity was severe enough that most residents were evacuated to the southern part of the island. On 6 May, the volcano started emitting steam, and the Rabaka Dry and Wallibu rivers became torrents of muddy water. At 2:00 pm on 7 May, Soufrière generated a pyroclastic flow, which destroyed most of the northern part of the island killing 2000 people. Fortunately, the evacuations avoided a higher deathtoll.

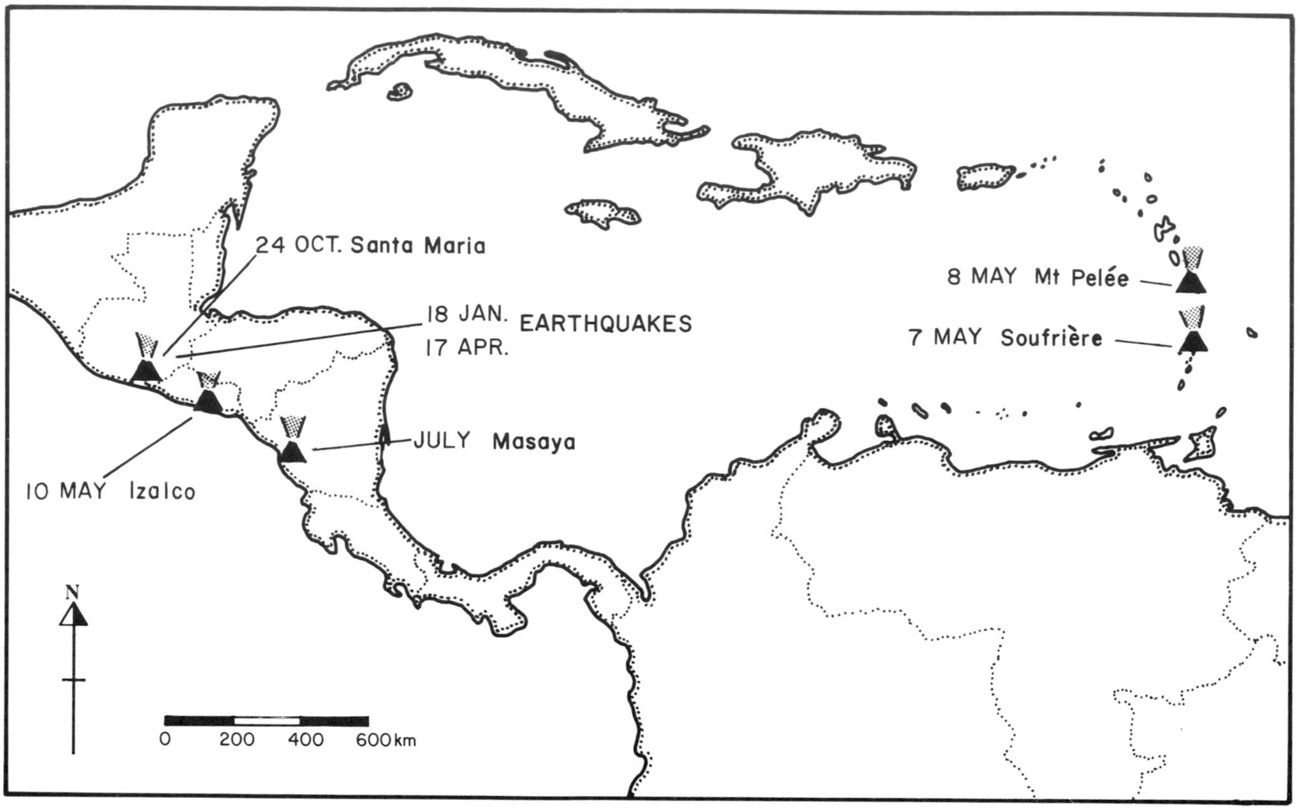

Fig. 12.5 Location map of volcanoes and earthquakes in the Caribbean region in 1902

Mount Pelée began to erupt on 20 April 1902. The mountain was conical in shape, with a notch on the southwest side that led into a topographic low called the Rivière Blanche (figure 12.6). This valley descends steeply for about 1.5 km, and then bends sharply westward. The town of St Pierre lies on the coast 3 km south of this bend. St Pierre was one of the most prosperous cities in the Caribbean. Its business centered on the export of rum made from local sugar cane. The eruption was preceded by a long series of warnings foreboding imminent disaster. Lava was found in the crater shortly after the initial eruption, and in the last week of April, the volcano belched ash and sulphurous fumes so intense that birds dropped dead from the sky. Reports were issued that the volcano was still safe, and that St Pierre was protected from any lava flows by the ridge between it and the Rivière Blanche. On the night of 3 May, Mount Pelée entered a new stage of eruption. The Roxelane River, running through St Pierre, turned into a torrent of mud, knocking out the power supply in the city. Fissures opened up on the slopes, and sent out steam and boiling mud that killed 160 people in the town of Ajoupa-Bouillon. About

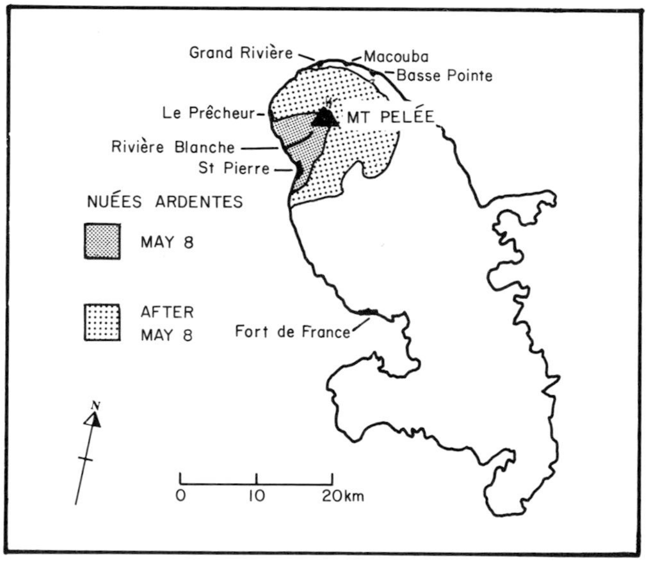

Fig. 12.6 Location map of Mount Pelée

10 000 people then fled the slopes and crowded into St Pierre. Biblical-type plagues then afflicted the city and countryside. Snakes and insects driven from the mountain by the heat and gas emissions invaded the city's suburbs. More than 100 deadly pit vipers entered the north of the city. The army

was called out to shoot them, but not before over 50 people and 200 animals had been bitten to death. At the sugar plantation at the mouth of Rivière Blanche, ants and deadly 0.3 m long centipedes invaded the mill buildings, swarmed up the legs of horses and the workmen, and bit them.

By 5 May, the crater at the top of the mountain had filled with muddy water. On 5 May, this was blasted out sending a large mud flow down the Rivière Blanche to the coast. The lahar killed 30 of the sugarmill workers who still remained after the insect attack, and drowned 159 workers in surrounding fields. When the 35 m high wall of mud hit the ocean, it generated a tsunami that crashed into the lower part of St Pierre killing 100 people there. All during these events, the government, headed by Governor Mouttet, issued press releases stating that there was no threat. Mouttet faced an election on 11 May and, because there was no absentee balloting, he feared losing power if the mainly white citizens of the largest city on the island fled. On the day of the mud flow, the government received a report from a committee of learned citizens, stating that the activity of Mount Pelée did not warrant evacuation. Following this statement, in what must be rated as one of the most foolhardy acts in history, Mouttet and his wife moved from the capital, Fort de France, into St Pierre to allay any fears of impending doom amongst the citizens. The move was sanctioned by the church, which had seen continued panic as worshippers flocked to give their confessions.

On 6 May Mount Pelée issued tephra which was so voluminous that it crushed roofs and clogged streets. The city entered a stage of mass panic, and troops were dispatched to block off all roads leading out of the city to prevent evacuation. All the rivers around the mountain then flooded for the next two days as the mountain began roaring and blowing out pumice. Hot gases rose to the surface expelling groundwater from the water-table. On 7 May, Soufrière on St Vincent island exploded, and underwater earthquakes cut all but one of the telegraph lines to Martinique. Reports soon circulated that the eruption on St Vincent would release the pressures building up under Mount Pelée. Then Mount Pelée quietened, supporting this rumour.

At 7:50 am on 8 May, a series of four violent explosions shot dust to heights of 15 km. A second cloud was blasted as a basal surge horizontally southwest. This cloud immediately flowed down the Rivière Blanche. When it came to the bend in the river valley, it overrode the bank, and part of the flow, travelling at velocities in excess of 160 km hr^{-1}, swept through the city of St Pierre at 7:52 am. Eyewitness accounts, from the few people who had fled the city early that morning, describe the pyroclastic flow as a hurricane of flame. Temperatures within the gas cloud were as high as 1075°C. Walls of buildings were blown down, a three-ton statue of the Virgin Mary was tossed 12 m, 18 out of the 20 ships in the harbour were sunk, and most of the standing city set ablaze. The hot blast exploded a rum distillery and ignited rum, which then flowed through the streets completing the incineration of the city (figure 12.7). The dust cover only averaged 30 cm in thickness throughout the city, but it engulfed the area in complete darkness. Towards the mouth of the Rivière Blanche pyroclastic deposits thickened to 4 m. Up to 30 000 people, including Governor Mouttet, died within two minutes. Many had clothing stripped from them by the force of the blast. Others were grotesquely disfigured as their body fluids boiled and burst. There were only two survivors, one of whom, Auguste Ciparis, was a condemned prisoner in the local jail. He suffered severe burns and later had his sentence commuted. Subsequent blasts on 19 May and 20 August swept across most of the northern and western slopes, causing further destruction. The last explosion killed 2000 people in five mountain villages. Of all historical descriptions of volcanic disasters, the residents of the city of St Pierre and the island of Martinique appear to have suffered one of the most devastating eruptions known.

FIG. 12.7 The city of St Pierre, Martinique, after the 8 May 1902 eruption of Mount Pelée (photograph from a book by A. Lacroix, courtesy of the Geological Museum, London). The pyroclastic flow which destroyed the city swept over the ridge at the top of the photograph. The waterfront had been wrecked by a tsunami caused by a lahar several days previously.

Krakatoa (26–27 August 1883) (Bolt et al., 1975; Cornell, 1976; Whittow, 1980; Simkin and Fiske, 1983; Blong, 1984; Myles, 1985; Blong and Johnson, 1986)

Krakatoa was one of the largest explosive eruptions known to humanity. The volcano lies in the Straits of Sunda between Sumatra and Java, Indonesia (figure 12.8B). It had last been active in 1681, and during the 1870s underwent increased earthquake activity. In May of 1883, one vent became active throwing ash 10 km into the air. By the beginning of August, a dozen Vesuvian eruptions reverberated across the island. On 26 August loud explosions recurred at 10-minute intervals, and a dense tephra cloud rose 25 km above the island. The dust particles formed nuclei for water condensation and muddy rain fell on adjacent islands. The explosions could be heard throughout the islands of Java and Sumatra. That evening small tsunami waves 1–2 m in height swept the straits. On the morning of 27 August, three horrific explosions occurred. The first explosion disintegrated the 130 m peak of Perboewetan, forming a caldera, which immediately infilled with sea water. Within half an hour, the 500 m peak of Danan exploded and collapsed, sending more sea water into the molten magma chamber of the eruption. The third blast at 10:02 am tore the remaining island of Rakata apart. A caldera, 6 km in diameter and 300 m deep, formed where the central island had once stood (figure 12.8C). The final blast was the largest sound ever heard by humanity, and was heard 4800 km away on the island of Rodriguez in the Indian Ocean, and 3200 km away at Elsey Creek, Northern Territory, Australia. Windows 150 km away were shattered. The atmospheric shock wave travelled around the world seven times. Barometers in Europe and the United States measured significant oscillations in pressure over nine days following the blast.

The blasts were accompanied by a cloud of ash, which rose to heights of 80 km. The debris from this cloud formed deposits with a volume of 18 km^3. On the island remnants, 60 m of ignimbrite were deposited on top of 15 m of tephra. Large rafts of pumice, consisting of magma from inside the

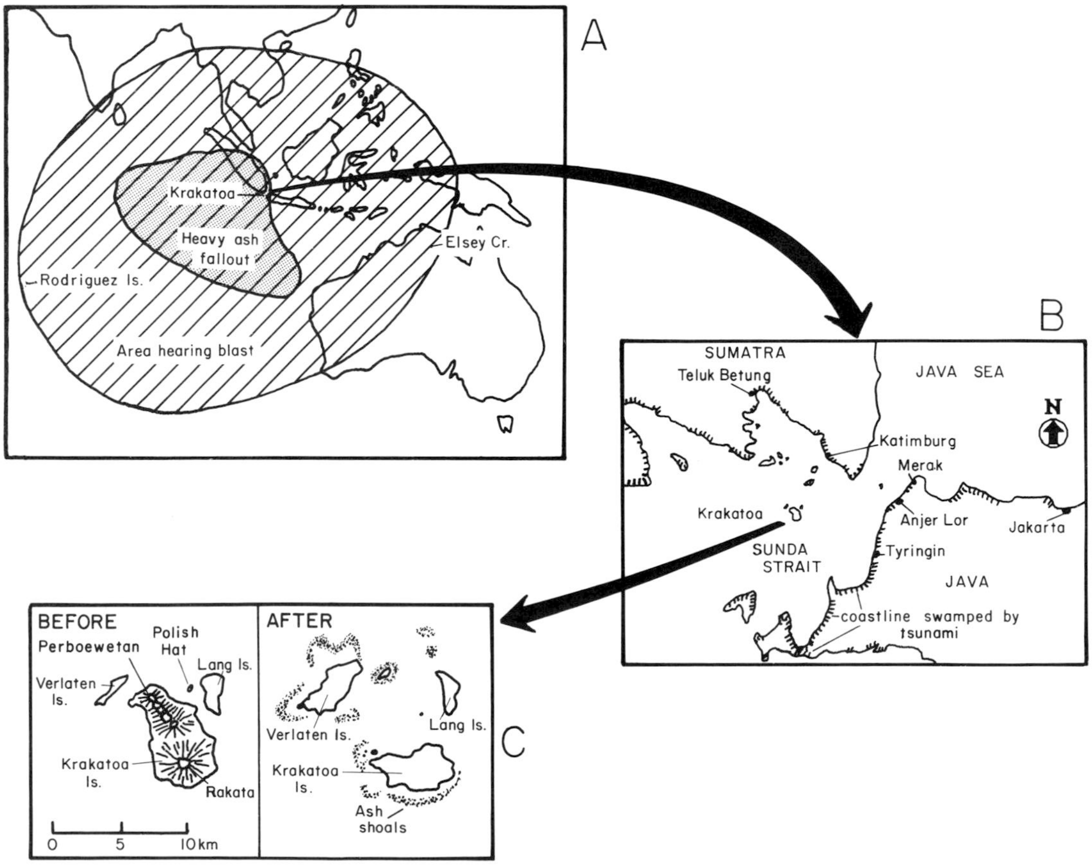

Fig. 12.8 Location map of Krakatoa (after Judd, 1888; Francis, 1976; Whittow, 1980)

volcano rather than remnants of the volcano itself, blocked the Sunda Straits. Tephra fell over an area of 300 000 km^2 and turned day into night 200 km downwind for forty-eight hours (figure 12.8A). The ash that was shot into the stratosphere reduced solar radiation reaching the Earth's surface by 13 per cent. Two years later solar radiation in France was still 10 per cent below normal. After six weeks, sunlight reflecting off the dust particles, produced spectacular sunrises and sunsets that continued for two years around the world. In the eastern United States, sunsets triggered false alarms to fire departments.

Each caldera collapse, plus the pyroclastic surge and shock waves from the eruption generated enormous tsunami that engulfed the adjacent coastline of Java and Sumatra. The third blast-induced wave was cataclysmic. Over the next 32 hours, this shock wave was responsible for the measurement of tsunami waves in such distant and disassociated bodies of water as the English Channel, the east Pacific Ocean at San Francisco and at Lake Taupo in the center of New Zealand. Nine hours after the blast, 300 riverboats were swamped and sunk at Calcutta on the Ganges River 3000 km away. The coastline adjacent to the eruption was struck by waves with a maximum height of 42 m, that rolled 5 km inland over lowlying areas within 30–60 minutes (figure 12.8B). The largest wave struck the town of Merak; the town of Anjer Lor was swamped by an 11-meter high wave, the town of Tyringin by one 23 m in height, and the towns of Katimburg and Teluk Betung were each struck by a wave 26 m high. In the latter town, the Dutch warship *Berouw* was carried 2 km inland and left stranded 10 m above sea-level. Between 5000 and 6000 boats in the straits were sunk. In total, 36 000 people in these major towns and 300 other villages died because of this final tsunami.

In recorded history, only the Tambora (Indonesia) and the Santorini (Aegean Sea) eruptions were bigger. Following Tambora, 5000–6000 died in pyroclastic flows, and 4000–5000 in the resulting tsunami; however, 82 000 died of starvation from the crop failures caused by over 1 m of tephra fallout on adjacent islands. Only disaster relief organized by the Dutch in 1883 prevented a similar deathtoll in the aftermath of Krakatoa. These eruptions reinforce the view that this heavily populated section of Indonesia from Sumatra to Timor represents one of the most hazardous zones of volcanic activity in the world.

Mount St Helens (18 May 1980) (Hays, 1981; Lipman and Mullineaux, 1981; Keller, 1982; Blong, 1984; Coates, 1985)

The Mount St Helens eruption is notable for several reasons. Firstly, it was the first large explosive eruption to occur in the world for several decades. Secondly, it heralded a period of substantial volcanic activity at the beginning of the 1980s. Thirdly, it had associated with it several hazard phenomena. Finally, it was the largest eruption to occur in the conterminous United States in recorded history. Mount St Helens is situated in the Cascade mountains of Washington State (figure 12.9). It had

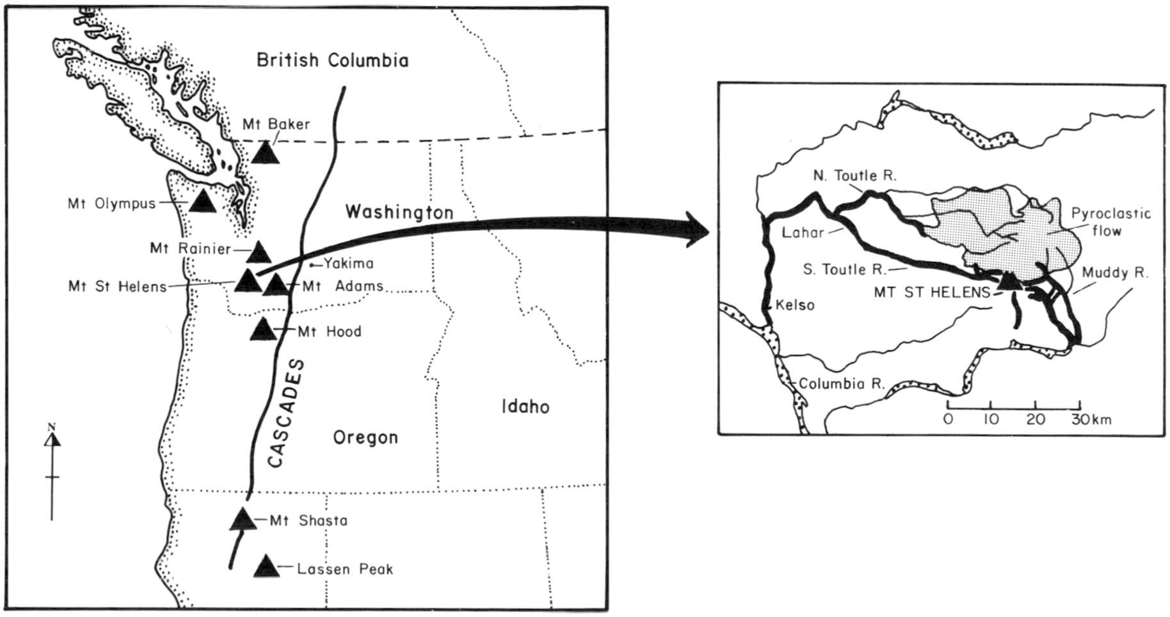

FIG. 12.9 Location map of Mount St Helens (adapted from Blong, 1984)

not had a major eruption for 123 years, and was considered a relatively minor volcano next to Mount Rainier. However, research in the mid-1970s began to indicate that ash found in the region originated mostly from Mount St Helens. On 20 March 1980, a swarm of micro-earthquakes around Mount St Helens signalled renewed activity, which culminated in a small eruption on 27 March sending clouds of ash 6 km skyward. Harmonic tremors began on 3 April, signalling the movement of magma into the magma chamber below the volcano. By this time, over 70 million m^3 of tephra had been ejected. Prior to the eruption, the north side of the volcano bulged more than 150 m at a rate of 1.5 m per day.

At 8:30 am on 18 May, an earthquake of magnitude 5.1 on the Richter scale started landslides in the bulge region. Immediately, the volcano erupted and sent out a lateral explosion, which was felt 425 km away. A dark cloud of ash, containing 2.7 km^3 of material, rose to a height of 23 km for the next nine hours. The height of the mountain was reduced by 500 m (figure 12.10). Over 500 km^2 of forest were devastated by the resulting base surge that reached temperatures of 260° C. The blast of debris overrode ridges and travelled over 10 km in distance. In some places, debris deposits accumulated to depths of 150 m. Snow and ice on the mountain were melted by the blast, and formed lahars that rushed into Spirit Lake, filling the lake to depths of 60 m. The lahars then moved over 50 km down the north and south forks of the Toutle River, reaching Kelso on the Columbia River. Lahars also flowed down into the Muddy River to the east. In all, over 300 km of roads and 48 road bridges were extensively damaged. The lahars and fine ash that settled in the surrounding area were soon carried into the Columbia River, where the 180-metre wide shipping canal was reduced from a depth of 12 m to one of 4.3 m. Only 60 people lost their lives, mainly because the government had ordered evacuations as the intensity of the eruption increased. However, some of this deathtoll included people who were permitted back into the area only days before the eruption.

Tephra fallout became the worst nuisance after the initial eruption (figure 10.13). It fell to depths of 10–15 mm, 150 km downwind of the eruption. It also made driving conditions in and around the area treacherous for five days because of low visibility and slippery roads. Speed limits had to be reduced to 8–10 km hr^{-1} to prevent accidents. The deposition of ash proved a major clean-up problem in urban centers. The small town of Yakima, Washington, with a population of 50 000, took ten weeks to remove half a million tonnes of ash, a task that cost $US2 million. The fine dust clogged air filters on cars, got into brakes and wrecked motors. Dust fallout on electrical equipment caused equipment failure, while crop yields noticeably decreased because the ash coated leaves and lowered the efficiency of photosynthesis.

FIG. 12.10 The north side of Mount St Helens after the 18 May 1980 eruption (photograph courtesy of James Ruhle and Associates, Fullerton California). The crater has not formed by collapse, but by a lateral blast. The material in front represents the pyroclastic debris from that blast.

Given the magnitude of the event, the media-hype, and the reaction by the public and government, the true economic effects of the disaster tended to be overestimated. Washington State estimated the damage at $US2700 million, and Federal Congress appropriated just under $US1000 million for disaster relief. In effect, the eruption caused $US844 million dollars damage. Clean-up cost $US270 million; agricultural losses amounted to $US39 million; property damage, mainly to roads and bridges, cost $US85 million; and commercial

TABLE 12.1 Volcanic Events since 1980

Volcano	*Date*	*Deathtoll*
Mount St Helens, United States	18 May 1980	60
Galunggung, Indonesia	4 April 1982	[a]
El Chichon, Mexico	28 March 1982	[a]
Lake Monoun, Cameroon	1984	37
Nevado del Ruiz, Columbia	13 November 1985	23 000
Lake Nios, Cameroon	25 August 1986	1 500

[a] Minor but unknown.

timber losses amounted to $US450 million. The eruption sparked renewed research interest into other volcanoes along the Cascade and Sierra Nevada ranges. Mount St Helens also highlighted the fact that the previous 60 years worldwide were generally quiescent volcanically (figure 10.13). Since Mount St Helens in 1980, renewed volcanic activity and associated phenomena have taken more than 24 000 lives (table 12.1).

References

Blong, R. J. 1984. *Volcanic hazards: a sourcebook on the effects of eruptions*. Academic Press, Sydney.

Francis, P. 1976. *Volcanoes*. Penguin, Harmondsworth.

Hays, W. W. 1981. 'Facing geologic and hydrologic hazards: Earth-science considerations'. *United States Geological Survey Professional Paper* 1240–B pp. 86–109.

Judd, J. W. 1888. 'On the volcanic phenomena of the eruption, and on the nature and distribution of the ejected materials'. In Symons, G. (ed.) *The eruption of Krakatoa and subsequent phenomena*. Royal Society of London Krakatoa Committee Report, pp. 1–46.

Perrett, F. A. 1935. 'Eruption of Mt Pelée 1929–32'. *Carnegie Institute of Washington Publication* No. 458.

Whittow, J. 1980. *Disasters: the anatomy of environmental hazards*. Pelican, Harmondsworth.

Further reading

Blong, R. J. and Johnson, R. W. 1986. 'Geological hazards in the southwest Pacific and southeast Asian region; identification, assessment, and impact'. *Bureau of Mineral Resources Journal of Australian Geology and Geophysics* v. 10 pp. 1–15.

Bolt, B. A., Horn, W. L., MacDonald, G. A. and Scott, R. F. 1975. *Geological hazards*. Springer-Verlag, Berlin.

Coates, D. R. 1985. *Geology and Society*. Chapman and Hall, NY, pp. 103–127.

Cornell, J. 1976. *The great international disaster book*. Scribner's, NY.

Fielder, G. and Wilson, I. 1975. *Volcanoes of the Earth, Moon and Mars*. Elek, London.

Holmes, A. 1965. *Principles of Physical Geology*. Nelson, London.

Keller, E. A. 1982. *Environmental Geology* (3rd edn.) Merrill, Columbus, Ohio, pp. 169–97.

Lipman, P. W. and Mullineaux, D. R. (eds). 1981. 'The 1980 Eruptions of Mount St Helens, Washington'. *United States Geological Survey Professional Paper* No. 1250.

Myles, D. 1985. *The Great Waves*. McGraw-Hill, NY.

Sigurdsson, H., Carey, S., Cornell, W. and Pescatore, T. 1985. 'The eruption of Vesuvius in 79 AD' National Geographic Research v. 1 pp. 332–87.

Simkin, T. and Fiske, R. S. 1983. *Krakatau: the volcanic eruption and its effect*. Smithsonian Institution Press, Washington, DC.

Symons, R. B., Rose, W. I. and Reed, M. H. 1988. 'Contribution of Cl- and F-bearing gases to the atmosphere by volcanoes'. *Nature* v. 334 pp. 415–18.

Tazieff, H. and Sabroux, J. C. (eds). 1983. *Forecasting volcanic events*. Lange and Springer, Berlin.

Wood, R. M. 1986. *Earthquakes and Volcanoes*. Mitchell Beazley, London.

13

Land Instability as a Hazard

Introduction

One of the most widespread natural hazards is the unexpected, and sometimes unpredictable movement of unconsolidated weathered material (*regolith*), or weathered rock layers near the Earth's surface. Landslides and avalanches, while historically not renowned for causing as large a deathtoll as other natural disasters such as tropical cyclones or earthquakes, have had just as dramatic an impact on property and lives. The sudden movement of slope material is as instantaneous as any earthquake event but it is a more widespread problem. In any moderate-to-high relief region subject to periods of high rainfall, slippage of part, or all of the regolith downslope is probably the most common hazard. Nowhere is this problem more prevalent than in cold regions underlain by *permafrost* or ground ice. Of a slower nature, and just as widespread a hazard, is land *subsidence*. While much of a land surface may be stable or even flat, human activity beneath the surface can cause structural failure of the surface and collapse. It is beyond the scope of this text to go into human-induced subsidence; however, there are a wide range of natural processes which can generate ground collapse. Another important aspect concerning land instability is the multitude of factors that can trigger ground movement. Almost all of the hazards presented in this book can generate secondary land instability problems. The large deathtoll from earthquakes and cyclones has in many cases been caused by associated landslides. Even droughts can exacerbate ground instability through the process of repeated drying and wetting of expansive clays. This can lead to surface deformation and eventual destruction of structures with insufficient foundations.

In this chapter an attempt will be made to present a broad overview of the wide range of land instability hazards. Firstly, the basic principles of soil mechanics will be described to show how surface material becomes unstable, and to point out what factors exacerbate ground failure. Secondly, each of the main types of land instability will be examined in turn, together with a description of some of the major disasters. In order to limit the coverage of such a wide topic as land instability, direct human-induced and ice-related (*cryogenetic*) factors will only be discussed at a cursory level. This is not to say that they are unimportant. For instance, cryogenetic landscapes dominate up to 20 per cent of the world's landmass.

Soil mechanics (Young, 1972; Chowdhury, 1978; Finlayson and Statham, 1980; Goudie, 1981; Bowles, 1984)

Stress and strain

Consider a body of soil with mass ω sitting on a slope of angle, α. This mass is affected by gravity and tends to move downslope. The effect of gravity is directly related to the angle of the slope as follows:

$$\text{effect of gravity on a slope} = \omega \sin \alpha \quad (13.1)$$

where ω = mass
α = the slope angle

Any force (gravity is by far the most important) that tends to move material downslope is termed a *stress* (figure 13.1). If a building is built on the slope, the weight of the building adds more stress to the soil. Additional weights on a slope are termed 'stress increments'. Stress increments can consist of rain, soil moved from upslope, buildings or any other mass. While gravitational stress is most obvious, there are other stresses which can exist in the regolith. Molecular stress arises with the movement of soil particles or even individual molecules. It is associated with such phenomena as swelling and shrinking of *colloids* (clay soil particles) on wetting and drying, thermal expansion and contraction, and the growth of ice crystals. Biological stress refers to the stress exerted on a soil body by the growth of plant roots or the activities of animals.

The effect of stress upon a soil or a regolith is called *strain*. Strain is not directly equated with stress but incorporates the additional factor of slope angle. Strain may not uniformly occur in the soil body, but may be restricted to joints where fracturing will eventuate. Strain can affect inter-particle movements, or act on the overall soil column. In combination, these strains result from what is labelled 'net shear stress'.

Friction, cohesion and coherence

One of the forces of resistance against the movement of material downslope is friction. Friction is the force which tends to resist the sliding of one object over, or against another. For regolith material sitting on a firm base, friction is due to irregularities present between the contact of these two materials no matter how smooth the contact may appear to be. The irregularities tend to interlock with each other and prevent movement. Generally, the greater the weight of an object pressing down on the contact, the greater the amount of friction. While friction is dependent on the degree of roughness of a surface, it is independent of the area of contact between a body of regolith and the underlying substrata. Small areas of soil material tend to fail at the same angles as larger areas. Figure 13.1 illustrates the application of stress to an object on a slope. There is a resisting force against movement due to friction. Sliding will commence when the applied stress exceeds maximum frictional resistance.

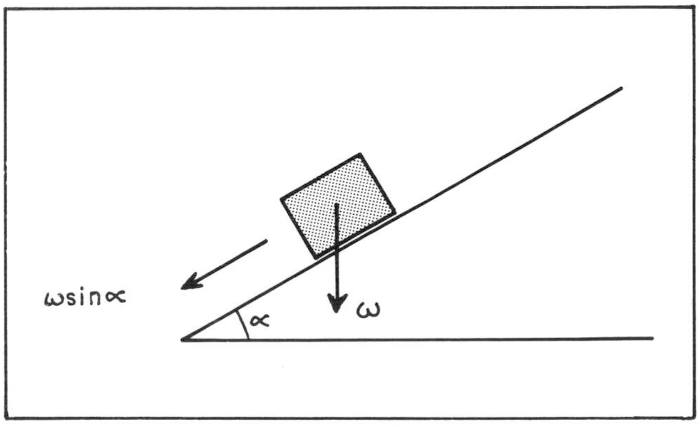

FIG. 13.1 Effect of gravity on an object with mass ω on a slope with angle α

Friction is expressed as a coefficient μ and this coefficient is equal to the the tangent of the slope angle at which sliding just begins, $\tan \alpha_\mu = (\omega \sin \alpha_\mu)\,(\omega \cos \alpha_\mu)^{-1}$. This critical slope angle is termed the 'angle of plane sliding friction'. The equations of movement of objects can be expressed in terms of critical frictional resistance ($R_{f\,\text{crit}}$) and critical applied force (F_{crit}). Failure commences when the critical applied force exceeds the critical frictional resistance as follows:

$$F_{\text{crit}} > R_{f\,\text{crit}} \qquad (13.2)$$

where $R_{f\,\text{crit}} \sim \omega \tan \alpha_\mu$

This equation implies that as the weight of an object approaches zero the force required to move it downslope approaches zero. For non-rigid objects (unconsolidated material) this is not so, because as the weight of the regolith approaches zero, there is still an additional force resisting downslope movement. This force is termed *cohesion*, which is defined as the bonding that exists between particles making up the soil body. The above relationship, now including cohesion, can be expressed as follows:

$$F_{\text{crit}} > R_{f\,\text{crit}} \qquad (13.3)$$

where $R_{f\,\text{crit}} \sim \omega \tan \alpha_\mu + c$
and c = cohesion

Strictly, the term cohesion refers to chemical or physical forces between clay particles. The term *coherence* is used to describe the binding of soil particles of all sizes, as a group or mass, due to capillary cohesion by water at the interface between individual grains, by chemical bonds or by *cementation*. In cementation, the chemical bonds are considered primary and, thus, are very strong. Common cementing agents include carbonates, silica, alumina, iron oxide and organic compounds. Cementation can also occur as a result of compaction, especially if material comes under pressure from above. The compaction process gives stability to materials on slopes. With certain grain shapes, especially with clays, it is possible to realign grains so that they interlock effectively to increase coherence and the resisting force to movement. Chemical bonds consist mainly of oppositely charged electrical fields that develop on the surfaces of large molecules, especially clay minerals. These attracting charges are called *Van der Waals bonds*, which for clay minerals remain active even when the clay particles or colloids are moved relative to each other. For clays, this gives rise to plasticity which will be described later.

Figure 13.2 presents the type of bonds that can exist depending upon particle size and the relative strength of bonding. Van der Waals bonds are restricted to material less than 0.03 mm or 30 microns in diameter. As the material gets smaller, the bonding strength increases considerably until it reaches values of 1 kg cm^{-2} for sizes less than 0.001 microns. As grain size increases, capillary cohesion mainly due to water becomes dominant. Capillary cohesion is also very important in bonding clay minerals, creating forces three orders of magnitude stronger than Van der Waals bonds. The only way that coherence of material coarser than 1 cm can be achieved is by compaction and cementation.

The effect of capillary cohesion can be demonstrated easily. Damp sand can be moulded into almost vertical walls without any sign of failure. As the sand dries out and the cohesive tension of water at sand grain interfaces is removed, the sand pile begins to crumple, eventually reaching the angle of repose for loose sand, which is only 33°. Cohesion in damp sand is totally produced by the effect of water tension between sand grains, and gives wet sand a remarkable degree of stability. Note, however, that the sand cannot be too wet otherwise liquefaction occurs. In other words, if more water is added to the sand, or any other material for that matter, such that the pore spaces in the material are filled and the capillary thickness of water at the grain interfaces is exceeded, then the cohesion of the material approaches zero and the material begins to behave as a fluid. This process will be discussed in more detail later.

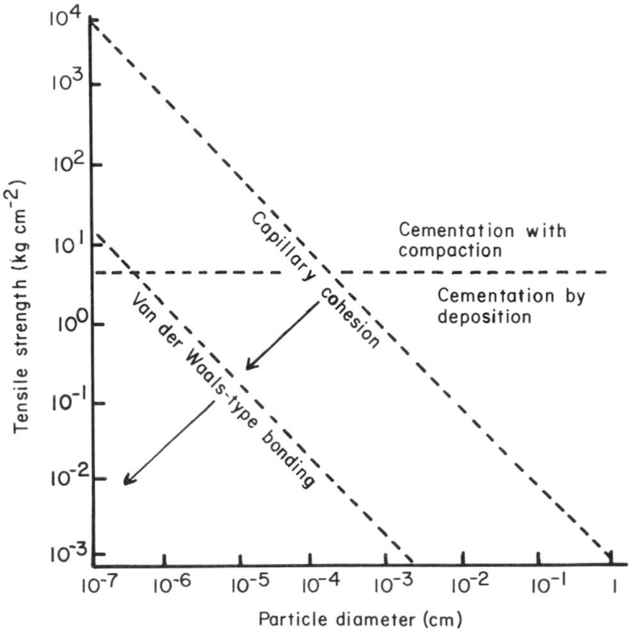

FIG. 13.2 Bonding strength on particles of different sizes (Finlayson and Statham, ©1980, with permission Butterworths, Sydney)

Shear strength of soils: Mohr-Coulomb equation

The way that soil particles behave as a group or mass (coherence) depends not only upon the inner cohesion of the soil particles but also upon the friction generated between individual soil grains. The latter characteristic is internal friction or shearing resistance. How much shear stress a soil or regolith can withstand is given by the following equation termed the *Mohr-Coulomb equation*:

$$\tau_s = c + \sigma \tan \phi \qquad (13.4)$$

where τ_s = the shear strength of the soil
c = soil cohesion
σ = the *normal stress* (at right angles to the slope)
ϕ = the angle of internal friction or shearing resistance

The Mohr-Coulomb equation is represented diagrammatically in figure 13.3. Note, that this equation is in a form similar to equation 13.3 except for two differences. Firstly, in the Mohr-Coulomb equation the critical force for movement is determined by the stress normal to the ground surface, rather than by the weight piled on a slope.

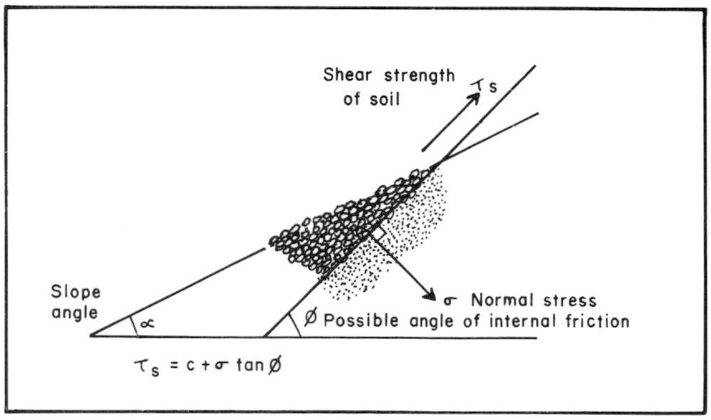

FIG. 13.3 Schematic representation of the Mohr-Coulomb equation. Note, that angle ϕ is not necessarily the slope angle, but the angle of internal friction within the slope mass.

Secondly, the angle of the slope has been replaced by the angle of shearing resistance, which represents the angle of contact between particles making up unconsolidated material. Loosely compacted material tends to have a lower angle of shearing resistance than very compressed or compacted material. All unconsolidated material tends to fail at internal angles less than the slope angle upon which it is resting. The Mohr-Coulomb equation can also be used to define the *shear strength* of a unit of rock resting on a failure plane and the susceptibility of that material to landslides. In this case, the formula must be modified to include the characteristics of bedding planes. If the stress applied to the soil exceeds the shear strength, then the material will fail and begin to move downslope.

Pore water pressure (see also Aune, 1983)

Almost all material, no matter what its state of consolidation, contains pores or voids which may be filled with air or, more often, water. If the water does not solely adhere to individual particles as capillary water, but completely fills voids forming a water-table, then it will flow freely under a pressure head. If the soil is incompressible, a rise in the water-table will cause the pressure of water in voids at depth to increase. Furthermore, some of the added pressure may be taken up by the grain-to-grain contacts in the soil. If an external load is applied to a soil mass on a slope in the form of additional water, or overburden, then the pore water pressure will build up in that mass and water will be expelled at weak points. These weak points occur along fracture lines, bedding planes or at the base of the regolith, where pore water pressure is highest. As water drains from a soil body, the pore

The increase in pore water pressure in the Mohr-Coulomb equation reduces the effective resistance of the soil body or regolith as shown by the following equation:

$$\tau_s = c + (\sigma - \xi) \tan \phi \qquad (13.5)$$

where ξ = pore water pressure

In this case, normal stress, σ, is reduced by the pore water pressure defined by ξ. If overburden is added to a soil, it will immediately increase pore water pressure. Unless the excess pore water pressure is reduced through drainage of water, the critical stress on that soil will exceed the critical resistance of the soil, causing slope failure. If, on the other hand, wet material is added upslope, there is no change in the stress being applied to the soil. However, the water in this material can drain into the soil downslope, increasing progressively over time its pore water pressure. Failure of the slope may happen sometime after the overburden was emplaced upslope. A slope that has been stable under existing load and water-table conditions can also become unstable if drainage patterns are changed in the surrounding area. Dams raise water-tables locally, a fact that may cause subsequent slope failure in adjacent areas some time after the dam was filled. Construction of buildings on slopes with septic systems can also increase the pore water pressure, leading to subsequent failure. The source of water need not come solely from septic tanks, it could also come from lawn watering and other domestic discharges.

Earthquakes can increase the pore water pressure of a soil with each passage of a compressional shock wave. As the pore water pressure increases, it may not be reduced fast enough by discharge of water from the ground before the next compressional wave sweeps past. The effective stress in the material thus decreases with each shock wave until the pore water pressure is equal to the normal stress in the soil (figure 11.9). At this point liquefaction will result. In unconsolidated sediment, the smaller the particle size, the more water movement is inhibited because of capillary cohesion. Very fine silt and clays should, thus, be susceptible to liquefaction during earthquakes. However, depending upon the degree of consolidation, weathering or other factors, cohesion in these materials may exceed the increases in pore water pressure during the passage of earthquake shock waves. Thus, liquefaction tends to occur best in medium-to-fine grained sands which have not completely compacted. Because these types of sediments are widespread, especially in marine environments or on river floodplains, liquefaction is an almost universal feature of moderate-to-large earthquakes.

There is an exception to the above occurrence of liquefaction. Some sodium cation-rich clays, termed *quick clays*, will liquefy if salt is leached by fresh water. This scenario often occurs with glacially deposited marine clays that are subsequently elevated above sea-level and allowed to be flushed with fresh water. The salt or sodium chloride deposited with the clays acts as an electrolytic glue, adhering to the clay particles and providing cohesiveness and structure to the clay matrix. If the sodium is leached out or replaced by calcium—a procedure which can even be performed by the mobilization of free calcium in cement foundations into the surrounding clay subsoil—the clay dramatically loses its cohesiveness. These clays then become very responsive to shock waves, and will liquefy under moderately intensive earthquakes. Such a process occurred in the Bootlegger Cove clay underlying Anchorage during the Alaskan earthquake of 1964 (see figure 13.4 for the location of major place names mentioned in this chapter). The process can also be instigated by the flushing of evaporite lake clays or highly weathered clays such as montmorillonite, and is certainly exacerbated by irrigation, pipe leakage or simple lawn watering. The latter circumstances exist today in many parts of southern California including the San Joaquin valley, southern coast ranges, Mojave desert, and Los Angeles basin—all regions which are seismically active.

Rigid and elastic solids

Depending upon how stress and strain are related, soil bodies may behave in four ways—as rigid solids, as elastic solids, as plastics, or as fluids. The process of liquefaction determines the point at which a soil body behaves as a fluid. The concept of a rigid solid is also easily explained. No matter how much stress is applied to a solid, that solid is considered rigid until the stress exceeds the strength of the material. At this point, the solid either deforms or fractures. The rate at which the stress is applied determines how the solid will be behave. For instance, toffee candy, when given a sharp jolt, will fracture. In this case the toffee can be considered a rigid solid. If, however, gentle pressure is applied to the toffee, it may simply deform. In this case the toffee is not rigid under this type of stress. A solid may be considered elastic if a stress is applied which results in slow, continuous deformation proportional to the applied stress before fracturing. The deformation at this stage is reversible, hence the description of the body as an elastic. It is the capability for reversal of the deformation that defines *elasticity*. Earth materials behave elastically as long as the stress changes applied are small enough.

Plastic solids

The relationship between the moisture content of a soil and its volume permits three factors to be determined—shrinkage, plasticity and liquid limits. These three terms are labelled *Atterberg limits*. They refer mainly to properties of clays and form a continuum, which is illustrated schematically in figure 13.5. The shrinkage limit is defined as the moisture content of a soil, at which point the soil stays at a

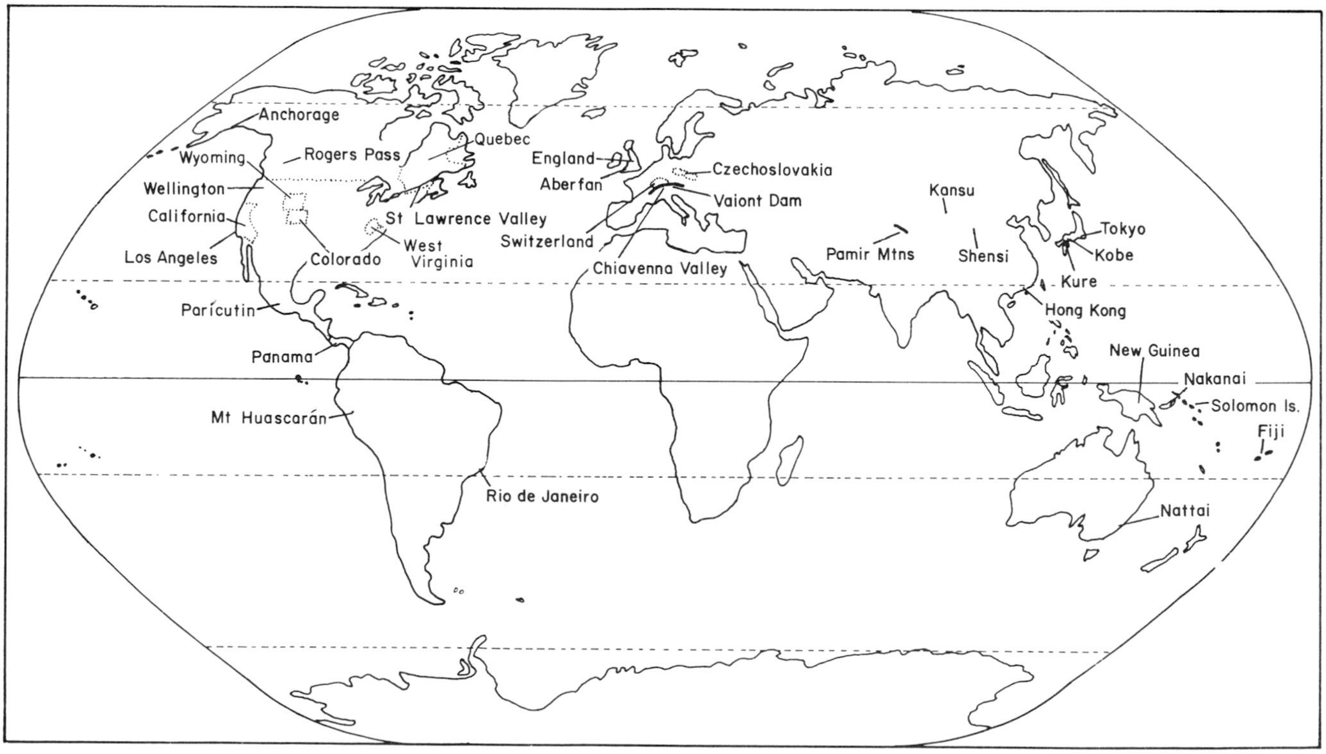

Fig. 13.4 Location map

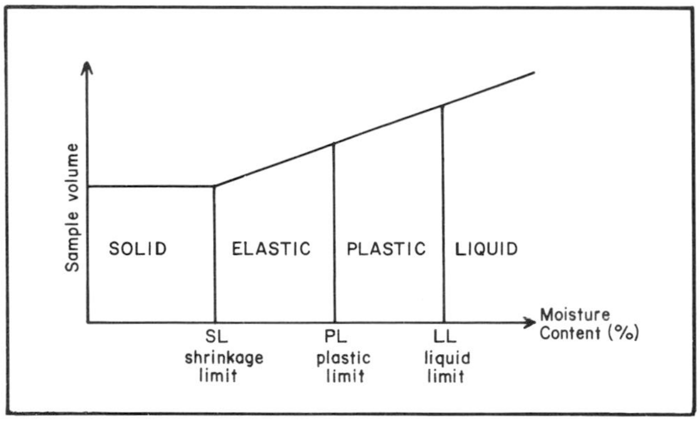

FIG. 13.5 Schematic representation of Atterberg limits (after Goudie, 1981)

constant volume upon drying. Above the shrinkage limit, soil material will behave as a rigid or elastic solid because it consists of interlocked assemblages of grains and particles that possess a finite strength termed the *yield limit*. However, if a force is applied to that material such that the rate of deformation is proportional to the amount of applied stress above this yield limit, then the material is termed a 'plastic solid'. The difference between an elastic and plastic solid concerns how the material reacts after the stress is released. In an elastic solid, *yield strength* has not been exceeded, and the solid will tend to return to its original configuration after deformation. In a plastic solid, the deformation is irreversible because the yield limit has been exceeded. The yield or plastic limit is defined by the minimum moisture content, at which point that material, usually clay, can be moulded. It is also the point where the angle of shearing resistance in clays approaches zero as moisture content increases. This limit defines *plasticity*, and for clays this state can be maintained over a wide range of moisture contents. As long as a clay has a moisture content below the plastic limit, it will support objects. However, when the plastic limit is reached, then the bearing strength of the clay is greatly reduced and it will begin to deform. The liquid limit defines the moisture content at which the clay flows under its own weight. When the liquid limit is reached, clay behaves like a fluid and easily flows downslope. This point also defines when liquefaction occurs, because the material has no shear strength.

Plastic and shrinkage limits are very much a function of the type of clay material. There are three main types of clays, dependent upon the degree of weathering of feldspars or other easily weatherable minerals. Feldspars progressively weather to montmorillonite, *illite* and finally *kaolinite*. Montmorillonite $[(Mg,Ca)O.Al_2O_35SiO_2.nH_2O]$ is composed of one alumina layer sandwiched between two silica layers. The layers are bound together by Van der Waals bonds, which are weak and permit water to enter. The clay minerals have large residual negative charge because of a charge imbalance in their structure. This leads to a high capability for the extensive substitution of water molecules into the molecular structure of the clay. As a result, the cation exchange, plasticity and swelling capacity of such clays are high, making them especially susceptible to deformation and swelling. As the clays hydrate through progressive chemical weathering, they begin to lose these characteristics. Illite $[KAl_2(OH)_2(AlSi_3(O,OH)_{10})]$, a daughter product of montmorillonite, contains a higher proportion of silicon atoms, which induces a high, net negative charge between layers. Potassium ions are attracted to these sites in the crystal lattice, and bind the silica layers together, thus reducing their swelling capacity and plasticity. Kaolinite $[Al_2O_3.2SiO_2.2H_2O]$ is a highly weathered clay mineral in which all potassium ions have been stripped from the lattice. These clay minerals consist of one silica and one alumina layer bonded together by Van der Waals bonds or hydrogen ions. Both bonds are strong. There is little substitution of other atoms or molecules possible in the mineral; hence, it has low plasticity and low swelling capacity.

Classification of land instability

Introduction (Sharpe, 1968; Varnes, 1978; Finlayson and Statham, 1980; Crozier, 1986)

Land instability can best be described by setting the various types of failure into a general classification. At present, land subsidence can be considered as a special case of land instability because it is more related to the behaviour of the substratum than it is to the stress applied to material on a slope. Most classifications of land instability are based upon the type of material and the type of movement. There are five types of movement—falls, topples, slides, lateral spreads and flows. Material can consist of bedrock, consolidated soil and regolith, loose debris, various mixtures of sediment and water, and pure water in the form of snow or ice. Unfortunately, many of the classifications diverge in their emphases. Some are based on geotechnical aspects, reflecting an engineering orientation; while others center around processes and morphology, reflecting a geomorphological

TABLE 13.1 United States Highway Research Board Landslide Committee Classification of Mass Movements

Type of movement		Type of material			
		BEDROCK →			→ SOIL
Falls		Rockfall			Soilfall
		ROTATIONAL →	→ PLANAR ←		← ROTATIONAL
Slides	No. of Units Few ↓	Slump	Block glide		Block slump
	Many		Rockslide	Debris slide	Lateral spread
		ALL UNCONSOLIDATED			
		ROCK FRAGMENTS →	SAND OR SILT →	MIXED →	PLASTIC
Flows	Moisture Content Dry ↓	Rock fragment flow	Sand run; Loess flow		
			Rapid earth flow	Debris avalanche	Slow earth flow; Creep
				Rubble flow: dirty snow avalanche	Solifluction; Mud flow
	Wet		Debris flow: sand or silt		

Source: from Leopold et al., 1964.

perspective. Each classification has tended to bring in different terminology that at times is confusing and contradictory.

A useful classification in terms of material and movement was that proposed originally by Varnes in 1958 and shown in table 13.1. This classification has since been modified to include topples and lateral spreads. This scheme emphasizes the composition of the material being moved, but it and its modification clearly do not include time. From the hazard point of view in terms of warning, human response and prevention, these classifications can be limited. A temporal classification which gives some idea of the speed of movement of land instability can be more attractive. In table 1.1 (in chapter 1), where hazard characteristics and impacts mentioned in this textbook were ranked, some emphasis was placed on the suddenness of the event. One such temporal classification is shown in figure 13.6. It is immediately obvious that this classification includes expansive soils, which are ignored in most engineering and morphometric classifications. Figure 13.6 is used in this chapter, not because it is better than the others, but because it emphasizes time and includes expansive soils, which in the long term form the most expensive type of land instability. Nor should the classification used here be considered definitive. Readers wanting to pursue land instability hazards should certainly peruse the extensive engineering and earth science literature, especially the United States Transport Research Board report cited under Varnes (1978) in the references at the end of this chapter.

Expansive soils (Hays, 1981)

Expansive soils annually cause more than $US3000 million damage (1989 dollar value) to roads and buildings in the United States. This cost exceeds the combined, annual damage bill from all climatic

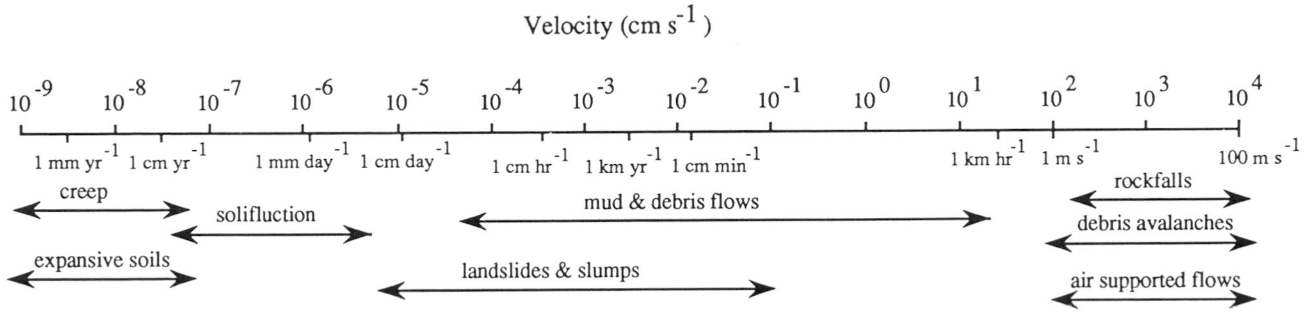

FIG. 13.6 Classification of land instability based upon rate of movement (derived from Finlayson and Statham, 1980)

hazards in this country. Because the process works so slowly, damage is not always obvious, and the actual annual cost to society may in fact exceed \$US10 000 million. Fifty per cent of the damage occurs to highways and streets, while 14 per cent occurs to family dwellings or commercial buildings. Of the 250 000 homes built in the United States each year on expansive clays, about 10 per cent will undergo significant damage and 60 per cent will undergo minor damage in their lifespan. Figure 13.7 maps the extent of expansive soils in the conterminous United States. Over one-third of the country is affected. The extent of the problem is not unique to the United States, but occurs on a similar scale in other countries such as Australia and the Soviet Union, where soil and geological conditions are similar.

Expansive soils are produced mainly by clays derived from two major groups of rocks. The first group consists of aluminum silicate minerals in volcanic material which decompose to form montmorillonite. The second group consists of sedimentary rocks, mainly shales, containing this clay mineral. In Australia both these rock types are very prevalent because large areas of the eastern half of that continent were subject geologically to volcanism. Additionally, slow evolution of the Australian landscape has permitted the widespread

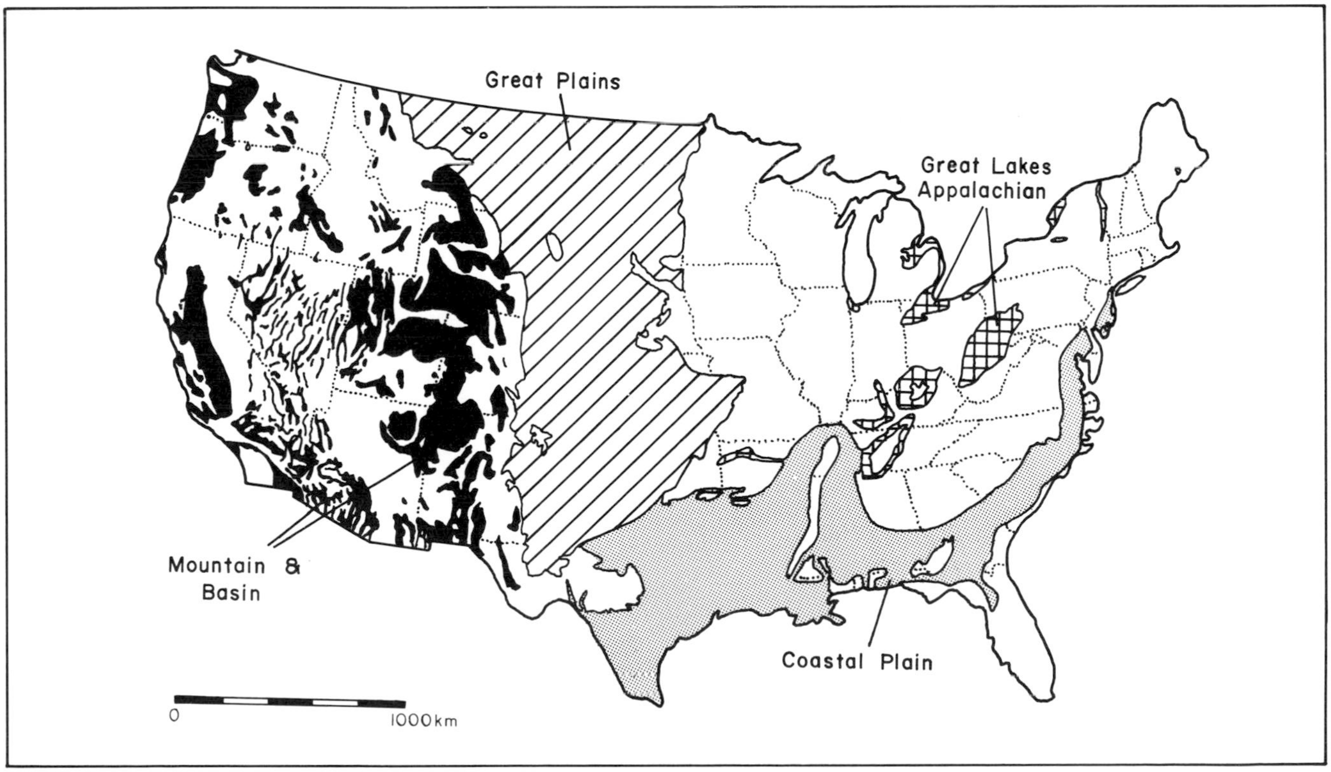

FIG. 13.7 Extent of expansive soils in the United States (adapted from Hays, 1981)

accumulation of large quantities of montmorillonite clay as cracking clays on inland river systems.

Expansion usually takes place when water penetrates the lattice structure of clay minerals. Two factors must be met before swelling can take place. First, it must be possible for a volume change to occur and, second, water must be present. An increase in soil moisture content of only 1–2 per cent is sufficient to cause expansion. In order for the volume change to be made, sufficient clay must be present. A thick layer of expansive clay has the potential for greater volume change than a thin layer. If the load on the clay is high, then compressive forces may exceed the expansive force exerted by clay upon wetting. Because this compressive load decreases towards the ground surface, the presence of expansive clays in the upper soil layers increases the effect of swelling and shrinking. Large buildings positioned on expansive clays will usually not be affected because the weight of the structure can prevent expansion. Extensive surficial cracking due to desiccation is also indicative of expanding clays. In Australia the presence of *gilgai* or undulations in the soil surface with wavelengths of 1–2 m is indicative of the problem.

People can enhance the effect of expanding clay by disrupting ground drainage and modifying vegetation. Drainage from a house gutter or septic tank into an area of expansive clays can cause that material to expand, while the drier material under the house does not. This differential expansion can affect foundations closest to the wetted soil. The removal of vegetation can also result in increased soil moisture, because evapotranspiration from trees or shrubs is negated. In built-up areas where rainfall variation is high, expanding clays can undergo repetitive cycles of swelling and shrinking, hence putting unequal stress on foundations over time. The process is quite common in Australia and is illustrated schematically in figure 13.8 for my house. During droughts, the soil underneath a house will dry out more slowly than that in the open. The exposed soil shrinks and the outer foundations of the house tend to settle before the inner ones. After rainfall, the exposed soil is wetted and expands. A wetting front then proceeds under the house causing expansion over time towards the center of the house. Trees around a house can accelerate the drying effect during drought because of their higher evapotranspiration, while lawn watering can accelerate the wetting effect. Cracking of interior walls, especially during a change from drought-to-wet or vice versa is very common. Floors can become creaky as pilings sink differentially due to variations in clay or moisture content.

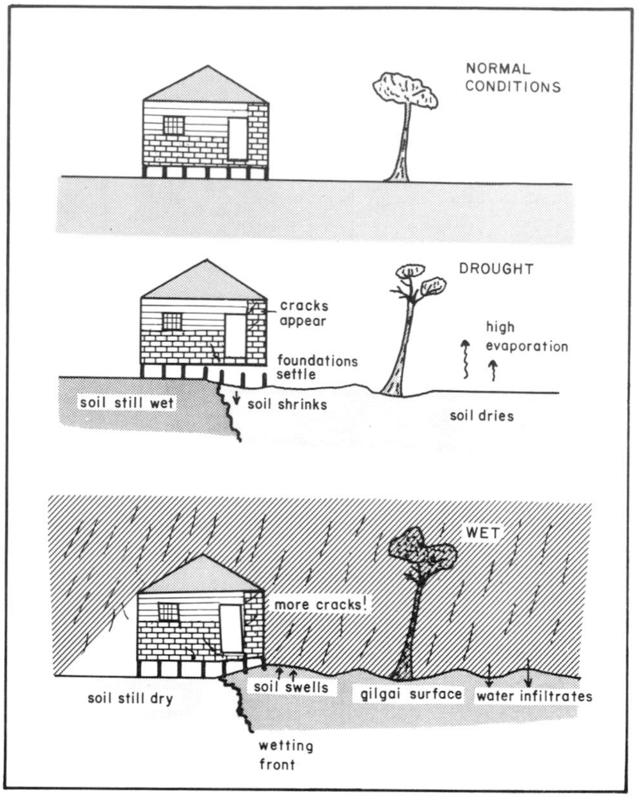

FIG. 13.8 Schematic representation of the effect on a house of alternate wetting and drying of expansive soils

The most efficient way to reduce damage because of expansive soils, is to avoid them. In Australia, however, this is almost impossible. Other methods to minimize damage include removing the expanding soil, especially at the surface, applying heavy loads to the surface, preventing water access to the site, prewetting the soil and preventing its drying out forever afterwards, or ensuring that foundations are sunk deep enough to minimize the effects of near-surface swelling and shrinking. It is also possible to change the ionic character of the soil by adding hydrated lime, $Ca(OH)_2$, to the surface of an expanding clay. The lime lowers the exchange capacity of the clay. Water, thus, cannot be substituted into the internal lattice structure of this clay mineral as efficiently, and expansion is reduced. Note, that the addition of lime to a soil subject to cracking is not recommended, if that soil is also a quick clay. Liming dramatically decreases this latter soil's cohesiveness.

Creep and solifluction (Leopold et al., 1964; Sharpe, 1968; Young, 1972; Finlayson and Statham, 1980)

Under sustained shearing stresses, all soil and rock materials on slopes exhibit viscous behaviour, which is termed *creep*. Where the melting of ground ice exacerbates the movement, the process is termed *solifluction*. Rates of creep are not substantial and rarely exceed 1–2 mm yr^{-1}, although velocities over 100 mm yr^{-1} have been measured on slopes as steep as 40 degrees. Because creep is more active towards the surface of a soil profile, the rate of movement and the resulting displacement decrease with depth. This is partly due to compaction and increased loading at depth. If material has some long-term instability, shearing planes or surfaces will develop in the region of maximum shear stress. This point usually occurs at the base of the soil profile between the B and C horizons, or along fractures (bedding planes) in the underlying weathered bedrock. It is often evidenced by a stone line, above which banding in the soil occurs, and below which imperceptible movement takes place. Material in this zone will shear until its strength is reached. If further shearing eventuates upwards in the profile, the displacement becomes cumulative and, hence, greatest towards the surface.

Creeping of a regolith downslope can also be accomplished by alternate expansion and shrinking of a soil. The most obvious mechanisms for this process is the presence of expanding clays under the influence of seasonal inequalities in rainfall. The process can also be accomplished by freezing and thawing of water in the soil, and by salt crystallization. Each time that expansion happens, the soil tends to be pushed upwards at right angles to the slope (figure 13.9). This process weakens the coherence of the soil mass. When shrinkage or thawing eventuates, the soil settles back to its original position; however, gravity will tend to move the material slightly downslope. Creep rates should be proportional to the sine of the angle of the slope, and all material should move downslope. In practice, it has been found that the rates of movement on a constant slope can vary greatly over small distances, to the point of being random. There are also measurements indicating that creep movement can occur upslope. This result implies that expansion and contraction do not operate normal to the slope. In this regard, a slope must be considered three-dimensional, and it is possible for movement in any direction including upslope and laterally. This aspect has already been recognized in periglacial environments, where lateral sorting of sediment dominates the micro-morphological evolution of the landscape. In environments where a distinct thawing season exists after the soil has been frozen, or where rainfall is seasonal, creep rates will vary seasonally. The process is fastest in spring, when temperatures are still cool enough for the ground to re-freeze at night, and during the wet season, when distinct seasonal inequalities in rainfall exist.

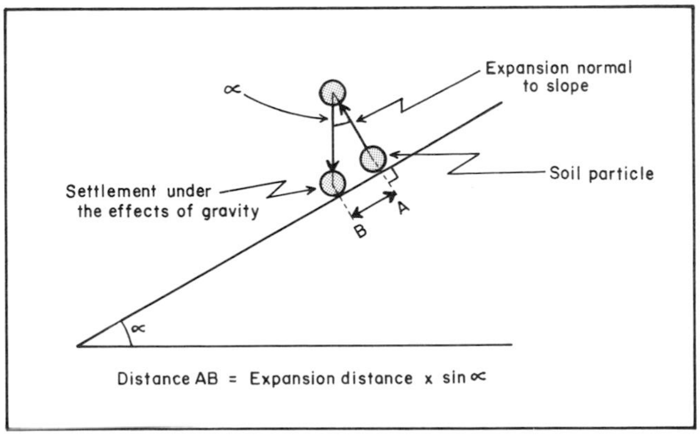

Fig. 13.9 Mechanism for the downhill movement of material on a slope caused by soil creep

Creep in clay does not always depend upon expansion and contraction. If clay is wet enough to become plastic, it will deform under load in the downslope direction. As the rate of creep depends upon the weight of overburden, deformation may occur only at depth in the profile and not at the surface. Creep, in clays at depth in a particular zone, can cause the clay minerals to realign themselves, reducing shear strength in this layer, which subsequently becomes the location for development of a shear plane. The cohesiveness of clay decreases with increasing moisture content; so, continual creep in clays has its highest rate where the clay is moist for the longest duration.

Creep processes are not restricted solely to clays or to locations with expansive soils. Because creep operates basically under the effect of gravity, any disturbance to a soil on a slope will result in downhill movement of material. Animal burrowing, plant-root penetration, tree collapse, and simple trampling by animals or people can be the cause of this disturbance. While the short-term disturbance seems minor, aggregated over time these effects can have significant influence on near-surface creep rates.

The process of solifluction involves additional factors. Solifluction is restricted to environments where ground freezing takes place. The expansive force initiating creep is due to moisture in the soil freezing, because frozen water occupies a greater volume than the equivalent weight in liquid form. Solifluction can produce accelerated rates of downslope movement of material on slopes as gentle as 1 degree. Freezing in soil can also concentrate water at a freezing front and, hence, form ice lenses, which can constitute up to 80 per cent by volume of the soil mass. The melting of all or some of this ice can increase pore water pressures, and form shear planes for slope failure. In areas underlain by permafrost, where seasonal melting prevails in an active surface zone, solifluction can totally dominate a landscape such that material on all slopes appears to be in a continual state of rapid movement downslope. The discussion of solifluction and its geomorphic ramifications are beyond the scope of this book; but it should be realized that ground ice and solifluction processes represent a serious hazard to construction and transport in Arctic regions of Canada, Alaska and the Soviet Union. Most mountain regions are also cool enough, because of their high elevation, to be affected by solifluction. The actual areas susceptible to these deleterious agents represent about 20 per cent of the world's landmass.

Mud and debris flows (Sharpe, 1968; Cornell, 1976; Bolt et al., 1975; Whittow, 1980; Innes, 1983)

A flow is any slope failure where water constitutes a major component of the material and a major controlling factor in its behaviour. Flows can usually be subdivided into two categories, based upon the type of material. If clays dominate, the flow is termed a 'mud flow' or 'earth flow' in the United States and a 'mudslide' in the United Kingdom. Note that this term incorporates lahars mentioned in the previous chapter. The term 'lahar' is more restrictive in that it is a mud flow originating in volcanic ash. If the range in particle size is highly variable, then the flow is termed a *debris flow*. Generally, the material in a flow is water-saturated and unconsolidated. Water not only can permeate through the material, but it can also be absorbed by the sediment. For this reason, most flows represent the further downslope movement of material that has already undergone some sort of failure. For instance, a rockfall that disintegrates into smaller pieces while falling will produce an unconsolidated debris mantle when it comes to rest. Over time, this debris can chemically weather, breaking down into smaller particles. When wetted, this material may absorb water, increase in weight and begin to fail again at lower slope angles.

Flows appear to behave more like rivers than they do glaciers. The high water content increases pore water pressure in the deposit, a process that decreases shearing strength. Flows usually begin moving along a basal shear plane. Rates of movement can be as low as $1-2\,m\,yr^{-1}$, or as high as 600 m or more per year. Fastest movement occurs during the wettest months. Mud flows do not necessarily move at consistent rates and commonly surge. In southern California, surge rates of $3-4\,m\,s^{-1}$ have been observed in large mud flows. Typically, even with lahars, velocities average under $20\,km\,hr^{-1}$. Slow-moving earth flows have caused property damage, but minimal loss of life, in Czechoslovakia, England, Switzerland and Colorado, United States. In periglacial environments dominated by solifluction, mud flows are a common occurrence.

Coarse debris flows are much more difficult to move. Fine material must be present to aid water retention. The debris flow may consist of very low volumes of sediment moving downslope at rates of several kilometers per hour. Coarser debris tends to get extruded to the surface and to the sides during movement, such that coarse-sized levee banks are built to the sides and front of the moving deposit. Typically, debris flows require slopes of 30–40° for movement to start, but flow continues over slopes of 12–20° (figure 13.10). Note that this type of flow does not include large, catastrophic events moving coarse debris at higher velocities—such an event is termed *debris avalanche* and will be discussed in detail subsequently. When mud flows and debris flows stop, excess water tends to drain from the deposit, forming either lobate or planar fans. In some cases where subsequent filtering out of sediment has happened, flow deposits can be

Fig. 13.10 Widespread debris flows on slopes of 12–30° in the Kiwi valley area, Wairoa, New Zealand, in February 1977 (photograph courtesy of the Hawke's Bay Catchment Board, Napier, New Zealand). These were initiated by heavy rainfall, with intensities of 127 mm in 2 hours, and 252 mm in 5.5 hours being recorded.

mistaken for alluvial fan deposits. The thickness of the deposit is inversely proportional to the fluidity of the deposit, and to the slope angle at rest.

Under extreme wetness, both mud flows and debris flows can liquefy, whereupon the weight of material is borne by pore water pressure and not grain-to-grain contacts. This problem is particularly severe in marine clays laid down following the retreat of the last major glaciation. Subsequent desalinization forms quick clays, which are very susceptible to liquefaction. The St Lawrence valley area in eastern Canada was effectively flooded by the ocean during the waning phases of continental glaciation, with the subsequent deposition of extensive, thick clay beds susceptible to earthflow and liquefaction. Mud flows with volumes in excess of 20 million m^3 have been quite common in this region. The St Jean-Vianney, Quebec, failure in 1971 took 35 lives. Similar types of deposits have been recorded in Scandinavia and the Soviet Union. In 1893 at Vaerdael, Norway, 112 people were killed by one such failure.

Debris flows can also be caused by people dumping unstable mine tailings. The Aberfan, Wales, disaster of 21 October 1966 was one of the more tragic disasters of this nature to happen this century. A 250-meter high, fine-grained, rain-soaked coal-tailings pile failed and turned into a debris flow, which swept through a school killing 116 children, the entire juvenile population of the town. A similar flow in a coal-mining waste dump in West Virginia killed 118 people in 1972. The Aberfan disaster illustrated the unsafe nature of coal tailings in the United Kingdom: over 25 per cent were subsequently found to be in a similar condition of instability.

Landslides and slumps

Mechanics (Leopold et al., 1964; Whittow, 1980; Chowdhury, 1978)

Slope failure in terms of landslides can be modelled using the Mohr-Coulomb equation. Because many slides occur along a distinct shear plane, this equation can be modified to include the effects of shear planes within an unconsolidated deposit, bedding planes and joints in weathered or unweathered bedrock, and the presence of a water-table. Landslides, therefore, tend to happen in two different types of material, consisting of either bedrock or unconsolidated sediment, usually clay. In the simplest case, where the Mohr-Coulomb equation is easily applied, landslides move down-slope parallel to the line of failure. However, landslides can undergo rotation as well. In this case, the failure develops slowly as a *slump*. Figure 13.11 illustrates this process of rotational sliding or slumping. While the mechanics of rotation are the same as a simple planar slide, the movement of the material is determined by the arc of rotation centered at a point outside the slip material. Rotational slides can be discussed in terms of: a head scarp, where the material has separated from the main slope; a slip surface, upon which the material rotates; and a toe, which consists of the debris from the failure. Rotational slides may also generate transverse cracks across the body of the slipped mass, and tensional cracks above the head scarp. These tensional cracks can be lines of failure for future slides. The toe of the slide controls the terminal point of the failure. Interference with the toe of the slide can also initiate further failure.

Causes (Sharpe, 1968; Finlayson and Statham, 1980; Crozier, 1986)

With landslides and slumps, not only is the stress applied to the area of potential failure important but the behaviour of the toe also becomes crucial to the timing of the slide. One of the major causes of landslides is undercutting of the basal toe in some manner. The causes and circumstances for landslides in any form are thus multitudinous. Geological or topographical factors condition the location of a potential slide. If weathered or unweathered bedrock contains inherent lithological or structural weakness, then these areas will favour slide failure. Lithologically favourable material includes leached, hydrated, decomposed, chloritic

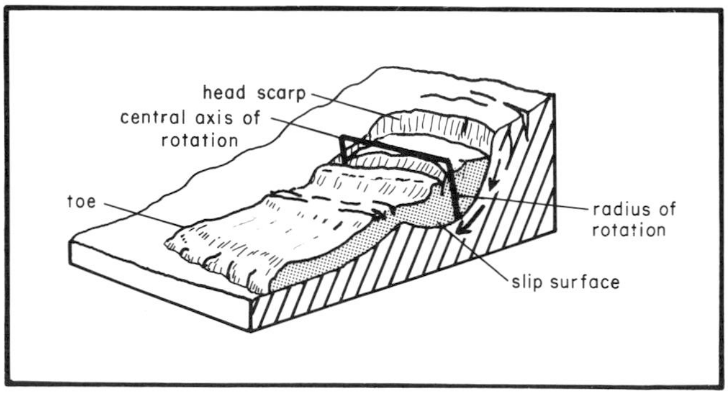

Fig. 13.11 Process of rotational sliding (Whittow, 1980, © and reproduced with the permission of John Whittow, Department of Geography, University of Reading)

or micaceous rocks, shales, poorly cemented sediments or unconsolidated material. Additionally, if this material is well-rounded or contains clay, then it will flow more freely under pressure. Stratigraphically favourable conditions for landslides include massive beds overlying weaker ones, alternating permeable and impermeable beds, or clay layers. Structurally favourable conditions include steeply or moderately dipping foliations, cleavage, joint, fault or bedding planes. Rock that is strongly fractured, jointed, or contains parallel alignment of grains because of crushing, folding, faulting, earthquake shock, columnar cooling or desiccation is also likely to fail. Internal deformation structures resulting from tectonic movement, solution or underground excavation will aid slope failure, especially if such features do not reach the surface. Any water-saturated, porous lens within a regolith will also tend to form a zone of preferred shearing. Topographically, any cliffed or steep slope caused by stream, glacier, wind or wave erosion is also liable to landsliding, as are areas of block faulting, previous landslides or subsidence, and artificial excavation. Finally, any area which has been denuded of its soil-retaining vegetation because of deforestation, overgrazing, cultivation, fires or climatic change has the potential for sliding.

Initiating causes of landslides are just as plentiful. The removal of basal support provides the easiest way to start a landslide. Natural agents such as undercutting by running water, waves, wind and glaciers are prime candidates. Extrusion of material from the base of a slope is also effective. This can take the form of outflow of plastic material within the deposit, washout of fines, and melting of ice. The material at the base could also change its characteristics through water absorption due to flooding, solution of soluble material such as limestone or salt, or chemical alteration. In developed areas, people have now become the major cause of toe instability on slopes—through mining, excavation, quarrying, foundation construction, and road, rail and canal works. Overloading of the slope material is also a common means of initiating failure. Overloading can occur by saturating the slope with rainwater or water from upslope streams or springs, or by loading the slope with snow or debris from upslope instability. Humans can also initiate overloading by dumping spoil or by building foundations and structures on the slope.

Slope failure can also be induced by reducing internal coherence in the slope material. Increased lubrication due to heavy rainfall, or runoff filtering through a regolith is a common mechanism for triggering failure, either immediately, or some months afterwards. Cracking of the slope material through partial failure, desiccation, earthquakes, or internal movement can also permit easier water penetration into the slope material. Blockage of drainage, usually by raising the base of the water-table, can reduce shear resistance internally. As already mentioned, the activities of humans through drainage modification, deforestation, overgrazing, or water discharge from septic systems and domestic usage also reduce internal coherence through lubrication. Earthquakes and volcanoes can either crack slope material, or reduce coherence through the process of liquefaction. Even thunderstorms can trigger rockfalls and landslides because of the vibrations that are transmitted through slopes. Similarly, humans now perform activities which vibrate the ground and could trigger landslides. These activities include vehicle movements, blasting, pile-driving, drilling and seismic work.

There are a number of processes which can cause slope failure due to prying or wedging. Water freezing in cracks, increased pore water pressure after heavy rain, expansion of soil material because of hydration, oxidation, carbonation or the presence of swelling clays, tree-root growth, and temperature changes are the main mechanisms. Finally, strains in the earth due to sudden changes in temperature, to atmospheric pressure or the passage of earth tides are also factors which could trigger landslides.

The above conditions favouring landslides may act in concert to trigger instability. As a result, slope failure can happen instantaneously, irregularly but accumulatively over time, or with considerable temporal lag after the occurrence of a triggering event. Figure 13.12 illustrates this complexity in landslide initiation. Ultimately, failure will eventuate when the shear strength of the soil is finally exceeded by a critical shear stress. The critical shear stress can fluctuate slowly or dramatically over time, while the shear strength of the material can oscillate because of seasonal factors, or decrease slowly because of any of the above conditions. Slope failure may happen at a number of points in time. If stress dramatically increases, for instance because of heavy rainfall, and exceeds soil strength, then the cause of the landslide is obvious. However, soil strength could be decreasing, or the stress on a slope increasing slowly and imperceptibly over time, to the point that the cause of the landslide is not definable. Thus, with any landslide, there is a randomness as to the timing and cause of the failure.

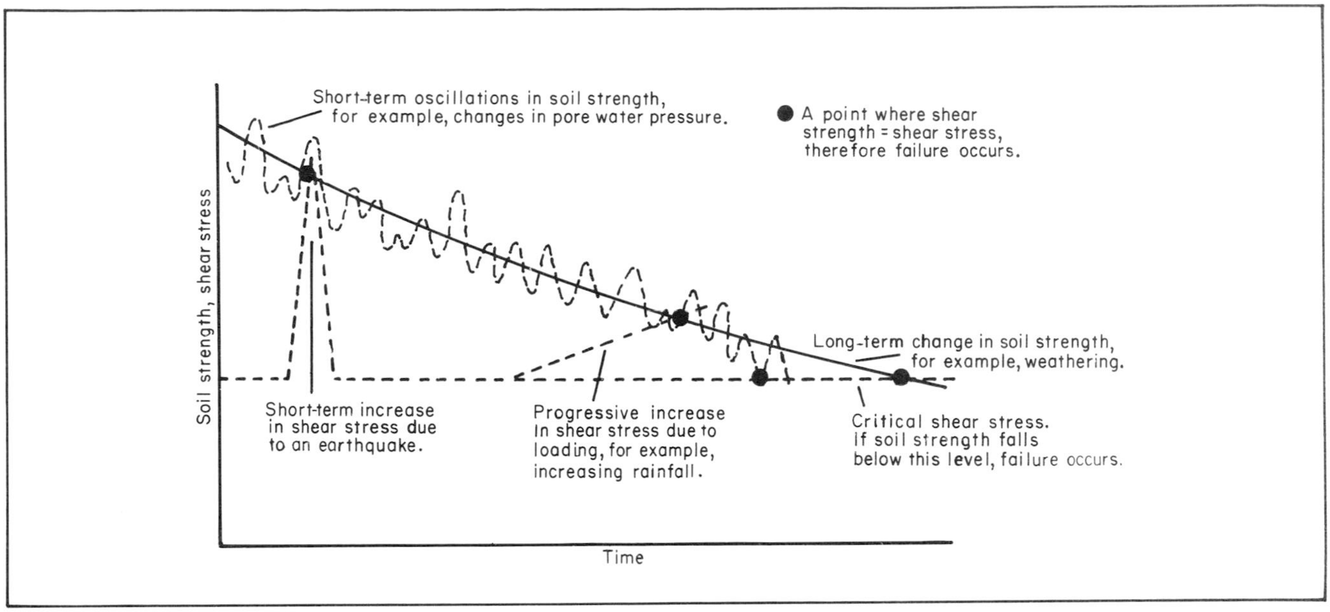

FIG. 13.12 The complexity of landslide initiation over time (Finlayson and Statham, ©1980 with permission Butterworths, Sydney)

Landslide disasters (Cornell, 1976; Bolt et al., 1975; Whittow, 1980; Coates, 1985; Blong and Johnson, 1986)

Landslide disasters are difficult to separate from other types of land instability, and from some of the larger disasters that trigger the event. For instance, tropical cyclones, apart from drowning people, usually bring very heavy rainfalls; this not only increases pore water pressure in potentially unstable material but, through added weight, also increases stress in slope deposits. Such a situation occurred dramatically during Cyclone Namu, which struck the Solomon Islands in May 1986. Well-vegetated mountain slopes, unaccustomed to the torrential rain dropped by the cyclone, failed over a wide area, triggering landslides that flowed into rivers and turned into mud flows that flooded over fertile cultivated coastal plains. The landslide-initiated mud flows were responsible for the damage to about 50 per cent of the homes on the islands. It is difficult to ascertain if people died from the landslide, drowning or some other effect in these circumstances. Some of the worst natural disasters have also occurred as landslides triggered by earthquakes. The Chinese earthquakes in Shensi province in 1556, and at Kansu in 1920 caused failure in loess deposits, collapsing homes dug into the cliffs.

Urban areas seem to be most affected by landslides. In Rio de Janeiro in 1966 record-breaking rainfall in January and March caused catastrophic landslides, which struck shanty towns around the mountains in the city. Over 500 people were killed in the slides and another 4 000 000 were affected by disrupted transportation and communications. In the following two-year period, over 2700 people were killed in the Rio de Janeiro area by landslides and other slope instability events which afflicted over 170 km^2. In Hong Kong a tropical cyclone in 1976 dropped 500 mm of rain in two days, triggering landslides on steep slopes, which killed 22 people. Similar slides there in 1966 killed 64 people. Many of the slides occurred where dense urban sprawl encroached upon steep slopes and undermined the toe of unstable sediments. Japan is also particularly vulnerable to typhoon-generated slides near urban areas. Kobe was hit in 1939 by rain-induced landslides that killed 461 people and damaged 100 000 homes. In 1945 Kure was affected by similar landslides that killed 1154 people. And in 1958 Tokyo was struck by a typhoon that generated over 1000 landslides and killed 61 people. Even in the United States, urban landslides have caused substantial death at times. For example, landslides in the Los Angeles area killed 200 people on 2 March 1938.

In underdeveloped, heavily vegetated regions, especially in the tropics, mega-landslides play a major role in the downslope movement of material. In 1935 New Guinean landslides cleared 130 km^2 of vegetated slopes. The Bialla earthquake of 10 May 1985, on the island of New Britain in

Papua New Guinea, triggered a mega-landslide in the Nakanai range that dramatically infilled the Bairaman River (figure 13.13). About 12 per cent of slopes in New Guinea are subject to landslide denudation every century. Elsewhere in the tropics, an area of $54\,km^2$ was cleared on slopes in Panama in 1976. Tropical Cyclone Wally, which struck Fiji in April 1980, and Cyclone Namu, which passed over the Solomon Islands in 1986, both generated hundreds of landslides with an average volume of material moved in excess of $500\,m^3$ per hectare. Such landslides had not been witnessed in these areas in the previous 50 years. The cyclones effectively destabilized the landscape enough to ensure continued mass movements triggered by less severe rainfalls for years to come.

The ubiquitous nature of landslides is illustrated with reference to the conterminous United States in figure 13.14. By far the most prevalent zone for landslides occurs in the Appalachian mountains in the eastern part of the continent. Second in importance are the Rocky Mountains, particularly in the states of Colorado and Wyoming. The coastal ranges along the Pacific Ocean form the third most hazardous zone. In all, about one-seventh of the country is affected by landslides in some form. While most of the failures happen in mountainous areas, it is not restricted to these zones. The

Fig. 13.13 Mega-landslide on the northern slope of the Nakanai Range, New Britain, Papua New Guinea (photography by Dr Peter Lowenstein, courtesy of C. O. McKee, Principal Government Volcanologist, Rabaul Volcanology Observatory, Department of Minerals and Energy, Papua New Guinea Geological Survey). This slide, which was triggered by the magnitude 7 Bialla earthquake on 10 May 1985, occurred in Miocene limestone and extensively infilled the Bairaman River valley.

Mississippi valley has a low, but significant risk, as do plateaux on the southern Great Plains, where weak shales are found. The United States is not unique in this respect: the landslide is a common geomorphic process in most countries with steep or raised terrain.

Rockfalls (Scheidegger, 1975; Voight, 1978)

Rockfalls form one of most rapid land-instability events. They are generally restricted to unvegetated, vertical, rock cliff-faces but can occur in certain unconsolidated sediments which permit vertical slopes to develop. Generally, the vertical faces must be continually maintained by erosion of talus and screes, which form at the base of cliffs. If this debris is permitted to accumulate, then the vertical face becomes buried. The debris is removed through chemical breakdown, washing out of fines and slippage. There is, thus, a balance between the rate of cliff retreat controlled by the occurrence of rockfalls, and the rate of removal of debris at the base. The production of talus does not have to occur catastrophically, but instead debris can simply accumulate through the continual addition of debris as it is loosened from the cliff face by weathering, rainfall, small shock waves or frost action.

The preservation of a vertical face is dependent upon the characteristics of the bedrock. Generally, less resistant rock such as shales will not be able to develop cliff faces, because the rock is susceptible to weathering and slippage. Situations where there are massive beds of sandstone or limestone, or where these resistant beds overlie weaker beds, are ideal for the formation of cliffs. However, massive beds without much jointing do not erode easily, and hence do not produce rockfalls. For example, the escarpments along the east Australian coastline have faces up to 100 m or more in height in places, and appear to be retreating at rates as low as $1\,mm\,yr^{-1}$ without significant rockfalls. Significant joints or vertical failure planes should also be present for the occurrence of large rockfalls. Even jointing is not a guarantee that rockfalls will eventuate. Blocks separating from cliffs along vertical lines of weakness must also undergo disintegration in order for rockfalls to take place. Most of the major rockfalls on vertical faces in Australia appear to be occurring where humans have undercut the base of the cliff, where vibration is significant or where mining tunnels are collapsing near cliffs (figure 13.15).

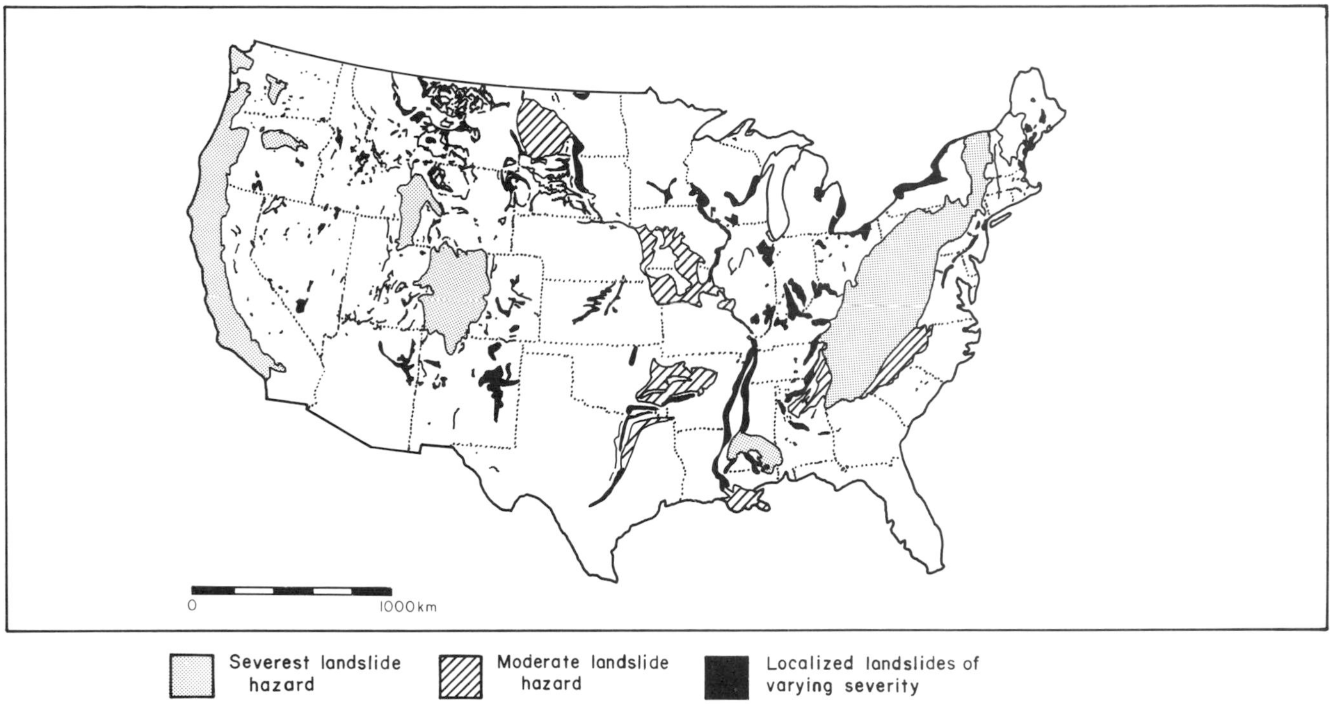

FIG. 13.14 Extent of landslides in the conterminous United States (after Hays, 1981)

FIG. 13.15 Rockfall along cliff face undercut by N. Nattai coal mine, Illawarra, New South Wales, Australia (photograph by Dave Cunningham, NSW National Parks). Note the lack of fresh talus elsewhere along this escarpment.

Globally, rockfall is the major process for the removal of debris from cliffs in scarped or steep mountainous regions affected by periglacial activity. While rockfalls are initiated by many of the factors which trigger landslides, the majority occur because of earthquakes. The Yungay disaster in Peru started out as a large-scale rockfall from Mount Huascarán triggered by an earthquake. Many of the landslides discussed above, especially in the Alps also began as rockfalls. In deeply snow-covered mountains, rockfalls are a major triggering mechanism for avalanches.

Debris avalanches (Varnes, 1978; Voight, 1978)

Debris flows and mud flows can also take on catastrophic proportions several orders of magnitude greater in size and speed than described above. The term 'debris avalanche' is more appropriate to these types of flows. Many of these events begin as landslides or rockfalls. All appear eventually to entrain a variety of particle sizes and behave as a fluid flow. The high velocities appear to be achieved by lubrication provided by a cushion of air entrapped beneath the debris. The flow literally travels like a hovercraft. In some instances there is little evidence of water involved in the avalanche, and in other cases masses of debris appear to have travelled with minimal deformation or re-alteration.

By far the most destructive debris avalanches in recent times occurred around Mount Huascarán, Peru, in 1962, and again in 1970. The 1962 event started as an estimated 2 million m^3 of ice avalanched from mountain slopes. This ice mixed with mud and water, and turned into a much larger mud flow with a volume of 10 million m^3. It swept down the Rio Shacsha valley killing 4000 people, mainly in the town of Ranrahirca. Material, some of which consisted of boulders over 2000 tonnes in weight, travelled down the valley at 100 km hr^{-1}, and 100 m up valley slopes. Having survived that event—and believing that another catastrophe like it was unlikely—residents of the area then experienced an earthquake on 31 May 1970, which caused the icecap of Mount Huascarán, together with thousands of tonnes of rock, to cascade down the valley. This debris flow entrained boulders the size of houses; as it travelled down the valley at speeds of 320 km hr^{-1}, it picked up more debris each time it crossed a glacial moraine. At the bottom of the valley, the flow had incorporated enough fine sediment and water to become a mud flow with a 1 km wide front. Eye-witness accounts describe the flow as an enormous 80 m high wave with a curl like a huge breaker as it approached the town of Yungay. Ridges over 140 m in height were overridden, and boulders weighing several tonnes were tossed up to 1 km beyond the rims of the flow. Within four minutes, Ranrahirca and the much larger town of Yungay were completely obliterated by 50–100 million m^3 of debris. The flow then continued down the Rio Santa reaching the Pacific Ocean 160 km downstream. Over 25 000 lives were lost. Only those who had raced to the top of ridges survived. The disaster ranks with the Mount Pelée eruption of 1902 in the completeness of its destruction.

Many large landslides described as such in the literature should in fact be classified as debris avalanches. Almost all have originated in steep mountain areas. For instance, the Bialla slide referred to above as a mega-slide, and shown in figure 13.13, is technically a debris avalanche. Historically, many large landslides fit into the same category. At Elm, Switzerland, on 11 September 1881 about 10 million m^3 of mountain slipped onto the town, burying it to a depth of 7 m and killing 115 people within 55 seconds. A debris avalanche of similar size at Goldau in Switzerland, in 1806, buried four villages and killed 457 people. On 4 September 1618, over 2400 people were killed by debris avalanches in the Chiavenna valley in Italy. Larger disasters have also taken place. Reportedly, the total population of 12 000 in the town of Khait, Tadzhikistan, Soviet Union was killed by converging avalanches triggered by earthquakes in the Pamir mountains in 1949. The largest amount of debris ever involved in a recorded failure occurred in the same mountains at Usoy in 1911. Approximately 2.5×10^9 m^3 of rock failed, damming the Murgab River and forming a lake 284 m deep, and 53 km long. The debris avalanche was triggered by an earthquake registering 7 on the Richter scale, and destroyed the town of Usoy, killing 54 inhabitants. A failure one-tenth this size, containing 0.25×10^9 m^3 of soil and rock, fell into a reservoir across the Vaiont valley of Italy in October 1963. The impact sent a wave of water 100 m high over the top of the dam and down the valley, killing 3000 people.

Air-supported flows (Bolt et al., 1975; Scheidegger, 1975; Whittow, 1980)

Avalanches are commonly thought of in terms of ice or snow cascading down slopes; however, they are not restricted just to frozen water. Snow avalanches are part of a phenomenon of debris movement involving the interstitial entrapment of air. Any material may be involved, including hot gasses entrained within volcanic tephra (pyroclastic flows), mixtures of sediment and water in debris avalanches as described in the previous section, and collapsing loess (the Shensi, China, earthquake of 1556). The material in an avalanche flow is supported by air rather than by grain-to-grain contacts. Thus, avalanches are analogous to the liquefaction process whereby air, rather than water, fills voids and sustains the weight of the flow. The process of liquefaction has been described in detail in both this chapter and chapter 11. Even if the avalanche makes contact with the ground, or contains a high proportion of liquid water, the internal air content ensures that coherence is minimal, and friction so low that the avalanche can obtain high ground-velocities. Without turbulence, it is possible for velocities to reach speeds in excess of 1200 m s^{-1}. However, avalanches rarely exceed 300 m s^{-1} because of the disruption to movement by turbulence. Unfortunately, turbulence permits most avalanches to entrain loose debris up to the size of boulders, a process that can cause substantial erosion along part of its path. The swift movement ensures that such flows are very powerful, and that they are preceded by a shock wave, which for snow avalanches may exert forces greater than 0.5 tonnes m^{-2} as shown in figure 13.16. The internal turbulence within a snow avalanche tends to generate swirls of debris with velocities in excess of 300 km hr^{-1}, greater than the movement of the avalanche itself. Forces of 5–50 tonnes m^{-2} can be generated internally, and are of sufficient magnitude to uproot trees and move large structures.

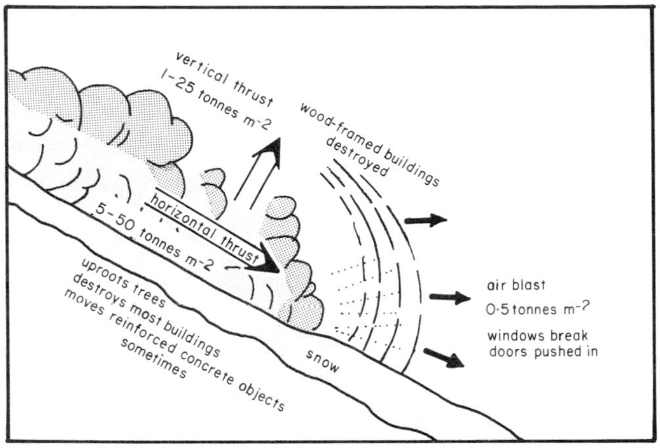

FIG. 13.16 Process and impact forces generated by a snow avalanche (Whittow, 1980, © and reproduced with the permission of John Whittow, Department of Geography, University of Reading)

If the debris in the flow has liquefied, velocities are high and the flow thin, then it is possible for shear sorting to occur internally in the flow. Shear sorting tends to produce internal sorting of debris such that larger objects are expelled to the top and sides of the flow. Basal surges in volcanoes behave in this fashion, and produce ash deposits which are well sorted with the coarsest material towards the top of the deposit. The debris avalanche which wiped out Yungay in Peru in 1970 preferentially moved boulders towards the top and sides of the flow as it underwent shear sorting; however, it is not certain whether water or air was involved in the dynamics of the sorting process. The results of shear sorting enhance the prospect of survival for people caught in snow avalanches. Because human bodies are relatively large compared to the rest of the debris in the flow, they tend to move to the top of the avalanche under shear sorting. Many people are found and rescued alive from snow avalanches by rescuers simply probing the upper layers of the avalanche deposit.

The word 'avalanche' has always conjured up images of snow cascading down mountains at high speed and wiping out unsuspecting alpine villages. This image is based upon accumulative disasters in the European Alps. Snow avalanches are also a common hazard in the Himalayas, Andes, and Rocky Mountains, especially in spring. One of the more recent disasters happened in 1978 at Col des Mosses, Switzerland, when an avalanche overwhelmed a ski-lift and killed 60 people. Even the Prince of Wales was almost killed by an avalanche that overtook his skiing party in 1987. Two of the largest disasters have occurred during wars. When Hannibal decided to invade Rome in 218 BC, by crossing the Italian Alps, his efforts were doomed by the huge losses inflicted on his army by snow avalanches. It is estimated that 15 000–18 000 of Hannibal's men were killed on this one trip. Between 1916 and 1918 in the First World War, snow avalanches killed around 40 000 Italian and Austrian troops fighting in the Italian Alps. Other disasters have been less costly, but just as spectacular. Between 1718 and 1720 in Switzerland, four separate snow avalanches took over 250 lives. The largest Swiss avalanche took 300 lives in 1584. Other Alpine areas have suffered similar tragedies more recently. In 1910 an avalanche at Wellington, Washington, in the Rocky Mountains, swept through two trains killing 96 people. In the same year at Rogers Pass, British Columbia, 62 people were killed.

In recent years, snow avalanche disasters have decreased with the advent of monitoring techniques and remedial measures designed to negate the hazard. For instance, most zones where snow avalanches are likely to occur have been mapped in populated mountainous areas. The fact that avalanches tend to recur in the same place has aided this mapping. Sharp increases in temperature after a winter of heavy snowfall, or during a spring accompanied by rainfall, produce ideal conditions for avalanches. Studies of the dynamics of snow and ice on slopes have allowed these conditions to be forecast to a high degree of accuracy. In the North American Rockies and the European Alps, army field cannon or explosives are used to trigger snow avalanches before they can build up to catastrophic dimensions. These techniques have proved effective in keeping passes free of large disruptive flows, and in protecting ski slopes in resort areas. Where road and train routes must be kept open, avalanche chutes are used to divert flows where they recur frequently. These chutes simply consist of roofs which permit the avalanche to pass harmlessly over the road or railway. Near settlements, where these techniques are impractical, other techniques for braking, deflecting or disrupting the flow have been established. Upward-deflecting barriers can be constructed along preferred avalanche pathways to convert a ground-based avalanche into a slower, less forceful, airborne one. Concrete blocks can be arranged on slopes to divert avalanches away from settlements. The judicious alignment and reinforcement of buildings facing avalanche-prone slopes can also negate damage and prevent loss of life. These techniques are presently so successful that the risk of death due to avalanches has been virtually reduced to zero in most of Europe and North America. Rather than being common events at the end of each winter, deadly avalanches have become rare enough to be unusual.

Subsidence (Sharpe, 1968; Bolt et al., 1975; Coates, 1979, 1985; Whittow, 1980; Hays, 1981)

Land subsidence is a major type of land instability that does not depend upon many of the basic mechanisms outlined in this chapter so far. Rather, it is due to factors that include the effects of humans through water and oil extraction, and mining activities. The proper discussion of these latter factors is beyond the scope of this text. In this section only the natural causes and mechanisms of land subsidence will be discussed.

The importance of land subsidence can take place at several scales of magnitude, spatially and temporally. For instance, a small sinkhole caused by limestone solution is not significant except to a local landowner; however, uniform subsidence of several square kilometers can become severely disruptive to a much wider community. The Alaskan earthquake of 1964, in a few minutes, produced regional, tectonically induced subsidence along the south coast of Alaska, which had a profound effect upon the landscape. This sudden change in the relative location of land and sea resulted in severe economic problems in harbours. At the same time, there are locations, such as the south coast of England, where long-term subsidence of the crust has been occurring since the demise of the last continental ice sheets, and as far back as the Pliocene. While coastal erosion and flooding can result from this regional subsidence, it is possible for humans to adjust to these changes.

The mechanisms for land subsidence are as varied as those for landslides, and can be divided into natural or artificial categories. Natural factors fall into three groupings: chemical, mechanical and tectonic. Chemical agencies of land subsidence involve mainly solution by ground water of limestone, halite, gypsum, potash or other soluble minerals. Because limestone is very extensive, the development of *karst* topography can lead to sudden ground subsidence over large areas. This is a particular problem in the southeastern and mid-western part of the United States, which is underlain by extensive limestone deposits. In Australia subsidence features are widespread across the Nullarbor Plain; however, the region is neither densely settled nor intensely farmed, so that karst subsidence has little effect on people. A rarely recognized chemical cause of subsidence is the natural ignition (or, more recently, by industrial vandalism) of coal seams such as has occurred at Burning Mountain, north of Scone, New South Wales, Australia.

Mechanical factors involve the removal of material at depth in the soil by extrusion. This can include the ejection of quick clays. However, by far the most effective process is the melting of ice lenses in permafrost areas. Because living and decayed vegetation provides an insulating blanket against the penetration of heat into permanently frozen ground, its removal will often result in catastrophic and irreversible melting of ground ice and subsequent surface collapse. The emplacement of heat conduits, such as foundations and telephone poles, will also permit the transfer of heat from the surface to depth where melting will occur. Even heat from a building can conduct into permafrost to cause melting and subsidence unless it is prevented from doing so by an intervening layer of insulation. The effect of subsidence in permafrost regions has major implications for landform development and the survival of fragile, cold-climate ecosystems.

Tectonic factors involve warping of the Earth's crust. The Earth is an elastic body which can be deformed by the weight of glaciation, and which will return to its original shape after the ice load is removed. Glaciation causes forebulging some distance in front of the ice front. In these regions following glacial retreat, the crust settles instead of rebounding. Large sections of the southern coastline of England, including London and northern Europe, are affected by this process, and are presently undergoing subsidence at rates in excess of $2-3\,mm\,yr^{-1}$. This is also being exacerbated by the long-term downwarping of the southern North Sea basin. If there is a threat of sea-level rise in the next century (discussed in chapter 4), then it is possible for this rise to be exacerbated by tectonic subsidence in these regions. Subsidence can also be caused by ground loading. This is an obvious factor where major rivers are presently depositing large volumes of sediment on shallow continental shelves. Much of the coastline surrounding the Mississippi delta is subsiding because of continual, long-term deformation of the crust due to the deposition of sediment debouched from the Mississippi River.

Concluding comments

Land instability hazards encompass the most diverse range of processes and phenomena of any hazard discussed in this book. They also occur over the greatest timespan, from soil creeping downslope at rates of millimeters per year, to the various types of avalanches reaching speeds in excess of 300 km hr^{-1}. In many respects land instability is the most commonly occurring hazard. For instance, solifluction affects 20 per cent of the Earth's landmass, while expansive soils and landslides cover 33 per cent and 14 per cent respectively of the conterminous United States. These aspects imply that this hazard should be intently studied and any examination of the number of books devoted to the topic on library shelves would tend to back up this statement. However, land instability hazards are not an easy topic to cover. The fact that they have attracted equal attention from engineering- and geology-related disciplines has led to a confusion of terms and a discipline-specific emphasis. This has meant that some important phenomena such as expansive soils are relegated in treatment to disciplines, in this case soil science, that do not always attract the attention of natural hazard researchers. It has also made exchange between disciplines difficult, not only because of the clash over terminology, but also because engineers tend to center discussion around mathematics while geologists and geomorphologist are more interested in processes and morphology. Despite the wide research attraction, many aspects about land instability remain under-explained. For instance, little is known about the mechanics of debris avalanche movement. This has meant that our ability to predict the occurrence of many land instability hazards is far from effective. For example, research in the Wollongong region of Australia would indicate that landslides should be prevalent whenever the monthly total of rainfall exceeds 600 mm. This condition was met for the month of April both in 1988 and 1989. While numerous artificial embankments failed, the number of landslides to occur on natural slopes did not reach expectations.

Finally, land instability hazards are frequently associated with other natural hazards. Flows, landslides and avalanches can all occur as secondary phenomena following tropical cyclones and other large-scale storms, earthquakes, and volcanic eruptions. Even drought affects soil stability because of its control on the frequency of soil contraction and expansion. For these reasons, it is often tempting to begin a textbook of this nature by describing land instability first. This has not been done for two reasons. Firstly, the diversity of the topic and its treatment in the literature by various disciplines precludes a simplistic presentation that students can easily grasp when first approaching the topic of natural hazards. And secondly, land instability hazards cannot be ranked as importantly as floods, storms, earthquakes or volcanoes. It is more logical to treat the processes and causes of these latter events before discussing what are basically secondary phenomena.

This chapter concludes the treatment of major climatic and geological natural hazards discussed in this textbook. Before summarizing some of the major themes to be drawn from the text, it is appropriate to discuss human reactions to the types of hazards that have been described. This will be covered briefly in the following chapter.

References

Finlayson, B. and Statham, I. 1980. *Hillslope analysis.* Butterworth, London.

Goudie, A. (ed.). 1981. *Geomorphological techniques.* Allen and Unwin, London.

Hays, W. W. 1981. 'Facing geologic and hydrologic hazards: Earth-science considerations'. *United States Geological Survey Professional Paper* 1240–B pp. 54–85.

Leopold, L. B., Wolman, M. G. and Miller, J. P. 1964. *Fluvial processes in Geomorphology.* Freeman, San Francisco.

Whittow, J. 1980. *Disasters: the anatomy of environmental hazards.* Pelican, Harmondsworth.

Further reading

Aune, Q. A. 1983. 'Quick clays and California's clays: no quick solutions'. In Tank, R. W. (ed.) *Environmental Geology.* Oxford University Press, Oxford, pp. 145–50.

Blong, R. J. and Johnson, R. W. 1986. 'Geological hazards in the southwest Pacific and southeast Asian region; identification, assessment, and impact'. *Bureau of Mineral Resources Journal Australian Geology and Geophysics* v. 10 pp. 1–15.

Bolt, B. A., Horn, W. L., MacDonald, G. A. and Scott, R. F. 1975. *Geological hazards.* Springer-Verlag, Berlin.

Bowles, J. E. 1984. *Physical and Geotechnical properties of Soils* (2nd edn). McGraw-Hill, Sydney.

Chowdhury, R. N. 1978. *Slope analysis.* Elsevier, Amsterdam.

Coates, D. R. 1979. 'Subsurface influences'. In Gregory, K. J. and Walling, D. E. (eds) *Man and environmental processes.* Dawson, Folkestone, pp. 163–88.

——. 1985. *Geology and Society.* Chapman and Hall, NY, pp. 153–82.

Cornell, J. 1976. *The great international disaster book.* Scribner's, NY.

Crozier, M. J. 1986. *Landslides: causes, consequences and environment.* Croom Helm, London.

Innes, J. L. 1983. 'Debris flows'. *Progress in Physical Geography* v. 7 pp. 469–501.

Scheidegger, A. E. 1975. *Physical aspects of natural catastrophes.* Elsevier, Amsterdam.

Sharpe, C. R. S. 1968. *Landslides and related phenomena: a study of mass-movements of soil and rock.* Cooper Square, NY.

Varnes, D. J. 1978. 'Slope movement types and processes'. In Schuster, R. L. and Krizek, R. J. (eds) *Landslides: analysis and control.* National Research Council Transportation Research Board Special Report No. 176 pp. 11–33.

Voight, B. (ed.). 1978. *Rockslides and avalanches 1: Natural phenomena.* Elsevier, Amsterdam.

Young, A. 1972. *Slopes.* Oliver and Boyd, Edinburgh.

III

Social Impact

14

Personal and Group Response to Hazards

Introduction

The mitigation and survival of any natural hazard ultimately depends upon the individual, family or community. An individual has the choice to heed warnings, to prepare for an impending disaster, and to respond to that event. In the end, the individual, family unit or local community bears the brunt of most of a disaster in the form of property loss, injury, loss of friends or relatives, or personal death. If these small social units could recognize the potential for hazards in their environment, and respond to their occurrence before they happened, then there would be minimal loss of life and property. Unfortunately, not everyone heeds advice or recognizes warning signs, not because of stupidity, but because of very important personal and socio-economic reasons. The reaction of individuals, families and community groups after a disaster also determines their ability to survive physically and mentally. Some individuals and families can overcome all bureaucratic obstacles in rebuilding, others end up with shattered lives despite all the support available to them. This chapter will describe and account for some of these personal and group reactions.

Before the event

Warnings and evacuation

People react differently in their perception of a hazard. While some may not know that an earthquake is about to occur until they hear the rumbling and feel the shaking of the ground, most will know that they live in an earthquake-prone area. For example, there are 15 000 000 people living along the San Andreas fault zone in southern California. The bulk of this population knows that a major earthquake is overdue, but continues to live there despite the risk. People are faced with the same decision in Wellington, New Zealand, where the occurrence of a major earthquake could destroy most of the city at any time. Wellington is expanding as a city along its fault zone. In both cases, it is the 'sense of place' or 'home' that overrides all common sense about the threat of the hazard. People born in an area, tend to want to stay there. Home is familiar, it has people one knows and associates with at a personal level. There is some sort of historic identity with the area which is difficult to give up, and which is defended against change. There is a need to maintain links with the past or with ancestors, to sustain one's roots. People require an association with place and history. No threat of a hazard will make them leave, and they will rationalize the threat to minimize its occurrence. Residents of San Francisco refer to the 'Great Fire of 1906' rather than the 'Great Earthquake of 1906'. In these people's minds, great fires are rarer than great earthquakes and, hence, a similar disaster would be less likely to recur. If a place is attractive climatically, has an alluring lifestyle or promises better economic opportunities, then it will continue to grow despite any calamity which can be associated with it. That is why people continue to move to California, that is why Mexico City has grown so rapidly during the 1960s and 1970s, and that is why Wellington, New Zealand, continues to grow.

In the short term, even in the above locations, a warning of imminent disaster can also invoke a nonchalant reaction bordering on disregard. For example, during the Ash Wednesday bushfires in Australia in 1983, some people nonchalantly sat around pubs escaping the heat of the day, while they watched one of the worst conflagrations witnessed by humanity bearing down on them (figure 14.1). A non-evacuation reaction may be machoistically based bravado, a public display of fearlessness, a normal psychological response of a low-anxiety individual, a religious taboo, or even the sensible thing to do. An example of a religious taboo occurred in Eastern Pakistan (Bangladesh) during the 1970 cyclone surge. At the time the cyclone struck many women were prohibited, by established religious practice, from going outside for the period of a month. Even if residents had been given warning of the storm surge, few women would have evacuated and broken that taboo. Non-reaction to imminent disaster can occur for other reasons. Warnings of disaster imply evacuation, and that may appear too radical a departure from everyday life to be acceptable. Peer-group pressure also may be at work. If you are the only one in your neighbourhood to evacuate, then you may be ridiculed or made to look silly. This has always been a human response to warnings of doom. For example, in 1549, a cadi (judge) in eastern Persia predicted an earthquake for his city. After trying to persuade his friends to leave their homes for the open, and after spending time alone in the cold outside, he eventually returned home only to be killed along with 3000 others in the ensuing earthquake (Michaelis, 1985).

Impending disasters are also times of worry, fear and confusion. The overt decision to stay put may represent a subliminal mechanism for coping with the unknown and the unexpected. Generally, those of low socio-economic status, ethnic minorities and women respond to warnings the least. Older people find such change difficult to handle physically and mentally. For example, one old man died in the Mount St Helens eruption because he refused to move despite all the warnings. He had lived there all his life, and nothing was going to make him move. During any tropical cyclone evacuation in the United States, people have had to be evacuated from coastal areas at gunpoint by the militia. Old couples died in the Australian bushfires of 1983 along the coast south of Melbourne, because they had retired to those communities and there was nowhere else for them to go. If their retirement home was to be destroyed by fire, they would stay behind to save it, or go down with the ship.

Fig. 14.1 Drinkers at a pub in Lorne, Victoria casually watching a bushfire bearing down on them on Ash Wednesday, 16 February 1983 (photograph © and reproduced courtesy of the *Age*, Melbourne). This fire and others turned into one of the greatest conflagrations ever witnessed.

This reaction to imminent disaster can be explained in psychological terms. High-anxiety people perceive events occurring around them as confirming their anxiety. Thus, high-anxiety people may in fact be psychologically conditioned for survival. The same reaction could be produced by someone who feels they are in control of their life, as opposed to someone who is fatalistic and feels that their life is controlled by a divine direction or other external factors. Usually, the two aspects are one and the same thing. Low-anxiety people often are the fatalistic ones who believe their life is controlled externally. This difference also relates to variation in socio-cultural groups. For example, as mentioned in chapter 3, Sims and Baumann (1972) found that there was a higher incidence of death due to tornadoes in the south of the United States, than in the north. People in Alabama distrust the Weather Bureau and its warnings. They await their fate and as a result there is a greater incidence of death there because of tornadoes. On the other hand, people in Illinois believe they are in control of their own lives, are technologically orientated, and take precautions against the threat of tornadoes. They survive the risk better as a result.

For some, the decision to evacuate represents an economic decision. Any evacuation, no matter who organizes it, or carries it out, costs money. An evacuation involves loss of income for most of the community. For example, fishermen in the Caribbean do not evacuate during cyclones. Staying behind with their boat is more acceptable than evacuating, knowing that it is going to cost them money to move, and loss of income for the time they have moved. Loss of income may be unacceptable if your income is paying for an expensive mortgage, or keeping you from the moneylender. In that case, evacuation will only be contemplated if the hazard is a certainty.

Others get lulled by the 'cry wolf' syndrome. For example, even though the residents of Darwin in 1974 had up to three days warning of the arrival of cyclone Tracy, only 30 per cent took precautions. This was because a cyclone had approached Darwin in the previous week, and had veered away. False cyclone threats are always happening in Darwin. Tracy also had the misfortune of occurring just before Christmas. Preparing for Tracy meant interrupting Christmas celebrations for a cyclone that might never come. Evacuation meant going inland, where there were no facilities, and the possibility of being trapped by heavy rains. Similarly in the United States, tornado warnings are ignored because they are issued for very large areas, when their probability of occurrence in part of that area is small. The Weather Bureau assumes that the total population will respond to a warning. However, so few people actually experience a tornado, that the majority of the population fail to take any precautionary measures. Civil defense groups in the United States now make more dramatic pleas for evacuation. For instance, after a newscaster has just broadcast a tropical cyclone warning in the weather report, the head of the civil defense organization may interrupt further broadcasting, identify who he is, and in an authoritative voice state that the cyclone threat is now real, and people should take immediate action to evacuate for their own personal safety.

There are also those who are quite willing to evacuate, even though the signs of impending disaster are not obvious to the general public. Many people in touch with nature, such as North American Indians, have left areas of disaster weeks in advance, because of subtle signs shown by insects and animals. For example, some Indians in the Mount St Helens area packed up and left before the eruption, because they noticed animals doing the same thing. Indian tribes in southern California have left flood-prone areas in the midst of drought, because they noticed insects and ground burrowing animals seeking higher ground. Several weeks later flash flooding occurred when the drought subsequently broke.

Evacuation can also be dangerous. It may seem advantageous to flee the landfall of an approaching cyclone, and certainly if warning is given of an earthquake, it is wise to leave buildings and head for open spaces. The same type of evacuation might also seem appropriate if a wildfire is swooping towards your home at $20\,km\,hr^{-1}$, as happened in the Ash Wednesday bushfires of 1983 in Australia. However, work by Wilson and Ferguson (1985) in a study of attended houses at Mount Macedon during this bushfire, showed that it was better not to flee. Only 44 per cent of unattended houses survived, compared to 82 per cent of attended houses. Six people lost their lives at Mount Macedon; however, this could not be attributed to the fact that people remained at home to fight the fires. Evacuation during the fire was downright deadly. Roads were clogged by other evacuees, smoke made driving impossible, and some cars ended up trapped in exposed areas, where death from heat radiation was a risk. Similarly, it is unwise to try and flee approaching tornadoes in cars. Studies in the United States have shown that 50–60 per cent of deaths and injuries from tornadoes are caused by flying glass from broken windscreens.

In some cases, where the disaster occurs suddenly, evacuation may not be a viable option. During most earthquakes or landslides, individuals have only a few seconds to react. Personal decisions must be deliberately made against a background of seemingly organized flight by tens or hundreds of people. The decision to buck the trend becomes mentally more difficult, and in certain cases physically impossible. However, most individuals, be they sane or not, evidence remarkable composure. While panic may arise in any disaster, it is not that common a response. Panic may appear evident simply because people in a rapidly developing disaster situation, such as a fire, spend critical time confirming the threat, thus leaving little time to flee.

Preparedness, if warned

In many of the above cases, people who stay put and do not prepare for disaster are the exception. For instance, for every person who resists cyclone evacuation in the United States, there are hundreds of thousands who undertake it in an orderly fashion. The threat of personal loss of life, or loss of close family, outweighs the risk of staying. For most inhabitants of cyclone-threatened areas in Australia, the approach of the cyclone means tying down loose objects in the home and yard, stocking three days supply of food and water, and taping windows. Such precautions are taken seriously and initiated individually.

Preparation for imminent disaster by individuals centers around four areas: self, family, property and community—in that order. Organizing oneself can range from packing one's clothes and toothbrush to mental preparation. The latter usually includes some sort of catharsis such as conveying apprehension or fear to oneself, family or friends, or taking a sedative or alcoholic drink. In some cases, it may take the form of prayer for divine aid or intervention. Mental preparation appears to be a first priority in meeting any hazard. For those with immediate family, the second priority is ensuring their safety, preparedness and mental stability. Attempts may be made to check the safety of close family living in the same threatened area, or communicating assurances of safety to distant relatives.

Next there is a direct concern for property. Usually, this involves the immediate home either well before, or immediately preceding any disaster. For example, McKay (1983) found that the degree of preparation for bushfires in the Adelaide Hills, following the 1983 Ash Wednesday fires, varied throughout the area. Fewer than 20 per cent of people did nothing. Fifty-four per cent were involved in either temporary or permanent measures. Temporary measures involved such preventative tasks as clearing gutters of leaves, and cutting away undergrowth around buildings. Permanent measures involved structural changes to the house, landscape modifications, or the construction of a sprinkler system. These remedies required little maintenance, but were more expensive. There was no socio-economic differentiation in the number or types of measures people utilized. However, people who rented their accommodation tended to perform fewer measures. About 60 per cent of people carrying out preparations did so after seeking advice from neighbours, friends or relatives. Only 9 per cent obtained information from public bodies and 11 per cent from the country fire brigade. Drabek (1986) also found a similar trend for many other hazards worldwide. Obviously, people trust neighbours or relatives more for advice, than they trust government bodies that have the expertise. Significantly, people who had neighbours experiencing bushfires, or people who themselves were affected in the last few years by fires, took greater precautions. In Australia people's enthusiasm for implementing preventative measures against bushfires may wane as the memory of the last fire fades. Large segments of the community may be unprepared for bushfires as the threat increases. This fact may require an annual advertising program to make people aware of the continuing threat. A similar pattern has been found in the United States with people's degree of preparedness to cyclones.

During any tragedy, victims usually made some attempt at saving their home. If there is little hope that the house can be saved, then personal belongings or mementos such as photographs, trophies or presents are rescued first. If there is time, then more expensive transportable property is saved. Reaction to pets can also be included in this behaviour. Sometimes considerable risk can be expended with people dying in attempts to save pets or livestock. For example, floods in Australia have seen farmers sheltering complete dairy herds in their house. Evacuation of 40 000 families in Missassauga, Ontario, Canada, following a gas leak in 1979 also involved 30 000 pets. If preparation for an imminent hazard can be completely carried out, and the victim perceives that such preparation has enhanced their chance of survival, or preserved personal property, the shock of the event is lessened and the recovery after the event is more easily dealt with.

Dealing with the event and its aftermath

Response during the event

Response during the event also centers around the same four areas of interest involved in preparation. It is doubtful if anyone going through a tropical cyclone will be worrying about the state of the American or Australian economy. Almost all descriptions of human behaviour during a natural disaster refer to a lack of panic. People may be scared but, in the face of adversity, most remain outwardly calm. Hysteria is rarely evidenced. In an extensive search of the literature, Drabek (1986) could find no proof that disasters generated mass fear or panic. Generally, the media overtly reinforced a panic myth by exaggerating incidents of non-rational behaviour. If panic exists, it usually represents a rapidly reinforced, collective response to flee some very real threat.

It is a human instinct to try and protect one's family during disaster. This instinct tends to override all fear. One mechanism to ensure maximum protection is for family members to remain close or huddled together during life-threatening situations. Tropical cyclone victims in Australia, usually find themselves huddling together for refuge in the toilet, because in new houses this room is best designed to withstand high winds. During Cyclone Tracy in 1974, huddling environments were not always conducive to survival. One family who had huddled together in bed in the main bedroom found that they were exposed to the full force of the storm as the house disintegrated around them. The bed began to drift across the floor downwind towards the edge of the house and only came to rest when one of the legs fell through the floor. Earthquake victims are often found huddled together in the crush of fallen buildings. Such closeness often permits an adult to save a child in miraculous situations. For instance, a father, fleeing the debris flow of the Mount Huascaran earthquake in Yungay, Peru, in 1970, managed to throw his two children up a slope to safety just as the flow overtook and killed him. Following the Nevado del Ruiz lahar of 1985 in Columbia, a child was found trapped in a building in mud up to her neck propped up by her father who had suffocated to death.

The above examples illustrate another basic response during disaster, and that is the innate instinct for survival. The miracle babies of the Mexican earthquake of 1985 survived in the maternity ward of a collapsed hospital for up to two weeks after the earthquake because their bodies went into a natural state of hibernation that conserved fluids and energy. Victims, some severely wounded, have been dug out of collapsed mines, earthquake rubble, bombed buildings and avalanches after similar periods of incarceration. Almost all survived because of a determination to live, a feeling that it was too early to die, a feeling that there was still more to be done in life. The Tarawera eruption of 10 June 1886, in New Zealand, buried Maori villages under more than 2 m of ash and mud, killing 156 people. A 100-year-old survivor, Tuhoto of the Arawa tribe, was found alive and in good health after being buried in his house for four days. He, in fact, told his rescuers to go away because the gods were taking care of him. He died ten days later under hospital care. The girl trapped in the lahar mud in Columbia in November 1985 met a similar fate. She remained cheerful and survived for several days stuck in the mud, only to die just before being pulled free. The physical devastation greeting the first outsiders to arrive in Darwin after cyclone Tracy, led many to expect more than the 64 deaths that actually occurred. Over 90 per cent of homes were totally destroyed and many families endured Tracy completely exposed to the elements. The small number of casualties underscores the ability of people to endure much more hardship, deprivation, undernourishment and shock than is normally believed possible. It is only after the event, that survivors must cope with the psychological trauma which inevitably sets in.

Finally, there is one other characteristic of human behaviour that often gets noted during a disaster: curiosity to see what is going on. It is not to be confused with sightseeing, which involves people who have not experienced the disaster wanting to see what has happened. For example, during the passage of the eye of a hurricane, almost everyone wants to get outside and take a look around. You and your family appear to be uninjured, but you wonder, 'What does the house look like, where's the dog, how are the neighbours doing?'. For example, whenever a flood crest affects the Mississippi River system in America, people who could be flooded are always wandering down to the river to look at the river's height. Mount Vesuvius during eruptions has always attracted local sightseers curious to find out what is going on. Even residents of St Pierre journeyed up to Mount Pelée in 1902 to see what was happening a few days prior to the eruption. Ten thousand people flocked to San Francisco beaches after the Alaskan earthquake to see the tsunami coming in. The deathtoll in America during that tsunami event, and the one which hit Hawaii in 1946, resulted from mainly curious, and probably uninformed, residents returning home too soon to evaluate the damage before the last waves had hit the coast.

Maybe there is some euphoria to living through natural disasters or record-breaking events. The news has just reported the heaviest one-day rainfall on record and you wander outside to stand in it. For instance, I can still remember as a child waiting for days the approach of Hurricane Hazel in southern Ontario, because no one in the area had ever experienced a tropical cyclone before. When Hazel finally struck, there was a sense of relief that it had arrived at last. The same thrill of expectation goes through anyone who is witnessing for the first time the initial stages of a ground tremor. But then reality sinks in. The tremor becomes a major earthquake, the nonchalant attitude to a cyclone becomes a nightmare as the house disintegrates around you, the flood hits home as rumours of friends being wiped out become facts. The disaster has struck, and new emotions and needs arise. Mechanisms for cleaning up after the disaster must be carried out, and reconstruction and recovery instigated. At this stage, the emphasis shifts slowly away from the individual towards social groups.

Death and grief

One of the most immediate aspects of a disaster that must be faced is coping with the injured and the dead. Often the search for survivors and the rescuing of the injured becomes a major component of large-scale natural disasters, requiring international co-operation. While rescue operations tend to take precedence over coping with the dead, dead bodies pose special problems; both hygienically and legally. The health aspect of dead, but unburied, bodies is obvious. Decaying corpses, human or animal, breed disease and must be cleared away within hours or days if epidemics are to be avoided. The more involved problem occurs in trying to define people as legally dead. The problem is minor in the case of unidentifiable bodies. Forensic science can be used to match bodies against reports of missing people. An extra body poses no problems. After time, it will simply be classified as a 'John Doe' and buried as such. The more difficult legal problem occurs when a person is reported as missing, and a body cannot be found. Western society presumes that such a person is still legally alive. They could be wandering around with amnesia; they may have taken the opportunity of fleeing an unbearable family or work situation in the confusion of the disaster; or they might be responsible for the disaster and are fleeing conviction. Unless there is a body, the person is considered still alive for up to seven years. A will cannot be read, or an estate settled within this statutory period. In extreme cases, where a will has not been written, it is possible for a person's estate to be sealed off. A dead person's car must not be moved from a driveway, a house is locked up, business assets frozen. Not only can such action be financially trying for the surviving relatives, but the presence of the dead person's possessions in the immediate vicinity is also a constant reminder that they are missing.

In all societies, the trauma of the dead is resolved by a burial service. It allows family members and close friends to publicly show feelings of grief. Mentally, it permits an individual to accept the death and get on with life. The administration of last rites in the Catholic faith provides a good example of a procedure for ensuring this transition. Devout family members know that the person administered last rites can move into an afterlife, and that the prospects of seeing the person again after death are good. Psychologically and sociologically, funeral services and death rites are essential to overcome the problem of fatalities following a disaster. If the disaster is large scale and the deathtoll significant, then society organizes mass burials and memorial services. An official period of national mourning may also be instigated following large disasters. Total strangers who have never known any of the dead, or had any association with the affected communities, find that they have to attend some sort of funeral. Even in Australia, where less than 20 per cent of the population attends regular church services, special services had to be arranged to fulfill this need following both Cyclone Tracy and the Ash Wednesday bushfires.

Possessions and homes

While the death of close family may seem to be the worst outcome of a disaster, the destruction of property may leave longer scars. No funeral service is held for a wrecked home, or a burnt photograph. People following the Ash Wednesday bushfires stated that the worst aspect was losing, firstly, personal photographs; secondly, prized trophies or awards; thirdly, property which had some sort of link to dead parents, children or friends. Some even vocalized that the loss of possessions was worse than losing close family. Nothing can replace old family photographs or a melted down trophy following a bushfire. Almost everyone has seen pictures of people returning home after a natural disaster to pick through the ruins to find some trace of remembrances. The pictures often show people sitting afterwards in complete despair or shock when nothing can be found. The personal belongings give some sense of continuity to life. They also provide a sense of history—a connection with past events or ancestors. Survivors have gone to extreme lengths to try and recover some of these possessions. Relatives are contacted to provide copies of significant photographs, and antique shops may be searched to obtain replicas of prized objects.

Many people immediately commit themselves to rebuilding in the exact same location. Often a duplicate of the original house is built. This was the case for many people rebuilding after the Ash Wednesday bushfires. Homes were re-established in their entirety to designs which offered little protection to repeat bushfires. The need to rebuild exact replicas can be taken to extremes. For example, following the San Francisco earthquake of 1906, the city rejected a modern design in favour of a street layout and a building program which would, as quickly as possible, recreate the original city. Sometimes it is investment in assets that conditions reconstruction. For instance, following the Hawke's Bay earthquake of 3 February 1931 in New Zealand, the town of Hastings was rebuilt to its original, narrow street plan, even though total destruction permitted an opportunity to rebuild wider streets. The decision to rebuild in this manner was made at the adamant insistence of one councillor, who owned the only shop with any part of its structure still standing. The councillor wanted to save money by not having to rebuild a new shop, which he would have to do if the streets were widened.

Feelings about property can be quite intense. For example, following the autumn coastal storms of 1974 around Sydney, Australia, I was walking with colleagues along remnants of the beach in front of a series of houses propped up by makeshift supports. As we dodged waves surging up amongst the rocks, one resident furiously insisted that we were on private property and he would get the police after us for trespassing. This incident highlights a second stage of response following a natural disaster. After the initial impact, and after all one's resources have been utilized to ensure that safety of the family, the anger sets in. For some, the response manifests itself in long-term psychological shock that alters one's life forever. This latter aspect will be examined in a subsequent section.

Anti-social behaviour

No group of people is so disdained as the sightseer or the looter. Looting, while rare, does occur. However, it also tends to be exaggerated by the media. In the United States the prospect of looting traditionally has been so great following disasters, that the National Guard is often mobilized to block off affected areas and ordered to shoot looters on sight. Residents have to show evidence that they live in an area, and must restrict their activities to the street where they live. In Australia looting has been reported, and confirmed, in some incidences. People were arrested following both Cyclone Tracy and the Ash Wednesday bushfires but no one was ever jailed. The Lismore flood of 1954 saw looters taking whatever they could carry from the business district, including a rescue boat which was used openly to transport stolen goods.

A real problem for rescue and relief organizers after large disasters is coping with the sightseers. While people may just want to see what a natural disaster such as a cyclone or bushfire can do, the victims often perceive sightseeing as an intrusion on their privacy by the nosey, or as an opportunity by those who survived unscathed, to gloat over their luck and the victim's misfortune. In Australia sightseeing can take on mammoth proportions, not only by the general public but also by the politicians and bureaucrats. Almost every cabinet minister in the Whitlam government made a fact-finding journey to Darwin after Cyclone Tracy. The same planes that evacuated survivors returned nearly as full with government sightseers. Following the Ash Wednesday bushfires, which occurred in the middle of the 1983 election campaign, almost all party leaders made the journey to affected areas—not, as they took pains to state in front of the cameras, for political reasons, but to see if they could offer genuine help.

People's misfortune can also breed extraordinary ghoulish behaviour. For example, an old lady was robbed in busy downtown Toronto, Canada, in broad daylight, after being blown to the ground in January 1978 by wind gusts in excess of 160 km hr^{-1}. As she lay pinned to the sidewalk, a more agile motorist, in what at first appeared to be an act of chivalry, stopped his car and ran to assist her. Before stunned onlookers could react, he snatched her purse, hopped back into his car and sped away. Michaelis (1985) reports that after the Naples earthquake of November 1980, smartly dressed people in flashy cars were seen leaving the scene with orphaned babies and young children, abducted for themselves or to be sold for adoption. Relief goods in Naples were also pilfered and resold on the black market. Holthouse (1986) also reports that during the Mackay cyclone and subsequent massive flood of 1918 in Queensland, boat owners were seen offering to rescue people stranded on rooftops and in trees on payment of a substantial fee—no fee, no rescue. In the longer term, disasters attract strangers and hangers-on, the type of people you were warned about as a child. As Hauptman (1984) so aptly puts it, 'The sort of trashy people who follow disasters . . . people who could not hold a job anyplace else.' Just like vultures flocking to encircle a dying body, these people flock to victimize survivors, who often let their guard down following a disaster. They have to because they are now reliant upon total strangers for assistance. The worse their financial plight, the more reliant they become. And for every honest stranger dispensing cash and materials for relief, there is a dishonest one disguised as a contractor or household merchant attempting to steal it back.

Resettlement

So far the impression has been given, especially with the example of Cyclone Tracy, that a significant number of people may not, or are not able to, go back to the scene of the disaster, and re-establish their lives. Nothing could be further from the truth. In almost all cases, people want to go back and resettle in the same place they were living before the disaster. This may seem odd if the place happens lie in a major earthquake zone such as the San Andreas faultline, or a coastal area subject to recurring tsunami. There are a number of reasons why people return home again, even when that environment has been shown to be hazardous.

There may simply be no alternate place to live or occupation to undertake. If people are employed in agriculture or fishing, they cannot go elsewhere because they know no other career. Additionally, they are experts in farming or fishing in that particular area, and cannot afford to lose that expertise. For instance, people farming the rich volcanic soils surrounding volcanoes may want to return after an eruption, because they know that the tephra will immediately increase crop yields. The residents of the town of Sanriku on the east coast of Japan, which has been struck repetitively by large tsunami, have always returned because they know the fishing conditions along that section of coast. The latter example also illustrates the point that in some cases, people return to the same place because there may be nowhere else for them to go that is less hazardous. The citizens of Managua, Nicaragua, after the 1972 earthquake also found themselves in that situation. The whole country is seismic and no other location is necessarily safer. In Nicaragua or Japan, movement from one area of the country to another does not necessarily increase a person's sense of security.

Evacuation can also mean loss of assets. By returning, people can at least salvage something. Residents at Terrigal, north of Sydney, were caught in that predicament following the Australian coastal storms of 1974. The value of houses, perched precipitously at the edge of dunes eroded by storm waves, collapsed. Residents returned to these homes to try and protect them from further erosion. Fortunately for some, housing prices climbed back to normal and they were able to sell out with minimal loss. Unfortunately for others, storms in 1978 encroached further upon properties, and land had to be resumed for token sums by the local council. Change after a disaster also raises legal complications, especially if governments attempt to restructure cities and cut across existing private or commercial property boundaries. For people outside western culture, a high value is placed upon traditional land rights. All of these examples show that humans are basically territorial, and do not want to surrender property no matter what has happened to it.

There are also psychological benefits in reconstructing at the site of a disaster. By rebuilding as fast as possible, people can forget the actual event. They also keep busy at something they are familiar with, and can master. This keeps their minds off the disaster, builds up self-esteem and overcomes the despairing feeling of hopelessness. It has already been pointed out that the city of San Francisco was rebuilt almost exactly as it had been before the 1906 earthquake and, after the 1983 bushfires in Victoria, some home-owners rebuilt exact replicas of the original home. In the case of San Francisco, the decision to rebuild in a hazardous area leads to outright denial of the hazard. Residents of San Francisco will point out that San Francisco is as safe as any other area in southern California.

A loyalty to local history, and a commitment to a locality, also draws people back to a disaster scene. Rebuilding signals a loyalty to friends or relatives who might have been killed. Their lives were not lost in vain, and their remembrance will be maintained. For example, following the Columbian lahar tragedy, the whole affected area was declared a national shrine by the government in memory of the 20 000 people buried beneath the mud. Within months resettlement of the area by survivors was occurring. This sense of loyalty also includes the concept of ancestor attachments. Many people want to return to their homeland. In some cases, this relates to ethnicity; in other cases, to respect for ancestor links. Indian tribes in the Americas and the Aboriginal groups in Australia have equally strong attachments to the land. To be removed, or not to return to an ancestral homeland, makes life meaningless.

Religious beliefs can be a significant control on resettlement. Many people believe that the disaster is due to the wrath of a revengeful or angry god. Disasters do not occur because areas are hazardous, but because the people living there did something wrong. For instance, the Lisbon earthquake of 1755 was perceived at the time to be punishment for the wickedness of its citizens, to be God's answer to the cruelty of the Portuguese Inquisition. Accounts of the destruction of the town of St Pierre, Martinique, after Mount Pelée erupted, mention the fact that the town had a reputation for being permissive. Prophets of doom have predicted the imminent destruction of San Francisco by an earthquake because of its large homosexual population. In Papua New Guinea, some members of the Orokaivan group rationalized the 1951 Mount Lamington eruption as payback for the killing of missionaries. An angry god who can cause disaster can also be placated by hymns, gifts, offerings and prayers. Hazardous environments in Indonesia and the Philippines are often viewed as being imbued by a divine spirit that can be quietened in this fashion.

Similarly such beliefs can also dictate against resettlement. For example, Maoris in New Zealand attributed the 1886 eruption of Mount Tarawera to an act of revenge by an alleged sorcerer. This supposed sorcerer was, in fact, Tuhoto—the same man referred to earlier who was found alive after being buried for four days. Tuhoto had visited some friends close to the site of the eruption earlier in the year and, following his visit, a child of a chief had died. The child's grandmother accused Tuhoto of bewitching the child; when Tuhoto heard about the accusation, he was so angered that he called upon the god of volcanoes to wipe out the whole village. Following the eruption, no Maori would have anything to do with Tuhoto. The site of the eruption and the surrounding devastated area were considered taboo, and to this day are still avoided by Maoris.

Resettlement may also be rationalized by a belief that modern-day technology can prevent destruction in future disasters. In southern California, great faith is put into reinforcing older buildings and expressway overpasses against collapse. Minor earthquakes in 1971 and 1989 have shown that not all this technological strengthening works. New Zealand is pursuing the idea of building large buildings on rubber bungs, that can absorb earthquake shock waves, and minimize destruction. Certainly, in areas where this has been done earthquakes are not as destructive as they are in regions where buildings are not built to earthquake standards. The large deathtolls of both the Agadir, Morocco, earthquake of 1960 and the Armenian earthquake of 1988 attest to this latter fact even though these events were of lesser magnitude than recent Californian ones.

Moving elsewhere after a disaster may also be too disruptive to a wider segment of society. Evacuation of large numbers of people represents a major task that can put pressure on housing, government, and consumer supplies in other parts of a country. By containing the survivors, efforts can be efficiently directed to a single area. For example, the evacuation of Darwin for such a long time-period was not effective. It caused disruption to the people who were evacuated. Had the numbers been larger, it would have severely disrupted the economy in the south. There is an inertia effect resisting the relocation of so many people. Large communities attract a communications and business infrastructure that represents a significant capital investment that cannot be easily abandoned or relocated. For instance, Messina, Italy, after the earthquake of 1908, rather than being abandoned, was rebuilt in the same area because it was a significant port. Tangshan, China, plays an important link between Peking and Tienshen on the coast. For this reason, it had to be quickly rebuilt following the 1976 earthquake. The only concession to the disaster was the rebuilding of the city in a series of satellite cities spread out to minimize damage in any future earthquake. Some countries such as Japan and China have more centralized economies, where the government decides the movement and location of people. There is no personal choice to leaving or resettlement. The government makes the decision, and either enforces it or encourages it.

Finally, the disaster may appear to be a one-off or exceedingly rare event. In this case, alternatives to resettlement in the same area do not arise. For example, tropical cyclones affecting southern Ontario, Canada, are relatively rare events. Hurricane Hazel in 1954 represented a 1:200-year event. While areas prone to such infrequent flooding have been set aside in Toronto as open space, and planning takes account of this event, no plans for massive evacuation have been established in case another cyclone strikes. The event is just too rare to be of significant concern. The same is true of some earthquakes and most volcanoes. Historically some of the largest earthquakes in the United States have occurred at locations such as New Madrid, Missouri, and Charleston, South Carolina, which are not considered seismic. An earthquake disaster movie set in South Carolina just would not be believable. Nor does the disaster have to be infrequent to be dismissed. It just has to be perceived as infrequent. The slopes of Mount Vesuvius in Italy, and the town of St Pierre, Martinique, have been resettled despite the fact that the local volcanoes are still considered dangerously active. This perception of infrequency can form very quickly, because people have short memories of disaster events. For example, within six months of the Brisbane River floods of 1974 in Australia, property values in flooded areas had returned to their pre-flood levels.

Myths and heroes

The myth that people panic during disasters has already been dispelled in this chapter. There are a number of other myths about human response during disasters that should also be dismissed. The first myth is that people experiencing the calamity appear confused, are stunned, and helpless. If anything, disasters bring out the strength in people. Survivors, far from being helpless, are almost immediately back into the rubble trying to save people and property. At this stage in disaster relief, the victims are more than capable of organizing themselves and performing the rescue work. The only things they may lack are the tools to carry out the tasks efficiently. Secondly, it is generally believed that disasters bring out anti-social behaviour—stories of mothers dropping their babies to go after a few seeds of grain spilled from a passing truck, of men pushing children from life boats on sinking ships. Such behaviour is rare, or else it is often the product of media distortions and bears little resemblance to reality. While there may be anti-social behaviour—for example the ghoulish conduct of baby-snatchers in Naples, mentioned above—survivors are generally very sociable. Disasters kindle community spirit and altruistic acts in people.

Another aspect about disasters, which may or may not be mythical, is the role of heroes. In the United States forest fires made heroes. For example, the actions of the crew on the St Paul and Duluth train to Hinckley, Minnesota during the fire of 1894 was certainly heroic. Despite decreasing visibility due to smoke, the train casually approached the town only to be met by 150 fleeing townspeople, closely pursued by a wall of flame. The engineer decided to evacuate people to a marsh 7 km back up the line, and throttled in reverse at full speed. However, the fire overtook the train, consuming one passenger car after another. As flames licked through cracks and windows, the conductor calmly put out fires on woman's dresses with a fire extinguisher. In the cab, the situation was worse. Both the engineer's and fireman's clothes caught fire, and the cab began to burn. The glass in the cab window burst from the heat cutting the engineer's jugular vein, the controls caught fire, the seats began to smolder and the cabin lamp melted. Several times both men collapsed in the heat, but managed to keep the train moving. As the train reached the marsh and erupted into flame, both passengers and crew flung themselves into the mud where they sheltered for four hours. The fireman, however, after carrying the bleeding engineer to safety, returned to the train, uncoupled the engine and moved it to safety, while the conductor struggled back to the nearest station to warn the next train coming through. The crew's actions saved over 300 people.

In Australia, no one would doubt that General Stretton's work in the first weeks after Cyclone Tracy was heroic. His organizational skills stand out as superhuman. He was able to mobilize evacuation and ensure that the city was cleaned up, laying the foundations for the reconstruction that was to follow. He made decisions which appeared innovative and effective. He was given an impossible task—to make order out of chaos; and he succeeded without any additional casualties. Stretton rose to the occasion, but then that is what was expected of him. He had a military background, and was in charge of the National Disasters Organization in Canberra. His training, experience and position aided him in performing his assignment.

And yet, there have been situations where common people with few of these attributes have also spontaneously risen to the occasion. Following the 23 January 1973 eruption of Heimaey in Iceland, the fire chief from Keflavik found himself saving the town of Vestmannaeyjar. He had only flown in to offer firefighting advice. He found himself in charge of salvage operations, evacuation of household goods off the island, the distribution of relief food, the organization of workmen to strengthen houses against collapse from ash, the dispatch of troops and volunteers to clear ash from the town, and the efforts to spread sea water onto the lava to slow its movement. Similarly, when rescuers arrived two days later at the scene of the 4 February 1976 earthquake in Guatemala, they found an old man in charge. He had organized rescue, reassured survivors and, when the aid came in, he took charge of the distribution of food and supplies. The other survivors were deferring to his wisdom whenever a decision or action was required. In a way, hero stories reinforce society's sense of security. People know that if they ever have the misfortune of being caught in a natural catastrophe and require help, there will be someone to save them.

Additional impacts

Emotional problems

> Why did it have to happen to me? I've worked hard all my life and the house is gone. There stands George's house next door. He built it after winning the lotto, the shrubs are stolen, and he was at the pub all through the fire. Look, his fence isn't even scorched.

This scenario describes well the reaction that begins to set in after the disaster. While humans are quite resilient during disaster, a state of shock or numbness that borders on self-pity sets in afterwards.

These feelings inevitably turn to anger. Maybe anger at oneself for not preparing better for the event, maybe anger at a neighbour who survived the disaster better, maybe anger at some larger organization such as the Weather Bureau for not giving enough warning about the approach of a cyclone, maybe anger at some unidentified group such as the conservationists who prevented prescribed burning. The object may be real or imaginary; it makes no difference. One becomes just angry. The anger is vocalized to anyone who will listen. It is a common trait of people that they must talk about the event to anyone who will listen. The press prey upon this characteristic to get pictures and material on the disaster, and often record this anger stage. In some cases, they become the object of the resentment. Relief and social workers are aware of the fact that if they remain around the scene of a disaster long enough, they must expect to be on the receiving end of some of the verbal attacks.

For some, a disaster elicits a *victim–rescuer syndrome*. Large natural disasters such as earthquakes can completely destroy the physical fabric of a community. Survivors become totally dependent upon outside aid for their daily requirements such as shelter, clothes, food and services. Often the victims have no resources left for reconstruction. They may be suffering from injury or shock after losing family members. These people become victims of the disaster. Without anything which is familiar, without any ability to cope for themselves or power to direct their own lives, they succumb to lethargy and establish a complete dependence upon the rescuer. This disaster syndrome is not that common, being most pronounced following traumatic catastrophes for only a minority of people. In some cases, unless aid is directed to re-establishing a normal existence, the victims can become permanently locked into a dependency syndrome. They become disaster refugees. For example, the Ethiopian government at the end of the Sahel drought of 1984, forcibly repatriated refugees in relief camps back to their homelands in northern provinces, because refugees had become so dependent upon emergency aid that they were hesitant to return to what they perceived as still drought-ravaged farms. A significant number of evacuees of Cyclone Tracy, after a year's living in the southern States of Australia, where the standard of living was better, also found it difficult deciding to return home to Darwin.

The evacuation of survivors causes other psychological problems. For instance, after Cyclone Tracy, most residents were initially evacuated. However, husbands were permitted back to rebuild. Families were often separated for up to a year. When finally reunited, parents had drifted apart socially, and a gulf had arisen between fathers and children who had not seen each other for months. In this case, the resulting tragedy of Darwin was not the initial deathtoll from the cyclone; it was the fact that 50 per cent of families who had experienced the disaster separated, or divorced within a decade. Nor was this phenomenon restricted to evacuees. Families in the south who had volunteered their homes as temporary billets were faced with increased stress, or for some, increased temptation to become involved in an extra-marital affair with younger guests. Evacuees, especially those people who were self-employed, also experienced loss of income. This was not necessarily a problem for government workers or employees of benevolent national companies, who were relocated in southern offices. Those not so lucky faced unemployment or found temporary work. For some, this alternative employment became so attractive that they were reluctant to return to Darwin and face an uncertain future.

Single-parent families left in the south during the reconstruction also faced difficulties with children. Because of the trauma and changed surroundings, younger children underwent regressive behaviour. Without a father figure, rejected by their peers as odd, and considered transient by the community, school-aged children found it difficult adjusting to a strange environment and fitting into the community's social activities. Many developed antisocial behavioural traits, became delinquent, or took up drugs and alcohol to cope. The problems were especially acute for younger children as evidenced in table 14.1. Behaviour of these non-returnee children significantly differed from similarly aged children who had either returned to Darwin, or for some reason, had not left.

Within Australia, this situation is not unique to Cyclone Tracy. The drought of 1982–83 drove many families off the land into rural towns and the capital cities, where society was more permissive than generally experienced on isolated farms. Many children found the availability of drugs, which were beginning their dramatic spread at the time beyond the capital cities, too irresistible. At the time of the Live-Aid Concert for Ethiopia, it was pointed out by Molly Meldrum, the Australian organizer of that relief campaign, that we were more concerned in Australia with people's plight in the Sahel than we were about the unrecognized,

TABLE 14.1 Frequency of Psychological Problems amongst young Survivors of Cyclone Tracy, Darwin, 1974

Trait	*Stayers* %	*Returnees* %	*Non-Returnees* %
fear of rain and wind	19.8	24.2	29.3
fear of dark	10.8	13.2	11.5
fear of jet noise	4.5	7.9	15.5
clinging to mother	6.3	5.3	12.6
bed-wetting	2.7	6.8	7.6
thumb-sucking	0.0	1.6	2.0
temper tantrums	2.7	1.1	7.2
fighting	0.9	3.2	4.9
deliberate vandalism	2.7	1.1	3.4
injuries	0.0	2.6	5.5
diseases	9.9	5.8	10.9

Source: Data from Western and Milne, 1979.

or deliberately ignored drug problems of children driven into cities by the drought in Australia. Significant adolescent problems still existed in country towns in 1989, six years after that drought, with many young people drifting to capital cities or resort towns and contributing to delinquency rates in those places.

The shock of a natural hazard and the trauma of death also leads to psychiatric problems as time passes. The Aberfan, Wales, disaster, while not strictly a natural disaster, illustrates this phenomenon well. Failure of a 250-metre high, coal-tailings pile killed 116 children, the entire juvenile population, in a manner of seconds. The whole population personally was affected and underwent psychological trauma. Over 158 residents subsequently sought psychiatric treatment. A similar, but less severe situation, arose in communities badly affected by the Ash Wednesday bushfires of 1983 in Australia. The number of people requiring psychiatric help to cope with the disaster and its after-effects increased significantly. Following a particularly bad tornado in Wichita Falls, Kansas, in 1979, the number of suicides rose in the following year.

Just as work stress over time can lead to increased illness and absenteeism in workers, so the culmination of life changes, such as those resulting from a natural disaster, can lead to psychosomatic illness. Psychosomatic illness is any physical disease, such as ulcers, depression, or arthritis, that owes its origin to a psychological problem. Scales have been constructed foreshadowing the likelihood of an individual undergoing a psychosomatic illness. Death in the immediate family, a household move, the birth of a child, a change in job, or a divorce all contribute equally to the chance of physical illness in the following year. Again, non-returnees from the evacuation of Darwin in 1974 suffered a significant increase in stress-related illness. Further study has shown that it was not necessarily non-returnees that suffered more, it was the people who wanted to return but could not (Kearney and Britton, 1985). In this group, there was a 50 per cent increase in the number of people who were nervous, depressed or worried about the future, and a 100 per cent increase in the number of people complaining about bowel trouble or who were using sedatives.

The prolonged prominence of psychosomatic illness can not only destabilize a family unit in the short term but also have profound long-term ramifications for society if not resolved. Where one or more family member suffers from a psychological disorder, the chance of that disorder appearing in offspring is significantly increased. Natural disasters can also breed a cycle of sociological and psychological disorder, affecting a future generation who have never actually experienced the event. Ten years after Cyclone Tracy in Brisbane, where some of the victims had been evacuated, social workers were able to recognize cyclone-traumatized victims as a distinct sub-group.

The impression should not be given that all natural disasters produce copious psychological problems. A natural disaster may simply apply stress to an already stressed person or family. The disaster may simply be the straw that broke the camel's back. The evacuees from Darwin who did not return, and who developed psychological problems, may have been those people in Darwin already suffering problems, who would have left the city permanently some day even if Cyclone Tracy had not struck. Even if the families were stable units, the long-term isolation in unfamiliar surroundings and crowded accommodation was an abnormal social pressure. The study by Luketina (1986) of the Southland Floods in New Zealand in 1984 indicated that people evacuated for similar periods of time as the Darwin evacuees also underwent exceptional stress. However, after life returned to normal, very few admitted to, or sought help for psychological problems attributed to the floods. The exceptions to this occurred with the elderly, some of whom became confused by the events; with those already going through the stages of marriage breakdown; and with children. Immediately after the floods, 7 per cent of children from flooded homes were exhibiting behavioural changes of concern to parents and teachers. However, the majority of these problems disappeared after several months as life normalized.

Basically, there is a sequence of psychological adjustment following a disaster. Foremost, the majority of disaster victims appear to be resilient and adaptive to the disaster. For example, in the Brisbane floods of 1974 the community basically shunned disaster relief. Less than 2 per cent of the victims sought help subsequently from a counselling service. Of course, in many of these events there was not a large loss of life or substantial injury. A significant criteria, for disaster-induced psychosis appears to be the culmination within a short period of time of major life-changes. If the disaster generates these changes, or adds to ones which have already occurred, then an individual or family will be more likely to be psychologically affected by that disaster.

Critical incidence stress syndrome

A more worrying problem with disasters is the fact that anyone who comes into contact with a tragedy spawning death or injury can be affected by that tragedy for the rest of their lives. The experience may be as innocuous as being a bystander to the tragedy. For instance, during the Waterloo, London, subway station fire of 1987, passengers on incoming trains saw their train stop with the doors remaining closed. As people fleeing the inferno desperately banged on windows and doors trying to get into the train to escape, it pulled swiftly away. For the stunned passengers, the stop lasted only 30 seconds, but the experience will be with them for the rest of their lives. The closer one is to the disaster, the more permanent and damaging the effect. Sadly, the closest people are the heroes. Whereas a hero may have been awarded society's everlasting gratitude, and should be basking in glory, many in fact suffer from extreme depression and eke out shattered lives. The pattern is now known as *Critical Incidence Stress Syndrome*, and is just beginning to be recognized as a severe, long-term psychological response to disaster by rescue workers and heroes. The pattern is characterized by extreme depression and unfounded guilt feelings. For example, the policeman who repetitively risked his life to save commuters in the King's Cross subway fire, in London in 1987, cannot ascertain why he has survived while others still died; he now lives a timid life, afraid to go outdoors. The passenger, who became a human bridge permitting fellow travelers to crawl to safety after the *Herald of Free Enterprise* ferry sunk at Zeebrugge, Belgium, in 1987, lives a life without direction, even though he has been given every bravery award his country can offer. Another policeman, who dragged a burning man from the Bradford stadium fire in 1985, finds himself so emotionally weakened that he cannot help crying over trivial matters years after the incident. Even General Stretton was affected by guilt and depression because he believed his evacuation orders for Darwin after Cyclone Tracy had shattered survivors' lives.

As with the victims of many disasters who receive psychiatric counselling within minutes of surviving, rescue workers are now being debriefed by professional counselors following their rescue work. In New South Wales, Australia, these debriefing sessions take the form of team counselling and, depending upon the magnitude of the disaster and the response, can be followed up for months afterwards. The policy is not to let rescue workers discuss the disaster clannishly amongst themselves in an unstructured environment such as the local pub. Instead, counselling is directed towards relieving initial feelings of anger and guilt, which may not only destroy a good rescue worker but wreck their lives. Where death is involved, there is no rosy conclusion to a disaster. Anyone, be they victim, bystander, rescue worker or hero, who has witnessed the tragedy, potentially can have their lives irreparably harmed by the experience.

References

Drabek, T. E. 1986. *Human System Responses to Disaster: An Inventory of Sociological Findings*. Springer-Verlag, NY.

Hauptman, W. 1984. 'On the dryline'. *Atlantic Monthly*, May 1984, pp. 76–87.

Holthouse, H. 1986. *Cyclone: a century of cyclonic destruction*. Angus and Robertson, Sydney.

Kearney, G. E. and Britton, N. R. 1985. 'Insurance response in disaster'. *Report of Proceedings of Research Workshop on Human behaviour in disaster in Australia*. Natural Disasters Organisation, Mt Macedon, pp. 300–24.

Luketina, F. 1986. 'The Psychological Effects of floods'. *Soil and Water* v. 22 No. 1 pp. 20–24.

McKay, J. M. 1983. 'Community adoption of bushfire mitigation measures in the Adelaide Hills'. In Healey, D. T., Jarrett, F. G. and McKay, J. M. (eds) *The economics of bushfires-the South Australian experience*. Oxford University Press, Melbourne, pp. 116–31.

Michaelis, A. R. 1985. 'Interdisciplinary disaster research'. *Report of Proceedings of Research Workshop on*

Human behaviour in disaster in Australia. Natural Disasters Organisation, Mt Macedon, pp. 325–46.

Sims, J. H. and Baumann, D. D. 1972. 'The tornado threat: coping styles of the north and south'. *Science* v. 176 pp. 1386–92.

Western, J. S. and Milne, G. 1979. 'Some social effects of a natural hazard: Darwin residents and cyclone "Tracy"'. In Heathcote, R. L. and Thom, B. G. (eds) *Natural hazards in Australia*. Australian Academy of Science, Canberra, pp. 488–502.

Wilson, A. A. G. and Ferguson, I. S. 1985. 'Fight or flee?—a case study of the Mount Macedon bushfire' Pt 2. *Bushfire Bulletin* v. 8 No. 1&2 pp. 7–10.

Further reading

Burton, I., Kates R. W. and White, G. F. 1978. *The environment as hazard*. Oxford University Press, Oxford.

Cornell, J. 1976. *The great international disaster book*. Scribner's, NY.

Dalitz, E. R. 1979. 'Personal reactions to natural disasters'. In Heathcote and Thom (eds) *Natural hazards in Australia*. pp. 340–51.

Reser, J. P. 1980. 'The Psychological reality of natural disasters'. In Ollier, J. (ed.) *Response to disaster*. Centre for Disaster Studies, James Cook University, Townsville, pp. 29–44.

15

EPILOGUE

This book has been concerned with natural hazards that are within the present realm of possibility, that we have lived with, will continue to experience, and hopefully can survive. The basic purpose of the book was to present the reader with a clear description of the mechanisms producing climatic and geological hazards. An attempt was made to enliven that presentation with historical descriptions and supportive diagrams. There are however four themes which were indirectly addressed throughout the text. It is worthwhile here briefly summarizing those themes.

Firstly, natural hazards are predictable. This is especially applicable to climatic hazards. With satellite monitoring, short-range computer-forecasting models, and the wealth of experience of trained personnel and scientists, most climatically hazardous events can be forecast early enough to give the general population time to respond to any threat. Whether or not those warnings are believed is most relevant in today's world. The fact that farmers in Bangladesh cannot really do anything to enhance their chances of survival should a tropical cyclone strike, that the British Meteorological Office ignored an extra-tropical storm warning issued by their counterparts in France, or residents in Darwin were more interested in partying for Christmas when Cyclone Tracy was barrelling down on them, does not negate the fact that most climatically hazardous events are being predicted. It only illustrates that a society, its organized bureaucracy or its individuals have a large influence in determining how severely a natural disaster affects that community.

We have also pointed out in this text that the degree of prediction is strongly influenced by regular astronomical cycles, such as the 11- and 22-year sunspot and the 18.6-year lunar cycles. This belief, especially in relation to sunspot cycles, has been strongly criticized, and it will undoubtedly be one of the major criticisms against this book. Yet, frequencies equating with the return period of sunspots appear in enough climatic and geophysical data that sunspot cycles cannot be ignored as a precursor for natural hazard events. The excellent work by Currie, in relating the 18.6-year lunar cycle to the occurrence of drought and extreme rainfall worldwide, is convincing. I seriously believe that drought will afflict the United States Great Plains in the early 1990s, coincidentally with a maximum in the 18.6-year lunar cycle. If this prediction comes true, and remembering that many of the world's grain-producing areas will have surplus rainfall at this time, then the 18.6-year lunar cycle will be established as the best long-term precursor for defining the quantity and timing of regional rainfall. The economic and sociological implications of such a relationship will be profound.

Secondly, the occurrence of natural hazards is increasing in frequency. The Munich Reinsurance Company has shown, through payout claims, that natural hazards are on the increase, especially since 1960. This point is well illustrated for volcanic activity this century. After the Caribbean eruptions of 1902, volcanism leading to large deathtolls virtually ceased for half a century. A similar quiescence is evident in climatic records. The 1930s and 1940s were exceedingly passive as far as storm events were concerned. Air temperatures globally tended to be warm, and weather patterns did not fluctuate strongly. This does not imply that there were no climatic disasters. After all, this period witnessed the dust bowl years on the Great Plains of the United States, and Hitler's armies experienced a

Russian winter that was only matched in severity by that which afflicted Napoleon in 1812. Beginning about 1948, climate globally began to become more extreme. This variability has increased substantially since 1970. For example, over the past 20 years Australia has experienced the worst drought on record sandwiched between two of the largest regional flood events experienced. In the United Kingdom, in the late 1980s, temperatures seasonally have tended to oscillate from the coldest to the warmest on record. This enhanced variability is prominent in both climatic and geological phenomena. In terms of natural hazard events, we presently live in interesting times.

Thirdly, natural hazards are pervasive in time and space. A simple plot of the location of the hazards mentioned in this book will illustrate this fact. Normally, we do not expect droughts to strike England, tornadoes to sweep through Edmonton, Canada, or snow to fall in Los Angeles. However, such events have happened in the last 20 years. Part of the reason so-called unusual events are occurring in places where they are not expected may lie in the fact that we have lived through a passive period for hazards, as mentioned above. Another explanation is the fact that, as we develop longer geophysical records, extreme events are more likely to appear in those records. Additionally, technological advances in communications have dramatically shrunk the size of the Earth to the point that today's unusual weather, or geological event, is everyone's experience. Such developments have heightened people's awareness of natural hazards. I believe that the long-term prognosis, at any location, is for the occurrence of a wide range of hazards that have not necessarily been detected in the historical record. It is not unusual for an earthquake of 7.2 on the Richter scale to occur in the supposedly aseismic area of Tennant Creek, Australia (September 1988), nor for tornadoes to sweep through New York City (July 1989). These extreme events, while not common in these areas, are certainly within the realm of possibility. There is a pressing need for the serious assessment of extreme hazards in many parts of the world. For example, volcanic activity is possible in Australia even though there are no active volcanoes, and tropical cyclones could affect an area like southern California. In the former case, volcanoes have erupted in the past 10 000 years; in the latter case, southern California lies close to an area generating tropical cyclones. Natural hazards, and in some cases very unexpected hazards, are simply more common than perceived.

Finally, the pervasiveness of so many hazards, plus ones that we have not discussed, should not be viewed pessimistically. When a natural disaster occurs, and is splattered across our television screens, we tend to believe that few survive. In fact, it is surprising that so few people die or are injured. There have only been a few disasters in the past century that have wiped out all but a handful of people. Mount Pelée in 1902 left two survivors out of 30 000; the Mount Huascarán debris flow of May 1970 killed most of the population of Yungay, Peru, while the November 1985 eruption of the Nevado del Ruiz volcano in Columbia killed most of the people in the path of its subsequent lahar. Consider the odds of survival for other large natural disasters mentioned in this text: Cyclone Tracy killed 64 people out of a population of 25 000 in Darwin on 25 December 1974; only one out of every four people died in the horrific Tangshan earthquake in China in 1976; the death-toll in the Tokyo earthquake of 1923 was 140 000 people out of a population well in excess of 1 million. Human beings are very resilient and good at surviving and, in particular, are gifted at coping with disasters. These aspects were emphasized in chapter 14. Any textbook on natural hazards tends to emphasize, as this one does, the sensational, negative or gruesome aspects of disasters. However, it is only apt to end this book by noting people's remarkable ability to survive and recuperate from calamity.

SELECT GLOSSARY OF TERMS

ADIABATIC LAPSE RATE. The rate at which air temperature decreases with elevation because of the decreasing effect of gravity. The result is an increase in air volume without any change in the total amount of heat. The normal adiabatic lapse rate is $-1°C$ for each 100 meters of elevation: but the saturated adiabatic lapse rate, when air cools more slowly, is $0.4-0.9°C$ per 100 meters. If a parcel of air is moved towards the ground the reverse process occurs and the air begins to warm: this is adiabatic warming.

AGGRADE. To build up or accrete.

ALBEDO. The degree to which shortwave solar radiation is reflected from any surface or object.

ANDESITE. Volcanic rock containing the minerals andesine and pyroxene, hornblende or biotite. In a melted state, such material contains a high amount of dissolved gas. Volcanoes producing magma with a high content of this rock are classified andesitic.

ANGULAR MOMENTUM. The force that an object exerts as it spins around a central axis as affected by the angle that the object makes to the axis of rotation.

APERIODIC. Recurring but not with a fixed regularity. For instance, the 1st of each month occurs aperiodically because there are not the same number of days in every month.

ASEISMIC. Characterizing a region which historically has not experienced earthquakes.

ATTERBERG LIMITS. These limits define the moisture content of a soil at the points where it becomes elastic, plastic and liquid.

BARRIER ISLAND. A sandy island formed along a coastline where relative sea-level is rising or has risen substantially in the past. Shoreward movement of sediment cannot shift the island landward fast enough, so the island becomes separated from the mainland by a lagoon or marsh.

BASALT. The most common liquid rock produced by a volcano. Containing mainly magnesium, iron, calcium, aluminum and some silicon, it forms a hard, but easily weathered, fine-grained dark grey rock. Volcanoes producing magma with these constituents are classified basaltic.

BASE SURGE. The sudden eruption of a volcano laterally rather than vertically.

BATHYMETRY. Depths below sea-level.

BEARING STRENGTH. The amount of weight that a soil can support, including its own weight, before the soil deforms irrevocably.

BENIOFF ZONE. The zone at a depth of 700 km associated with earthquakes along a subduction zone.

BIFURCATION THEORY. A system of mathematics which describes a process, event or form subdividing suddenly into two, sometimes completely different entities.

BIOMASS. The total weight of all living organisms.

BIOTA. The animal (fauna) and plant (flora) life of a region.

BISTABLE PHASING. Droughts and wet periods tend to alternate every 9.3 years in many locations over two to three centuries. In some cases a drought suddenly has been followed 9 years later by another drought before the alternating cycle is re-established.

BLIZZARD. Any wind event with velocities exceeding $60\,km\,hr^{-1}$, with temperatures below $-6°C$ and, often, with snow being blown within tens of meters of ground level.

BLOCKING HIGH. A high-pressure system whose normal eastward movement stalls, blocking the passage of subsequent low- or high-pressure systems and deflecting them towards the pole or equator.

BLOCKS. A type of lava—angular, boulder-sized pieces of still solid rock blown out of an erupting volcano.

BOILING. A process whereby the extra elevation of water during a flood may generate sufficient water pressure in the spaces between particles in any raised embankment to force water through the bank, weakening its structure and winnowing out fine material.

BOMBS. (1) *See* EAST-COAST LOWS. (2) A type of lava—boulder-sized pieces of melted rock blown out of an erupting volcano.

BOREAL FOREST. Northern forest, mainly consisting of conifers and birch.

BRUUN RULE. A relationship stating that a sandy beach retreats under rising sea-level by moving sand from the shoreline and spreading it offshore to the same thickness that sea-level has risen.

BURNING OFF. *See* PRESCRIBED BURNING.

CALCAREOUS. Any soil or sedimentary deposit containing a preponderance of calcium carbonate.

CALDERA. The round depression formed at the summit of a volcano caused by the collapse of the underlying magma chamber, which has been emptied by an eruption.

CAPILLARY COHESION. The binding together of soil particles by the tension that exists between water molecules in a thin film around particles.

CATASTROPHE THEORY. A modern mathematical theory that may explain sudden changes that occur in nature, such as earthquakes, tsunami, and people's reactions under stress.

CATASTROPHISTS. A group of people, little challenged before 1760, who believed that the Earth's surface was shaped by cataclysmic events that were Acts of God.

CAVITATION. A process, highly corrosive to solids, whereby water velocities over a rigid surface are so high that the vapour pressure of water is exceeded and bubbles begin to form.

CEMENTATION. The process whereby the spaces or voids between particles of sediment become filled with chemical precipitates, such as iron oxides and calcium carbonate, which effectively weld the particles together into rock.

CENOZOIC. The last of four basic geological time eras; it dates from 65 million years BP to the present.

CHAOS THEORY. A modern theory that states that any process, governed by as few as three non-linear functions or mechanisms, will behave in basically an unpredictable manner even though, in the short term, repetitive patterns may be apparent.

CHAPARRAL. The name given to dense, often thorny thickets of low, brushy vegetation, frequently including small evergreen oaks, which is found in the Mediterranean climate of the United States southwest.

CHERNOZEM. A very black soil—rich in humus and carbonates—forming under cool, temperate, semi-arid climates.

CLUSTERING. The tendency for events of a similar magnitude to be located together in space or time.

COHERENCE. (1) Of soils: the binding together of soil particles due to chemical bonding, cementation or tension in thin films of water adhering to the surface of particles. (2) Of time series: the tendency for a phenomenon to be mirrored at another location over time. For instance, wet periods in eastern Australia often occur simultaneously with a heavy monsoon in India.

COHESION. Of soils: the capacity of clay soil particles to stick or adhere together because of physical or chemical forces independent of inter-particle friction.

COLLOIDS. Fine clay-sized particles 1–10 microns in size, usually formed in fluid suspensions.

COMPARTMENTALIZED. Describing beaches where escape of sediment alongshore is completely obstructed, usually by headlands.

CONFLAGRATION. A particularly intense fire with a heat output of 60 000–250 000 kW per meter along the fire front.

CONTINENTAL SHELF. That part of a continental plate submerged under the ocean, tapering off seaward at an average slope of less than 0.1 degrees, and terminating at a depth of 100–150 m in a steep drop to the ocean bottom.

CONVECTION. The transfer of heat upwards in a fluid or air mass by the movement of heated particles of air or water.

CONVECTIVE INSTABILITY. A parcel of air continuing to rise when lifted if its temperature is always greater than the surrounding air—a process that is guaranteed if condensation occurs releasing heat into the parcel.

CORIOLIS FORCE. The apparent force caused by the Earth's rotation, through which a moving body on the Earth's surface deflects to the right in the northern hemisphere and to the left in the southern hemisphere. The degree of deflection increases poleward.

CORRELATION. The statistical measure of the strength of a relationship between two variables.

CRACKING CLAYS. Soils consisting of clays which have a high ability to expand when wetted, and to shrink and, hence, crack when dried.

CREEP. The imperceptibly slow movement of surface soil material downslope under the effect of gravity.

CRETACEOUS. The geological time period 135–65 million years BP, which was characterized by sudden, large extinctions of many plant and animal species.

CRITICAL INCIDENCE STRESS SYNDROME. The condition affecting rescuers who perform superhuman, selfless feats of heroism or are involved with death during a disaster, when they suffer severe guilt, become reclusive and generally broken in spirit for months or years after the event.

CROSS-CORRELATION. The degree of correspondence between two variables which can be plotted over time.

CRUST. The outermost shell of the Earth, 20–70 km thick under continents, but only 5 km thick under the oceans. It consists of lighter, silica-rich rocks, which flow on denser liquid rock making up the mantle.

CRYOGENETIC. Any process involving the freezing of water within soil or rock.

CUMULO-NIMBUS CELL. A small center of rising air in which condensation and the formation of cumulus cloud takes place, usually in association with thunder and lightning.

CYCLIC. Recurring at regular intervals over time—for instance, sunspot numbers, which recur at 10–11-year intervals.

CYCLOGENESIS. The process forming an extra-tropical, mid-latitude cell or depression of low air pressure.

DEBRIS AVALANCHE. A flow of debris of different particle sizes that takes on catastrophic proportions. Volumes of 50–100 million cubic meters travelling at velocities of 300–400 km hr^{-1} have been measured.

DEBRIS FLOW. Any flow of sediment moving at any velocity downslope in which a range of particle sizes is involved.

DEFLATION. The removal by wind, over a long time period, of fine soil particles from the ground surface.

DEGASSING. The sudden or slow release into the atmosphere of gas dissolved in melted rock.

DEGRADATION. The slow loss from a soil of humus, minerals and fine particles essential to plant growth; or the contamination of a soil by salts or heavy minerals which are toxic to plant growth.

DESERTIFICATION. The process whereby semi-arid land capable of supporting enough plant life for grazing or extensive agriculture becomes infertile because of increased aridity, degradation or overuse.

DIATOMS. Mainly marine unicellular plant life, secreting silica.

DILATION. Rock in the Earth's crust will tend to crack and expand in volume when subjected to forces capable of fracturing it.

DUST BOWL. Part of the American southern Great Plains, centered on Oklahoma, which experienced extensive dust storms during the drought years of 1935–36 at the height of the great depression.

DUST VEIL INDEX. A subjectively defined index which attempts to measure the relative amounts of dust injected into the stratosphere by volcanic eruptions. The index is referenced at 1000 to the eruption of Krakatoa in 1883.

DYNAMO EFFECT. An electric current passing through a coil will generate a magnetic field in an iron bar placed in the center and, similarly, a fluctuating magnetic field in the iron bar will generate a current in the coil. A similar effect can be generated with saline ocean currents replacing the coil and the Earth's magnetic field replacing the iron bar.

EAST-COAST LOWS. Very intense storms that develop in mid latitudes on the seaward (eastern) side of continents. Such lows develop over warm ocean water in the lee of mountains and are dominated by intense convection, storm waves and heavy rainfall. Events where wind velocities reach in excess of 100 km hr^{-1} within a few hours are called 'bombs'.

EASTERLY WAVE. In the tropics pressure tends to increase parallel to the equator with easterly winds blowing in straight lines over large distances. However, such air flow tends to wobble, naturally or because of some outside forcing. As a result, isobars become wavy and the easterly winds develop a slight degree of rotation, which can eventually intensify over warm bodies of water into tropical cyclones.

EL NIÑO. The name given to the summer warming of waters off the coastline of Peru at Christmas time (from Spanish, meaning 'The Child').

ELASTICITY. A property of any solid that deforms under a load and then tends to return to its original shape after the load is released.

ELECTRIC RESISTIVITY. The degree to which a solid body can conduct an electric current.

ELECTROSPHERE. The zone of positively charged air at 50 km elevation, the strength of which determines the frequency of thunderstorms.

ELECTROSTATIC. The generation of electrical charges on stationary particles.

ENERGY. The capacity of any system to do work. Energy can be described as 'kinetic', which is the energy any system possesses by virtue of its motion; as 'potential', which is the energy any system possesses because of its position; and 'thermal', which is the energy any system possesses because of its heat.

ENSO. Abbreviation for the words 'El Niño-Southern Oscillation', referring to times when waters off the coast of Peru become abnormally warm and the easterly trade winds blowing in the tropics across the Pacific Ocean weaken or reverse.

EPICENTER. The point at the surface immediately above the location of a sudden movement in the Earth's crust generating an earthquake.

EPISODIC. The tendency for natural events to group into time periods characterized by common or loosely connected features. Weak cyclicity or regularity may be implied in this term.

EQUINOX. The time—occurring twice yearly, around 21 March and 21 September—when the apparent motion of the sun crosses the Earth's equator, which results in all parts of the Earth's surface receiving equal amounts of daylight.

ERG DESERT. Deserts formed of sandy material.

EUCALYPTS. A genus of tree which is mainly evergreen,

tends to shed its bark annually, is rich in leaf oil, flammable, drought resistant and capable of growing in nutrient-deficient soil.

Eustatic. Worldwide sea-level change.

Evapotranspiration. The amount of moisture lost by plants by evaporation to the atmosphere mainly through leaf openings or stomata. This amount is always greater than that loss from equivalent, unvegetated surfaces.

Expansive soils. Soils consisting of clay particles, mainly montmorillonite, that absorb water when wetted, thus, expanding in volume.

Extra-tropical depression. A mid-latitude, low-pressure cell with inwardly and upwardly spiralling winds. Unlike tropical cyclones, the cell can develop over land as well as water, usually in relation to the polar front and with a core of cold air.

Eye. The center of a tropical cyclone (or tornado), where air is descending, surrounded by a wall structure created by a sharp decline in pressure and characterized by intensely inwardly spiralling winds.

Fault. A planar zone where one part of the Earth's crust moves relative to another. A fault is categorized according to the direction of movement in the horizontal or vertical plane (*See* figure 10.10). Transcurrent faults are faultlines approximately at right angles to major lines of spreading crust in the oceans.

Feedback. The situation where one process reinforces (positive) or cancels (negative) the effect of another. For instance, a microphone placed close to an amplifier catches its own sound waves and rebroadcasts them louder—this is positive feedback.

Fetch. The length of water over which wind blows to generate waves: the longer the fetch, the bigger the wave.

Firestorm. Under intense burning, air can undergo rapid uplift and draw in surrounding air along the ground to replace it. This movement can generate destructive winds at ground level in excess of $100\,km\,hr^{-1}$.

Flood. A large discharge of water flowing down (if not outside) any watercourse in a relatively short time period compared to normal flows.

Flood lava. Lava flows occurring over a very large area.

Floodplain. Flattish land adjacent to any watercourse built up by sediment deposition from water flowing outside the channel during a flood.

Föhn wind. Air rising over a mountain and undergoing condensation releases latent heat of evaporation and, thus, cools adiabatically at a slow rate of about 0.5°C per 100 m. When this air then descends downslope on the leeward side, it tends to warm at the faster, dry adiabatic lapse rate of 1°C per 100 m. The resulting wind is much warmer and drier than on the windward side.

Foraminifera. Mainly marine unicellular animals that secrete or build grains of calcium carbonate.

Foreshocks. The small seismic shocks preceding a major earthquake or volcanic event.

Freezing rain. Light rain falling through a layer of cold air or upon objects with a temperature below freezing will tend to freeze or adhere to those objects; over time it weighs down and breaks power lines and tree branches.

Friction. The force due to the resistance of one particle sliding over another because of a weight pressing irregularities on the surface of both particles together.

Fumarole. A hole or vent from which volcanic fumes or vapours issue.

Galactic disk. The Milky Way galaxy consists of a mass of stars spread out in spiralling arms in a disk-shape from a central bulge. Our sun lies along one such arm, rotating around the galactic disk axis every 33 million years.

General air circulation. Equatorial regions have an excess of heat energy relative to the poles because incoming solar radiation exceeds outgoing longwave radiation. To maintain a heat balance over the Earth's surface this excess heat flows, mainly in air currents, from the equator to the pole.

Geochemical. Involving chemical reactions on or inside the Earth.

Geo-electric activity. Processes involving the transmission or induction of electrical currents along the surface of, or through the Earth.

Geomagnetic activity. Fluctuations in the Earth's magnetic field over space and time. The strength of this field and its variation can be changed by solar activity.

Geophysical. Dealing with the structure, composition and development of the Earth, including the atmosphere and oceans.

Geosyncline. An area of sediment accumulation so great that the earth's crust is deformed downwards.

Gilgai. The tendency for soil, affected by alternating wetting and drying, to overturn near the surface. The process can result in local relief up to 2 meters.

Glacier burst. The sudden release of huge volumes of water melted by volcanism under a glacier and held in place by the weight of ice until the glacier eventually floats (also called Jökulhlaups).

Greenhouse effect. The process whereby certain gases in the atmosphere, such as water vapour, carbon dioxide and methane, transmit incoming shortwave solar radiation, but absorb outgoing longwave radiation, thus raising surface air temperature.

Gumbel distribution. The plot of the magnitude of an

event on a logarithmic scale against the average length of time it takes for that event to occur again.

GYRES. Large rotating cells of ocean water which often completely occupy an ocean basin. The Gulf Stream forms the western arm of one such gyre in the north Atlantic.

HADLEY CELL. Heated air at the equator rises and moves poleward in the upper atmosphere. It cools radiationally and begins to sink at about 20–30° North and South of the equator, whereupon part of the air returns back to the equator. This forms a large, semi-permanent vertical cell, first postulated by George Hadley in 1735, on each side of the equator.

HALE SUNSPOT CYCLE. Sunspot numbers peak every 10–11 years. Every alternate peak is larger forming a 22-year cycle, noted first by the American astronomer George Hale. The Hale cycle is, in fact, the full magnetic sunspot cycle.

HECTARE. Standard international unit measuring area. 1 hectare = 1000 m^2 = 2.471 acres.

HECTOPASCAL. Standard international unit, abbreviated as hPa, measuring air pressure: 1 hPa = 1 millibar.

HINDCASTING. The procedure whereby wave characteristics can be calculated from the wind direction, strength and duration derived from weather charts.

HOLOCENE. The most recent part of the Quaternary, beginning 15 000 years BP, when the last major glaciation terminated.

HOWLING TERRORS. An Australian term for small, destructive, tropical cyclones with an eye diameter of less than 20 km; also called 'kooinar' by the Aborigines.

HYDROSTATIC. Describing pressure in ground water at a location exerted by the weight of water at higher levels in a slope.

HYDROTHERMAL. Describes the processes associated with heated or hot melted rock rich in water.

ICE-DAMMED LAKES. Lakes formed by glaciers or ice-sheets completely infilling a drainage course or flowing against higher topography, thus blocking the natural escape of meltwater.

IGNIMBRITES. Very hot ash can be blasted laterally across the landscape; when deposited, the ash particles fuse together to produce a hard, welded rock.

ILLITE. A clay mineral slightly more weathered than montmorillonite, containing hydroxyl radicals and having less iron and silica. Chemical formula $[KAl_2(OH)_2(AlSi_3(O,OH)_{10})]$.

INTERFLUVES. The hill slopes between streams flowing in the same direction.

INTER-TROPICAL CONVERGENCE ZONE. Air rises because of heating in the tropics beneath the seasonal position of the sun. Winds flow from the north and south and converge on this zone of heating, effectively producing a barrier to the exchange of air between hemispheres.

INTRA-PLATE. Locations lying near the center of a plate or, at least, away from tectonic activity associated with the margins.

INVERTED BAROMETER EFFECT. The height of sea-level inversely relates to the pressure of the atmosphere above, such that a decrease of 1 hectopascal in air pressure results in a rise in sea-level locally of 1 centimeter.

ISLAND ARC. A series of islands located on the continental plate side of a deep ocean trench, under which gas-rich crust is being subducted to produce andesitic magma.

ISOBARS. Lines plotted on a map joining points of equal pressure, usually at a fixed spacing of 4 hectopascals.

ISOSTATIC. Describing local changes in sea-level induced by regional tectonic variation.

ISOTHERMAL. Having equal degrees of heat.

JET STREAM. A fast moving flow of air at the top of the troposphere, usually linked to large-scale uplift or subsidence of air. The two most common jet streams are: the polar jet stream, attached to uplift along the polar front; the subtropical jet, centered over the poleward and trailing side of Hadley cells.

JURASSIC. The geological period 180–135 million years BP, dominated by the dinosaurs.

KAOLINITE. A common, clay mineral which is resistant to further weathering, containing only metals of aluminum and silica. Chemical formula $[Al_2O_3.2SiO_2.2H_2O]$.

KÁRMÁN VORTEX STREET. This describes the pattern of wind movement behind an obstacle. In going round an obstacle, winds tend to generate vortices with alternate clockwise and counterclockwise wind rotations.

KARST. The process in which solution is a major mechanism of erosion, or landscape developed by that process. Karst most commonly occurs in limestone, but in some instances can involve sandstone and volcanic rocks.

KATABATIC WINDS. Winds formed by the flow down topographic drainage lines of denser air cooled by longwave emission under clear skies and calm conditions. The resultant wind can reach speeds in excess of 100 km hr^{-1} and is prevalent in winter, mountainous terrain or adjacent to icecaps.

KELVIN WAVE. A water wave trapped along the boundary of a coast, channel, embayment or two bodies of water differing in some physical characteristic. The wave moves slowly parallel to the boundary and has a height that decreases rapidly away from the boundary.

LA NIÑA. The opposite of an ENSO event, during which waters in the west Pacific are warmer than normal, trade winds or Walker circulation is stronger and, consequently, rainfalls heavier in Southeast Asia.

LAHAR. A mudslide induced by volcanic eruption either at the time of the eruption (by the mixing of hot gases, melted ice or water, and ash) or years later (by the failure of volcanic ash deposits in the presence of heavy rain).

LANDSLIDE. The general term given to movement of material downslope in a mass.

LAPILLI. Volcanic fragments, about 2–60 mm in diameter.

LATENT HEAT. (1) of evaporation: each kilogram of water converted to water vapour at 26°C requires 2425 kilojoules of heat energy. Because this heat energy is not felt, it is termed 'latent'. (2) Of condensation: if this air is cooled and moisture condenses, the latent heat energy is released back to the atmosphere, warming the air.

LAVA. Molten rock. Different terms are used to describe the nature of the lava, mainly as determined by viscosity.

LENGTH OF DAY (LOD). Refers to the time, measured to a precision of a thousandth of a second, that it takes for the Earth to rotate once upon its axis.

LEVEE. When a river overflows its banks, there is an immediate decrease in velocity. This results in deposition of suspended mud and, eventually, the buildup of an embankment that can contain the river above the elevation of its adjacent floodplain.

LIQUEFACTION. If the weight of a sediment deposit or flow is not supported by the touching of individual grains against each other, but by the pressure of water or air entrapped in voids between particles, then such sediment will liquefy—that is, behave as a fluid. When liquefaction occurs the sediment loses its ability to support objects and is able to travel at fast velocities.

LITHOLOGY. Refers to the physical composition of any rock.

LITTLE ICE AGE. The period between AD 1650 and 1850, when sunspot activity was minimal, global temperatures were 1°C cooler and temperate mountain glaciers advanced.

LOESS. Dust picked up from barren ground in front of icecaps by dense, cold winds blowing off the icecap and deposited hundreds of kilometers downwind.

LONGITUDINAL. Any movement which occurs in a north-south direction.

LONGSHORE DRIFT. The movement of water and sediment within surf zones due to waves approaching shore at an angle, or causing small changes in sea-level parallel to the coastline.

LUNAR CYCLE. *See* M_N LUNAR CYCLE.

MAGMA. Liquid rock derived from the mantle.

MAGNETIC REVERSAL. The Earth has a magnetic north-south polarity centered within 1000 km of its present axis of rotation. The intensity of this magnetic field can vary, and at intervals of hundreds of thousands of years, weaken and reverse polarity very rapidly.

MAGNITUDE-FREQUENCY. The intensity or magnitude of all geophysical events such as floods occur at discrete time intervals in such a way that the lower the magnitude of the event, the more frequent its occurrence. The magnitude of an event can be plotted against how often that event occurs, to produce a magnitude-frequency diagram.

MANTLE. That part of the Earth from the crust to depths of 3000 km, which consists of molten, dense rock made up of silicates of magnesium, iron, calcium and aluminum.

MARGINALIZATION. The process whereby the fertility, water capacity or other characteristics of a soil in a region are reduced to the point where agricultural production is not sustainable permanently, but subject to periodic collapses.

MASS EXTINCTIONS. The dying out or disappearance of large numbers of plant and animal species in short periods of time, supposedly because of a catastrophic event that has affected the climate of the Earth in the prehistoric past.

MAUNDER MINIMUM. Refers to the lack of sunspots between AD 1650 and 1700 at the height of the Little Ice Age.

MEANDERING. The process whereby a flow of air or water tends to become unstable and travel in a winding path.

MEDITERRANEAN CLIMATE. A seasonally dry climate having rainfall in winter months. Its average temperature for at least four months exceeds 10°C, with no monthly average falling below −3°C. Rainfall of the driest summer month is less than one-third that of the wettest winter month and less than 40 mm.

MERIDIONAL AIR FLOW. Movement of air in a north-south direction.

MICROSEISMS. Very small earthquakes that can only be detected by sensitive instruments.

M_N LUNAR CYCLE. A periodicity of 18.6 years that exists in the orbit of the moon about the Earth because of the orientation of the moon's orbit to the equatorial plane of the sun.

MOGI DOUGHNUT. In some seismic areas, the epicenter of a large earthquake is ringed in the prior decade by moderate earthquakes forming a doughnut-like pattern. Named after its discoverer.

MOHR-COULOMB EQUATION. An equation in which the shear strength of a soil is related to soil cohesion, a force applied to the soil at right angles to the slope, and the angle at which particles move relative to each other within the soil.

MONSOON. Defines any region characterized by a distinct change of 180° in wind direction between summer and winter, resulting in alternating copious rain and aridity seasonally.

MONTMORILLONITE. A group of clay minerals with an aluminum layer sandwiched between two silica layers bonded with magnesium or calcium cations. Such soils expand easily because they can absorb water into the layered structure when wetted. Chemical formula $[(Mg,Ca)O.Al_2O_35SiO_2.nH_2O]$.

MULCHING. Any process which causes soils not to compact, but to become more friable towards the surface.

NOMOGRAM. Any graph that depicts numerical relationships.

NORMAL STRESS. Any force which is applied at right angles to a soil surface (*See* figure 13.3).

NUÉE ARDENTE. A cloud of hot ash that, blown upwards during an eruption, collapses under the weight of gravity and begins to flow downslope at great velocity suspended by the gases expelled in the eruption (*See* figure 12.3). Also called pyroclastic flow.

OORT BELT. A zone of billions of comets surrounding the solar system 4–13 trillion miles from the sun. These comets are perturbed by the gravitational influence of other stars and become the source of comets that orbit within the solar system.

OROGRAPHIC. Any aspect of physical geography dealing with mountains.

OROGRAPHICALLY CONTROLLED. Refers to the control of the Tibetan plateau and Rocky Mountains upon the spatial location of the polar jet stream in the northern hemisphere.

OVERBANK DEPOSITS. Occur when sediment-laden water floods out of a stream channel and, because of a drop in velocity, immediately deposits sediment on the sides of the channel and floodplain.

OVERWASHING. The process whereby sandy beaches or barrier islands are overriden by waves and high seas during storms or by the arrival of tsunami at a coastline.

OZONE HOLE. Ozone in the stratosphere decreases in concentration each year in the Antarctic spring because it reacts with chlorine ions in the presence of sunlight at very cold temperatures on ice particles. This seasonal depletion has been amplified in the 1980s because of the presence of more chlorine, attributed to the breakdown of artificial chlorofluorocarbons (CFCs).

PALAEOZOIC. The geological era 600–270 million years BP.

PARENT CLOUD. A large cloud characterized by turbulent air flow, and subdividing into separate areas of convection that can give rise to tornadoes, intense rainfall and hail.

PARTIAL PRESSURE. The contribution made by a gas to the total pressure of the atmosphere.

PERMAFROST. Permanently frozen ground.

PERMEABILITY. The degree to which water can flow through the voids in sediment deposits or microfractures in rock.

PERSISTENCE. The tendency for an abnormal climatic situation to continue over time.

PHOTOSYNTHESIS. The process whereby plants convert carbon dioxide and water in the presence of sunlight and, aided by chlorophyll, into carbohydrates or biomass.

PIEZOMAGNETISM. The generation of magnetism in iron-rich magma by increases in pressure or stress.

PLASTICITY. A property of any solid which permanently deforms without breaking up or rupturing when subject to a load.

PLATES. The Earth's crust is subdivided into six or seven large segments or plates, which move independently to each other at rates of 4–10 centimeters per year over the underlying mantle. These plates can collide (converge) with, or separate (diverge) from each other.

PLEISTOCENE. The geological age in the Quaternary, beginning approximately 2–3 million years BP; it was characterized by periods of worldwide glaciation.

PLIOCENE. The last period in the Tertiary, 12–2 million years BP. The Pliocene terminated when large-scale glaciation began to dominate the Earth.

PLUVIOMETERS. Instruments used to measure the rate at which rain falls over very short timespans.

POLAR FRONT. Dense air, cooled by longwave emission of heat in the polar regions and flowing towards the equator, differs substantially in temperature and pressure from the air it displaces. This difference occurs over a short distance, termed the 'polar front'.

PORE WATER PRESSURE. The pressure that builds up because of gravity in water filling the spaces or voids between sediment particles.

POWER. The rate at which energy is expended.

PRE-CAMBRIAN. Geological time greater than 600 million years BP.

PRECURSORS. Variables which give an indication of the occurrence of some future event.

PRESCRIBED BURNING. The deliberate and controlled burning of vegetation growing close to, or on the ground to minimize the fuel supply for future bush or forest fires. Also called (in Australia especially) 'burning off'.

PROBABILITY OF EXCEEDENCE. The probability or likelihood, expressed in the range of 0–100 per cent, of an event with a certain magnitude or size occurring.

PUMICE. Volcanic ash which is filled with gas particles and, hence, able to float on water.

PYROCLASTIC EJECTA. Pyroclastic material blown out into the atmosphere. *See also* TEPHRA.

PYROCLASTIC FLOW. *See* NUÉE ARDENTE.

PYROCLASTIC MATERIAL. The fine, hot ash produced from the fragmentation of magma in the atmosphere by the force of a volcanic eruption and the gases contained within.

QUATERNARY. The most recent geological period, beginning approximately 2–3 million years BP and consisting of the Pleistocene and the Holocene.

QUICK CLAYS. Clays which have been deposited with a high degree of sodium cations that tend to glue particles together, giving the deposit a high degree of stability. However, sodium cations are easily leached by groundwater, with the result that the clay deposit progressively becomes unstable over time.

RECURRENCE INTERVAL. The average time between repeat occurrences of an event of given magnitude.

REG DESERT. The residual, stony desert formed after clay and sand-sized particles are blown away.

REGOLITH. The layer of weathered or unweathered, loose, incoherent rock material of whatever origin, forming the surface of the Earth.

RESONANCE. A process involving the development of air or water waves in which the wave length is equal to or some harmonic of the physical dimensions of a basin or embayment.

RIFT SYSTEM. A zone of near-vertical fracturing of the Earth's surface that causes one segment to drop substantially below another.

ROSSBY WAVES. Any wave motion that develops with a long wavelength of hundreds or thousands of kilometers in a fluid moving parallel to the Earth's surface and controlled by Coriolis force. Generally restricted to the upper westerlies attached to the polar jet stream at mid-latitudes.

RUN-UP. When any wave reaches the shoreline, its momentum tends to carry a mass of water landward. The distance landward of the shoreline is the run-up distance, while the elevation above sea-level is the run-up elevation.

SAHEL. The seven African countries—Gambia, Senegal, Mauritania, Mali, Niger and Chad—at the southern edge of the Sahara desert.

SALTATION. The tendency for sediment particles to move from the bed into air- or water-flow and back to the bed in zig-zag paths, where the distance beneath upward movement is much shorter than the distance beneath downward movement.

SCORIAE. Fused volcanic slag or ash that results from a pyroclastic flow and contains a high degree of gas.

SEAMOUNT. A volcano which has formed in the ocean, been planed off at the top because of close proximity to the ocean's surface, and then lowered below the sea as the Earth's crust moves away from the center of volcanic activity.

SEASAT ALTIMETER. The SEASAT satellite, in orbit between July and October 1978, carried an altimeter that could measure the distance between the satellite and the surface of the Earth to an accuracy of 10 centimeters.

SECONDARY AIR CIRCULATION. The initial tendency for air movement across the Earth's surface is from the pole to the equator. Coriolis force causes any air movement to form secondary, circular cells hundreds to thousands of kilometers in diameter.

SEICHING. Excitation of a regular oscillation of water waves within a basin or embayment caused by an earthquake, the passage of an atmospheric pressure wave or the occurrence of strong winds.

SEISMIC. Related to sudden and usually large movement of the Earth's crust.

SEISMIC GAP. That part of an active fault zone which has not experienced moderate or major earthquake activity for at least 30 years.

SEISMIC RISK. The probability of earthquakes of given magnitude occurring in a region.

SET-BACK LINE. A line established shoreward of a coastline, incorporating 50–100 years of postulated shoreline retreat, and in front of which no development should proceed.

SET-UP. The enhanced elevation of sea-level at a coastline mainly due to the process of wave-breaking across a surf zone, but also aided by wind, shelf waves, Coriolis force, upwelling and current impinging along a coast.

SHALLOW FOCUS. Earthquakes occurring within 60 km of the surface.

SHEAR SORTING. Any process which causes coarse particles to be transported more efficiently in a liquid or air flow that is thin, full of sediment and travelling at high velocity.

SHEAR STRENGTH. The internal resistance of a body of material to any stress applied with a horizontal component.

SHEAR STRESS. The force applied to a body of material which tends to move it parallel to the contact with another solid.

SHEARING. A stress caused by two adjacent parts of a solid tending to slide past each other parallel to the plane of contact.

SLUMP. The tendency for some landslides to undergo rotation, with the greatest failure of material at the downslope end.

SOLAR FLARES. Large ejections of ionized hydrogen from the sun's surface accompanied by a pulse of electromagnetic radiation.

SOLAR LUMINOSITY. The brightness and, thus, the energy intensity of the sun.

SOLIFLUCTION. The slow movement of water-saturated material downslope, usually in environments where freezing and thawing are the dominant processes affecting the surface.

SOUTHERLY BUSTER. The term given in Australia to a cold front moving rapidly northwards along the eastern coastline.

SOUTHERN OSCILLATION. The usual movement of air in the tropical Pacific is for strong easterly trades reinforced by low pressure over Indonesia-Australia and high pressure over the southeast Pacific. Every two to seven years this low and high pattern weakens or even reverses, causing the trades to fail or become westerlies. It is this switching from one pattern to another that is the Southern Oscillation.

SPOT FIRES. Small fires ignited by flying embers carried on updrafts generated by heated air from the main core of a forest or bush fire.

STERIC. Referring to the change in volume of sea-water due to heating or cooling.

STORM SURGE. The elevation of sea- or lake-levels above predicted or normal levels mainly resulting from a drop in air pressure, or from the physical piling up of water by wind associated with a low-pressure disturbance or storm.

STRAIN. The deformation or movement caused by applying a force to material on a slope.

STRATOSPHERE. That part of the atmosphere lying 12–60 kilometers above the Earth, and characterized by stability and an increase in temperature with altitude.

STRATOSPHERIC FOUNTAINS. Little air is exchanged between the troposphere and stratosphere. However, intense convection, which is generated by thunderstorms in the tropics at the west side of the Pacific, breaches this barrier and, like a fountain, injects air originating from the ground into the stratosphere.

STRESS. Applied to soil, it is any force which tends to move material downslope.

STRIP FARMING. An agricultural practice used to conserve soil moisture whereby crops are alternated in strips with fallow land.

SUBCRITICAL FLOW. Tranquil river flow in which velocities relative to the channel depth are low enough that sediment transport is not excessive.

SUBDUCTION ZONES. The zone down which an oceanic plate passes beneath a continental plate to be consumed into the mantle.

SUBSIDENCE. The downward failure of the Earth's surface brought about mainly by removal of material from below.

SUNSPOTS. Areas 20 000 km in diameter on the sun's surface of strong magnetic disturbance, so strong that convective movement of heat to the surface is curtailed. Hence, the spots are darker because they are cooler.

SUPERCELL THUNDERSTORM. A particularly large, convective thunderstorm cell with a diameter of 50–100 kilometers. Such a storm has a greater chance of producing hail, tornadoes, strong wind or copious rainfall.

SUPERCRITICAL FLOW. River flow in which velocities relative to the channel depth are so high that the flow begins to shoot or jet. Such flow is very erosive.

SURCHARGING. The escape of stormwater through access-hole covers, which are designed to lift off under excess pressures caused by flooding. This prevents stormwater pipes from bursting underground, which would necessitate costly repairs.

SUSPENSION LOAD. That portion of sediment being carried in air or water flow by turbulence.

SYNTHETIC APERTURE SIDE SCAN RADAR. A satellite technique for measuring surface ocean wave characteristics by bouncing short radar pulses at an oblique angle to the sea-surface.

TAIGA. Swampy, coniferous, evergreen forest of subarctic Siberia.

TECTONIC. The process characterizing the deformation of the Earth's crust because of earthquakes.

TECTONIC CREEP. Deformation of the Earth's crust which occurs in such small increments that it is virtually undetectable, except over time.

TELECONNECTION. The linkage of a measured climatic time series to another one at some distance across the globe. For instance, rainfall in Florida can be related to sea surface temperature in northern Australia through the Southern Oscillation.

TELLURIC CURRENTS. Electric currents moving through the Earth.

TEPHRA. Volcanic ash which is disaggregated and blown by the force of an eruption, usually vertically, into the atmosphere.

TERTIARY. The geological period from 65 to 2–3 million years BP.

TERTIARY AIR CIRCULATION. This refers to localized air movement—for example sea breezes, gravity winds or convection in a thunderstorm.

THERMAL EXPANSION OF ICE AND WATER. Water is unusual in that it has its greatest density at 4°C. Thus, it expands readily when heated. Ice however has its greatest volume around the freezing point. It shrinks and cracks around −10°C. If water can enter the cracks and refreeze, such ice will expand and buckle as it warms.

Thermocline. The abrupt change in water temperature with depth within a large body of water.

Thixotrophy. The property of becoming fluid when stirred or agitated especially by compressive shock waves. *See also* liquefaction.

Tidal inlet. The opening that forms across a barrier island in which there is a free and substantial exchange of saline water between the ocean and a landward lagoon during each tide cycle.

Tidal waves. A misnomer applied to tsunami. In a truer sense it refers to the amplification of a tidal cycle because of the effects of seiching or resonance induced by the passage of a low-pressure storm.

Tilt steps. Tilting of the Earth's crust which occurs, not smoothly, but in a series of discrete steps.

Topographic enhancement. Any process which is exaggerated by the effect of topography. For example, wind blowing across a series of ridges becomes more turbulent because the hills present a rougher surface for air to flow over.

Tornado. An extremely violent vortex or whirlwind of air measuring only hundreds of meters in diameter and with wind speeds up to 1300 km hr^{-1}.

Trade winds. Winds on the equatorial side of Hadley cells, blowing from east to west within 20–30° of the equator. Their steadiness for three to four months of the year greatly benefited trading ships in the days of sail.

Trench. An area of oceanic crust which has been dragged downwards by the movement of oceanic crust beneath a continental plate.

Tropical cyclone. A large-scale vortex of air hundreds of kilometers in diameter. It is formed over the tropical oceans, characterized by copious rain and a central area of calm, and surrounded by clockwise-spinning winds blowing at speeds in excess of 200–250 km hr^{-1}.

Tropopause. The band separating the troposphere (where temperature decreases with altitude) from the stratosphere (where temperature increases with altitude). It forms a barrier to the ready exchange of air between these two zones.

Troposphere. The lower part of the atmosphere 10–12 km in altitude characterized by cloud formation and a drop in temperature with elevation.

Tsunami. A Japanese word for 'harbour wave'. It is used to define a water wave generated by a sudden change in the sea bed resulting from an earthquake, volcanic eruption or landslide.

Tundra. The treeless plain of Arctic regions, characterized by permanently frozen ground.

Turbulence. Flow (also called turbulent flow) in which individual particles or molecules travel unpredictable paths apparently unaffected by gravity.

Ultrasound. Sound with a frequency above 20 000 vibrations per second, and generally not perceivable by the human ear.

Uniformitarianism. A geological explanation developed in the eighteenth century to counteract the Catastrophists. It involves two concepts: the first implies that geological processes follow natural laws applicable to science (methodological uniformitarianism); the second implies that the type and rate of processes operative today have remained constant throughout geological time (inductive reasoning).

Van der Waals bonds. The weak attraction exerted by all molecules to each other, caused by the electrostatic attraction of the nuclei of one molecule for the electrons of another.

Vertical accretion. The deposition of sediment in the vertical direction.

VHF. Abbreviation for 'very high frequency'. That part of the electromagnetic spectrum with a wavelength between 30 and 300 megahertz, characteristic of radio and television transmission.

Victim-rescuer syndrome. In disasters where survivors are at the mercy of outside support for survival or their daily requirements, a strong dependence can develop between a victim and those who are providing the aid. This dependency may not be easily terminated when the necessity for relief has ended.

Viscosity. The resistance of a fluid to the motion of its own molecules because they tend to stick together.

Voids. The spaces or gaps that exist between particles in a sedimentary deposit and filled with air or water.

Vortex. A spinning or swirling mass of air or water that can obtain considerable velocity.

Walker circulation. The normal movement of air across the tropical Pacific is from the east between a stationary high-pressure cell off the coast of South America and a low-pressure zone over Indonesia-India. In the 1920s Gilbert Walker discovered that the strength of this air movement determined the intensity of the Indian monsoon.

Wave diffraction. The tendency for wave energy to spread out along the crest of a wave. Hence, when a wave passes through a narrow harbour entrance, the wave will spread out to affect all the coastline inside a harbour.

Wave refraction. The tendency for the velocity at which a wave travels to slow down over shallow bathymetry. Hence, a wave crest tends to wrap around headlands and to spread out into bays.

Wave rider buoy. An instrument, consisting of sensitive accelerometers encased inside a buoy and floating on the ocean surface, used for measuring the height and energy of a wave.

WIND-CHILL FACTOR. The factor by which the temperature of a surface is reduced because of the ability of wind to remove heat from that surface more efficiently by convection (*See* figure 7.11).

WIND SHEAR. The tendency for winds to move at different velocities with elevation because of strong temperature differences. The wind, as a result, may accelerate in speed over short altitudes or even change direction—a particularly hazardous situation near the ground for airplanes trying to land or take off.

YIELD LIMIT. The limit beyond which a soil can no longer resist a force without deforming permanently.

YIELD STRENGTH. The maximum resistance of a solid to a force before it can be permanently deformed.

ZONAL. Restricted to an area bounded by parallels of latitude on the Earth's surface.

INDEX

Note: References to figures and tables are not distinguished in this index.